Drugs and a Methodological Compendium

Vikrant Singh Rajput • Ashish Runthala
Editors

Drugs and a Methodological Compendium

From bench to bedside

Springer

Editors
Vikrant Singh Rajput
Department of Biomedical Engineering
Central University of Rajasthan
Ajmer, India

Ashish Runthala
Koneru Lakshmaiah Education Foundation
Guntur, Andhra Pradesh, India

ISBN 978-981-19-7951-4 ISBN 978-981-19-7952-1 (eBook)
https://doi.org/10.1007/978-981-19-7952-1

This Springer imprint is published by the registered company Springer Nature Singapore Pte Ltd.
The registered company address is: 152 Beach Road, #21-01/04 Gateway East, Singapore 189721, Singapore

Preface

Biological sciences and its interlinked biomedical research are a gruesome exercise which is heavily dependent on several intensive literature survey followed by intensive experimental protocols involving complex methodologies to computational algorithms, which are constantly being explored and refined in search of novel, safe and effective drugs and regimens in laboratories. The cell's molecular network has been exploited to decipher the essential drug targets for developing and delivering the potential drug candidates against various conditions, which remains at the interface of hottest topics of the modern drug discovery and development fundamentals. Many books/volumes independently put forth intricate details of drug discovery without scrutinizing the framework of methodologies and techniques in use. Therefore, with the support of various authors with expertise in different domains of drug discovery and development, we have substantially streamlined and bundled the complicated methodologies involved at various stages of a drug discovery and development process, along with several case studies. Aimed at researchers/students, the book shall provide a compendious and perspicuous guide to the techniques deployed for the biological and biomedical research pertaining to the intricacies of techniques involved during the journey of a potent drug molecule. It will present detailed methodological insights from bench to bedside. The emphasis will be on basic principles, highlighting the paramount significance of each step during drug discovery till the delivery of drug molecule to the clinics. The text shall focus on approaches of modern high-throughput screening methods, computational modus operandi, pharmacological optimization, medicinal chemistry approaches for drug design, formulation and drug delivery, clinical trial procedures and others. The book also encompasses specific case studies, regulatory approval proceedings, and industrial viewpoint alongside the conceptual layout. The volume will integrate medical, biological, medicinal, pharmacological, and computational streams and would be of prime interest to a wide audience including molecular biologists, biochemists, pharmacologists, medicinal chemists, toxicologists, drug discovery and development researchers, and all other students interested in these disciplines.

Ajmer, India Vikrant Singh Rajput
Guntur, India Ashish Runthala

Acknowledgments

We do not achieve anything by ourselves;

We achieve it with the help of the community of people around us.

Venky Ramakrishnan

Nobel Prize in Chemistry, 2009

It feels so good to have our first edited book published with Springer. It is the concerted efforts of many exceptional brains, without whom it would have been close to impossible to complete this book. Although it is just our name on the cover, I acknowledge every single individual who actively/passively helped us in their own way.

First and foremost, we greatly appreciate the creative liberty we have received through several dedicated minds of authors, who have written various chapters in this book. Their ideas, and our timely constructive inputs, have consistently steered it in the right direction.

We are also grateful to all our teachers and mentors for their support and benevolence. Further, we also thank our colleagues for stimulating novel ideas during the various stages of this book and enriching our research aptitude.

Finally, the last mentions are always important. We thank our families for being a pillar of support and literally pacifying our nerves during the tough times, managing various things and deadlines. Often, we find no words of gratitude to those who have been everything in our lives, which is our families—"We are, what we are, which could not have been possible without them."

Contents

Editors and Contributors

About the Editors

Ashish Runthala graduated with an M.Sc. in Biological Science (2005) from BITS Pilani, Rajasthan, India. Meanwhile qualifying the National Eligibility Test (NET), he pursued his M.E. in Biotechnology (2008) here and completed his Ph.D. in Computational Biology in 2016. At the Indian Institute of Science, he was awarded the Ucchayatar Avishkar Yojana postdoctoral fellowship in 2017. Starting his academic responsibilities at BITS Pilani from 2005, he has held several positions viz. Teaching Assistant, Assistant Lecturer, Lecturer, and Visiting Faculty. He has also worked at Presidency University Bangalore as an Assistant Professor. He is currently works as Associate Professor at KL University, Vijayawada, and is doing research on developing the functionally improved enzymes of the DXP pathway for enhancing the overall yield of the pathway.

Areas of Interest: Computational algorithms; Bioinformatics; Protein structure prediction; Enzyme design; Decoding active site; Designing thermostable enzyme

Vikrant Singh Rajput graduated with a B. Tech in Biotechnology (2010) from Amity University, Noida, UP, India, and a Ph.D. in Biological Science from CSIR-Indian Institute of Integrative Medicine, Jammu, J&K, India (2017). He was awarded postdoctoral fellowship from Indian Institute of Technology (IIT) Roorkee and Young Scientist Award/Research grant from the Department of Health Research (Govt. of India). He has also held different positions like Adhoc Faculty at National Institute of Technology (NIT) Raipur, Project Scientist at Biosafety Support Unit, Regional Centre for Biotechnology, Faridabad and Scientist at Foundation for Neglected Diseases Research, Bengaluru, India, primarily working on TB and other bacterial pathogens for drug discovery against them. He currently works as Assistant Professor at Department of Biomedical Engineering, Central University of Rajasthan, Ajmer.

Areas of Interest: TB biology; TB and antibacterial drug discovery; Combatting drug resistance; Novel drug targets; Computational drug discovery and immunoinformatics.

Contributors

Saleem Akbar Department of Pharmaceutical Chemistry, School of Pharmaceutical Education and Research, Jamia Hamdard (Deemed to be University), New Delhi, India

Aditya Anand School of Biochemical Engineering, Indian Institute of Technology (Banaras Hindu University), Varanasi, India

Killivalavan Asaithambi Molecular Biophysics Unit, Indian Institute of Science, Bangalore, India

Somya Asthana Food Toxicology Division, Food, Drug and Chemical Toxicology Group, CSIR-Indian Institute of Toxicology Research, Lucknow, India
Department of Biotechnology, Manav Rachna International Institute of Research and Studies (MRIIRS), Faridabad, India

Aggarapu Chandana Department of Biotechnology, KoneruLakshmaiah Education Foundation, Guntur, India

Shivanandappa Kukkaler Channappa Pertussis Vaccine Section, Pasteur Institute of India, Coonoor, India

Deepak Das Department of Chemistry, School of Chemical and Life Sciences, Jamia Hamdard, New Delhi, India

Rikeshwer Prasad Dewangan Department of Pharmaceutical Chemistry, School of Pharmaceutical Education and Research, Jamia Hamdard (Deemed to be University), New Delhi, India

Ravindra Dhar Dubey Specialty Formulation Department, Alembic Pharmaceutical Ltd., Hyderabad, Telangana, India

Rupita Ghosh Department of Biotechnology, KoneruLakshmaiah Education Foundation, Guntur, India

Ankita Gupta Department of Zoology, Gauhati University, Gauhati, India

Parul Gupta Department of Respiratory Medicine, King George's Medical University, Lucknow, Uttar Pradesh, India

G. N. S. HemaSree Pharamacological Modeling and Simulation Centre, Faculty of Pharmacy, M. S. Ramaiah University of Applied Sciences, Bangalore, Karnataka, India

Chandrashekhar Kadam Specialty Formulation Department, Alembic Pharmaceutical Ltd., Hyderabad, Telangana, India

Nitin Pal Kalia Department of Biological Sciences (Pharmacology and Toxicology), National Institute of Pharmaceutical Education and Research, Hyderabad, Telangana, India

Saravanan Kandasamy Molecular Biophysics Unit, Indian Institute of Science, Bangalore, India

Harvinder Kour Khera Tata Institute for Genetics and Society - Centre at inStem, Bengaluru, Karnataka, India

Ajay Kumar Government Pharmacy College, BRD Medical College Campus, Gorakhpur, India

Sanjay Kumar Glocal College of Paramedical Science and Research Centre, The Glocal University, Mirzapur Pole, Saharanpur, U.P., India
School of Life and Allied Health Sciences, The Glocal University, Mirzapur Pole, Saharanpur, U.P., India

Anchala Kumari Indian Council of Medical Research, International Health Division, New Delhi, India

Shalini Kumari CSIR-Institute of Genomics and Integrative Biology, New Delhi, India

Aman Kumar Mahto Department of Pharmaceutical Chemistry, School of Pharmaceutical Education and Research, Jamia Hamdard (Deemed to be University), New Delhi, India

Saima Shakil Malik Centre of Biotechnology and Genomic Medicine, Medical College of Georgia, Augusta University, Augusta, GA, USA

Sarada Prasanna Mallick Department of Biotechnology, Koneru Lakshmaiah Education Foundation, Guntur, India

V. Lakshmi Prasanna Marise Department of Pharmacy Practice, Faculty of Pharmacy, M. S. Ramaiah University of Applied Sciences, Bangalore, Karnataka, India
Pharamacological Modeling and Simulation Centre, Faculty of Pharmacy, M. S. Ramaiah University of Applied Sciences, Bangalore, Karnataka, India

Nosheen Masood Department of Biotechnology, Fatima Jinnah Women University, The Mall Rawalpindi, Rawalpindi, Pakistan

S. R. Krishna Motukuri Department of Agricultural Biotechnology, College of Agriculture, Koneru Lakshmaiah Education Foundation, Guntur, India

Manikanta Murahari Department of Pharmacy, Koneru Lakshmaiah Education Foundation, Vaddeswaram, Guntur District, Andhra Pradesh, India

Mamatha Krishna Murthy Department of Pharmacy Practice, Faculty of Pharmacy, M. S. Ramaiah University of Applied Sciences, Bangalore, Karnataka, India
Pharamacological Modeling and Simulation Centre, Faculty of Pharmacy, M. S. Ramaiah University of Applied Sciences, Bangalore, Karnataka, India

Seema Negi Central Research Station, Subharti Medical College, Swami Vivekanand Subharti University, Meerut, U.P., India

S. L. Neha Department of Pharmaceutics, Delhi Institute of Pharmaceutical Sciences and Research (DIPSAR), Delhi Pharmaceutical Sciences and Research University, New Delhi, India

Arzoo Pannu Department of Pharmacology, Delhi Pharmaceutical Sciences and Research University, New Delhi, India

Shweta Paroha Department of Pharmaceutics, Delhi Institute of Pharmaceutical Sciences and Research (DIPSAR), Delhi Pharmaceutical Sciences and Research University, New Delhi, India

Burra V. L. S. Prasad Centre for Advanced Research and Innovation in Structural Biology of Diseases, K L E F University, Vaddeswaram, Andhra Pradesh, India

Vijayakumar Rajendran Molecular Biophysics Unit, Indian Institute of Science, Bangalore, India

Vikrant Singh Rajput Department of Biomedical Engineering, Central University of Rajasthan, Ajmer, India

Chitra Rani University of Connecticut Health Center, Farmington, Connecticut, USA

Laxmi Rani Department of Pharmaceutics, Delhi Institute of Pharmaceutical Sciences and Research (DIPSAR), Delhi Pharmaceutical Sciences and Research University, New Delhi, India

Ashish Runthala Koneru Lakshmaiah Education Foundation, Vijayawada, India

Pravat Kumar Sahoo Department of Pharmaceutics, Delhi Institute of Pharmaceutical Sciences and Research (DIPSAR), Delhi Pharmaceutical Sciences and Research University, New Delhi, India

Ganesan Rajalekshmi Saraswathy Department of Pharmacy Practice, Faculty of Pharmacy, M. S. Ramaiah University of Applied Sciences, Bangalore, Karnataka, India
Pharamacological Modeling and Simulation Centre, Faculty of Pharmacy, M. S. Ramaiah University of Applied Sciences, Bangalore, Karnataka, India

Kamatchi Sundara Saravanan Department of Pharmacognosy, Faculty of Pharmacy, M. S. Ramaiah University of Applied Sciences, Bangalore, Karnataka, India

Kshreeraja S. Satish Department of Pharmacy Practice, Faculty of Pharmacy, M. S. Ramaiah University of Applied Sciences, Bangalore, Karnataka, India

Shashikanta Sau Department of Biological Sciences (Pharmacology and Toxicology), National Institute of Pharmaceutical Education and Research, Hyderabad, Telangana, India

Jagannathan Selvaraj Tissue Culture Anti-Rabies Vaccine Section, Pasteur Institute of India, Coonoor, India

Syed Shafi Department of Chemistry, School of Chemical and Life Sciences, Jamia Hamdard, New Delhi, India

Vibha Shukla Food Toxicology Division, Food, Drug and Chemical Toxicology Group, CSIR-Indian Institute of Toxicology Research, Lucknow, India
Academy of Scientific and Innovative Research (AcSIR), Ghaziabad, India

Ajeet Singh Department of Pharmaceutical Sciences, J. S. University, Shikohabad, Firozabad, U.P., India

Bhisham Narayan Singh Department of Ageing Research, Manipal School of Life Sciences, Manipal Academy of Higher Education, Manipal, India

Subhash Singh ICMR-RMRC, Bhubaneswar, Odisha, India

Amit Kumar Srivastava Indian Institute for Technology, Roorkee, Uttarakhand, India

Dheerendra Kumar Suman Department of Biotechnology, National Institute of Technology, Tadepalligudem, India

Anurag Tripathi Food Toxicology Division, Food, Drug and Chemical Toxicology Group, CSIR-Indian Institute of Toxicology Research, Lucknow, India
Academy of Scientific and Innovative Research (AcSIR), Ghaziabad, India

Shah Ubaid-ullah J & K Higher Education Department (Govt. Degree College, Pulwama), Srinagar, India
Department of Biotechnology, Islamia College of Science & Commerce (ICSC), Srinagar, India

Ajay Kumar Verma Department of Respiratory Medicine, King George's Medical University, Lucknow, Uttar Pradesh, India

Juhi Verma Product Development Cell, National Institute of Immunology, Aruna Asaf Ali Marg, New Delhi, India

Venkata Rajesh Yella Department of Biotechnology, Koneru Lakshmaiah Education Foundation, Guntur, India

RETRACTED CHAPTER: Computational and Informatics Methodologies in Drug Discovery, with Focus on Natural Products

Anchala Kumari and Vikrant Singh Rajput

Abstract In the field of biomedical research, innovations for developing novel therapeutic drugs are one of the distinguished assignments that are indispensable for scientific, economic, and social progress. The precision of drug discovery and development approaches can be enhanced by efficient computational methodologies. Developments in the computational biology and informatics algorithms have greatly augmented the efficiency of the drug discovery strategies at various levels. However, drug discovery rate, in general, has reduced immensely because of its dependence on the small molecules as the preliminary source of new and innovative hypothesis. Natural products like metabolites and immunological factors are a massive resource of bioactive molecules from varied origins. Some of them are supported by traditional medicines for thousands of years and are immensely uncoupled from the range of small molecules usually utilized for drug discovery. Nevertheless, natural products retain unique features that differentiate them from conventional and outdated small-molecule drug leads, demanding novel techniques and access for evaluating their therapeutic capability. In this chapter, we scrutinize a list of advanced techniques in computational biology, cheminformatics, bioinformatics, and informational engineering for data directed drug discovery from biological resources. Herein, we emphasize on the procedures that intend to saturate the void among the conventional outdated small-molecule drug leads and the various types of organic compound discovery methodologies. In order to completely clout the limitations of standard drug informatics practices, we explored the existing gaps and barriers that need to be overcome. At last, we concluded the chapter through a "road map" of the research significances, which pursues to apprehend this objective.

The original version of this chapter was retracted: The retraction note to this chapter is available at https://doi.org/10.1007/978-981-19-7952-1_17

A. Kumari (✉)
Indian Council of Medical Research, International Health Division, New Delhi, India

V. S. Rajput (✉)
Department of Biomedical Engineering, Central University of Rajasthan, Ajmer, India

V. S. Rajput, A. Runthala (eds.), *Drugs and a Methodological Compendium*,
https://doi.org/10.1007/978-981-19-7952-1_1

1 Introduction

Drugs are essential for the inhibition and mitigation of human diseases. Drug discovery is the technique for identifying new pharmaceutical interventions against a natural target. A drug advances through various stages including target validation, hit identification, lead optimization, and preclinical development, followed by clinical trials, and finally a newly discovered drug reaches the market for patients. It embraces one of the most significant accomplishments in pharmacological sciences. The complete procedure of discovering a novel drug is considered to be a laborious, hazardous, and an overpriced task. The US National Institutes of Health (NIH) funded the discovery and advancement of 210 novel molecular entities, which was roughly about 20% of the total budget for the 2010–2016 period as per 2018 analysis (Cleary et al. 2018). Most of the systematic drug inventions have concentrated on the small molecular entities like, for instance, the DrugBank database in which almost 86% of drugs including both experimental and approved ones are small molecules, with the aim of advancing in the modern medical science (Wishart et al. 2018). The reason behind this includes the comparative effortlessness synthesis operandi available, frequent higher biochemical steadiness, and additional direct categorization of reactivity (Drews 2000). The Lipinski's "rule of five" helps to predict the extensiveness of small molecules in drug realization by defining the preeminent practical standards for screening probable orally effective drug molecules: the molecular mass of "good" compounds should be less than 500, hydrogen bond donors should be less than five, and hydrogen bond acceptors should be less than ten, among the other rules (Lipinski 2004).

For many recent years, the pervasiveness of computational methods and computers in science has stretched out to drug discovery (Sliwoski et al. 2014). For instance, cheminformatics is the utilization of computer knowledge for comprehensively representing the biomolecular features and biochemical activities of definite compounds. These strategies have produced vast small-molecule libraries for screening against explicit curative progressions (Blaney and Martin 1997). Whenever small molecules are large in numbers, cheminformatics strategies can be utilized to produce compound libraries that are chemically and structurally more likely to produce "hits," to enhance efficacy, kinetics, stability, and of course the discovery rate. Integrally, bioinformatics methods can be utilized to find potential drug candidates, leading to a remedial action against a disorder while anticipating the synergies among proteins and other drugs along with the possible effect on various biological functions and pathways and illustrating genomic variations that can change a drug reaction (Drews 2000).

Regardless of these hi-tech approaches in drug development, the consent of novel remedial drugs has lowered down abruptly in last few years. Like in 1996 and 2007, the amount of novel drug entities recommended by the United States Food and Drug Administration (US FDA) has decreased from 53 to 17 every year, showing the similar rate as to over 50 years ago (FitzGerald 2008; Munos 2009). This is mostly because of several factors mentioned below:

1. In terms of small molecules, the "lowest hanging fruits" as drug candidates have been broadly scrutinized, and computational objections interrupt addendum of traditional approaches to further convoluted structures. Scientists allude to "rediscovering the sweet spot" in the drug invention procedure (Brown and Superti-Furga 2003) and have dedicated a lot of exertion to delivering new, directed screening libraries that influence foreseen attributes of lead compounds (Welsch et al. 2010; Cheng et al. 2012).
2. Numerous human diseases in dire need for clinical interventions have exceptionally multifaceted etiologies, and therefore potent drug targets are harder to be validated (Ramsay et al. 2018).
3. Model living beings may not give satisfactory layouts for testing medicines against more perplexing infections, because of interspecies varieties that are vital to remedial activities (Hunter 2008; Ehret et al. 2017).

A way out to delimit the primary two challenges is to center on novel types of potent drugs, outside the purview of small molecules. Natural products (NPs) may assist in this cause by recurring to the basis of remedial composites that have cured sickness for several decades (Dias et al. 2012). Even though stringent pharmacological science is new in relation to the traditional usage of NP drugs, several contemporary advancements have occurred with the assurance of "modernizing" this area (Harvey 2008). Alongside a reestablished intriguing list of NP drugs within biomedical research and development fraternity, many novel NP have progressed significantly in the pharmacological industry. A global count by Newman and Cragg appears that 646 out of 1562, i.e., 41% of every unused drug authorizations from 1981 to 2020, are NPs or inspired through NPs (Newman and Cragg 2016). Some of the latest surveys have included the needful outlines of the NP drugs and the vast range of methods that have been utilized for their recognition and characterization (Katz and Baltz 2016; Rodrigues et al. 2016), especially as of the perception having practical research strategies and advanced improvements in biotechnology. In view of the previously mentioned patterns in modern informatics strategies and the propellors in traditional computation required for the translational research applications for these strategies, such surveys can be accompanied by a steadfast deliberation for drug discovery through NPs.

Additional drift in drug discovery empowered by computational and informatics strategies is an enhancing move toward a data-determined drug discovery (Tatonetti et al. 2012; Lusher et al. 2014). Conventionally, drug discovery has been accomplished as follows: standard researchers firstly predict a target three-dimensional structure in living organism associated with an illness or a disease, trailed by high-throughput virtual screening for the "potent" small molecules that represent binding affinity toward the target three-dimensional structure. Hence, the small molecules' list is narrowed down (utilizing a few of the strategies depicted in this chapter) to

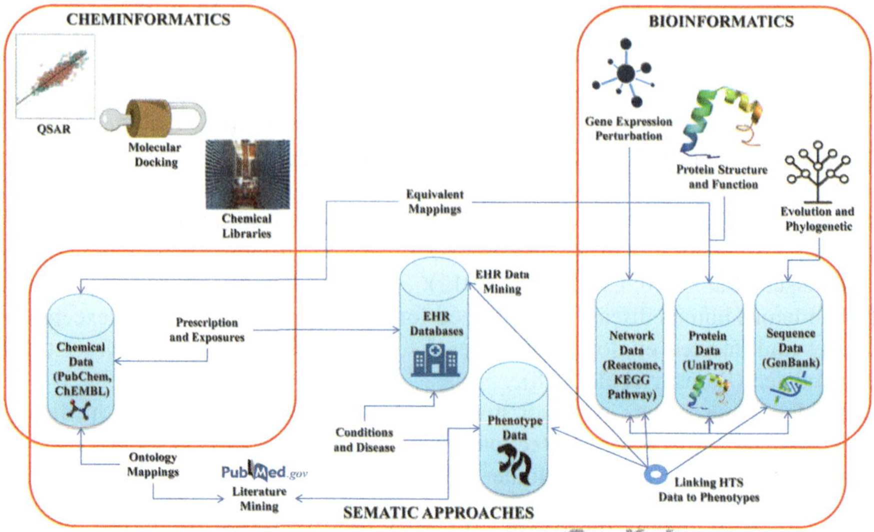

Fig. 1 Computational approaches for NP drug designing covered in this chapter

discover the foremost prominent hit(s), which at that point endure the advancement procedure to evaluate protection and adequacy in classical living beings and, inevitably, the human body. Failing to any phase in this pipeline can, and frequently does, require beginning from the start, which adds funding to the assessed budget of about 2.6 billion USD in order to convey a novel drug to retail and market (Avorn 2015).

Data-determined drug discovery fits the method on its bean, by utilizing data quarrying on big data depositories of lead small molecules and information of the disease to create innovative remedial theories methodically instead of trusting for one restorative speculation to convey actionable outputs. Apart from maintaining a strategic distance from orderly predispositions displayed within the hypothesis-driven model, data-determined drug discovery furthermore makes a difference to make strides in profit rate on ensuing bench-work experimentation and approval of lead small molecules, eventually leading to cost reduction and enhancing efficiency (Jorgensen 2004). Data-determined drug designing influences new knowledge-based information forms that were already blocked off and depends intensely upon super-computers and informatics methods to deliver progressively precise outputs (Butte and Ito 2012).

In this chapter, we investigate a few most important disciplines centered on informatics processing and computational methodologies – cheminformatics, bioinformatics, and semantic (or "knowledge-based") informatics processes – and their related strategies that can be utilized particularly for NP drug designing. These strategies are abridged graphically in Fig. 1. At last, we clinch with a review of the main crevices as of now confronting the arena of informatics NP drug designing and recommend activities for long term that may offer assistance to resolve these issues.

2 Cheminformatics Strategies

Cheminformatics strategies can for the most part be categorized in agreement to the forms of features they explore: any straight events of biochemical action (e.g., a biochemical constants, ADME estimations, or reactive sets) or unintended events (e.g., conformational motifs, small-molecule-type participation, or additional complex perceptions). These approaches can be later subcategorized; for example, structural correlation can be applicable either after or before auspicious chemical action is acknowledged (that we allude to here as imminent and review conformational mining, correspondingly). Planned conformational mining is leading in a directed way, where recognized chemical actions of well-classified small molecules are matched to the conformations of query small molecules to foresee the remedial potency of the queries. Reflective conformational mining, on the other hand, is additionally close in its resemblance to unsupervised learning strategies, where additional screening procedures begin to distinguish an intriguing compound (alluded to as a "hit") and after that looks to extend the quota of lead small molecules by viewing for conforms that are comparative to the best lead small molecule.

2.1 Cheminformatics besides Natural Products

Numerous conventional cheminformatics strategies are thought provoking to acclimatize to definite types of NPs, especially when the NPs comprise big biochemical conforms (e.g., antibodies, venoms, or further protein-built NP drug leads). For instance, creating combinatorial reference libraries of big polypeptides is presently unmanageable, because of the enormous search space. In any case, extra characteristics that are one of a kind to these types of NPs empower any abridging conventions to be made or the innovation of totally innovative methods for anticipating bioactivity (Huang et al. 2016). Here, we distribute cheminformatics into three key sets of strategies that have been utilized for accomplishment with NPs, offering explanations of the cautions that must be deliberated for NPs in specific.

Natural Product and QSAR Study

Quantitative structure activity relationship (QSAR) study is a broadly used – if every so often vaguely demarcated – method in cheminformatics for anticipating a counter variable provided a demeanor of physical, biochemical, and/or structural input factors (well known as molecular descriptors). Mostly, the objective is to study a utility of the formula

$$\overline{y} = f(\mathrm{x}) + \varepsilon$$

where x = (x_1,. . . ., x_N), i.e., the vector of *N* input factors, $\bar{y}$ is the assessed counter (uninterrupted in the event of regression and integer assessed in the instance of characterization), and ε is an error factor. *f* can be an either suitable model; collective selections comprise support vector machines, logistic regression, artificial neural networks, random forest, and others. Lately, deep learning has appeared to be especially successful for calculating an extensive range of responses, comprising probe-likeness, solubility, and others (Korotcov et al. 2017). Several commercial and freely available software applications of QSAR are offered for a range of requirements (Benfenati et al. 2011; Tosco and Balle 2011), and methods for acclimatizing genetic arithmetical and machine learning prototypes or models for QSAR are also promptly accessible (Lavecchia 2015).

QSAR has been implemented legitimately broadly to distinctive types of NPs, where particular types contribute to manage the selected molecular descriptors. Classic selections for non-NP functions comprise representative (1D or 2D) descriptors, 3D structural association, advanced order (e.g., time reliant or ligand destined) structural features (Polanski 2009), investigational estimations (partition coefficient, refractivity, polarizability, etc.), and several others. For a point-by-point survey of these and comparable descriptors, readers are directed to a detailed review by Cherkasov et al. (Cherkasov et al. 2014). Further features that can be utilized for NP small molecules comprise absolute ("one-hot") factors showing class participation (e.g., alkaloid, terpenoid), species of ancestry (or added common taxonomic clades), and further organic characteristics. Macromolecular NPs are significantly highly limited, in particular, of the classes of descriptors that can be utilized viably. Mostly, the 3D structural descriptors and interaction knowledge-based information work preeminent for these NPs and produce great outcomes (Mladenović et al. 2017; Dhiman et al. 2018). QSAR has accomplished satisfactorily for anticipating interacting affinity of antibodies to proteins, as Mandrika et al. illustrate a model comprising 26 physicochemical descriptors (comprising polarity, electronegativity, hydrophobicity, etc.) at every amino acid place in a database comprising one-chain monoclonal antibodies (Mandrika et al. 2007). Whereas such a model has not been implemented to NP drug designing, however, it appears to be a feasible approach.

Molecular Docking and Molecular Dynamics Simulations

QSAR is a valuable arithmetical strategy for anticipating possible remedial contacts, but then it is regularly alluring to precisely model the physical or chemical interface that is supposed to be examined. Molecular docking is a method that investigates to determine if and in what way two structures (generally a receptor and a ligand) substantially bind with each other. This is generally implemented in the following two stages: (1) looking for probable structural pocket binding poses and subsequently (2) scoring those binding poses. Molecular dynamics simulation is a specific simulation procedure that can be functional to molecular docking and is well known in drug advancement. Since an extraordinary grade, molecular dynamics simulation accomplishes an informatics simulation of the atoms along with molecules (usually with solvents) shown in a presumed response and let the molecules bind for a definite tenure. The methodological specifics and algorithms for molecular docking and molecular dynamics simulation are very well defined in previous studies (Karplus and McCammon 2002; Pagadala et al. 2017) – we will mainly focus on the varied limitations, disputes, and novelties in relation to particular NPs.

The variety of NP small molecules contribute to guide the appearance (receptor vs. ligand) that the small molecule represents in molecular docking simulations. Characteristically, small-molecule NPs and moderately small polypeptides (e.g., venom elements and peptide contaminants) act as ligands, whereas protein complexes and bigger proteins act as receptors (in spite of the fact that special cases are communal). This difference is essential, particularly once the objective is to filter numerous lead small molecules: generally, the receptor is detained stationary, whereas the ligand can be fetched from databases of numerous small-molecule repositories. In this manner, computationally, it is attainable to achieve docking simulation of numerous molecules once a particular molecular receptor is well recognized beforehand (Khan et al. 2009; Lee et al. 2011; Ma et al. 2011). On the contrary, in the event that a macromolecular NP is assumed of binding through endogenous compound metabolites, for example, in human cancerous cells, molecular docking simulation can be utilized to pit the compound metabolites that seem to interact with the NP molecule (Pithayanukul et al. 2009). In the event that both a receptor and a ligand are as of now anticipated by different ways like QSAR (or, preferentially, other strategies defined in this chapter), molecular docking is usually utilized as a subordinate confirmation approach. Despite their high molecular weight, the antibodies are moderately simple to filter in huge quantities by means of molecular docking, because of their particular conformational and interacting constraints that can significantly decrease the computational intricacy of simulations (Walls and Sternberg 1992; Abagyan and Totrov 2001).

Molecular dynamics simulation is an essential method for illustrating substantial binding of acknowledged drugs with known receptors; however, because of computational hassles, it cannot be utilized with existing innovations in an information-determined way to screen exceptionally huge numbers of NPs against correspondingly expansive numbers of probable receptors at the same time (Salmaso and Moro 2018). In any case, it has been demonstrated inconceivably profitable in revealing particular remedial components of NPs like in few venom proteins. A primary and compelling illustration of this appeared in 1995 by Albrand et al., whereby merged molecular dynamics simulation alongside NMR to elucidate in what way toxin FS2 obstructs L-type calcium channels was elucidated, triggering strong cardiological toxic impacts (Albrand et al. 1995). Furthermore, there are remarkable achievement stories that developed from screening comparatively trivial NP databases contrary to particular drug receptors. For instance, the compound ellagic acid has displayed both antioxidant and antiproliferative characteristics which was recognized by Moro et al. by filtering a restrictive library of 2000 small-molecule NPs contrary to oncoprotein casein kinase 2 (Cozza et al. 2006). Correspondingly, Fu et al. recognized jadomycin B – one more compound with anticancerous impacts—by filtering 15,000 microbial small-molecule metabolites contrary to oncoprotein Aurora-B kinase (Fu et al. 2008). These cases outline the possibility of molecular dynamic simulation studies for finding modern remedial NP molecules and propose that incapacitating related informatics intricacies can empower their extensive function in varied and information-determined perspectives.

Computational Mutagenesis and Library Creation

The foremost common methods for determining drug leads is to create huge databases of small molecules, which can be filtered simultaneously, amid the considerate that only a minor section will give output in "hits" (showing potent remedial actions). There are numerous methods where such databases are created, several of which come beneath the canopy name of combinatorial chemistry (i.e., identifying chemical conformations utilizing combinatorics) (Terrett et al. 1995). NPs offer a few benefits over conventional (non-NP) types of lead compounds, specifically that such "databases" already occur in nature. Universal drive online databases of chemical compounds (like ChEMBL and PubChem) (Li et al. 2010; Gaulton et al. 2017) comprise numerous NPs that are classified by compound type, whereas other, more particular databases (like VenomKB, ArachnoServer, and the Dictionary of Marine Natural Products) give indeed additional particulate interpretations for accumulating NP databases with numerous utilizable features (Pineda et al. 2018; Romano et al. 2018).

Computational mutagenesis is an associated type of method that has revealed adequacy in definite types of NPs. This strategy includes indicating a layout (e.g., a definite antibody having acknowledged remedial action that entails optimization) and, thereafter consecutively mutating positions within the layout's conformation to create a database of lead small molecules. These databases can at that point be filtered in silico (like through molecular docking simulations as explained in Part 1.2) to discover conformations, which can be synthesized within the laboratory. Antibodies, in specific, are especially compatible to in silico mutagenesis, by modulating residues in interacting section(s) (Sivasubramanian et al. 2009; Wollacott et al. 2019). The possibility of mutagenesis procedures within the perspective of NP drug designing was illustrated by Chen et al., who created a database of equivalents of the seven-residue NP peptide HUN-7293 to enhance its hindrance impacts on cell adhesion (Chen et al. 2002).

It ought to be eminent that one such benefit of research with NPs is the prospective of evading database filtering totally, supporting the supposition that nature has enhanced it for natural action. This opinion is further extended in Sect. 4.1.3.

3 Bioinformatics Strategies

Bioinformatics strategies for drug discovery incorporate everything associated with the organic characteristics of probable drug leads, comprising sequence-centered features, binding with conformational entities (cells, tissues, metabolites, proteins, etc.), pathway disturbances, and noxiousness, among others. Multiomics along with high-throughput virtual screening are correspondingly the main applications in bioinformatics concerning drug discovery. Many subcategories of bioinformatics can be practical in certain methods to the drug designing procedures (Wishart et al. 2018; Thomford et al. 2018).

3.1 Bioinformatics besides Natural Products

Within the framework of NPs, scientists have been quite successful in collating the methods associated with generating efficacious small molecules. In specific, evolution and phylogenetics offer numerous ways for different drug discovery actions. Closely associated organisms frequently offer comparable metabolites and proteins; thus once one natural compound having desired action has an inappropriate remedial catalogue for human utilization, databases of comparative small molecules can be effectively developed by probing the creatures in the similar genus. In any case, these strategies must be practical and ought to be carefully executed. For instance, individuals of certain collections of biological small molecules (like venom proteins) may augment a specific natural specialty, but the members of the similar species may display totally exclusive metabolic delineation with regard to the same small molecules in question. *Crotalus oreganus helleri*, a rattlesnake species, established a noticeable illustration of such case where individuals of this species existing on diverse edges of a mountain range created completely different venom patterns (Sunagar et al. 2014).

Gene Expression Concernment

The growth of multiomics methods in revealing our preparedness for the epidemic has driven to hoards of means to evaluate the impact that acknowledged medicines have on cells. In specific, gene expression concernment—enumeration utilizing transcriptomics and RNA sequencing – has driven to a sum of inventive developments in drug designing for infections based upon gene disarticulation, comprising cancers and several other infections with multifaceted hereditary diagnosis (Sirota et al. 2011; Subramanian et al. 2017). Alongside ecological acquaintances, structural variations from the norm, and other impacting variables, these infections frequently can be ascribed in section to variations from the norm in gene expression, comprising the system layer impacts of expression concernment in the greater framework of metabolic networks and cell signaling (Nica and Dermitzakis 2008; Cookson et al. 2009). More precisely, differential expression can be evaluated as a phenotypic indication that emerges from fundamental infection diagnosis. Consequently, drugs and drug leads that successfully alter such pernicious impacts are probable treatments for these infections.

This method is especially suited for utilization in NP drug designing, as a tremendous number of small molecules from every type of NPs are particularly augmented to have contributions in cell signaling or metabolic networks and are as of now recognized to be moderately naturally steady (Lewis and Garcia 2003). Compounds utilized in traditional Chinese medicine (TCM) have been especially well used in such research arenas. Analysts in 2014 revealed the probable contrivances by which the TCM small-molecule berberine shows anti-cancerous action, utilizing freely accessible expression information for berberine concerned human cells occupied from the Connectivity Map (CMap) project (Lee et al. 2014). An additional essential latest case by Lv et al. offers differential gene expression patterns in reaction to 102 distinctive TCM small molecules, displaying an outline for forthcoming efficient studies on the actions of TCMs (Lv et al. 2017).

An independent but associated method includes investigation of differential expression in the living beings generating NPs (instead of the living beings that NPs operate upon). An examination by Amos et al. identified formerly undiscovered NPs – along with presumed mechanisms portraying their characteristics – by correlating transcriptome patterns of distinctive bacterial species within the genus *Salinispora* (Amos et al. 2017), underlining the variety of intensifying multiomics methods that can be utilized in NP drug designing.

Molding Protein Conformation and its Role

In spite of the fact that the dimension and intricacy of proteins is frequently exorbitant to structure-centered investigations planned for compounds, other drug designing methods influence the exclusive features of proteins and additional macromolecules to achieve findings that are somewhat eccentric. Subsequently various types of protein content in NPs help these strategies that frequently are adjusted to NP drug designing having comparative effortlessness.

A few strategies utilize supervised machine learning algorithms accomplished on protein conformations (and motifs) having well-recognized response to foresee activity in novel, uncategorized proteins – this is fundamentally conventional QSAR planned to execute on proteins. The FEATURE outline (Halperin et al. 2008) ensures this utilizing three-dimensional spatial coordination of particles to identify action at various "microenvironments" in a bigger macromolecule and is consequently taken in a broader view to varied proteins having preserved practical action. Other analyst groups have planned comparable outlines grounded on additional machine learning prototypes, comprising deep learning prototypes resembling convolutional neural networks (Torng and Altman 2017; Thomas et al. 2018). For additional information on knowing protein characteristics from its conformation, we allude the reader to Pérez et al. (Pérez et al. 2018).

However, new protein practical modeling methods depend on initial factors that perform resembling "abstractions" of raw molecular features, containing residues or DNA conformation (laterally through sequence alignment algorithms) (Vyas et al. 2012), ontology interpretations (see sematic strategies segment for additional points of interest) (Mutowo et al. 2016), and biomarker reaction (Frank and Hargreaves 2003).

Utilizing Evolution to Identify Drug Lead

The point that NPs are determined from organisms suggests that they in one or the other way serve a particular reason in the perspective of that living being, or they are a derivative of an imperative procedure (Stone and Williams 1992). Subsequently, we can utilize taxonomy and evolution to identify novel small molecules and their impacts, along with libraries' creation of comparable biological products (Maplestone et al. 1992).

The easiest and utmost communal usage of phylogenetics in NP drug designing rotates over the precept that intimately associated species generate related NPs. This could be utilized to anticipate the conformations of NPs; specified conformations for comparable NPs in associated species are as of now well known (Ziemert and Jensen 2012). Succeeding an outline similar to QSAR modeling (explained in Sect. 2.1.1), phylogenetics can similarly be repurposed to foresee other features of thoroughly associated NPs, comprising compound types, harmfulness, solidness, and others, where rather than utilizing molecular descriptors as perceived characteristics of the NPs, one utilizes metamorphic features to construct a extrapolative model. Discriminant function analysis (DFA) was used as a notable illustration by Malhotra et al. specifying it to categorize and anticipate the utilities of above 250 phospholipase A2 proteins obtained from viperid snakes, where only the aligned residue sequences were used to build the initial characteristics of the DFA model (Malhotra et al. 2013).

Other practices of progression in drug designing apply phylogenomics to find relations over additional indistinctly associated species (such as among human and microbes). This comprises endeavors to set the complete breadth of several types of biological products to make inclusive NP-type libraries (Rønsted et al. 2012). Rudolf et al. in 2016 revealed that comparative genomics in varied microbial species could recognize 87 discrete gene collections through 78 bacterial species consistent to a type of acknowledged NP anticancerous medicines well recognized as enediynes (Rudolf et al. 2016). By discovery occurrences of NPs' coevolution in remotely associated species, investigations have revealed small molecules that perform vital tasks in metabolic procedures, driving to remedial resolutions in equivalent procedures in people. A notable and refined instance is revealed in the CSMNA strategy (Zhang et al. 2016), which is dependent upon the theory that resemblances between people and plant metabolic networks could be utilized to monitor phytochemical drug designing. The resemblances among the plant Foyer–Halliwell–Asada (HA) cycle and the human Nrf2–ARE pathway were used by the novelists to approve their respective drug discovery algorithm as they trigger antioxidant action of its HA cycle molecules on proteins within the Nrf2–ARE pathway.

A few cautions got to be saved in consciousness while utilizing developmental methods. Definite types of NPs are beneath metamorphic stresses that obscure phylogenetic investigation. Specifically, venom proteins could be exceedingly dissimilar indeed among species in the similar genus (Calvete et al. 2014), an anomaly recognized to the extraordinary metabolic rate of venom generation, and the profoundly directed behavior of numerous venom proteins to definite prey species.

4 Semantic (Knowledge-Centered) Strategies

Bioinformatics and cheminformatics both are the sectors of biomedical informational studies, which include the two essential disciplines involved in drug discovery and translational research. We presently swing our attention to a range of approaches developed from semiotics, library science, and linguistics; however, they have been adjusted to assist wide utilities in artificial intelligence and computational science – recognized as semantic (e.g., concerning to humans' illustratable importance) or knowledge-centered strategies. In common, these are strategies including the use of different information demonstrations, such as structured terminologies and ontologies. A few actions in this set contain rule-centered usual language handling, definite classes of medical records' mining, knowledge abstraction, semantic facts' standardization, and others. Particularly within the setting of drug discovery, knowledge-centered strategies are habitually practical in synchronization with cheminformatics and/or bioinformatics strategies and aid as solitary of the greatest methods to uniting and amalgamating discoveries and halfway outcomes spread through distinct research actions.

Conceivably the foremost well-used asset in knowledge-centered methods to drug designing is the Gene Ontology (Ashburner et al. 2000), which categorizes theoretical natural constituents to three sets: molecular utilities, cellular constituents, and biological progressions (all of them are vital in different phases of the drug discovery procedure). Analysts have generated hordes of data assets to help in drug designing, and several of these are recorded to the Gene Ontology to help through computational accumulation and preparatory approval of acknowledged theories. A few of these associated assets incorporate DrugBank (Wishart et al. 2018), UniProtKB/Swiss-Prot (and related explanation platforms like Tox-Prot) (Jungo et al. 2012), and ChEMBL (Gaulton et al. 2017); all of the inventory compounds herein will bestow certain remedial impact.

Quiet other implements have been made to outline formless data significant to drug discovery (like PubMed's journal article abstracts) to more organized illustrations. SemRep, Semantic Medline, and MetaMap from the National Library of Medicine, in addition to the NCBO Annotator from the National Center for Biomedical Ontology, recognize terminology and ontology terms in allowed content (frequently drawn from research papers) at different stages of perception. These implements have been utilized to effectively achieve ontological interpretation through numerous stages of confirmation for numerous finding errands, comprising drug designing. For added information, we allude the scholars to the authentic paper portraying Swanson's fish oil–Raynaud's syndrome theory (Swanson 1986) that elucidates how organized information and diagram algorithms could be utilized to determine descriptive correlations busted across alternatively irrelevant publications (Cameron et al. 2013).

Additional stages of knowledge illustration (like not officially measured at the perception level) correspondingly have imperative parts in drug designing; implements like OMIM can be utilized to outline recently revealed drug and gene affiliations to infections that are altered through that gene or range of genes. For inclusive entries of the several knowledge, ontologies' illustrations, and comparable implements with confirmed aspects in drug discovery, we allude the scholars to these reviews (Thomford et al. 2018; Gardner 2005; Vazquez-Naya et al. 2010).

4.1 Semantic Strategies besides Natural Products

Although the proportion of ontologies and comparative assets pertinent to drug designing are massive, progressive functions of these assets are moderately rare. This slant is indeed additionally outstanding in concerns to NP drug designing. Currently, maximum remedial correlations among NPs and infections are determined occasionally rather than over-organized, meticulous applications, even though previous segments of this chapter define eminent exclusions to this trend. In glare of the reality, which progressed utilization of semantic strategies is unusual in NP drug designing, we will also deliberate uses of terminologies and ontologies utilized for drug designing, which can be functional to NPs, dependent on existing information.

Literature Mining

One of the leading collective practices of sematic biomedical informational assets is literature mining as the process of accomplishing content mining on technical literature databases. Over 26 million biomedical content citations exist in the PubMed/MEDLINE database, of which thousands are associated with NPs and conceivably defining features of those NPs that offer directly or indirectly to indications of remedial action. Usually there are dual manners to naturally extricate corresponding information from the biological publications: (a) utilizing current terminology/ontology interpretations or (b) utilizing natural language processing (NLP) methods that find such interpretations.

One such terminology resource named Medical Subject Headings (MeSH) is designed to arrange the items of PubMed research papers and is implemented manually by skilled analysts at the US National Library of Medicine (NLM) to novel research papers soon afterward indexing in PubMed (Lipscomb 2000). MeSH terminology encompasses a varied collection of biomedical perceptions, orchestrated in a categorized design, and includes several types of NPs. The MeSH can be utilized to accumulate PubMed research papers portraying definite classes of NPs and can be advanced by utilizing extra terms (such as "drug discovery") or qualifiers (like "/therapeutic use"). The MeSH terminology could unite journal constituents to organize exterior libraries through one or the other utilizing cross mappings [comprising the NLM's Unified Medical Language System (UMLS)] or interpretations in exterior libraries straightforwardly to MeSH terminologies (Ruau et al. 2011). The MeSH terminologies have been utilized to review factors of plant genomes

(Beissinger and Morota 2017), indicating probable ways of advancing in finding novel NPs (instead of utilizing the terms to assemble information about well-recognized NPs).

A constrained number of libraries offer admittance to curated groups of research papers depicting NPs. VenomKB gives research papers interpreted to venom factors along with literature anticipations defining the acknowledged remedial impacts of those factors and mappings to other exterior databases (Romano and Tatonetti 2015). Additionally, NPASS grants chemical features of a wider collection of NPs and gives allusions to PubMed records defining manually curated organic activity calculations in a range of biological organisms (along with human beings) (Zeng et al. 2018). Other databases, containing NAPRALERT and MarinLit, offer marketable and compensated admittance to curated NP literature information.

Electronic Health Records' Mining

Essentially to literature mining, we could relate information reclamation methods to experimental information sources. To the extent that drug discovery is apprehensive, experimental information gives a technique for evaluating the impacts of compounds have on people in the context of lack of scrupulously organized clinical research readings. This fashion of statistical investigation propositions various key compensations over clinical prosecutions, comprising evasion of uncovering fresh patients to possibly hurtful medications and relieving definite sorts of bias related with qualification and patient assortment. Experimental statistics could frequently create bigger associates than clinical prosecutions. Several sources of experimental statistics could be used for drug discovery, but here we are concentrating on electronic health records (EHRs), because of their predominance and demonstrated usefulness for numerous translational research errands. In spite of the fact that confidentiality apprehensions, information disintegration, and normalization have conventionally hindered admittance to EHR statistics – predominantly for research standings deprived of clinical skill or association with an expansive scholastic therapeutic center, quickly growing endeavors, like Observational Health Data Sciences and Informatics (OHDSI) (Hripcsak et al. 2015) and the Electronic Medical Records and Genomics (eMERGE) set-up (McCarty et al. 2011), are flouting these obstructions in manners that will enhance admittance to information pertaining the extensiveness of the translational range.

EHR information are multimodal and multifaceted and focused on several exclusive predispositions and ethical/legal limitations (Weiskopf and Weng 2013). In expansion to free content (verified by healthcare suppliers), a quantity of organized information groups also exist (comprising assertions information, pharmaceutical remits, laboratory calculations, patient enumerations, and others). Currently, no chief uses of EHR information mining to NP drug designing have been revealed; however a sum of associated extents offer insights as to its possibility. Yao et al. in a review featured three particular illustrations by which EHRs can help in drug designing: (a) discovering connections among infections that drive drug repurposing, (b) assessing the utilization designs and protection of drugs and/or drug leads, and (c) determining genotype–phenotype affiliations that could lead to the innovation of novel drug receptors for particular infections (Yao et al. 2011). Significant cautions of all of these could be deliberated from the perception of NP drug designing, comprising particular points of interest and drawbacks that NPs offer relative to non-NP drugs and drug leads.

Drug repurposing includes captivating a prevailing drug and utilizing it to treat a diverse infection other than what it is presently envisioned for (Ashburn and Thor 2004). EHRs have been utilized for a numerous drug repurposing methods. The utmost mutual repurposing technique includes finding resemblances among illnesses and after that utilizing those resemblances to suggest innovative medications, usually dependent on the hypothesis that infection having comparable etiologies will create comparable indications in the EHR, which comparative etiologies may infer comparable treatments. A vital illustration by Rzhetsky et al. appeared unanticipated resemblance among breast cancer and bipolar disorder (Rzhetsky et al. 2007). Lately, it has been illustrated that the drug tamoxifen for breast cancer might be valuable for medicating the signs of bipolar syndrome (Kulkarni et al. 2006).

EHR information can likewise be utilized to evaluate the security of drugs (or putative drugs), by deciding if introduction to the drug increments the hazard of unfavorable impacts (Tatonetti et al. 2012; Schuemie et al. 2012). This can be accessible for affirmed drugs that have implicit illustrations within the EHR program (such as the ones with ATC codes or comparable – investigational and unaccepted drugs usually do not have an organized illustration in EHR libraries); however characteristic dialect dispensation can distinguish exploratory and acknowledged drugs with sensible viability (Björne et al. 2013). This recommends that NP drug lead security observation can be achieved on permitted content records within the EHR, particularly when preserved as natural contacts instead of doctor's recommended mediations. The achievability of such method was illustrated by Zhang et al., by showing that home-grown and biological appendages (that are ordinarily deliberated NPs) can be recognized in treatment records utilizing characteristic dialect transforming and evaluated the void among organized drug illustrations and such small molecules (Zhang et al. 2016). Two of the chief gaps in requirement of determination to understand this objective comprise postulating a consistent terminology for NPs (Dewick 2002) and recognizing where (topographically) hospital patients may be resolved to the NPs being explored.

Finding novel drug receptors is not entirely the similar aspect as drug designing; however, it does give a crucial beginning point for recognizing novel drug candidates. Last few years have seen a stable weakening within the discovery of novel receptors, and past surveys on the theme have aimed for novel and inventive techniques to report this concern (Lindsay 2005; Spedding 2006). The EHR information and clinical biobanks to carry out genome-wide association studies (GWASs) and phenome-wide association studies (PheWASs) are publicized as arrangements (Yao et al. 2011) by giving associative relations among illnesses and particular hereditary loci, which can then be utilized as receptors for novel exactness drug treatments (McCarty and Wilke 2010; Wilke et al. 2011). NPs, in specific, come into action when acknowledging their exclusive capabilities to target definite genes and gene items that are ineffectively focused on small molecules. The protein-centered remedies and monoclonal antibodies both are recognized for their capacity to object cell classes of a person, exclusively beneficial in cancers having particular hereditary marks (Adams and Weiner 2005; Cox et al. 2016). PheWAS and GWAS are comparatively novel for drug designing and advancement schedule; however we are possibly going to see numerous NP drugs arising from clinical prosecutions utilizing EHR and biobank-permitted examinations for object disclosure within the upcoming era (Thomford et al. 2018).

Associating HTS Statistics to Recognized Infectious Medication

Keeping in mind of the current research works, we have discussed about various means by which terminologies and ontologies could be utilized to recover and get conformational information. Nevertheless, another critical role that semantic strategies play in biomedicine is corresponding dissimilar information sources in manners that then necessitate enormous quantities of physical analysis and elucidation to administer at scale. This is essential for several causes, comprising observational approval, expanding arithmetical control and inferential aptitude, and indeed finding new information completely. A specific use that has knowledgeable quick development and key procedural progressions in drug designing is associating novel sorts of high-throughput sequencing (HTS) information to scientifically significant affiliations. Earlier familiar methods, like gene expression concernment (Part 2.1), abdicate outcomes comprising indications that have natural significance but no major association with clinical phenotypes. Previous vital illustrations of data-determined drug discovery from gene expression created remedial correlations among lung adenocarcinoma and cimetidine (Sirota et al. 2011), in addition to inflammatory bowel disease and topiramate (Dudley et al. 2011), but these illustrations needed manual analysis of numerous phenotype-associated expression patterns through which innovations could be accomplished. Information illustrations offer a strategy for creating these associations spontaneously, when accurately utilized.

Effective information amalgamation of this sort necessitates relations to be shaped among (a) groups of genes (or, more precisely, sets of inquest groups) and a metabolic pathway and (b) relations among phenotypes and pathways. A list of eminent and luxuriously clarified gene-pathway libraries (comprising KEGG and Reactome) (Fabregat et al. 2018; Kanehisa et al. 2017) exists and is utilized broadly through the biomedical research communal. Assets associating phenotypes to pathways are significantly not as much predominant (and comprehensively limited) due to their restriction to access important information, but the progressing determinations in translational bioinformatics communal are altering this. Assimilating contrasts in phenotypic response and gene expression at the cell and tissue level having pathway information has revealed specific assurance in this section (Hao et al. 2018; Hao and Tatonetti 2016). A latest assessment by Oellrich et al. summarizes developing and recognized implements for computational phenotyping (Oellrich et al. 2016).

Comparative studies are, though, closely lacking from the monarchy of NP drug designing. The exclusive features of diverse NP type (particularly those portrayed prior in this chapter) can enable the phenotyping procedure. Metabolomics information offers indications as to NPs' innovative characteristics in their source living beings that can regularly be expanded to their impacts when functional to people (Zhang et al. 2016; Xie et al. 2008; Yan et al. 2015). Phylogenomics can focus on resemblances among the genetic epidemiologies of multifaceted infections in people vs. model living beings, conceivably recommending species from which to mine leads that can cure these illnesses (Romano et al. 2015). Indeed the prey/predator alterations of NP-creating species can recommend the organic utility of NPs (de la Vega and Possani 2005; Miller et al. 2016); the revelation that the cone snail *Conus geographus* chases fish through discharging insulin toward the neighboring water (subsequent in quick hypoglycemic stun within the prey) driven to the recognizable proof of a capable insulin receptor interacting motif that has appeared impressive assurance for upcoming medications of diabetes (Menting et al. 2016). Certain latest studies centering on innovation from TCM information appeared promising: Cui et al. generated a TCM biochemical structural library, which they filtered in discrepancy of acetylcholinesterase (ACE) inhibitors, together through molecular docking simulations using the recognized conformation of ACE, in addition to resemblance to current ACE inhibitors recovered from BindingDB (Cui et al. 2015). Possibly, ontology assets can be utilized to acclimatize these strategies into a mechanized method for filtering several drug types through lesser to no labor-intensive curation.

Connecting HTS information to illness phenotypes is merely one use of semantic information assets that may be an advantage for NP drug designing. There are several additional possible employments for associating indication among medical information sets, drug terms, literature-extracted affiliations, and organismal biodiversity information, any of which might head toward possibly profitable revelations and enhanced indication for problematic theories.

References

Abagyan R, Totrov M (2001) High-throughput docking for lead generation. Curr Opin Chem Biol 5(4):375–382

Adams GP, Weiner LM (2005) Monoclonal antibody therapy of cancer. Nat Biotechnol 23(9): 1147–1157

Albrand J-P et al (1995) NMR and restrained molecular dynamics study of the three-dimensional solution structure of toxin FS2, a specific blocker of the L-type calcium channel, isolated from black mamba venom. Biochemistry 34(17):5923–5937

Amos GC et al (2017) Comparative transcriptomics as a guide to natural product discovery and biosynthetic gene cluster functionality. Proc Natl Acad Sci 114(52):E11121–E11130

Ashburn TT, Thor KB (2004) Drug repositioning: identifying and developing new uses for existing drugs. Nat Rev Drug Discov 3(8):673–683
Ashburner M et al (2000) Gene ontology: tool for the unification of biology. Nat Genet 25(1):25–29
Avorn J (2015) The $2.6 billion pill–methodologic and policy considerations. N Engl J Med 372(20):1877–1879
Beissinger TM, Morota G (2017) Medical subject heading (MeSH) annotations illuminate maize genetics and evolution. Plant Methods 13(1):1–8
Benfenati E et al (2011) CORAL software: QSAR for anticancer agents. Chem Biol Drug Des 77(6):471–476
Björne J, Kaewphan S, and Salakoski T. *UTurku: drug named entity recognition and drug-drug interaction extraction using SVM classification and domain knowledge*. in *Second Joint Conference on Lexical and Computational Semantics (* SEM), Volume 2: Proceedings of the Seventh International Workshop on Semantic Evaluation (SemEval 2013)*. 2013
Blaney JM, Martin EJ (1997) Computational approaches for combinatorial library design and molecular diversity analysis. Curr Opin Chem Biol 1(1):54–59
Brown D, Superti-Furga G (2003) Rediscovering the sweet spot in drug discovery. Drug Discov Today 8(23):1067–1077
Butte A, Ito S (2012) Translational bioinformatics: data-driven drug discovery and development. Wiley Online Library
Calvete JJ et al (2014) Omics meets biology: application to the design and preclinical assessment of antivenoms. Toxins 6(12):3388–3405
Cameron D et al (2013) A graph-based recovery and decomposition of Swanson's hypothesis using semantic predications. J Biomed Inform 46(2):238–251
Chen Y et al (2002) Solution-phase parallel synthesis of a pharmacophore library of HUN-7293 analogues: a general chemical mutagenesis approach to defining structure– function properties of naturally occurring cyclic (Depsi) peptides. J Am Chem Soc 124(19):5431–5440
Cheng T et al (2012) Structure-based virtual screening for drug discovery: a problem-centric review. AAPS J 14(1):133–141
Cherkasov A et al (2014) QSAR modeling: where have you been? Where are you going to? J Med Chem 57(12):4977–5010
Cleary EG et al (2018) Contribution of NIH funding to new drug approvals 2010–2016. Proc Natl Acad Sci 115(10):2329–2334
Cookson W et al (2009) Mapping complex disease traits with global gene expression. Nat Rev Genet 10(3):184–194
Cox N et al (2016) Integrin-targeting knottin peptide–drug conjugates are potent inhibitors of tumor cell proliferation. Angew Chem Int Ed 55(34):9894–9897
Cozza G et al (2006) Identification of ellagic acid as potent inhibitor of protein kinase CK2: a successful example of a virtual screening application. J Med Chem 49(8):2363–2366
Cui L et al (2015) Discovering new acetylcholinesterase inhibitors by mining the buzhongyiqi decoction recipe data. J Chem Inf Model 55(11):2455–2463
Dewick PM (2002) Medicinal natural products: a biosynthetic approach. John Wiley & Sons
Dhiman P, Malik N, Khatkar A (2018) 3D-QSAR and in-silico studies of natural products and related derivatives as monoamine oxidase inhibitors. Curr Neuropharmacol 16(6):881–900
Dias DA, Urban S, Roessner U (2012) A historical overview of natural products in drug discovery. Meta 2(2):303–336
Drews J (2000) Drug discovery: a historical perspective. Science 287(5460):1960–1964
Dudley JT et al (2011) Computational repositioning of the anticonvulsant topiramate for inflammatory bowel disease. Sci Transl Med 3(96):96ra76-96ra76
Ehret T et al (2017) Translational rodent models for research on parasitic protozoa—a review of confounders and possibilities. Front Cell Infect Microbiol 7:238
Fabregat A et al (2018) The reactome pathway knowledgebase. Nucleic Acids Res 46(D1):D649–D655

FitzGerald GA (2008) Drugs, industry, and academia. American Association for the Advancement of Science

Frank R, Hargreaves R (2003) Clinical biomarkers in drug discovery and development. Nat Rev Drug Discov 2(7):566–580

Fu D-H et al (2008) Jadomycin B, an Aurora-B kinase inhibitor discovered through virtual screening. Mol Cancer Ther 7(8):2386–2393

Gardner SP (2005) Ontologies in drug discovery. Drug Discov Today Technol 2(3):235–240

Gaulton A et al (2017) The ChEMBL database in 2017. Nucleic Acids Res 45(D1):D945–D954

Halperin I et al (2008) The FEATURE framework for protein function annotation: modeling new functions, improving performance, and extending to novel applications. BMC Genomics 9(2): 1–14

Hao Y, Tatonetti NP (2016) Predicting G protein-coupled receptor downstream signaling by tissue expression. Bioinformatics 32(22):3435–3443

Hao Y et al (2018) Tissue-specific analysis of pharmacological pathways. CPT Pharmacometrics Syst Pharmacol 7(7):453–463

Harvey AL (2008) Natural products in drug discovery. Drug Discov Today 13(19-20):894–901

Hripcsak G et al (2015) Observational health data sciences and informatics (OHDSI): opportunities for observational researchers. Stud Health Technol Inform 216:574

Huang P-S, Boyken SE, Baker D (2016) The coming of age of de novo protein design. Nature 537(7620):320–327

Hunter P (2008) The paradox of model organisms: the use of model organisms in research will continue despite their shortcomings. EMBO Rep 9(8):717–720

Jorgensen WL (2004) The many roles of computation in drug discovery. Science 303(5665): 1813–1818

Jungo F et al (2012) The UniProtKB/Swiss-Prot Tox-Prot program: a central hub of integrated venom protein data. Toxicon 60(4):551–557

Kanehisa M et al (2017) KEGG: new perspectives on genomes, pathways, diseases and drugs. Nucleic Acids Res 45(D1):D353–D361

Karplus M, McCammon JA (2002) Molecular dynamics simulations of biomolecules. Nat Struct Biol 9(9):646–652

Katz L, Baltz RH (2016) Natural product discovery: past, present, and future. J Ind Microbiol Biotechnol 43(2-3):155–176

Khan MTH et al (2009) Cholinesterase inhibitory activities of some flavonoid derivatives and chosen xanthone and their molecular docking studies. Chem Biol Interact 181(3):383–389

Korotcov A et al (2017) Comparison of deep learning with multiple machine learning methods and metrics using diverse drug discovery data sets. Mol Pharm 14(12):4462–4475

Kulkarni J et al (2006) A pilot study of hormone modulation as a new treatment for mania in women with bipolar affective disorder. Psychoneuroendocrinology 31(4):543–547

Lavecchia A (2015) Machine-learning approaches in drug discovery: methods and applications. Drug Discov Today 20(3):318–331

Lee K-H et al (2014) A gene expression signature-based approach reveals the mechanisms of action of the Chinese herbal medicine berberine. Sci Rep 4(1):1–9

Lee KW, Bode AM, Dong Z (2011) Molecular targets of phytochemicals for cancer prevention. Nat Rev Cancer 11(3):211–218

Lewis RJ, Garcia ML (2003) Therapeutic potential of venom peptides. Nat Rev Drug Discov 2(10): 790–802

Li Q et al (2010) PubChem as a public resource for drug discovery. Drug Discov Today 15(23-24): 1052–1057

Lindsay MA (2005) Finding new drug targets in the 21st century. Drug Discov Today 10(23-24): 1683–1687

Lipinski CA (2004) Lead-and drug-like compounds: the rule-of-five revolution. Drug Discov Today Technol 1(4):337–341

Lipscomb CE (2000) Medical subject headings (MeSH). Bull Med Libr Assoc 88(3):265

Lusher SJ et al (2014) Data-driven medicinal chemistry in the era of big data. Drug Discov Today 19(7):859–868
Lv C et al (2017) The gene expression profiles in response to 102 traditional Chinese medicine (TCM) components: a general template for research on TCMs. Sci Rep 7(1):1–10
Ma D-L, Chan DS-H, Leung C-H (2011) Molecular docking for virtual screening of natural product databases. Chem Sci 2(9):1656–1665
Malhotra A et al (2013) Predicting function from sequence in a large multifunctional toxin family. Toxicon 72:113–125
Mandrika I et al (2007) QSAR of multiple mutated antibodies. J Mol Recognit 20(2):97–102
Maplestone RA, Stone MJ, Williams DH (1992) The evolutionary role of secondary metabolites—a review. Gene 115(1-2):151–157
McCarty CA, Wilke RA (2010) Biobanking and pharmacogenomics. Pharmacogenomics 11(5): 637–641
McCarty CA et al (2011) The eMERGE Network: a consortium of biorepositories linked to electronic medical records data for conducting genomic studies. BMC Med Genet 4(1):1–11
Menting JG et al (2016) A minimized human insulin-receptor-binding motif revealed in a Conus geographus venom insulin. Nat Struct Mol Biol 23(10):916–920
Miller D et al (2016) Sex differences in defensive behavior and venom of the striped bark scorpion Centruroides vittatus (Scorpiones: Buthidae). Oxford University Press
Mladenović M et al (2017) Understanding the molecular determinant of reversible human monoamine oxidase b inhibitors containing 2 h-chromen-2-one core: structure-based and ligand-based derived three-dimensional quantitative structure–activity relationships predictive models. J Chem Inf Model 57(4):787–814
Munos B (2009) Lessons from 60 years of pharmaceutical innovation. Nat Rev Drug Discov 8(12): 959–968
Mutowo P et al (2016) A drug target slim: using gene ontology and gene ontology annotations to navigate protein-ligand target space in ChEMBL. J Biomed Semantics 7(1):1–7
Newman DJ, Cragg GM (2016) Natural products as sources of new drugs from 1981 to 2014. J Nat Prod 79(3):629–661
Nica AC, Dermitzakis ET (2008) Using gene expression to investigate the genetic basis of complex disorders. Hum Mol Genet 17(R2):R129–R134
Oellrich A et al (2016) The digital revolution in phenotyping. Brief Bioinform 17(5):819–830
Pagadala NS, Syed K, Tuszynski J (2017) Software for molecular docking: a review. Biophys Rev 9(2):91–102
Pérez A, Martínez-Rosell G, De Fabritiis G (2018) Simulations meet machine learning in structural biology. Curr Opin Struct Biol 49:139–144
Pineda SS et al (2018) ArachnoServer 3.0: an online resource for automated discovery, analysis and annotation of spider toxins. Bioinformatics 34(6):1074–1076
Pithayanukul P, Leanpolchareanchai J, Saparpakorn P (2009) Molecular docking studies and anti–Snake venom metalloproteinase activity of Thai mango seed kernel extract. Molecules 14(9): 3198–3213
Polanski J (2009) Receptor dependent multidimensional QSAR for modeling drug-receptor interactions. Curr Med Chem 16(25):3243–3257
Ramsay RR et al (2018) A perspective on multi-target drug discovery and design for complex diseases. Clin Transl Med 7(1):1–14
Rodrigues T et al (2016) Counting on natural products for drug design. Nat Chem 8(6):531
Romano JD, Nwankwo V, Tatonetti NP (2018) VenomKB v2. 0: a knowledge repository for computational toxinology. bioRxiv:295204
Romano JD, Tatonetti NP (2015) VenomKB, a new knowledge base for facilitating the validation of putative venom therapies. Sci Data 2(1):1–9
Romano JD, Tharp WG, Sarkar IN (2015) Adapting simultaneous analysis phylogenomic techniques to study complex disease gene relationships. J Biomed Inform 54:10–38

Rønsted N et al (2012) Can phylogeny predict chemical diversity and potential medicinal activity of plants? A case study of Amaryllidaceae. BMC Evol Biol 12(1):1–12

Ruau D et al (2011) Comparison of automated and human assignment of MeSH terms on publicly-available molecular datasets. J Biomed Inform 44:S39–S43

Rudolf JD, Yan X, Shen B (2016) Genome neighborhood network reveals insights into enediyne biosynthesis and facilitates prediction and prioritization for discovery. J Ind Microbiol Biotechnol 43(2-3):261–276

Rzhetsky A et al (2007) Probing genetic overlap among complex human phenotypes. Proc Natl Acad Sci 104(28):11694–11699

Salmaso V, Moro S (2018) Bridging molecular docking to molecular dynamics in exploring ligand-protein recognition process: an overview. Front Pharmacol 9:923

Schuemie MJ et al (2012) Using electronic health care records for drug safety signal detection: a comparative evaluation of statistical methods. Med Care 50(10):890–897

Sirota M et al (2011) Discovery and preclinical validation of drug indications using compendia of public gene expression data. Sci Transl Med 3(96):96ra77-96ra77

Sivasubramanian A et al (2009) Toward high-resolution homology modeling of antibody Fv regions and application to antibody–antigen docking. Proteins: Structure, Function, and Bioinformatics 74(2):497–514

Sliwoski G et al (2014) Computational methods in drug discovery. Pharmacol Rev 66(1):334–395

Spedding M (2006) New directions for drug discovery. Dialogues Clin Neurosci 8(3):295

Stone M, Williams D (1992) On the evolution of functional secondary metabolites (natural products). Mol Microbiol 6(1):29–34

Subramanian A et al (2017) A next generation connectivity map: L1000 platform and the first 1,000,000 profiles. Cell 171(6):1437–1452. e17

Sunagar K et al (2014) Intraspecific venom variation in the medically significant Southern Pacific Rattlesnake (Crotalus oreganus helleri): biodiscovery, clinical and evolutionary implications. J Proteome 99:68–83

Swanson DR (1986) Fish oil, Raynaud's syndrome, and undiscovered public knowledge. Perspect Biol Med 30(1):7–18

Tatonetti NP et al (2012) Data-driven prediction of drug effects and interactions. Sci Transl Med 4(125):125ra31-125ra31

Terrett NK et al (1995) Combinatorial synthesis—the design of compound libraries and their application to drug discovery. Tetrahedron 51(30):8135–8173

Thomas N et al (2018) Tensor field networks: rotation-and translation-equivariant neural networks for 3d point clouds. arXiv. preprint arXiv:1802.08219

Thomford NE et al (2018) Natural products for drug discovery in the 21st century: innovations for novel drug discovery. Int J Mol Sci 19(6):1578

Torng W, Altman RB (2017) 3D deep convolutional neural networks for amino acid environment similarity analysis. BMC Bioinformatics 18(1):1–23

Tosco P, Balle T (2011) Open3DQSAR: a new open-source software aimed at high-throughput chemometric analysis of molecular interaction fields. J Mol Model 17(1):201–208

Vazquez-Naya MJ et al (2010) Ontologies of drug discovery and design for neurology, cardiology and oncology. Curr Pharm Des 16(24):2724–2736

de la Vega RCR, Possani LD (2005) Overview of scorpion toxins specific for Na+ channels and related peptides: biodiversity, structure–function relationships and evolution. Toxicon 46(8): 831–844

Vyas V et al (2012) Homology modeling a fast tool for drug discovery: current perspectives. Indian J Pharm Sci 74(1):1

Walls PH, Sternberg MJ (1992) New algorithm to model protein-protein recognition based on surface complementarity: applications to antibody-antigen docking. J Mol Biol 228(1):277–297

Weiskopf NG, Weng C (2013) Methods and dimensions of electronic health record data quality assessment: enabling reuse for clinical research. J Am Med Inform Assoc 20(1):144–151

Welsch ME, Snyder SA, Stockwell BR (2010) Privileged scaffolds for library design and drug discovery. Curr Opin Chem Biol 14(3):347–361

Wilke R et al (2011) The emerging role of electronic medical records in pharmacogenomics. Clin Pharmacol Ther 89(3):379–386

Wishart DS et al (2018) DrugBank 5.0: a major update to the DrugBank database for 2018. Nucleic Acids Res 46(D1):D1074–D1082

Wollacott AM et al (2019) Structural prediction of antibody-APRIL complexes by computational docking constrained by antigen saturation mutagenesis library data. J Mol Recognit 32(7):e2778

Xie G et al (2008) Ultra-performance LC/TOF MS analysis of medicinal Panax herbs for metabolomic research. J Sep Sci 31(6-7):1015–1026

Yan T et al (2015) UPLC-MS/MS determination of ephedrine, methylephedrine, amygdalin and glycyrrhizic acid in Beagle plasma and its application to a pharmacokinetic study after oral administration of Ma Huang Tang. Drug Test Anal 7(2):158–163

Yao L et al (2011) Electronic health records: implications for drug discovery. Drug Discov Today 16(13-14):594–599

Zeng X et al (2018) NPASS: natural product activity and species source database for natural product research, discovery and tool development. Nucleic Acids Res 46(D1):D1217–D1222

Zhang B et al (2016) New strategy for drug discovery by large-scale association analysis of molecular networks of different species. Sci Rep 6(1):1–12

Ziemert N, Jensen PR (2012) Phylogenetic approaches to natural product structure prediction. Methods Enzymol 517:161–182

Structural Biology an Essential Tool for Drug Discovery and Development

Chitra Rani, Vikrant Singh Rajput, and Shah Ubaid-ullah

Abstract Deciphering the three-dimensional structures of biomolecules, in particular, enzymes/proteins using structural biology techniques like X-ray crystallography is an indispensable and powerful tool of present modern drug discovery programmes. Structure determination plays a key role in structure-based drug design (SBDD) approach, it is often considered as a benchmark of data defining the atomic structure of molecules like proteins and nucleic acids. Through crystal structure data one can easily map the active site residues present in the essential enzymes of pathogens which is used as blue print for designing of new drugs. Preparation of highly purified protein sample is the first step. The purified protein, thus obtained, is used for crystal preparation. Therefore, these important aspects of protein preparation, crystallization methodology and data collection are described in this chapter.

1 Introduction

It was roughly 30 years ago when the potential of structure determination by X-ray crystallography was realized when X-rays were used for the determination of 3-D structure of enzymes, globins and polypeptides hormones (Blundell et al. 1972; Beddell et al. 1976; Goodford et al. 1980). Identification of 'druggable' protein target is the first and primary step amongst all the steps involved in the drug

C. Rani (✉)
University of Connecticut Health Center, Farmington, Connecticut, USA

V. S. Rajput
Department of Biomedical Engineering, Central University of Rajasthan, Ajmer, India

S. Ubaid-ullah (✉)
Department of Biotechnology, Islamia College of Science & Commerce (ICSC), Srinagar, India

Present Address: J & K Higher Education Department (Govt. Degree College, Pulwama), Srinagar, India

V. S. Rajput, A. Runthala (eds.), *Drugs and a Methodological Compendium*,
https://doi.org/10.1007/978-981-19-7952-1_2

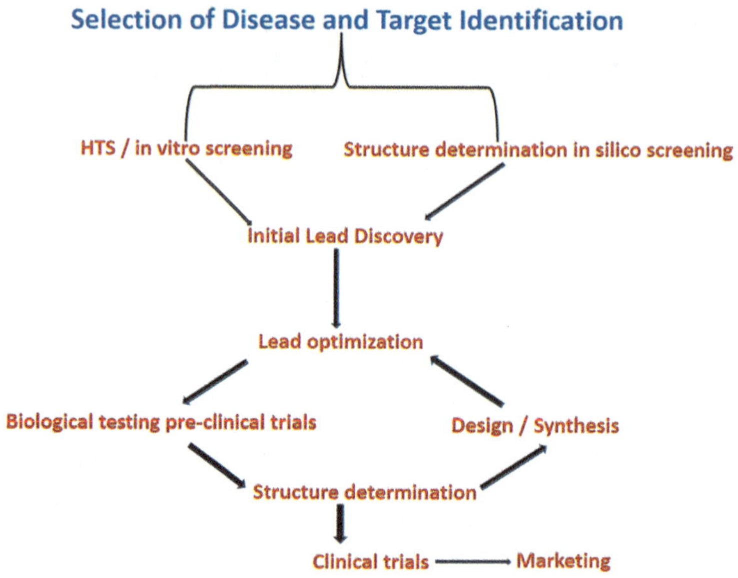

Fig. 1 Steps involved in drug discovery

discovery process (Fig. 1). A target is said to be *druggable* when there is a therapeutic effect after the modulation of its activity. In this context, 3-D structure information of a drug target serves as an initial template for the designing of drug-like molecules.

Antihypertensive drug, captopril which is an inhibitor for angiotensin-converting, was the first drug based on a structure-based approach, approved in 1981. This was followed by the approval of dorzolamide in 1995, an inhibitor for carbonic anhydrase used to treat glaucoma (Sugrue 2000). The crystallization and structure determination of macromolecules is not a single-step process but involves at least common five steps: (1) the generation of recombinant clones that produce a sufficient amount of protein in a suitable expression system, (2) optimizing and setting up an efficient purification methodology for the target protein and production of a good amount of purified protein to begin crystallization experiments, (3) optimization of crystallization condition using different salts and precipitants to develop quality crystals that give good X-ray diffraction pattern, (4) processing of this diffracted data into electron density maps, and lastly (5) building the model and its refinement. The crossing-point of any of these five steps can be a hurdle in the whole procedure. Since, all these steps are interrelated and follow each other, therefore without completing a particular step in an accurate way would not allow moving to the next step. The information generated through X-ray crystallography is used for generating 3-D models of the drug targets. In this chapter, the methodology used for the structure determination of druggable target proteins of Gram-negative bacteria *Acinetobacter baumannii* will be discussed.

2 *Acinetobacter Baumannii*: A Concern for Health System

The rate of bacterial infections is rising because of the development of multi-, extreme- and pan-drug resistant (MDR, XDR and PDR) forms (Eliopoulos et al. 2008). 'ESKAPE' is a term proposed by the Infectious Disease Society of America (IDSA) to describe bacterial pathogens that cause life-threatening infectious illnesses and are associated with high rates of drug-resistant strains. ESKAPE includes *Enterococcus faecium*, *Staphylococcus aureus*, *Klebsiella pneumoniae*, *Acinetobacter baumannii*, *Pseudomonas aeruginosa* and *Enterobacter* spp. (Abbo et al. 2005; Abbo et al. 2007; Rice 2008; Santajit and Indrawattana 2016). *Escherichia coli* is also included in the ESKAPE due to its capability to acquire and express plasmids carrying genes for extended-spectrum β-lactamases and carbapenemases. Drug-resistant infections are difficult to treat, with rising morbidity, death, and healthcare expenditures (HammoudiHalat and Ayoub 2020). The XDR strains of *K. pneumoniae* or *A. baumannii* are responsible for high mortality rates, which accounts for up to 50% and clinicians often used therapeutics such as colistin which possesses significant toxicity or side effects to treat these infections (Lee and Doi 2014). It has been reported that *A. baumannii* strains became resistant to colistin and gained cross-resistance to host antimicrobial peptides at the same time (Vila-Farres et al. 2012). The fear of post-antibiotic age was prophesied but never realized because of methicillin-resistant *S. aureus* (MRSA) or vancomycin-resistant Enterococcus (VRE). Antibiotic resistance in bacteria is caused by three fundamental mechanisms: (1) The bacterial system with plasmids which carries the genes which encodes for enzymes, such as a -lactamase that hydrolyze penicillin which results in that inactivation of the antibiotic (2) The enzyme's target site receptor or the ribosomal subunit is mutated, resulting in poor drug binding. (3) Protein transportation is disrupted, resulting in antibiotic entrance restrictions or active efflux from the cell (McDevitt et al. 2002). According to previous reports, more than 90% of the developed hit molecules were deemed unsuitable for drug-resistant diseases, whilst the rest were either ineffective against superbugs or had high cytotoxicity in humans (including Oritavancin and Dalvance). As a result, the WHO has labelled this era the antibiotic crisis era, implying that new tactics to identify and develop effective medicines to overcome widespread and developing antibiotic resistance are urgently needed in order to save millions of lives. One method to accomplish this is to find new targets for new drugs/small ligands. A cofactor can be a suitable or good target as a single cofactor may be required by several enzymes in a metabolic pathway, therefore cofactor production pathways provide a variety of possible therapeutic targets. Coenzyme A (CoA) is an example of such a cofactor plays a vital role in the metabolic activities of almost all organisms. CoA is used as an acyl group carrier, activating carbonyl group in a number of cellular reactions vital for metabolism and it also makes available in the −4'phospho pantetheine prosthetic group integrated by some carrier proteins which play an important role in fatty acid, polyketide and non-ribosomal peptide biosynthesis (Leonardi et al. 2005). It is well documented that several eukaryotes and prokaryotes depend on pantothenic acid present in the medium but some bacteria, fungi and plants, however, can synthesize pantothenic acid de novo by using intermediates like β-alanine and

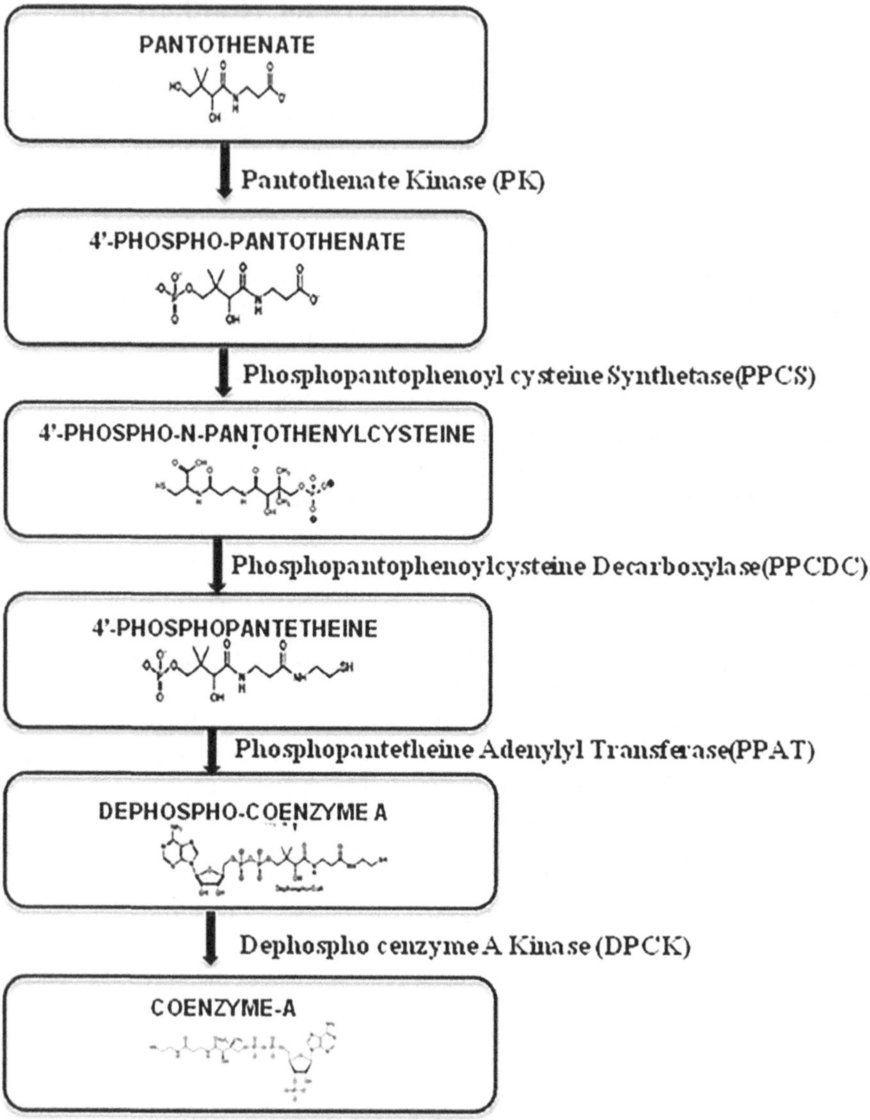

Fig. 2 Biosynthetic pathway of Co-enzyme A(Gupta et al. 2021)

α-ketoisovalerate. The pathway uses α-ketoisovalerate, which is an intermediate produced in the synthesis of branched-chain amino acid, and is converted to ketopantoate using the enzyme ketopantoatehydroxymethyltransferase (KPHMT). In the next step, ketopantoate is reduced to pantoate, a reaction catalyzed by ketopantoate reductase (KPR). Pantoate reacts with β-alanine using the enzyme pantothenate synthetase (PS) to give pantothenate (Gupta et al. 2021). In vivo, CoA biosynthesis is catalyzed by five enzymatic steps (Fig. 2).

In the first step, pantothenate is phosphorylated to 4′-phosphopantothenate with the involvement of enzyme pantothenate kinase (PanK). After this reaction, 4′--phosphopantothenate reacts with a molecule of cysteine to give 4′--phosphopantothenoylcysteine, catalyzed by phosphopantothenoylcysteine synthetase (PPCS). The third step is the decarboxylation of the cysteine moiety of 4′-phosphopantothenoylcysteine to yield 4′-phosphopantetheine involving enzyme phosphopantothenoylcysteine decarboxylase (PPCDC). The next step is the transfer of an adenylyl group to 4′-phosphopantetheine, to form dephospho-CoA. The final step is phosphorylation of dephospho-CoA at 3′ of ribose by the enzyme dephospho-CoA kinase (DPCK) to give Co-A. To understand the techniques used in structural biology, we will take the example of two enzymes involved in this pathway, i.e. PanK and PPAT as these two enzymes from *A. baumannii* are structurally and biophysically well characterized.

3 Cloning and Expression: A Technique for Scale up for Protein Production

The first and one of the most challenging and time-consuming step in understanding the structures of the selected proteins encoded by their genes is to produce these proteins in adequate amount and of high purity (1–10 mg). The purity of protein sample plays a crucial role in the process of crystallization. Any impurity present in the sample protein may be responsible for complete lack of crystal formation or may give rise to small crystals that may not be proper for diffraction studies or to development of crystals that do not scatter X-rays properly. The reason for this is the presence of impurities in the crystal which cause lattice strain thus decreasing the crystalline order leading to diffraction pattern which is poor both in terms of resolution as well as intensity(Giegé et al. 1986; Zulauf and D'Arcy 1992; Rosenberger 1996; Thomas et al. 1998; Structural Genomics Consortium 2008). The different strategies that are employed for optimizing protein expression and purification include using good DNA source, suitable transcription–translation system, appropriate protein-folding conditions, proper expression system, affinity tags for purification, protein modifications (if necessary), purification scheme, and quality assurance.

Bacterial expression systems are the first, easy and low-cost choices for protein expression. But, most eukaryotic or most human proteins involving post-translational modifications cannot be expressed in these systems in a soluble form (Edwards et al. 2000), thus expression systems are to be used. So, mammalian systems are used to produce these proteins in proper folded and modified form. Since mammalian cell culture is a complex technique often with a low yield of protein, therefore mammalian expression is not an economical choice. Baculovirus–insect cell expression system is used as an alternative approach to mammalian expression. For the expression of most soluble human proteins with proper post-translational modifications, different insect cell lines are used (Davies 1994). This system allows for the co-expression of numerous viral constructs encoding complementary proteins in the same cells, which help in protein complex formation or induce post-translational modifications. Whilst choosing expression vectors or systems, the nature and location of affinity tag used, localization and secretion signals have to be considered. Secondary structure modification due to mutations, deletion of a particular region or amino acid sequence that may promote protein degradation, changes in nucleotide to optimize codon usage are generally some of the factors that may improve transcription or translation (Salghetti et al. 2000). Structural data, if available, can also be used to design a better strategy. Structural data available of either homologous, orthologous or paralogous proteins can give clues about domains or unstructured loops present in a protein, which can be valuable whilst expressing and purifying the proteins for crystallization. For cloning of *A. baumannii Pank* and *PPAT* genes, the general methodology was used. Briefly, genes encoding PPAT and PanK proteins were amplified using *A. baumannii* genomic DNA as a template. The same restriction enzyme was used to digest purified PCR product and vector pET28a. In the next step, ligation reaction was set up to form recombinant vectors pET28a-*Ab*PanK and pET28a-*Ab* PPAT. These plasmids were used to express proteins in good quantity (Gupta et al. 2019; Singla et al. 2021).

3.1 *Expression and Purification of* Ab *PanK and* Ab *PPAT*

Several tags such as poly His, GST, MBP and CBP are used for co-expression and subsequent protein purification using resins bound with affinity tags, such as Ni^{2+}-nitrilotriacetic acid (Ni^{2+}-NTA), a strong ligand for the purification of His-tagged proteins under native conditions. Using this high-affinity chromatography, even proteins of interest whose concentrations may be as low as 1% in crude lysate can be purified in one step to more than 95% homogeneity (Janknecht et al. 1991). Expression and purification of *A. baumannii* PanK and PPAT proteins were done by using standard procedures. Briefly, recombinant vectors pET28a–*Ab*PanK and pET28a-*Ab* PPAT were transformed into *E. coli* BL 21 (λE3) cells for expression. LB media with kanamycin at a concentration of 50 μg ml^{-1} was inoculated with 1% of primary culture and allowed to grow at temperature of 37 °C at rpm of 200 till it attains 0.6 O.D at 600 nm. Induction was done by adding IPTG at a final concentration of 1 mM. The culture was centrifuged at 8000 rpm, 4 °C for 10 minutes to pellet down cells. This was followed by suspending cells in a small volume of ice-cold 50 mM tris-NaCl buffer having pH 8.0. For cell lysis, the suspension was sonicated at 40% amplitude for 10 minutes to release these proteins. Lysate was centrifuged at 15000 rpm for 45 minutes at 4 °C to collect the supernatant that contains our protein of interest. Proteins were purified from this supernatant using a Ni-NTA column. For washing three bed volumes of ice cold 20 mM imidazole, 50 mM Tris-Cl, 300 mM NaCl, 2 mM ATP, 1 mM MgCl, and pH 8.0 was used. For elution, 300 mM imidazole was used and all the fractions were analyzed by

SDS-PAGE. Fractions having our target protein were pooled and then concentrated using Amicon concentrator followed by desalting using PD10 columns (GE Health Care, USA) (Gupta et al. 2019; Singla et al. 2021). As crystallographic studies need protein free of any contamination, the purified protein samples were checked by different biophysical methods like SDS-PAGE, CD, IEF, MS and DLS (Bernstein et al. 1998). These techniques are used to access the homogeneity of the purified protein. Though, crystallization can be set up with 85% pure protein samples also, it is always good to set up crystallization experiments with pure and uniform protein samples of about 2–20 mg protein having a concentration of approximately 2–10 mg ml^{-1}.

4 Basics Methodology Used for the Crystallization

The two primary aspects of protein crystallography may be explained most easily: X-rays and crystals. When water is constantly withdrawn from a protein solution, the leftover water molecules become less mobile thus leading to crystallization. When protein-associated water molecules are lost, the entropy of the solvent increases, and if protein molecules get arranged in perfect crystal lattices achieving lower entropy. Crystal will grow as long as entropy change is positive. Nucleation, crystal development, and growth cessation are the three steps of crystallization(Chayen 2004). During nucleation, a good number of molecules arrange in 3-D to create a thermodynamically stable aggregate known as 'nucleus', which is the initial point for crystal formation. Growth stage follows nucleation; it involves particle diffusion to the surface of nuclei and their organized association with the developing crystal. It is to be noted that as the first nucleus forms and starts growing, the solution becomes depleted which prevents subsequent nucleation. It appears evident that water withdrawal should be gradual, stopping at a degree of supersaturation so that nuclei formation takes some days. Addition of precipitants slows down the loss of water as water molecules get associated with the precipitant rather than being lost from the protein solution. Salts, chiefly ammonium sulphate, or small molecular mass alcohols with the same vapour pressure as water, such as methylpentanediol, or inert polymers with high entropy, such as polyethylene glycol, are suitable precipitants used. The different buffers used for crystallization are given in Table 1 which is useful to access the protein's solubility curves.

A commonly used method used for crystallizing proteins is the hanging drop vapour diffusion method. Here, concentrated protein solution along with the suitable precipitant is allowed to concentrate by evaporation. Protein crystals will form under the right conditions in presence of a suitable precipitant. Small volume of protein sample and precipitant are put on a coverslip and mixed. This coverslip is then sealed over a well-holding precipitant solution in hanging drop vapour diffusion. As the concentration of precipitant on the coverslip is less than that present in the well, water evaporates from the hanging drop which increases the protein and precipitant concentration in it till equilibrium is reached. This increase in protein and precipitant

Table 1 List of buffers for protein solubility(Benvenuti and Mangani 2007)

Buffer	pH
Potassium/sodium phosphate	5.0
Potassium/sodium phosphate	6.0
Potassium/sodium phosphate	7.0
Sodium citrate	5.5
Bis–tris propane	6.5
HEPES	7.5
Sodium acetate	4.5
Sodium citrate	4.7
Sodium acetate	5.0
Sodium citrate	5.5
Cacodylic acid	6.5
Ammonium acetate	7.3
Imidazole	8.0
Bicine	8.5
Bicine	9.0
MES	5.8
MES	6.2
MES	6.5
HEPES	7.0
HEPES	8.0
TRIS	7.5
TRIS	8.0
TRIS	8.5

concentration in the drop favours crystal formation over precipitation (Jancarik and Kim 1991).

4.1 Setting-up Crystallization Experiment

Pre-greases 24-well crystallization trays are used for setting-up crystallization experiments. Before setting the crystallization trays, carefully organize all the buffers and record all the conditions tried for crystallization in each well.

The following steps are used generally:

- Pre-greased 24-well crystallization tray and 22 mm silicon cover slips are used. If setting up at a temperature of 4 °C temperature, it is important to equilibrate the material and the different solutions used before setting up the trials.
- First, the wells are filled with proper precipitant solutions.
- Coverslip to be used must be held only by edges.
- Pipette 2 μl of protein solution on the coverslip making sure no bubbles are formed whilst pipetting.

- Add an equal amount of precipitant and mix the two without any bubble formation.
- Cover the well with coverslip with a gentle press and turn 45° to ensure it is tight.
- The same procedure is repeated for the remaining wells.
- Immediately after preparing the plate, examine each drop for protein precipitation or foreign particles (glass shards, fibres or plastic bits) and make a proper note of the drops which may not be clear under a microscope.
- Leave the plate alone for at least 24 hours in a quiet location with the right temperature.

4.2 *Primary Screening of Crystallization Conditions*

Finding the right condition for crystallization is a complicated procedure in which different variables/conditions are tried and combined to achieve a satisfactory conclusion. A sparse matrix screen, in which a wide range of pH can be tested for a protein, salts and precipitants, is frequently used to find suitable protein crystallization conditions. There are great commercial screening kits available, so mixing your own initial screening chemicals is often unnecessary. In the order in which they should be used, the following commercial screens are recommended:

1. The Hampton Research Crystal Screen is the first of its kind. There are 50 reagents in this screen. As screen conditions #25 and #27 do not give satisfactory results, these may be skipped if the screen is run in two separate 24-well plates.
2. Hampton Research's Crystal Screen 2.

Evaluating Screens

After 24 hours, the plates should be checked under a microscope for protein crystallization, and then each day for 1 week. After this, plates may be checked weekly. Observations are recorded by using different abbreviations. The following are the abbreviations and their meanings:

- (C)—Clear drop; it showed no change in drop.
- (P)—Precipitate; it is usually light brown or granular in appearance. Precipitates of this type can sometimes form crystals. It' is unlikely that the precipitate will crystallize if it' is thick and swirly.
- (PP)—Precipitate/phase separation; it appears light black and granular, having little globules which resemble oil droplets. Crystals can develop at the intersection of phase separations.
- (MX)—Microcrystals; hard to differentiate from precipitate, but they are shiny not like precipitated protein, Better conditions may give good results.
- (NX)—Needle cluster; though these appear good crystals are not useful for X-ray diffraction. Optimum conditions could lead to fewer needle-like formations.

- (PX)—Plates; it might be beneficial for crystallography if it is not too thin. Better crystal shape may arise from optimization. If necessary, plate clusters can be broken down into single crystals.
- (RX)—Rod clusters; if this could be separated into single crystals, it could be valuable in crystallography.
- (X)—Single crystals; large, single crystals which are 3-D is the Holy Grail.

The crystallization conditions should be optimized if crystals formed during the screening phase are not appropriate for diffraction or are diffracted at low resolutions. Although it is always preferred to seek out the largest, most attractive crystals, there have been numerous stories of small, 'ugly' crystals producing the greatest data. It is worth noting that the production of crystals in protein crystal development is sometimes difficult to replicate, even when the original circumstances are strictly followed. As a result, optimization trials should be considered and used as a means of increasing reproducibility.

4.3 Optimizing Crystallization Conditions

Rescreening of the crystallization conditions should be considered when most of the drops are clear and, in this case, protein concentration has to be increased. In other conditions where the majority of the drops have copious precipitate, then lower the protein concentration and rescreening should be done.

It is always better to use only half of a screen at the beginning of a new experiment, so that protein wasted is minimized till the optimum conditions are worked out. Parameters that show good results in initial screening need to be further optimized so as to grow crystals of better form and size. To get ideal conditions for crystallization, pH, precipitant concentration and protein amount all can be changed. For playing with these parameters, it is necessary to have stock solutions of different solutions to be used for crystallization like buffer solutions of 1 M with different pH values. For salt solutions like, ammonium sulphate, saturated solutions are formed whose concentration can vary from 1 to 4 M depending on the salt. In addition to buffers and salts, different additives are tried to improve the process. Either different pre-formulated additive screens or 10X additive stocks commercially available from Hampton Research can be used at 10% concentration(Zaac et al. 2006). If the routine procedures fail to develop a crystal, the second option is the 'seeding technique' (McPherson and Gavira 2014). As we know, nucleation is the primary requirement for developing crystal, in the seeding technique ready-made nuclei referred to as 'seeds' are put in a crystallization drop equilibrated in metastable zone as shown in the phase diagram (Fig. 3) the zone which supports crystal growth but not nucleation (McPherson 2004). In metastable zone, nuclei is stabilized which can then form a large single crystal. Previously grown small crystals of either same protein known as homologous seeding or from a related protein like mutant or homologous protein known as heterologous can be used as seeds. Sometimes, even crystalline precipitate

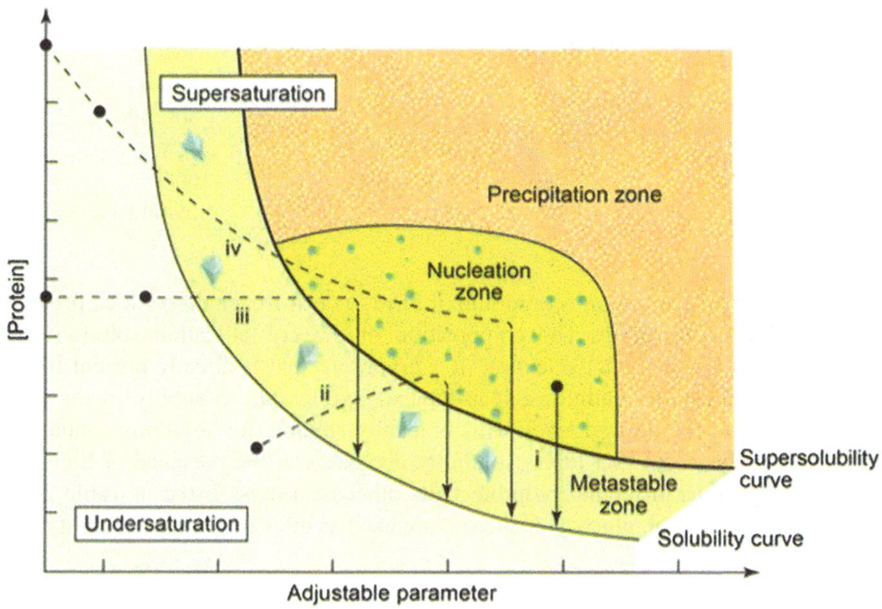

Fig. 3 Different phases during crystallization. The solubility curve which differentiates between the undersaturated with supersaturated. The superficial area is divided into three zones: (1) metastable zone, (2) the nucleation zone or labile zone and (3) the precipitation zone (McPherson 2004)

can be used to initiate crystallization. The frequently used technique used for seeding is 'micro-seeding' (Pusey et al. 2005).

Screening for a Suitable Cryoprotectant

Most of the crystal data is collected at a low temperature (usually 100 Kelvin) to avoid crystal weakening due to free radicals production by X-rays. When employing strong synchrotron X-ray sources, this is very crucial to use low temperatures. Prior to freezing, protein crystals must be treated with a cryoprotectant to prevent them from shattering.

The protein along with its thin layer of immediate molecules of mother liquid generate an amorphous glass in the presence of a cryoprotectant, which ensure production of a good diffraction pattern and least damage to the crystal. Usually, cryoprotectant is added to an artificial mother liquor solution, or a solution of artificial mother liquor having the required amount of cryoprotectant is prepared from scratch. Common cryoprotectants along with their concentrations used to guarantee good freezing protection are listed in Table 2. Mostly, the concentrations used are less than those listed in the table. In experiments having high concentrations of PEG (polyethylene glycol), a marginal concentration of cryoprotectant is needed. This can be estimated by taking around 10 μl of solution into liquid nitrogen; if the

Table 2 Cryoprotecantsalong with concentrations used for crystallization

Cryoprotectant	Concentration
Glycerol	30% v/v
Sucrose	30% w/v
Glucose	30% w/v
Ethylene glycol	30% w/v
2-Methyl-2,4-pentanediol (MPD)	30% v/v
PEG 400–20,000	25–40% (v/v or w/v)

cryoprotectant concentration is optimum, it forms a clear frozen drop which protects the crystal from breaking. The composition of the crystallization solution will influence the cryoprotectant selection. If a cryoprotectant is already present in the protein crystallization conditions, it is typically preferable to simply increase its concentration to the desired level. This is ideally suitable for solutions containing PEG. It is to be noted that PEGs solubility decreases in the presence of high salt concentration, thus different cryoprotectants other than those listed in Table 2 are used. Mostly, glycerol, glucose or sucrose are used as these have high solubility and thus offer a good choice.

4.4 *Soaking, Mounting and Freezing Protein Crystals*

Following the identification of an appropriate cryoprotectant to be used, the behaviour of protein crystals in these solutions should be studied to ensure that the crystals do not collapse, develop cracks or divide during freezing. If the situation is really challenging, 15% glucose and then 30% glucose can be used for soaking crystals before cryo-soaking. This allows proteins to freeze which otherwise was not possible. Cryoprotection can also be achieved without soaking crystals for long periods of time. All that is required is to soak the crystal in the cryoprotectant solution for a few seconds to replace the solution on the crystal's surface. Once the crystallization condition is optimized, in next step mounting of the crystals has been done for X-ray diffraction. Tiny loops (0.05–1.0 nm diameter) are used for crystal mounting. These loops are attached to hollow rods, which are fitted on magnetic caps, which can be easily put under liquid nitrogen on the X-ray diffractometer's goniometer head.

4.5 *Co-Crystallization*

Often it is required to solve a particular protein structure in the presence of a small molecule-like ligand. An easy way to form a protein–ligand complex is to saturate protein crystals in mother liquor having high concentration of ligand. In many cases, this is carried out at cryopreservation step. If dissociation constant (Kd) is known, the concentration of ligand employed should be 10–1000 X. Soaking for 10-–30 minutes is enough, for binding sites present in the crystal lattice. Co-crystallization with ligand should not be attempted if the ligand-binding site is covered by protein molecules in the crystal or if crystals cannot sustain prolonged soaking without splitting or dissolving (D'Arcy et al. 2014).

4.6 *Crystallization of* Ab*PanK and* Ab *PPAT*

For the crystallization of *Ab* PanK and PPAT proteins, crystal screens from Hampton Research were used for initial screening in order to find out appropriate buffer conditions.Concentrated *Ab* PanK and *Ab* PPAT protein solutions (approximately 12 mg ml^{-1}) were used. Protein was dissolved in 50 mM Tris HCl, pH 7.6 having 300 mM NaCl in a 24-well linbro crystallization plate at temperature of 298 K. As reservoir, a solution having 1 M tri-sodium citrate and sodium HEPES (pH 7.5) was used. The reservoir was filled with 1 ml of this solution. To 2 μl of protein solution, an equal volume of reservoir was added to make a final volume of 4 μl. The drops were put on coverslips of a 24-well crystallization hanging drop vapour diffusion setup. After 1 week, crystals of about 0.30 × 0.20 × 0.15 mm^3 size were formed. *Ab* Pank was crystallized in 20% PEG (M.W 3350) having 0.2 M sodium potassium tartrate (Gupta et al. 2019; Singla et al. 2021).

5 Basics of X-Ray Analysis and Data Collection

Basis of solving a structure by crystallography depends on the scattering of X-ray beam by the electrons of constituent molecules of macromolecule whose structure is to be solved. The fundamental building block of protein crystal known as 'unit cell' has a size ranging from ~30 Å to ~1000 Å, which is useful in determining molecular structure as this is comparable to classic length of a covalent bond which is about 1–2 Å. To study protein crystals, X-rays of wavelength in the range 0.5–2.0 Å are used. The crystal behaves as a diffraction grating leading to the interference of the scattered radiations constructively as well as destructively. The images that are formed by these scattered radiations have regularly spaced spots that correspond to different intensities superimposed on a low background. Crystals are made-up of any of the seven types of unit cells distinguished on the basis of lengths(a), (b) and (c) and angles α, β and γ between them.

These unit cells in macromolecules are repeated periodically in entire volume, it behaves as a 3-D diffraction grating and is specified by Bragg's law $n\lambda = 2d \sin\theta$. As per this law, incoming X-rays reflected by crystal lattice planes reflect photons known as 'reflections' which are identified by Miller indices (hkl). The intrinsic symmetry of the crystal transforms into a diffraction pattern. A diffraction experiment involves measurement of a large number of reflection intensities; the average number of measurements per individual is referred to as redundancy or multiplicity. As each measurement involves some error in measurement, more the number of measurements, more accurate will be the final estimated average. Due to the involved error in measurement, absorption or radiation damage, the measured intensities are not equal. Thus, comparing these symmetry-related reflections is a way to evaluate the reliability of a dataset. It is measured by a parameter known as R-factor, which is a total measure of the mean absolute difference between the intensity of a single reflection and the average intensity of its symmetry-related reflections, relative to that average intensity. R-factor lies between 8 and 10% for low-resolution data (8–10 Å), and has a value of 12–15% for high-resolution data (> 2.5 Å). Taking the square root of each measured reflection, I (hkl), it gets converted into structure factor, F (hkl) which is used to determine electron density, ρ(xyz) by Fourier transformation.

From this electron density map, atomic structure is built. Many programmes are available which are used to analyze X-ray diffraction data such as DENZO and SCALEPACK or MOSFLM and SCALA (Murshudov et al. 1997; Battye et al. 2011). The validation of the obtained structure can be checked by different software like PROCHECK and MolbrobityRCSB validation server(Chen et al. 2010; Laskowski et al. 1993).

5.1 Data Collection and Structure Determination of Ab *PPAT and* Ab *PANK*

Before mounting Ab PPAT crystals for diffraction, crystals were stabilized by soaking in 3 M tri-sodium citrate containing sodium HEPES buffer (pH 7.5) and 20% glycerol as cryoprotectant. Using a nylon loop and flash frozen in liquid nitrogen, the single crystal was mounted. Data was collected at a temperature of 100 K on MAR-225 CCD detector (Mar research Norderstedt, Germany) using Synchrotron beamline BM14 at the European Synchrotron Radiation Facility, Grenoble, France. Setting oscillation at 1° and exposure time of 4 sec per image with wavelength of 0.98 Å, a complete data set was collected. HKL 2000 was used for reducing data. Crystals were monoclinic having space group P212121 and values of a = 78.26 Å, b = 109.39 Å, c = 121.28 Å. Taking 18.3 kDa as molecular weight of *Ab* PPAT, it was calculated that the unit cell contains 24 protein molecules. Using the molecular replacement method using software MOLREP from CCP4i suite (Gupta et al. 2019), structure of *Ab* PPAT was determined. Prior to mounting, Ab PanK crystals were immersed in mother liquor having 20% glycerol as cryoprotectant and then immediately flash-frozen in liquid nitrogen. Crystals were diffracted at in-house beamline facility BL-21 from RRCAT, Indore, India. Even though, crystals diffracted to very low resolution, iMOSFLM was used for data processing to get preliminary X-ray data (Singla et al. 2021).

6 Lead Optimization with Structural Data

In the lead optimization step of structure-based drug design (SBDD), structural information plays an important role, by pointing to the chemical modifications needed in the lead compounds. The structural information available is of immense help in designing and synthesizing lead compounds with desired modification, thus immensely reducing the time and money investment in progress from a lead compound to a clinical candidate. It is a 'step-ahead' beforehand investing time and money in synthetic chemistry, if the crystal structure of the target protein/enzyme with its bound substrate or lead compound is available at the start of the drug discovery process. The available structural information can give clue about the interactions which must be conserved, interactions that must be eliminated, which group or atom to be modified so as to enhance the potency or selectivity. One such example is the synthesis and SAR of novel non-nucleoside Adenosine Deaminase (ADA) inhibitors (Terasaka et al. 2004). In this case, it was evident by X-ray analysis that lead compound binds to ADA with an induced fit that creates binding pockets dissimilar from those used by substrate-like inhibitors. The finding prompted the design and production of novel chemical compounds with much better inhibitory activity than usually used nucleoside analogues which have severe toxicity. Similarly, in case of phosphotyrosine phosphatase 1B (PTP-1B) structural details and molecular modelling which led to the development of potent PTP-1B inhibitors that showed less selectivity against TC-PTP, a related protein to tyrosine phosphatase (Lau et al. 2004). Structural details of novel designed compounds together with mutagenic studies directed to the identifications of selectivity at atomic level, thus confirming reliability on structure-based methods (Scapin et al. 2003).

7 Conclusions

Structure determination by X-ray is a fundamental and essential method for structure determination of macromolecules like proteins and thus understanding their biological functions. It is an essential and primary tool in the drug discovery process. The developments in the disciplines of structural and functional genomics, different algorithms available for high throughput screening and evaluation has paved the way for more and more structures being solved or predicted. With this, new and novel methods are used in target identification and lead compound discovery. However, it is to be noted that if there are some uncertainties or ambiguities in structure determination or predication, it may have a significant impact on structure-based drug design process or method. Use of docking, virtual and property-based screening, Lipinski's rule of five to evaluate drug likeliness of a compound, estimating ADME properties are important to check the reliabilities of these processes and methods. With each new structure being solved or determined, our understanding of the interactions between a macromolecule with a ligand increases, as are the chances for structure-based drug design.

Conflict of Interest The authors declare no conflict of interest.

References

Abbo A, Carmeli Y, Navon-Venezia S, Siegman-Igra Y, Schwaber MJ (2007) Impact of multi-drug-resistant Acinetobacter baumannii on clinical outcomes. Eur J Clin Microbiol Infect Dis 26(11):793–800

Abbo A, Navon-Venezia S, Hammer-Muntz O, Krichali T, Siegman-Igra Y, Carmeli Y (2005) Multidrug-resistant Acinetobacter baumannii. Emerg Infect Dis 11(1):22

Battye TG, Kontogiannis L, Johnson O, Powell HR, Leslie AG (2011) iMOSFLM: a new graphical interface for diffraction-image processing with MOSFLM. Acta Crystallogr D Biol Crystallogr 67(4):271–281

Beddell CR, Goodford PJ, Norrington FE, Wilkinson S, Wootton R (1976) Compounds designed to fit a site of known structure in human haemoglobin. Br J Pharmacol 57(2):201–209

Benvenuti M, Mangani S (2007 Jul) Crystallization of soluble proteins in vapor diffusion for x-ray crystallography. Nat Protoc 2(7):1633–1651

Bernstein BE, Michels PA, Kim H, Petra PH, Hol WG (1998) The importance of dynamic light scattering in obtaining multiple crystal forms of Trypanosoma brucei PGK. Protein Sci 7(2): 504–507

Blundell T, Dodson G, Hodgkin D, Mercola D (1972) Insulin: the structure in the crystal and its reflection in chemistry and biology by. Adv Protein Chem 26:279–402

Chayen NE (2004) Turning protein crystallisation from an art into a science. Curr Opin Struct Biol 14(5):577–583

Chen VB, Arendall WB, Headd JJ, Keedy DA, Immormino RM, Kapral GJ, Murray LW, Richardson JS, Richardson DC (2010) MolProbity: all-atom structure validation for macromolecular crystallography. Acta Crystallogr D Biol Crystallogr 66(1):12–21

D'Arcy A, Bergfors T, Cowan-Jacob SW, Marsh M (2014) Microseed matrix screening for optimization in protein crystallization: what have we learned? Acta Crystallograph Sect F: Struct Biol Commun 70(9):1117–1126

Davies AH (1994) Current methods for manipulating baculoviruses. Bio/Technology 12(1):47–50

Edwards AM, Arrowsmith CH, Christendat D, Dharamsi A, Friesen JD, Greenblatt JF, Vedadi M (2000) Protein production: feeding the crystallographers and NMR spectroscopists. Nat Struct Biol 7(11):970–972

Eliopoulos GM, Maragakis LL, Perl TM (2008) Acinetobacter baumannii: epidemiology, antimicrobial resistance, and treatment options. Clin Infect Dis 46(8):1254–1263

Giegé R, Dock AC, Kern D, Lorber B, Thierry JC, Moras D (1986) The role of purification in the crystallization of proteins and nucleic acids. J Cryst Growth 76(3):554–561

Goodford PJ, St-Louis J, Wootton R (1980) The interaction of human haemoglobin with allosteric effectors as a model for drug-receptor interactions. Br J Pharmacol 68(4):741

Gupta A, Sharma P, Singh TP, Sharma S (2021) PhosphopantetheineAdenylyltransferase: a promising drug target to combat antibiotic resistance. Biochimica et Biophysica Acta (BBA)-Proteins and Proteomics 1869(2):140566

Gupta A, Singh PK, Iqbal N, Sharma P, Baraigya HR, Kaur P, Umar MS, Ahmad F, Sharma A, Owais M, Sharma S (2019) Structural and binding studies of phosphopantetheine adenylyl transferase from Acinetobacter baumannii. Biochimica et Biophysica Acta (BBA)-Proteins and Proteomics. 1867(6):537–547

HammoudiHalat D, Ayoub MC (2020) The current burden of carbapenemases: review of significant properties and dissemination among gram-negative bacteria. Antibiotics 9(4):186.L

Jancarik JA, Kim SH (1991) Sparse matrix sampling: a screening method for crystallization of proteins. J Appl Crystallogr 24(4):409–411

Janknecht R, de Martynoff G, Lou J, Hipskind RA, Nordheim A, Stunnenberg HG (1991) Rapid and efficient purification of native histidine-tagged protein expressed by recombinant vaccinia virus. Proc Natl Acad Sci 88(20):8972–8976

Laskowski RA, MacArthur MW, Moss DS, Thornton JM (1993) PROCHECK: a program to check the stereochemical quality of protein structures. J Appl Crystallogr 26(2):283–291

Lau CK, Bayly CI, Gauthier JY, Li CS, Therien M, Asante-Appiah E, Cromlish W, Boie Y, Forghani F, Desmarais S, Wang Q (2004) Structure based design of a series of potent and selective non peptidic PTP-1B inhibitors. Bioorg Med Chem Lett 14(4):1043–1048

Lee CS, Doi Y (2014) Therapy of infections due to carbapenem-resistant gram-negative pathogens. Infection & chemotherapy 46(3):149–164

Leonardi R, Zhang YM, Rock CO, Jackowski S (2005) Coenzyme a: back in action. Prog Lipid Res 44(2–3):125–153

McDevitt D, Payne DJ, Holmes DJ, Rosenberg M (2002) Novel targets for the future development of antibacterial agents. J Appl Microbiol 92:28S–34S

McPherson A (2004) Introduction to protein crystallization. Methods 34(3):254–265

McPherson A, Gavira JA (2014) Introduction to protein crystallization. Acta Crystallograph Sect F: Struct Biol Commun. 70(1):2

Murshudov GN, Vagin AA, Dodson EJ (1997) Refinement of macromolecular structures by the maximum-likelihood method. Acta Crystallogr D Biol Crystallogr 53(3):240–255

Pusey ML, Liu ZJ, Tempel W, Praissman J, Lin D, Wang BC, Gavira JA, Ng JD (2005) Life in the fast lane for protein crystallization and X-ray crystallography. Prog Biophys Mol Biol 88(3): 359–386

Rice LB (2008) Federal funding for the study of antimicrobial resistance in nosocomial pathogens: no ESKAPE. J Infect Dis 197(8):1079–1081

Rosenberger F (1996) Protein crystallization. J Cryst Growth 166(1–4):40–54

Salghetti SE, Muratani M, Wijnen H, Futcher B, Tansey WP (2000) Functional overlap of sequences that activate transcription and signal ubiquitin-mediated proteolysis. Proc Natl Acad Sci 97(7):3118–3123

Santajit S, Indrawattana N (2016) Mechanisms of antimicrobial resistance in ESKAPE pathogens. Biomed Res Int 2016

Scapin G, Patel SB, Becker JW, Wang Q, Desponts C, Waddleton D, Skorey K, Cromlish W, Bayly C, Therien M, Gauthier JY (2003) The structural basis for the selectivity of benzotriazole inhibitors of PTP1B. Biochemistry 42(39):11451–11459

Singla A, Sharma P, Gupta A, Iqbal N, Rani C, Singh TP, Sharma S (2021) Biophysical characterization of type III pantothenate kinase (PanK) from Acinetobacter baumannii. Protein Pept Lett 28(4):450–458

Structural Genomics Consortium (2008) Protein production and purification. Nat Methods 5(2):135

Sugrue MF (2000) Pharmacological and ocular hypotensive properties of topical carbonic anhydrase inhibitors. Prog Retin Eye Res 19(1):87–112

Terasaka T, Kinoshita T, Kuno M, Seki N, Tanaka K, Nakanishi I (2004) Structure-based design, synthesis, and structure– activity relationship studies of novel non-nucleoside adenosine deaminase inhibitors. J Med Chem 47(15):3730–3743

Thomas BR, Carter D, Rosenberger F (1998) Effect of microheterogeneity on horse spleen apoferritin crystallization. J Cryst Growth 187(3–4):499–510

Vila-Farres X, De La Maria CG, López-Rojas R, Pachón J, Giralt E, Vila J (2012) In vitro activity of several antimicrobial peptides against colistin-susceptible and colistin-resistant Acinetobacter baumannii. Clin Microbiol Infect 18(4):383–387

Zaac A, Schall CA, Mueser TC (2006) Assessment of a preliminary solubility screen to improve crystallization trials: uncoupling crystal condition searches. Acta Crystallogr D Biol Crystallogr 62(7):833–842

Zulauf M, D'Arcy A (1992) Light scattering of proteins as a criterion for crystallization. J Cryst Growth 122(1–4):102–106

Bioactivity-Guided Fractionation and Identification of Bioactive Molecules: A Basic Method in Drug Discovery

Deepak Das and Syed Shafi

Abstract The process of discovering new medications is called as drug discovery. From the ancient period to modern age, there has been a significant improvement in the drug development. Different methodologies have been applied in drug discovery time to time.

Drug discovery has started with the usage of herbal extracts for different ailments, which were termed as traditional medicine. The knowledge, skills and practice of holistic health care, recognized and accepted for its role in the maintenance of health and the treatment of diseases, are defined as traditional medicine. Over the centuries, traditional medicine holds its prime place, and more than 80% of the population of developing countries are still depending upon traditional medicine. Later on, bioactivity-guided fractionation and identification of bioactive molecules (secondary metabolites) approach came into limelight and have been at forefront over the decades. Bioassay-guided fractionation and isolation are still the basic procedures for identifying new natural product-based scaffolds with defined biological activity. Historically, drugs were discovered serendipitously through the accidental identification of active ingredients from the traditional herbal extracts without knowing their targets. After the identification of an active substance, their targets and mechanism of action have been discovered. This approach of drug discovery is defined as classical pharmacology, forward pharmacology or phenotypic drug discovery. Natural products with their structural diversity and drug likeness have significantly contributed to drug discovery. More than 70% of the marketed medications are discovered either directly from natural sources or inspired by natural products. Natural products are still being treated as frontline ligands in the modern target-based drug discovery methods. As the natural products exert remarkable structural diversity, they have been the primary source of drug development. After being overlooked for twenty years, natural product research has regained its importance and is assuming new prominence.

D. Das · S. Shafi (✉)
Department of Chemistry, School of Chemical and Life Sciences, Jamia Hamdard, New Delhi, India
e-mail: syedshafi@jamiahamdard.ac.in

V. S. Rajput, A. Runthala (eds.), *Drugs and a Methodological Compendium*,
https://doi.org/10.1007/978-981-19-7952-1_3

In this chapter, a detailed study of the bioactivity-guided fractionation and identification of bioactive ligands, a basic approach in drug discovery, has been overviewed.

1 Introduction

Traditional medicine (TM) is built on beliefs, theories and experiences of indigenous peoples passed down from generation to generation and is recognized and accepted by the whole generation of healthcare knowledge, skills and practices for its role in health maintenance and disease treatment. Traditional medicine has always been a part of spiritual and cultural life and is still practised in many countries (Akerele 1983). Among the various forms of traditional medicine, herbal therapy is the most notable one, and as much as 80% relies on traditional remedies in developing countries (World Health Organization (WHO) 2002). The traditional ethnomedical information has guided to the isolation of biologically active compounds, which can be directly used in medicine. Historically, natural products have been used throughout in the form of remedies, traditional medications, concoctions, decoctions, tonics and oils (Butler 2004; Ivanova et al. 2018; Kinghorn et al. 2011).

Over the decades, natural products have been recognized as a rich and productive source of structurally diverse potential therapeutic agents (Butler 2004; Müller et al. 2000; Mishra and Tiwari 2011; Rey-Ladino et al. 2011; Cragg and Newman 2005; Haefner 2003) and thus have played a significant role in drug discovery.

However, the use of herbal formulations as extracts (in the crude form) originates numerous complications since the pharmacological action is affected by one or more bioactive ingredients present in the extract. The amount of these bioactive compound (s) varies with the habitat (location) of the plant and the season in which it is collected. In addition, these pharmaceutically active compounds (present in the extracts) are either highly poisonous when taken in excess or ineffective when they are present in sub-optimal levels. The medicinal values of many plants are also quickly lost on storage. In addition, the original extracts of many medicinal plants may contain other harmful components in addition to biologically active molecules. Therefore, isolation and identification of bioactive natural products in their pure form are very important. In addition, structural modifications of natural leads can improve efficacy and reduce the side effects (Malviya and Malviya 2017; Fabricant and Farnsworth 2001).

Ethno-pharmacological knowledge of folkloric medicine has prominently increased the pace of drug discovery in identifying the natural product-based leads for therapeutic development (Fabricant and Farnsworth 2001; Newman and Cragg 2012; Naman et al. 2017). Traditional herbs represent an extraordinary reservoir of bioactive compounds with the stake of about 25% of the currently prescribed modern medicine. The traditional knowledge of various healers has been transformed to modern scientific medicine and supports the experimental knowledge of "alternative" treatment. Among all the natural products isolated worldwide, at least

120 compounds have been derived from around 90 plant species. 74% of these substances were resulting from plants used in traditional medicine (Cragg and Newman 2005). As much as more than 50% of the 25 blockbuster drugs are directly associated with natural products (Phillipson 2001).

In natural product drug discovery, bioassay-guided isolation is a quick and authenticated method for the isolation of bioactive compounds in its pure form. This chapter describes about the "bioassay-guided isolation of bioactive compounds".

2 Bioassay-Guided Fractionation and Isolation of Bioactive Compounds

In natural product research, bioassay-guided fractionation approach is mostly used to isolate bioactive metabolites from a crude extract. When a medicinal plant is found to be used in folkloric medicine and its extracts are found to be active, the protocol used to recognize the compound accountable for the biological activity is bioassay-guided fractionation. It involves stepwise fractionation of extracted components based on differences in their physico-chemical properties and evaluating the biological activity, followed by purification and assaying.

This process comprised of the following steps: (A) identification and authentication of plant material, (B) extraction of plant metabolites from the plant material using different solvents, (C) biological screening of the extracts, (D) fractionation of the extract, (E) biological screening of each fractions and (F) isolation, characterization and biological screening of secondary metabolites (Newman and Cragg 2012; Bucar et al. 2013).

The step-by-step protocol has been depicted in the Fig. 1 given below.

A. **Identification and Authentication of Plant Material:** In the bioassay-guided fractionation protocol, the identification of correct plant material is most important as the quantity of secondary metabolites depends upon the locality of the plant and the season in which the plant has been collected and other geographical parameters (metals, nature of soil, water availability, etc.). Bioactive compounds may present in the whole plant or a part of the plant. Authentication of the plant can be done with the help of plant taxonomists.
B. **Extraction:** Extraction is pharmaceutically defined as the separation of bioactive components of plant/animal materials from the biologically inert portions by employing selective solvents. The impure extracts attained are present in the form of liquids, syrupy liquids, amorphous solids or powders envisioned for oral/topical use only. The extracts are recognized in several forms as powdered extracts, decoctions, infusions, fluid extracts and tinctures. The quality of the herbal drug depends upon the standardization of extraction procedures.

   The plant material will be shade dried, powdered and extracted with different solvents of variable polarities (water, n-butanol, t-butanol, ethanol, methanol,

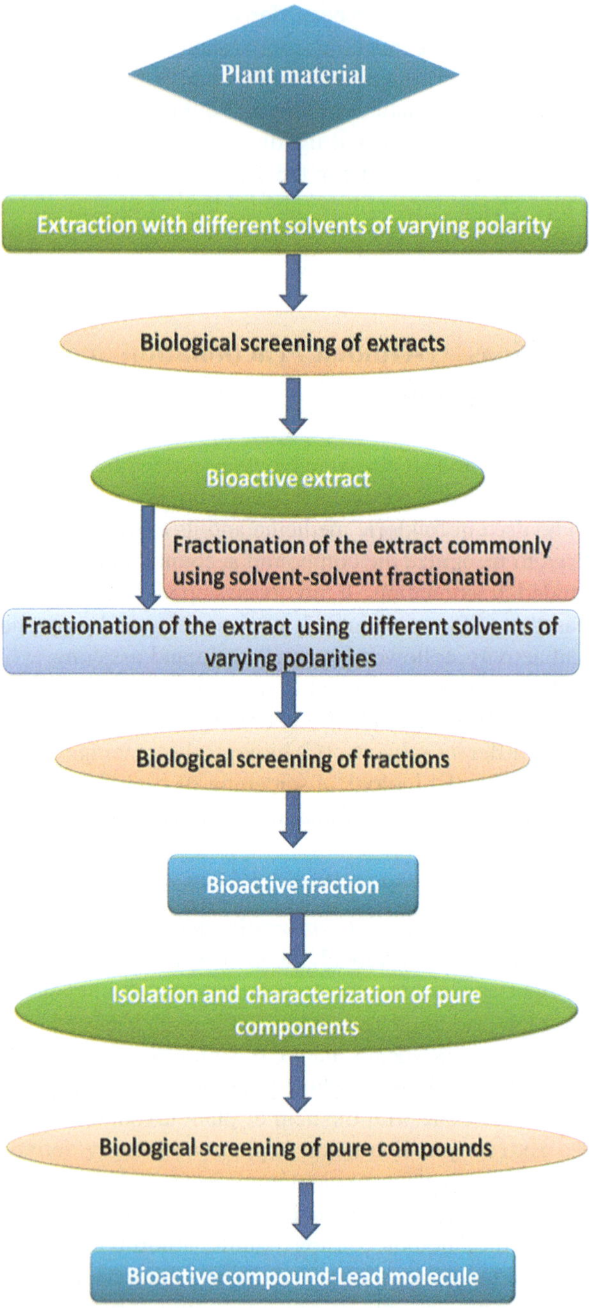

Fig. 1 Bioactivity-guided fractionation and identification of bioactive compounds

etc.) by employing various standard extraction methods including maceration, infusion, digestion, decoction, percolation, Soxhlet extraction, fermentation, counter-current extraction, sonication, supercritical fluid extraction and phytonics based on the nature of the constituents (Bucar et al. 2013).

Methods of Extraction

i. **Maceration:** In this procedure, finely powdered plant material is drenched in a stoppered container with menstruum (solvent) and kept at ambient temperature for three days. The solvent extract will be strained, and the marc (damp solid material) will be pressed. The combined liquid extracts are filtered or decanted after standing in order to get a clear extract.

   This is a perfect method for the extraction of phenolic compounds and other thermolabile components. However, it has the drawbacks of long extraction time and low extraction value.

ii. **Infusion:** In this procedure, secondary metabolites or flavours of plant material are extracted with water/oil or alcohol, by suspending the plant material in menstruum (cold or boiled) for the extended period of time (steeping). The resultant liquid extract is called as an infusion. These are the dilute solutions of readily soluble constituents of the plant material. The quantum of time and the constituents present in the solvent extract depend upon the type of infusion. Infusion times may vary from seconds to months.

   Examples of popular infusions include the following:

   - **Tea** is the common example of infusion. A variety of tea types were prepared by steeping the leaves in hot water. Many herbal teas including green tea, lemon tea, chamomile tea, turmeric tea, hibiscus tea, senna tea, apple tea, ginger tea, rooibos tea, etc. (individually or in combination) are prepared by infusion.
   - **Coffee** can also be made through infusion method.
   - **Herbal remedies** are generally prepared through infusions in water or oil.
   - **Flavoured oils:** Plants with desired flavours are soaked in an edible oil or vinegar for a prolonged period in order to extract the flavoured oils. The steeped oil or vinegar is generally used as a flavouring agent. Several flavoured oils are being produced from chillies, garlic, lemon and many other plants.
   - **Cucumber water:** A popular infusion of a mixture of sliced cucumber and citrus slices along with herbs (mint) in water. It is also known as "spa water" that is commonly served in spa centres and other personal care establishments.

iii. **Digestion:** This is maceration under moderate heating conditions. Under the heating conditions, the extraction efficacy of the solvent will be increased.

iv. **Decoction:** Over the centuries, decoctions of various plant materials are being used in traditional medicines of different cultures. Decoction is a method of extraction by boiling the plant material in a definite quantity of water for a specified time, followed by filtration. This process is appropriate for extracting water-soluble, thermally stable and non-volatile secondary metabolites. Generally polar compounds are extracted in this method. This procedure is generally used to prepare Ayurvedic extracts that are called "quath" or "kawath". Generally, water will be taken in excess compared to the plant material in the ratio of 4:1 or 16:1; the extract will be concentrated to one-fourth of its original volume through boiling. The reduced extract is filtered and either used directly as an Ayurvedic medication or further processed. Decoction method is also used to prepare herbal teas (tisane) and tinctures.

v. **Percolation:** In chemistry or material science, percolation (from the Latin word *percolare*, "to filter" or "trickle through") is defined as the movement and filtration of fluids through porous materials. This is the most frequently used procedure to extract active ingredients and herbal tinctures. The apparatus used in this method of extraction is called percolator.

   In a closed container, a finely powdered herb is imbibed with suitable amount of specified solvent and kept for approximately four hours. The percolator is packed with mass (filter paper, cotton, washed sand, etc.) above the plant material, and an additional menstruum is added on the top of the mass. The percolator will be closed from the top, and the mixture is allowed to stand for further 24 hours. The outlet of the percolator at the bottom is then opened, and the solvent present therein is allowed to trickle gradually. Some quantity of the solvent was added continuously, and the extraction takes place by gravitational force, pushing the solvent through the plant material. After collecting three-fourths of the required volume, the marc is then pressed and the released liquid is then added to the percolator. Adequate menstruum is further used to produce the required volume, and the combined solvent extract is filtered and concentrated under reduced pressure obtain the crude extract.

vi. **Reflux extraction:** It is a solid–liquid extraction procedure at a constant temperature with repeatable solvent evaporation and condensation for a definite period of time without losing the solvent. This method is widely used in herbal industries. Reflux extraction is more effective than maceration/percolation methods as it requires less solvent and extraction time, easy to operate and cost effective. However, reflex extraction method is not suitable for the extraction of thermolabile constituents.

vii. **Soxhlet extraction** (continuous hot extraction): The apparatus used in this extraction method is known as Soxhlet extractor, which contains a round bottom flask equipped with extraction chamber, siphon tube and condenser on the top. In this method, finely powdered plant material is placed in a porous bag (thimble) made up of strong filter paper or a clean cloth. The thimble is then placed in the extraction chamber of the Soxhlet apparatus. When the extraction solvent in the round bottomed flask is refluxed, the solvent vapours condensed in condenser drop down into the extraction chamber comprising the plant material in the thimble and extract it by contact. Subsequently when the solvent levels in the extraction chamber increases to the top of siphon tube, the solvent along with the extracted plant material will flow back into the round bottomed flask. The entire process is continuing repeatedly until the soluble contents are completely extracted.

The Soxhlet extraction method amalgamates the benefits percolation and reflux extraction, which exploits the principle of reflux and siphoning for continuous extraction of the plant material. This is a cyclic continuous extraction process with high efficacy and requires less solvent and extraction time compared to maceration or percolation.

This method is suitable to extract the constituents that are partly soluble in the chosen solvent. However, it is not a suitable method for heat-sensitive constituents.

viii. **Extraction by fermentation:** Fermentation techniques have been adopted for the extraction of some Ayurvedic preparations, viz. *asava* and *arista*. The technique involves the in situ generation of alcohol from the fermented plant material (powdered form or decoction), which was soaked in water for a specified time. The alcohol thus generated enables the extraction of the phytoconstituents from the plant material and also serves as a preservative. *Karpurasava, kanakasava* and *dashmularishta* are some of the examples of such preparations.

ix. **Counter-current extraction (CCE):** Counter-current extraction is a method of multiple liquid–liquid extractions (LLEs) that permits the separation of substances with different distribution coefficients. The apparatus used for this technique is called Craig apparatus. In CCE technique, two immiscible solvents flow in opposite directions in multiple stages; when they come in contact with each other, the soluble constituents will be dispersed into the extracted solvent. In this process, separation of components having variable solubility in two immiscible solvents is achieved. The ideal solvent used in this process should have high partition coefficient. This method has valuable applications in petrol industry including the separation of compounds with same boiling points and the recovery of aromatic compounds from the paraffin fraction.

In this extraction method, the wet pulverization plant material is directly used for the extraction in the form of fine slurry. In this extraction method, the material to be extracted is moved in one direction within the extractor and interacts with menstruum running in the opposite direction. Complete extraction is thus possible with the optimized amounts of solvent/plant material and their flow rates. This method of extraction is highly effective, requiring small volume of solvent and extraction time and should be well suited for the extraction of thermolabile constituents (as this method is processed at room temperature). Sufficient quantity of extract originates at one end of the extractor while the marc departs from the other end.

x. **Ultrasound extraction (sonication):** This method involves the use of ultrasounds of varied frequencies between 20 and 2000 kHz. The use of ultrasounds enhances the permeability of cell walls and generates cavitation. Since it is expensive for large-scale applications, this method has limited applications. The notable application of this method is in the extraction of Rauvolfia root. One disadvantage of the procedure is the influence of ultrasound energy (>20 kHz) on the secondary metabolites that may result in some chemical transformations.

xi. **Supercritical fluid extraction (SFE):** SFE is also termed as green method of extraction (an alternative extraction method), which reduces the use of toxic organic solvents and improves the sample output. In this procedure, liquid CO_2 is used as the fluid. To ease the polarity limits of CO_2, organic solvents are frequently mixed with liquid CO_2. Liquid argon has replaced CO_2 as a fluid in recent days as it provides more inert atmosphere and is economically inexpensive.

Advantages of SFE:

- The extraction of phytoconstituents at low temperature evades thermal degradation of secondary metabolites.
- No solvent residues.
- Eco-friendly extraction procedure.

Due to the rapid expansion of its applications, SFE has turned out to be one of the largest areas of growth. SFE method is extensively used in the extraction of essential oils, flavouring agents, pesticides, environmental samples, foods and fragrances, polymers and natural other products.

xii. **Phytonics process:** It is a recently developed eco-friendly technique for the extraction of plant materials using hydrofluorocarbon-134a. It is a non-hazardous and cost-effective method for the extraction of top-quality fragrant oils, essential oils, flavours and biological samples. The phytonics process has several applications in perfumery, essential oil, herbal drug and flavour industries. This method is also used in the extraction of antibiotics (produced by biotechnology methods) and the production of pharmacologically active products an.

C. **Biological Screening of Plant Extracts:** Different plant extracts obtained will be evaluated for their pharmacological activities, and the active extract will be taken forward for the further process.

D. **Fractionation of the Extract:** Fractionation is the process of separating a plant/animal extract into groups of chemicals having comparable physicochemical properties. Fractionation of an extract can be done using quite a few methods including solvent–solvent extraction, precipitation, distillation, dialysis, chromatography, electrophoresis, etc. based on the nature of the compound (solubility, size, shape, electrical charge) present in the extract, stability/application of the fraction and several other features (Houghton and Raman 1998). The molecules obtained from one fractionation method may not be the same as those obtained from a different fractionation method. When two fractionation methods are used in sequence, many fractions can be obtained, each containing only one or two components. In some cases, the extract will contain significant amounts of undesirable compounds that need to be removed, such as chlorophyll from leaf extracts, oils from seed extracts and sugars. The purification process that results in an extract enriched with the ingredients of interest is referred to as **clean-up procedure**. Most of the techniques used in fractionation are also used in cleaning procedures.

E. **Biological Screening of Fractions:** All the fractions obtained from the extract will be evaluated for their pharmacological activities, and the active fractions will be taken forward for the further process.
F. **Isolation and Characterization and Biological Screening of Pure Compounds**: Pharmacologically active fractions will be further processed for the isolation of the pure phytoconstituents using standard purification techniques. Pure compounds isolated will be characterized using different analytical techniques. Finally, pure secondary metabolites isolated will be evaluated for their biological activities; thereby the single phytoconstituent responsible for the biological activity will be identified.

Plants are an important source of structurally diverse natural products that are used in pharmaceuticals, agrochemicals, flavouring agents, fragrances, colours, bio-pesticides and food additives. Bioassay-guided fractionation followed by chromatographic purification techniques have led to the discovery of several drugs. This method has proven effective for the discovery of several pharmaceutically active compounds including taxol, artemisinin, vinblastine, quinine, morphine, paclitaxel, camptothecin, etoposide, mevastatin, artemisinin, etc. More than 75% of the marketed drugs are either directly isolated from the nature or the derivatives of natural products or the drugs developed based on natural product scaffolds due to its structural diversity and complexity.

The discovery of some of the pharmaceutically vital scaffolds from the nature is illustrated below:

3 Examples of Bioactive Compounds Isolated from Plants

(1) **Morphine:** Morphine is an analgesic agent of the opiate family that directly acts on the central nervous system to increase the feel of joy and warm relaxation and relieve pain. Naturally morphine is found in poppy plant (*Papaver somniferum*) as a dark brown viscous solid. Friedrich Sertürner, a German pharmacist, has first isolated morphine between 1803 and 1805 and is believed to be the first-ever active ingredient isolated from a plant.

The use of the opium poppy was known to the ancient Minoans.

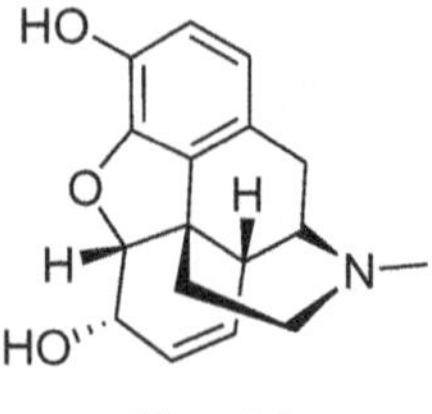

Morphine

The modern name of opium was obtained from the ancient Greek name *opion*, the sap prepared from the plant.

Isolation and Purification: Well-ground *Papaver somniferum* are mixed intimately with 10 ml. of a 10% aqueous sodium carbonate solution and extracted with 1:1 n-butanol-benzene solution; the collective organic layer is washed with aqueous sodium carbonate. The n-butanol-benzene extraction is shaken with 0.525N sulphuric acid and washed with water to remove morphine and other alkaloids to the aqueous phase. The aqueous solution thus obtained will be further processed for the isolation of morphine in pure form (Achor and Geiling 1954).

(2) **Artemisinin:** Artemisinin and its derivatives are frontline drugs for malarial chemotherapy. Artemisinin-based combination therapies (ACTs) are being used for the treatment of chloroquine-resistant *P. vivax* malaria and uncomplicated *P. falciparum*.

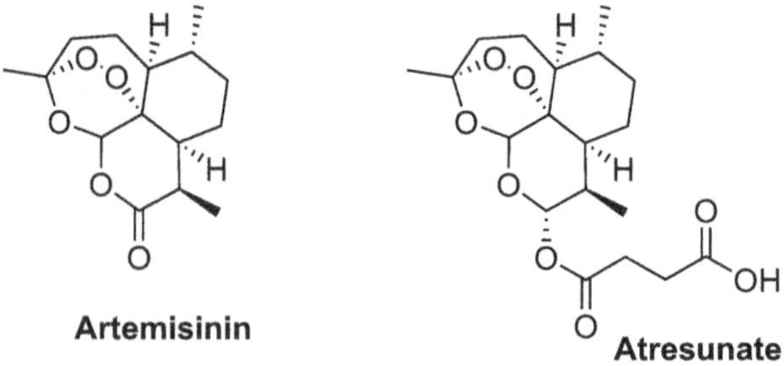

Artemisinin **Atresunate**

Artemisinin has been isolated from the plant *Artemisia annua*, an herb employed in traditional Chinese medicine (TCM) for centuries for reducing fevers and other ailments. Artemisinin-enriched extract has been obtained from the dried leaves of the plant through maceration in dichloromethane (DCM) and further purified by using a three-step crystallization process (Malwade et al. 2012).

(3) **Podophyllotoxin:** Podophyllotoxin is a natural lignan extensively found in *Podophyllum peltatum* with important antineoplastic and antiviral properties. Semi-synthetically modified analogues of podophyllotoxin, viz. etoposide and teniposide, are the marketed drugs for cancer chemotherapy.

Isolation: Powdered tissues were extracted with ethanol and water. Both the ethanolic and water extracts have been partitioned into ethyl acetate and water.

Podophyllotoxin-rich ethyl acetate fraction was evaporated and purified to obtain pure podophyllotoxin using different chromatographic techniques (Canel et al. 2001).

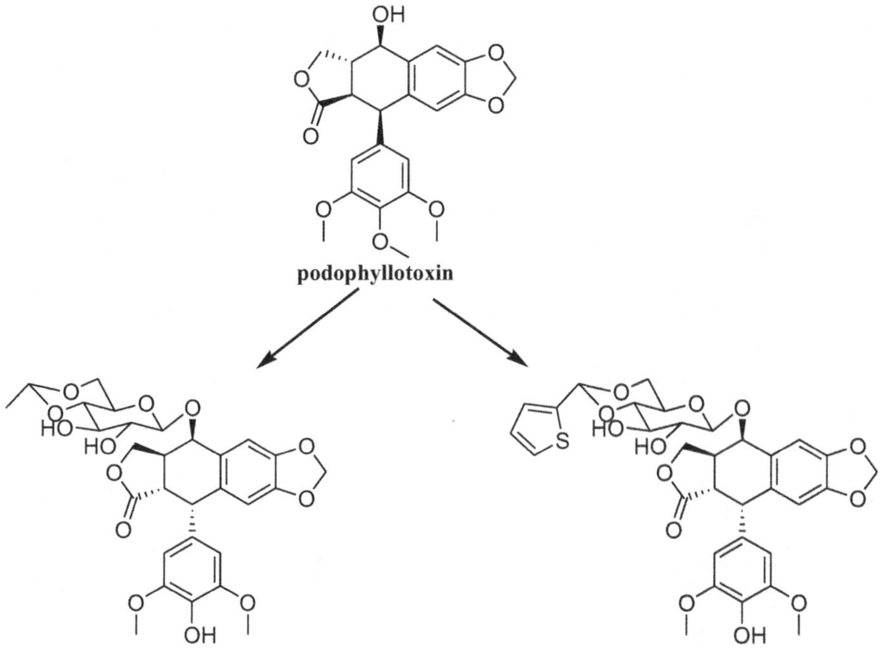

Etoposide **Tenoposide**

(4) **Paclitaxel: Paclitaxel (PTX)** is an anticancer drug sold under the brand name **Taxol**, used to treat several types of cancers including ovarian, breast, lung, oesophageal, Kaposi sarcoma, cervical and pancreatic cancer. Mechanistically, paclitaxel was found to stabilize the microtubule polymerization. It is naturally found in the trees of *Taxus* species including *Taxus brevifolia*, *Taxus chinensis*, etc.

Paclitaxel was first isolated from the bark of Pacific yew tree (*Taxus brevifolia*) in 1971 and approved for its medicinal use in 1993. It is one of the vital drugs listed on the WHO Model List of Essential Medicines. The leaves and shells of hazel tree (*Corylus avellana*) also contains paclitaxel along with 10-deacetylbaccatin III and baccatin III (Ottaggio et al. 2008).

Due to the low abundance of paclitaxel in natural sources, attempts were made to prepare it synthetically, semi-synthetically (from other natural precursors) and through cell cultures. Synthetic approaches for the total synthesis of paclitaxel result in the discovery of another anticancer drug docetaxel. Docetaxel is marketed under the brand name Taxotere that demonstrates similar clinical applications as paclitaxel.

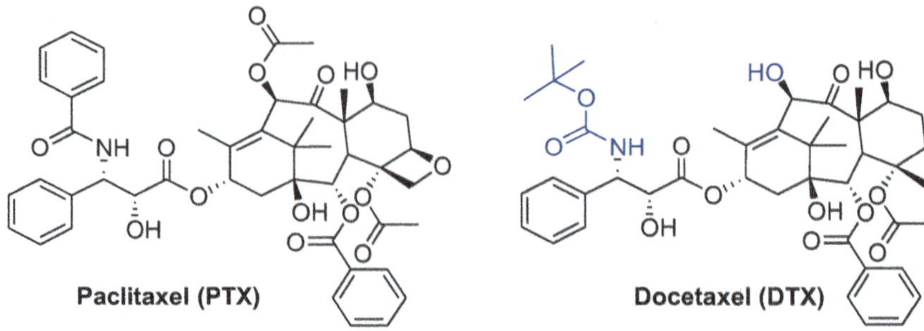

Paclitaxel (PTX) Docetaxel (DTX)

In 1988, Pierre Potier and his research group have reported the semi-synthetic approach for the production of paclitaxel from 10-deacetylbaccatin, a naturally abundant taxane isolated from the needles of *Taxus baccata* in large quantities (Denis et al. 1988). Later, in 1989, a new route for the synthesis of paclitaxel with better yields through the semi-synthetic modification of 10-deacetylbaccatin has been established by Holton's group.

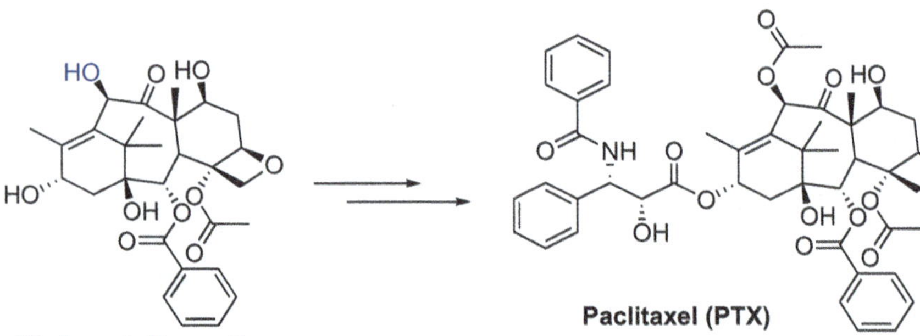

Paclitaxel (PTX)

10-deacetylbaccatin

In 1993, paclitaxel was discovered from an endophytic fungus *Taxomyces andreanae* living on the yew tree. Since then, several endophytic fungi that produce paclitaxel were discovered (Stierle and Strobel 1993).

Isolation: The plant material was extracted with methanol, and the extraction was repeated at least four times. Combined methanol extract was concentrated under vacuum to afford the methanolic extract. Further, it was processed for liquid–liquid extraction and subjected to chromatographic purifications to obtain pure paclitaxel (Pyo et al. 2004).

(5) **Reserpine:** Reserpine is an antihypertensive drug generally used along with a vasodilator. Reserpine was isolated in 1952 from the dried roots of *Rauvolfia serpentina* (*Sarpagandha*). It has been the part of India's traditional medical system over the years to treat fever, insanity and snakebites.

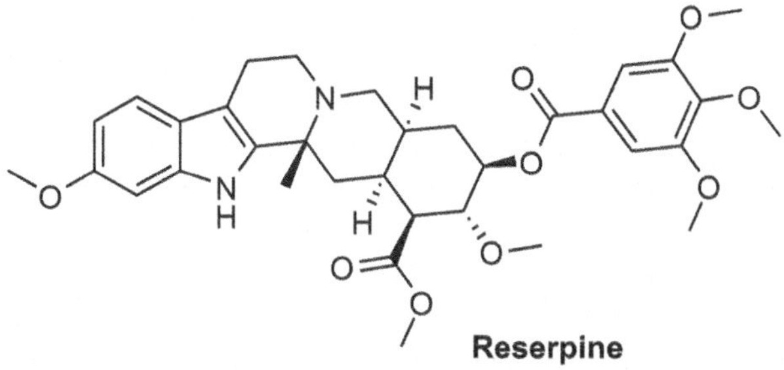

Reserpine

Isolation and Purification: A coarsely powdered plant material (Rauvolfia root) was extracted with chloroform using Soxhlet extraction method. Chloroform extract thus was further processed for the isolation of pure reserpine (Sharma and Gaikar 2012).

(6) **Caffeine:** Caffeine is the most widely consumed psychoactive agent. It is a natural stimulant of the methylxanthine class that is most commonly found in tea, coffee and cacao plants. Historically, the first brewed tea was supposed to be as far back as 2737 B.C. Many years later, coffee was reportedly accidentally discovered by an Ethiopian shepherd who noticed the extra energy it gave to his goats.

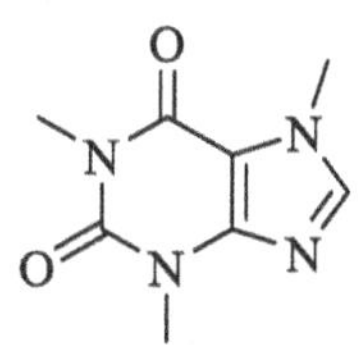

caffeine

Extraction: Well-ground plant material was stirred with sulphuric acid, and the solution was extracted with chloroform (sulphuric acid converts the tannins to their salts and makes them water soluble). The combined chloroform layer was evaporated under reduced pressure to yield the pure methylxanthine crystals (Vuong and Roach 2014).

(7) **Papain:** Papain is a proteolytic enzyme derived from the papaya plant's raw fruit. Proteolytic enzymes aid in the lysis of proteins into small peptides and amino acids. This is why papain is a popular meat tenderizer ingredient. Papain is used to be taken orally to relieve pain and inflammation, as well as to remove excess fluid after trauma or surgery. It is also used to treat parasitic worms, inflammation of the throat and pharynx, shingles symptoms, sore muscles, diarrhoea, runny nose, hay fever and psoriasis. Papain can also be taken orally to treat the side effects of radiation therapy, or it can be combined with other therapies to treat tumours.

Isolation: Freshly collected latex of the locally grown unripen *C. papaya* was stored at −20°C. Due to the greater influence of pH on papain partitioning, the pH of the latex was adjusted to the required pH using 6M HCl/6M NaOH. To the prepared latex, defined amounts of solid PEG and ammonium sulphate were added. After gently shaking the mixture for 15 minutes, the two phases were separated by centrifugation. Cathodic gel electrophoresis and FPLC were used to determine the presence of papain in aliquots of the phases. To obtain pure papain powder, the combined fractions exhibiting protease activity were dialyzed and lyophilized (Nitsawang et al. 2006).

(8) **Thymol and Carvacrol: Thymol** is a natural monoterpenoid derived from *Thymus vulgaris*, ajwain and other plants. It is a phenol derivative of p-cymene with a pleasing aromatic odour and strong antiseptic properties. It has been widely used to treat helminthic infections including tinea or ringworm and hookworm infection (Jannati et al. 2021).

Carvacrol (CV), a structural isomer of thymol, is widely distributed in essential oils of various plants including *Origanum vulgare* (oregano), *Citrus aurantium bergamia* (wild bergamot), *Lepidium flavum* (pepperwort), *Thymus vulgaris* (thyme), etc. Carvacrol has demonstrated a broad range of biological properties like antimicrobial, antioxidant and anticancer activities (Sharifi-Rad et al. 2018).

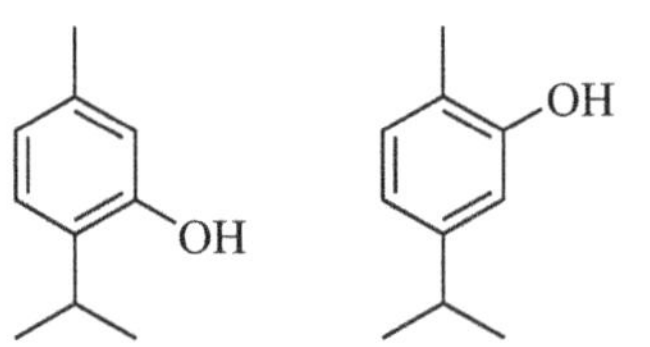

Thymol **Carvacrol**

(9) **Camptothecin:** Camptothecin (CPT) was isolated from the bark and stem of *Camptotheca acuminata* (Zeng et al. 2013), a native plant of China, and has been extensively used in traditional Chinese medicine for gastrointestinal tumours (Maz and Foetida 1972; Efferth et al. 2007). M. E. Wall and M. C. Wani have discovered CPT while screening the natural products in search of anticancer drugs. CPT was found to be a topoisomerase inhibitor that has demonstrated anticancer activity against colon, breast, ovarian, lung, and stomach cancers in preliminary clinical trials (Wang et al. 2019).

However, owing to its low solubility and adverse effects, CPT has limitations in therapeutic use. To overcome the limitations, various semi-synthetic analogues of camptothecin were developed, among which topotecan, irinotecan, belotecan and trastuzumab deruxtecan were approved for cancer chemotherapy (Wall et al. 1966).

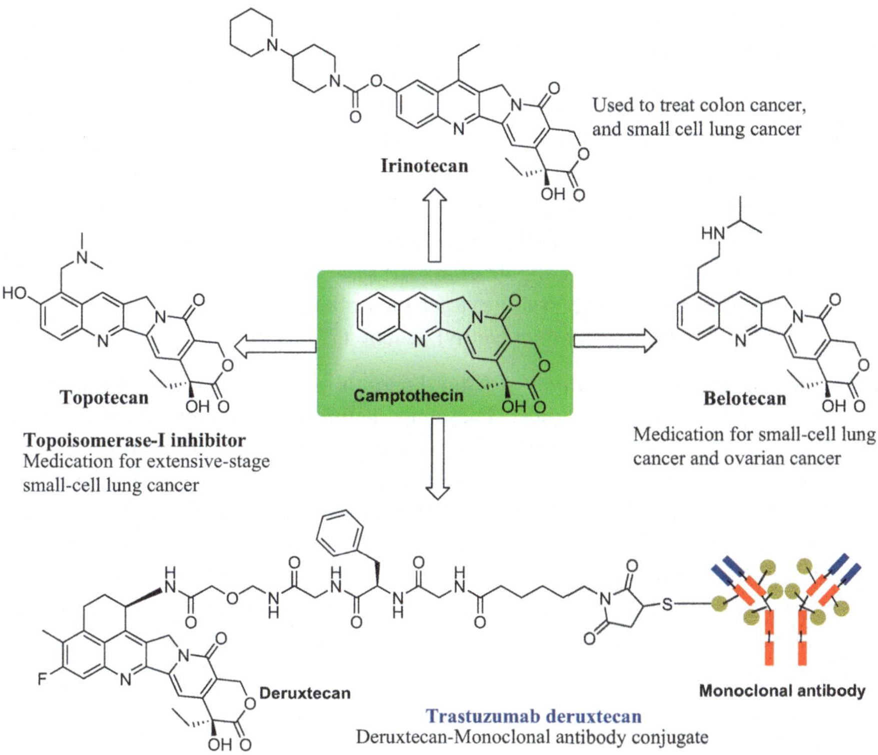

A monoclonal antibody medication used to treat breast cancer and gastroesophageal adenocarcinoma

(10) **Piperine:** Piperine is the chief pharmacologically active constituent of black pepper, which brought pungency and piercing taste to it. This natural alkaloid has demonstrated several health benefits and therapeutic applications. Piperine was isolated in 1819 by Hans Christian Ørsted from the fruits of *Piper nigrum*.

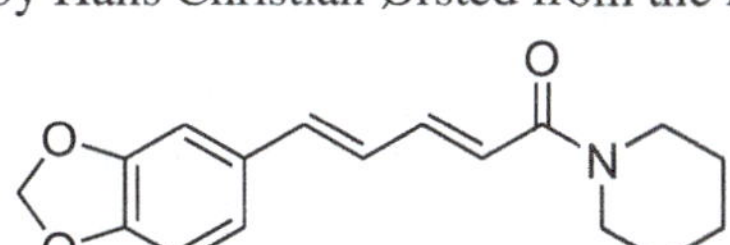

Piperine

Isolation

1. **Extraction with ethanol:** Finely ground black pepper (10 g) was extracted with 95% ethanol (150 ml) using Soxhlet extraction method for 2 h. The alcoholic extract was filtered and concentrated under vacuum, and the resultant crude extract was treated with 10% alcoholic potassium hydroxide (10 ml). The insoluble portion was filtered, and the filtrate was left overnight and the crystalline piperine obtained was filtered.
2. **Extraction with dichloromethane**: In a 100 ml round bottom flask, well-ground pepper powder (10 gm) was refluxed with dichloromethane (20 ml) for 20 min. The flask was gradually cooled to room temperature and filtered off. And the extract was treated with acetone and hexane to purify further.

3. **Extraction with glacial acetic acid**: Cold maceration was employed to obtain the acetic acid extract of black pepper (25 g) in glacial acetic acid (300 ml). The extract was then diluted with equivalent volume of water and extracted with chloroform. The chloroform layer was then washed with 10% $NaHCO_3$ solution followed by water. The organic layer was dried over anhydrous Mg_2SO_4 and concentrated under reduced pressure. Piperine-enriched resinous impurities were removed by washing with NaOH solution followed by water. Finally, piperine-enriched fraction was recrystallized by using diethyl ether (Shingate et al. 2013).

   Purification: The crude extract thus obtained was purified on conventional column using toluene/ethyl acetate (7:3).

(11) **Curcuminoids:** Derivatives of natural curcumin with varied functional groups are known as curcuminoids. These are naturally occurring phenolic linear diarylheptanoids responsible for turmeric's pronounced yellow colour. Curcumin is the chief constituent of the turmeric curcuminoids obtained from the rhizomes of turmeric plant *Curcuma longa* species, which is often used to colour foods and as medicines. Curcumin is generally used as a food ingredient, culturally used in cosmetics, and as a flavouring agent in foods and beverages (Wilken et al. 2011). Turmeric is commonly used for conditions involving pain and inflammation, such as osteoarthritis because of its antioxidants.

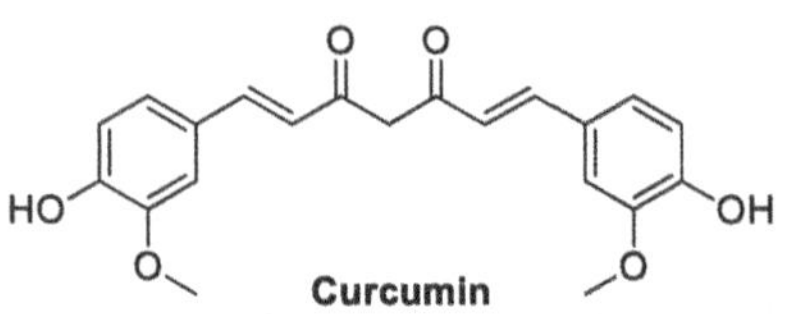

Curcumin

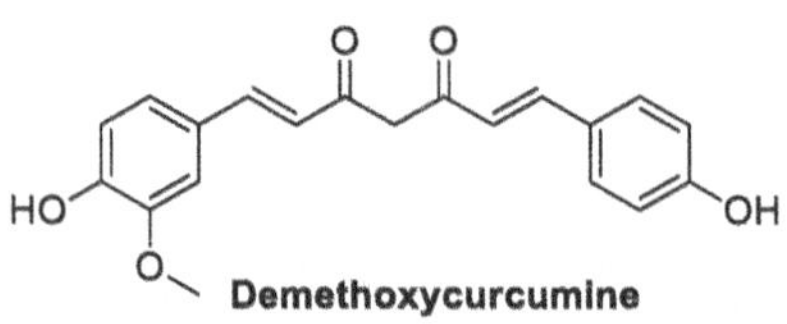

Demethoxycurcumine

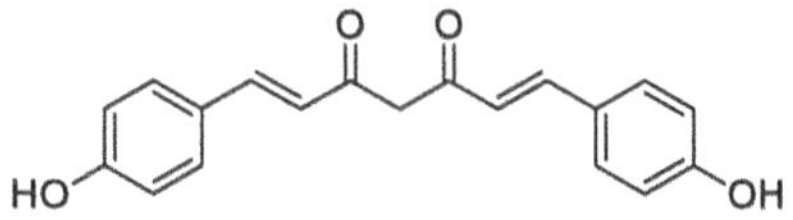

Bisdemethoxycurcumine

Isolation and Purification: Finely powdered rhizome was subjected to Soxhlet extraction (6 hours) using acetone. The acetone extract thus obtained was filtered off and evaporated under vacuum. The oleoresin obtained was precipitated with petroleum ether, filtered off and dried under vacuum to obtain the mixture of curcuminoids. The crude mixture consists of curcumin, demethoxycurcumin and bisdemethoxycurcumin was subjected to conventional column chromatography (chloroform followed by chloroform/methanol). The individual curcuminoids isolated were crystallized in methanol/ chloroform. The crystals obtained were finally washed with petroleum ether (Revathy et al. 2011).

(12) **Atropine and scopolamine:** Atropine and scopolamine are widely distributed in plants of several families. Plants of *Datura* species comprises of high concentrations of these two natural alkaloids. These phytoconstituents are

present in all parts of the plant that make the plant toxic. Over the years, herbal extracts of atropine and scopolamine have been used to treat nausea, intestinal cramping and motion sickness and as dilator of pupils for eye examination.

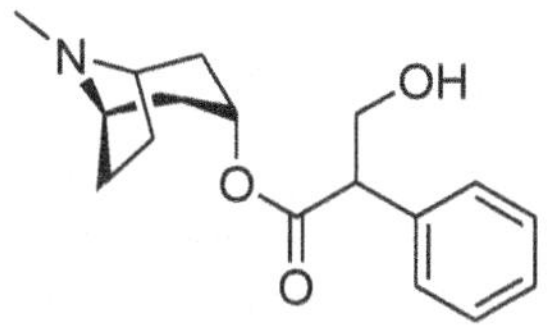

Atropine

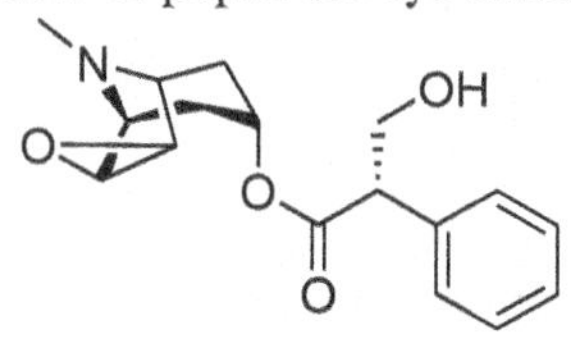

Scopolamine

Ultrasound-assisted extraction or liquid–liquid extraction (LLE) procedures was applied for the isolation of these alkaloids (Śramska et al. 2017).

(13) **Cocaine:** Cocaine, a powerful addictive stimulant drug, classified under a tropane class of alkaloids, is primarily isolated from the leaves of *Erythroxylum coca* and *Erythroxylum novogranatense* (plants of *coca* species) (Brachet et al. 2001). The leaves of many other species in the Erythroxylaceae family also contain cocaine in lesser amounts. South Americans use to chew and consume coca leaves (*Erythroxylum coca*) to get stimulated for millennia (Goldstein et al. 2009). Pure cocaine was isolated more than hundred years ago and in the early days of twentieth century; it was the chief constituent in many tonics and preparations developed to treat a wide variety of ailments. It was the ingredient of Coca-Cola in its initial formulations. Cocaine is generally used as a *recreational drug* and *euphoriant (*Pomara et al. 2012*)*. Cocaine is consumed in different ways including direct *snorting*, inhalation of the sublimated form (after heating) or in the form of injection into a *vein* (Zimmerman 2012). *Its ps*ychological effects include a penetrating sense of happiness, sexual provocation, hallucination, anxiety, etc. Physical symptoms include tachycardia, perspiration and enlarged pupils. High doses of cocaine can result in hypertension or high temperatures (Connors and Hoffman 2013). The activity of cocaine starts within the span of seconds to minutes of purpose and lasts for 5–90 min.

Prior to the invention of synthetic local anaesthetics, cocaine was used to block pain during surgeries (Calatayud and González 2003). Cocaine was topically applied as a local anaesthetic to handle the pain in the mouth or nose and is mainly used in nasal and lacrimal duct surgeries.

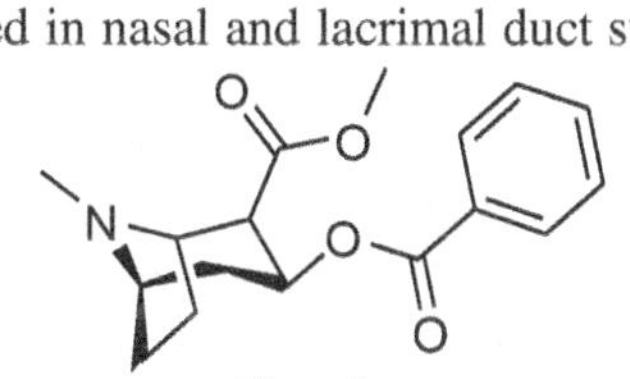

Cocaine

However, due to its influence on the reward pathway of the brain, cocaine is highly addictive and was declared as an abusive drug.

(14) **Berberine:** *Berberine* is widely distributed in nature and found in several plants including *Berberis vulgaris*, *Berberis aristata*, *Mahonia aquifolium*,

Hydrastis canadensis, *Coptis chinensis*, *Xanthorhiza simplicissima*, *Phellodendron amurense*, *Tinospora cordifolia*, *Eschscholzia californica* (Californian poppy) and *Argemone mexicana*. Berberine is generally distributed in roots, rhizomes, stems and bark. Owing to its natural yellow colour, plants of *Berberis* species were used to dye wool, leather and wood.

Medically, berberine is most commonly taken for diabetes, hyperlipidaemia and hypertension. It is also used to treat burns, canker sores, liver disease and many other conditions.

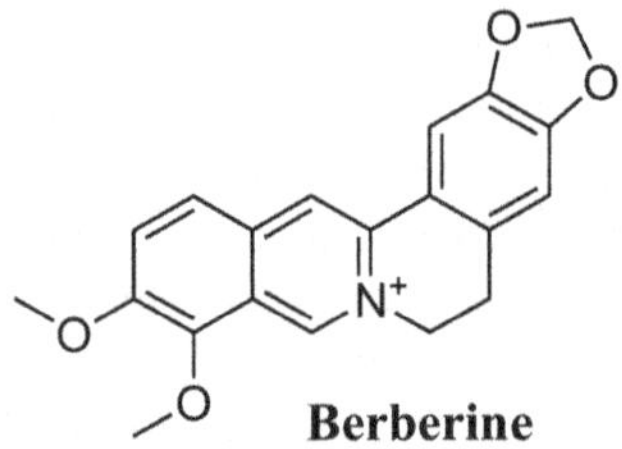

Berberine

Isolation and Purification: Fresh root of *Berberis vulgaris L.* were shade dried, and the air-dried roots were finely powdered for extraction. The aqueous extract was obtained by cold maceration. The extract was filtered, concentrated and dried under reduced pressure to get the crude extract that was further treated with 1% HCl, and the solid part that remained was filtered. The filtrate was basified with concentrated ammonium hydroxide till the pH reaches 8 and extracted with chloroform. The combined chloroform layer was evaporated to obtain tertiary alkaloids. The chloroform fraction was subjected to the conventional column chromatography to obtain the natural products in pure form (Pradhan et al. 2013).

(15) **Colchicine:** Colchicine, isolated from *Colchicum autumnale*, is a medication used to treat gout (Shekelle et al. 2017). *Colchicum autumnale* was described for the treatment of rheumatism and joint swelling in the ancient Ebers Papyrus, an Egyptian medical papyrus as early as 1500 BC (Graham and Roberts 1953). The utilization of the bulb-like corms of *Colchicum* to treat gout presumably goes back around 550 AD. *Colchicum* extract as a treatment for gout was first depicted in *De materia medica* by Pedanius Dioscorides in the first century AD.

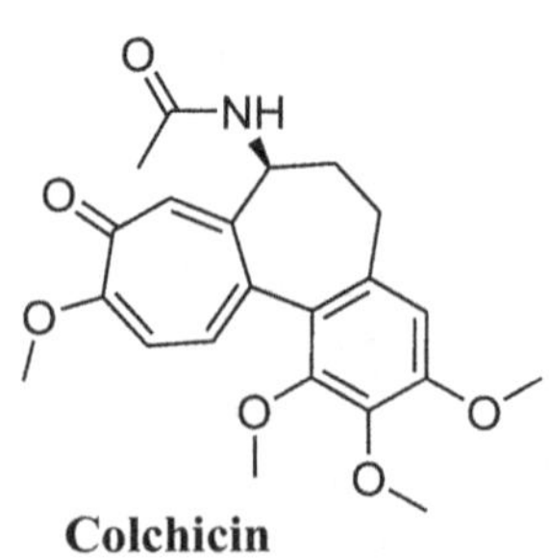

Colchicin

(16) **Combretastatin A-4:** Combretastatin A-4 is a well-known natural tubulin polymerization inhibitor, isolated from *Combretum caffrum* (African) and

Combretum leprosum. Combretastatin A-4 is mainly extracted from the stem-wood parts using dichloromethane–methanol solvent combination. Combretastatin A-4 and its derivative ombrabulin are in clinical trials for cancer chemotherapy.

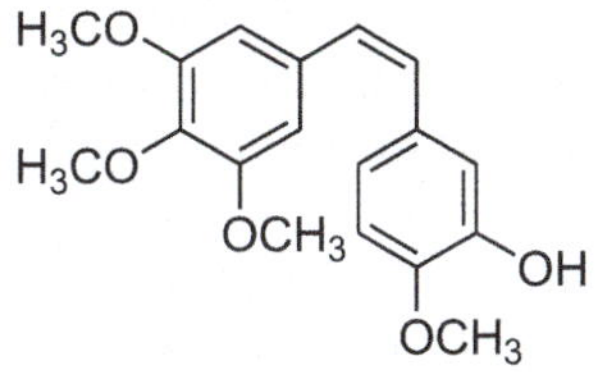

Combretastatin A-4

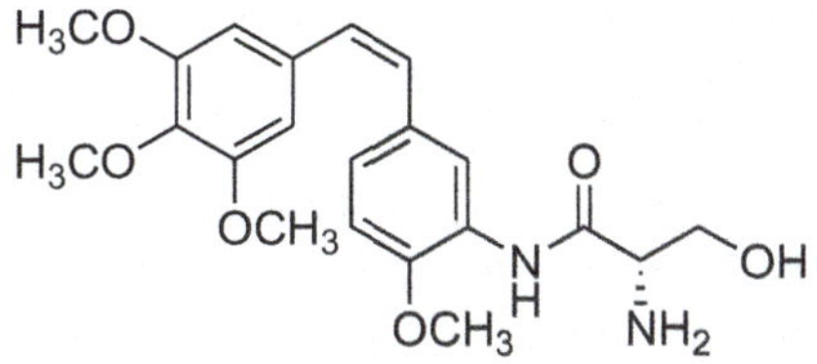

Ombrabulin

(17) **Phytol: Phytol** is an acyclic diterpenoid utilized in the manufacture of synthetic vitamins E (α-tocopherol) (Netscher 2007; Daines et al. 2005).

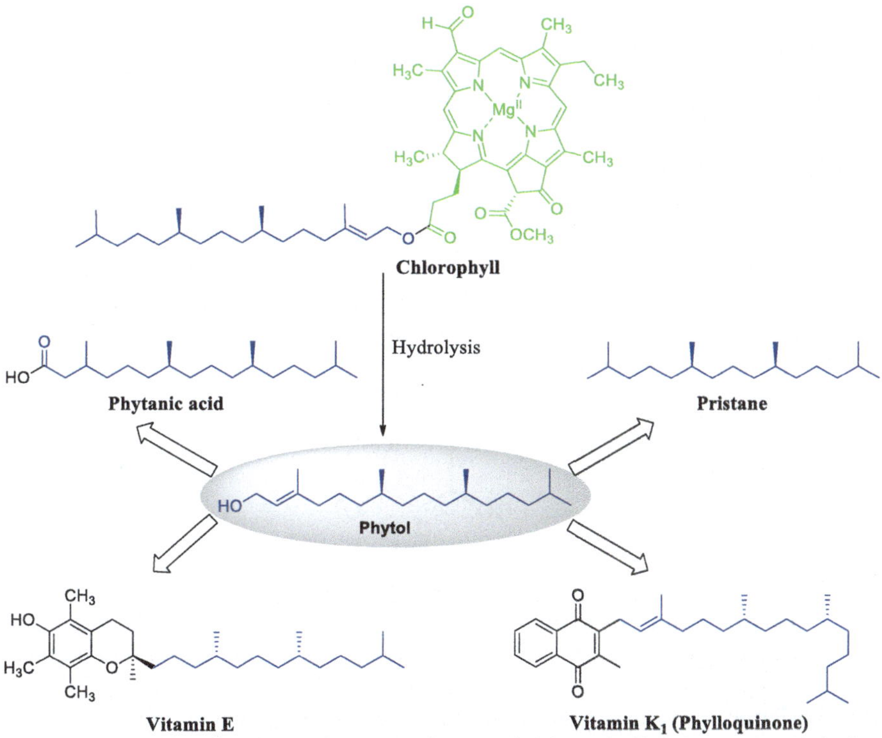

Naturally phytol is a phytoconstituent of chlorophyll. It is generated during the fermentation of ingested plant materials of ruminants in gut. Phytol is stored in the form of phytanic acid and stored in fats (biosynthetic products of phytol) (Van Den Brink and Wanders 2006) while it is transformed into pristane in shark liver.

Phytol was first obtained by hydrolysis of chlorophyll in 1909 by the German chemist Richard Willstatter, and its structure was determined by

F. G. Fischer in 1928. Phytol may be obtained in the process of separating chlorophyll from *Medicago sativa* (alfalfa/lucerne).

Phytol has been found to be a potent cytotoxic, antioxidant, anxiolytic, metabolism-modulating, autophagy, apoptosis-inducing, antinociceptive, anti-inflammatory, immune-modulating and antimicrobial agent (Islam et al. 2018).

Isolation: Phytol has been isolated from *Abutilon indicum* (Indian abutilon, Indian mallow) leaves using cold percolation extraction method (the hydromethanol extract) (Thakor et al. 2016).

(18) **Cynarine:** Cynarine (1,3-dicaffeoylquinic acid), a bioactive phytochemical of globe artichoke (*Cynara scolymus* L.), is known for its powerful hepatoprotective activity (Lattanzio et al. 2009). It is also reported to exhibit anti-HIV (Robinson et al. 1996), diuretic, choleretic and activities (Gebhardt 2005). Cynarine can be found in both artichoke leaves and heads, and about 1.5% is present in methanolic extracts (Lattanzio et al. 2009).

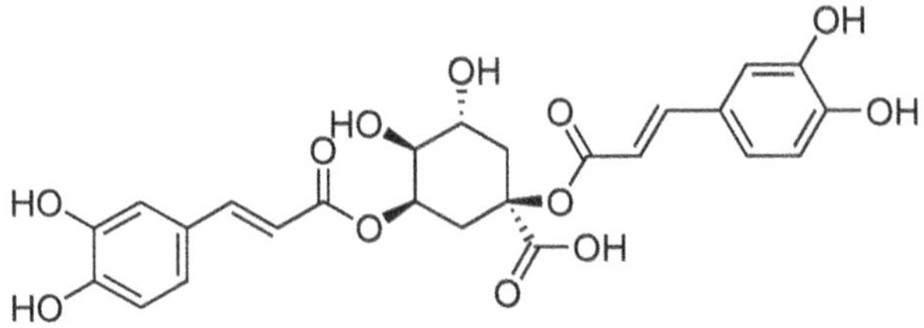

Cynarine

Isolation: Cynarine is commonly isolated from the methanolic extract of *Cynara scolymus* L. and *Onopordum illyricum* L. (Topal et al. 2016).

(19) **Pinitol:** Pinitol is an anti-diabetic agent derived from the leaves of *Argyrolobium roseum.*

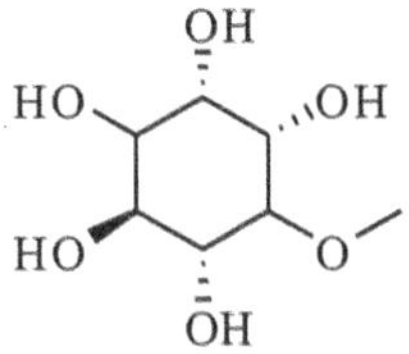

Pinitol

Isolation and Purification: The percolation method was used to attain the ethanolic extract of the plant material, which was further subjected to purification to get pinitol in pure form (Sharma et al. 2016).

(20) **Vasicine:** Vasicine is a quinazoline alkaloid mainly found in *Adhatoda vasica* (Acanthaceae). It is a well-known plant medication practised in Indian traditional Ayurvedic and Unani system of medicine over the years for the treatment of respiratory tract disorders and to induce abortions (Claeson et al. 2000).

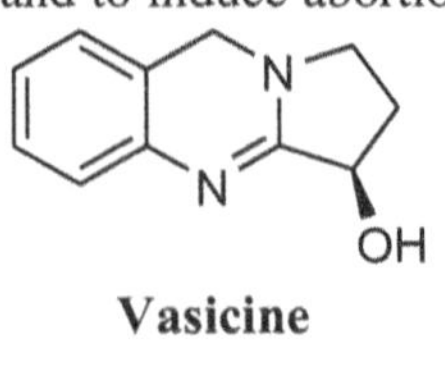

Vasicine

Isolation and Purification: Vasicine was first isolated in 1924 from the chloroform extract of *vasica* leaves (Sreelekshmi et al. 2021).

(21) **Betulinic and ursolic acid: Betulinic acid** is a triterpenoid, widely distributed in nature and demonstrated a wide range of biological activities including antiretroviral, antimalarial, anticancer (Chowdhury et al. 2002; Tan et al. 2003) and anti-inflammatory properties. Betulinic acid is commonly present in the bark of many plants including white birch (*Betula pubescens*) (Tan et al. 2003), ber tree (*Ziziphus mauritiana*), selfheal (*Prunella vulgaris*), *Triphyophyllum peltatum* and *Ancistrocladus heyneanus* (carnivorous plants), *Diospyros leucomelas*, *Tetracera boiviniana*, the jambul (*Syzygium formosanum*) (Zuco et al. 2002), flowering quince (Gao et al. 2003), rosemary (Abe et al. 2002) and *Pulsatilla chinensis (*Ji et al. 2002*)*.

Ursolic acid was identified in the epicuticular waxes of apples in 1920 and widely distributed in nature including in the peels of fruits, in rosemary (Abe et al. 2002) and thyme.

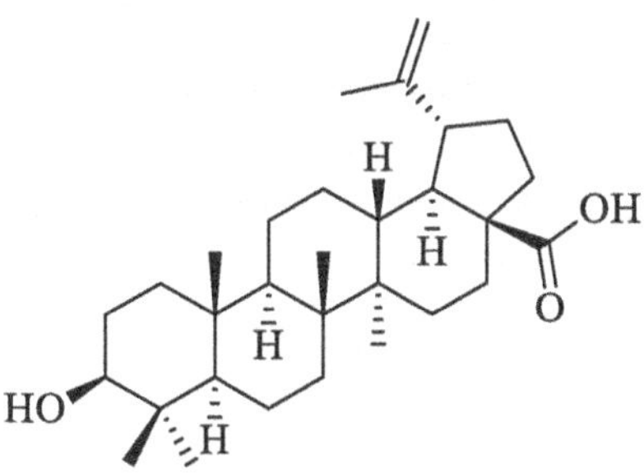

Beutilinic acid

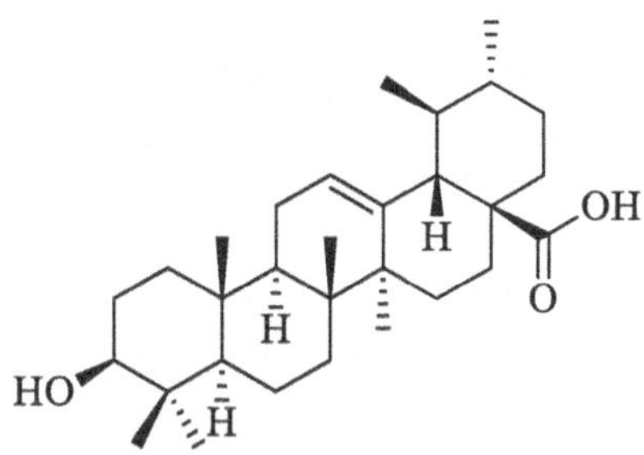

Urosilic acid

Both these bioactive natural products are found in the leaves of *Vitex negundo* L., a shrub also known as Chinese chaste tree.

Isolation and Purification: The defatted leaf powder of *Vitex negundo* L. was extracted with methanol. The methanolic extract was then evaporated under reduced pressure and purified on column chromatography to attain betulinic acid and ursolic acid (Chandramu et al. 2003).

(22) **Galantamine:** Galantamine is a marketed drug with the brand names Razadyne and GalantaMind™ used in Alzheimer's disease and other memory impairments (Lilienfeld 2003). US FDA has approved galantamine to treat mild to moderate dementia.

D. Paskov, a Bulgarian chemist, was the first to isolate galantamine in 1956 from the bulbs of *Galanthus nivalis* (common snowdrop) (Tewari et al. 2018). The active ingredient galantamine was found to be a potent *acetylcholinesterase* (AChE) inhibitor and is being treated as a modern-day frontline drug for Alzheimer's disease (Heinrich 2004). This alkaloid is also found in bulbs and flowers of different plants including *Galanthus caucasicus*, *Galanthus woronowii, Narcissus*, *Leucojum aestivum* and *Lycoris* including *Lycoris radiata*, etc.

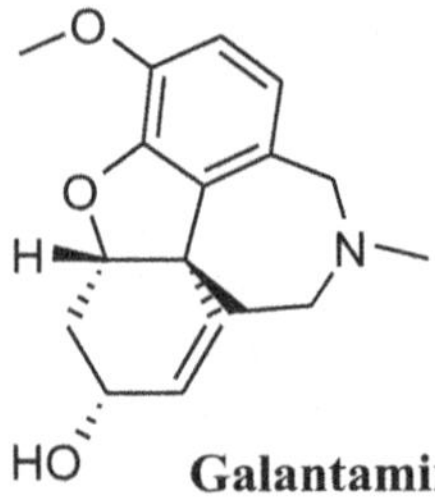

Galantamine

Isolation and Purification: Finely ground plant material (*L. aestivum* bulbs) was placed into a centrifuge tube and treated with 1% aqueous trifluoroacetic acid. Then the mixture was vigorously stirred for 35min and centrifuged. Two equal volumes of the supernatant were transferred into two fresh centrifuge tubes and basified with 1M NaOH. The organic matter was then extracted with ethyl acetate from each tube, dried with magnesium sulphate, filtered and reduced (~0.2 mL) under reduced pressure to afford the crude extract that was finally purified on flash column to obtain pure galantamine as a clear oil (Halpin et al. 2010).

(23) **Neoandrographolide:** Neoandrographolide is a colourless crystalline compound, isolated from the stem and leaves of *Andrographis paniculata. The herb that majorly contains* neoandrographolide and andrographolide has been used in several pharmaceutical preparations. These preparations are mainly employed for the treatment of acute bacterial dysentery, acute gastroenteritis, upper respiratory tract infection pharyngitis, acute tonsillitis, etc.

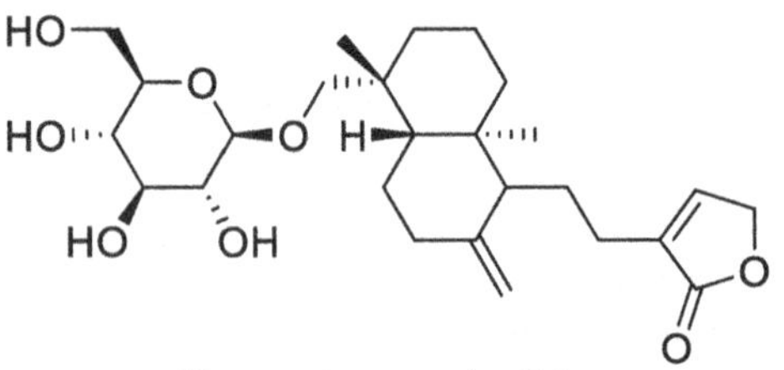

Neoandrographolide

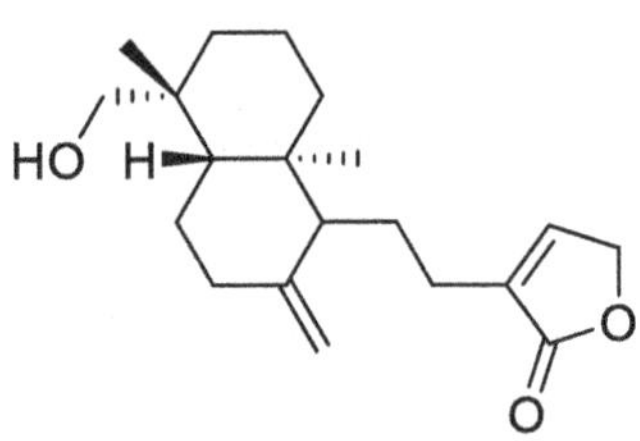

Andrographolide

Isolation: The plant material (507.60 gm) was soaked in ethanol for 10 days. Then the extracted material in ethanol was filtered and evaporated under vacuum to afford the ethanol extract. The sticky mass (62.60 g) thus obtained was fractionated into n-hexane, ethyl acetate and butanol soluble fractions. The ethyl acetate fraction was purified on column chromatography to obtain the required phytoconstituents (Al-Amin et al. 2012).

(24) **Nordihydroguaiaretic acid (masoprocol):** Nordihydroguaiaretic acid (NDGA) is an antioxidant phenolic lignin found in the genus *Larrea.* The primary source of NDGA is *Larrea tridentata* (creosote bush/chaparral), which is abundantly found in the deserts of Mexico and southwest USA (Page 1955). NDGA accounts for 10% of the weight of the dry leaves of the plant. Over the centuries, traditionally, chaparral leaves have been applied to treat a number

disorders including rheumatoid arthritis, diabetes, pain and inflammation by the locals of the mentioned areas (Arteaga et al. 2005).

NDGA was used as a food preservative in the 1950s and currently is used as a chaparral tea, a food supplement (Manda et al. 2020). It is best known for the ROS-scavenging activity, leading to the decrease of pro-oxidant effects of inflammation, and the lipoxygenases (5-LOX) inhibitor, leading to the reduction of lipid hydroperoxides (Mashima and Okuyama 2015).

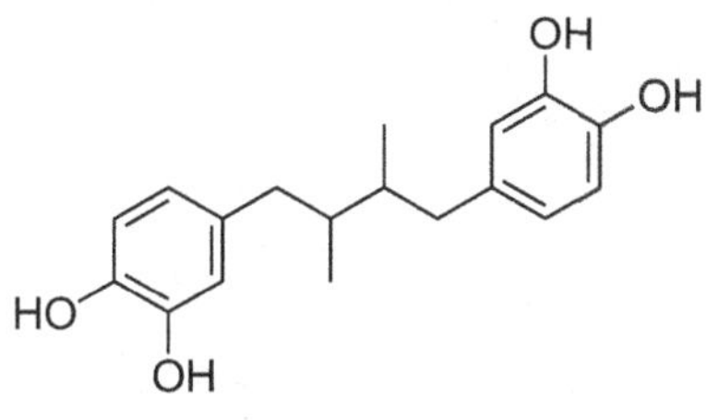

Nordihydroguaiareticacid

(25) **Protoveratrines:** In 1890, Salzberger first isolated the alkaloid protoveratrine from *Veratrum albumen*; later it was found to be a mixture of two closely related alkaloids, protoveratrine A and protoveratrine B.

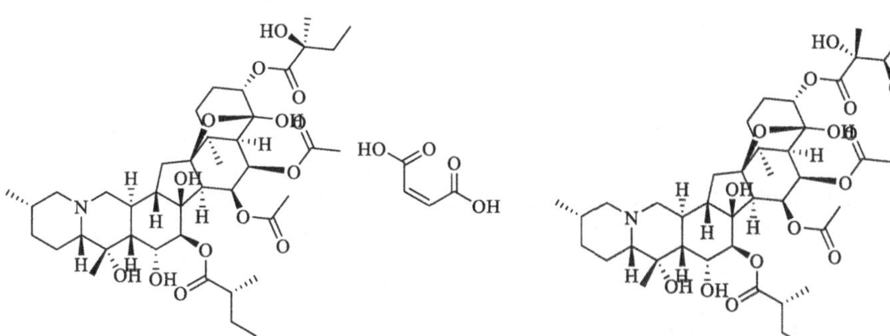

Protoveratrines A

Protoveratrines B

Protoveratrine A, the principal alkaloid of *Veratrum album*, has been used in the treatment of hypertension. Protoveratrine B is a vasodilator, and it can be administered at significantly higher doses before the patient begins to vomit. Dried rhizome extracts of *Veratrum album* were briefly used as a pesticide against the Colorado potato beetle (Aydin et al. 2014).

Isolation and Purification: The ground stalks and leaves were extracted by percolation method with 95% ethanol. The residue was dried and re-extracted with water. The alcohol was removed under vacuum to afford the syrupy alcohol extract. In the second experiment, the stalks and leaves were extracted with benzene and ammonia, followed by ethanol. Purified alkaloid fractions were prepared from dried ethanol and benzene extracts by a double partition between aqueous acid and chloroform phases. The final chloroform fraction was dried and purified to afford the pure alkaloids (Keeler and Binns 1964).

(26) **Sennosides:** Senna glycoside, also known as sennoside/senna, sold under brand names Ex-Lax and Senokot, is a medication used for episodic and chronic constipation and to empty the large intestine before surgery (Wald 2016). It typically starts its effect within 30 minutes when given by rectum and within twelve hours when given by mouth. It is enlisted in the WHO Model List of Essential Medicines (Organización Mundial de la Salud (OMS) 2019). Sennosides were found in the group of plants *Senna*). In plant form, it has been used since the 700s CE.

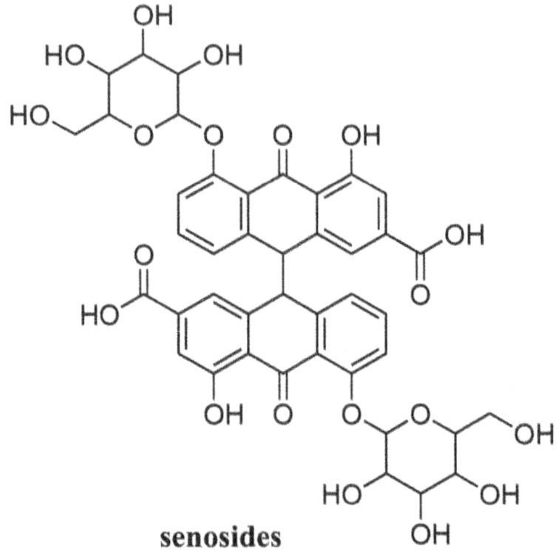

senosides

Isolation and Purification: Finely powdered leaves of *Senna alata* (also known as *Cassia alata*) were extracted with aqueous methanol followed by re-extraction with methanol and the powder was filtered. The combined filtrates were evaporated to obtain the methanolic extract. The most active ethyl acetate fraction was further processed for purification (Das et al. 2020).

(27) **Conessine:** Conessine is a steroidal alkaloid, obtained from the plant species of the family Apocynaceae.

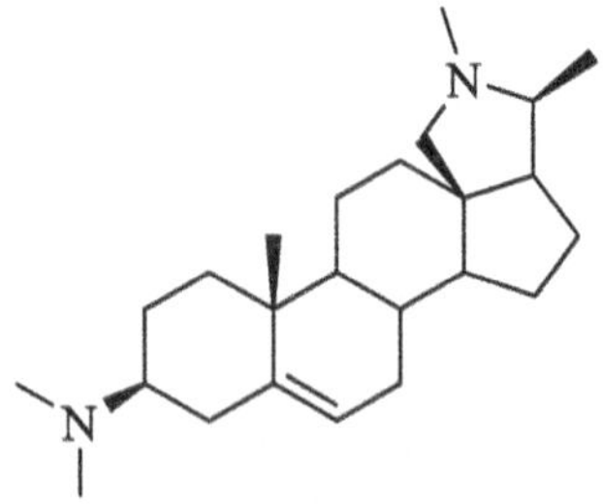

Conessine

It has shown a number of biological activities including antimalarial, H3-receptor antagonist activities, etc. It was also found to have the great ability of penetration through the blood–brain barrier and high affinity to the adrenergic receptors.

Isolation and Purification: The stem bark of *H. floribunda* was finely powdered and was macerated in methanol for 72h. The concentrated methanolic extract was then treated with 5% hydrochloric acid. The solid portion was filtered, and the aqueous solution was made alkaline with ammonia and extracted with ethyl acetate. The active ethyl acetate fraction was subjected to a bioactivity-guided fractionation by flash chromatography to obtain the pure conessine (Freeman et al. 2009).

(28) **Shatavarin: Shatavarin IV** (SIV), an anticancer steroidal saponin, is a chief bioactive phytoconstituent present in the root part of *Asparagus racemosus* (Liliaceae).

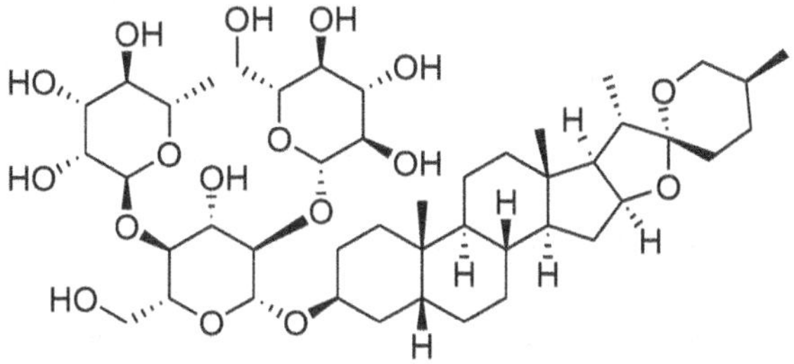

Shatavarin

Isolation: Roots of *A. racemosus* were pulverized in mechanical grinder, and the coarsely powdered material was extracted with hexane followed by methanol using Soxhlet apparatus. The methanolic extract was further processed for the purification to obtain shatavarin (Haldar et al. 2018).

(29) **Nicotine:** Nicotine is a chiral alkaloid, generally produced by the *Solanaceae* family of plants, mainly *Nicotiana tabacum* (Tobacco plant) and *Duboisia hopwoodii*. It is a commonly used recreational agent as a stimulant and anxiolytic agent. Nicotine was first isolated in 1828 from the tobacco plant by German chemists Wilhelm Heinrich Posselt and Karl Ludwig Reimann. Precise levels of nicotine are given to patients to deter their smoking dependency.

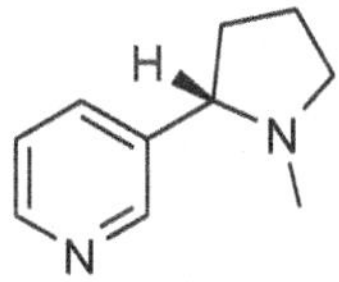

Nicotine

Isolation and Purification: The shade-dried leaves were ground, and the low alkaloid content was removed. By using a strong base, the alkaloid content was precipitated and further processed for purification to obtain nicotine in pure form (Jamin et al. 1997).

(30) **Cannabinoids:** Compounds found in cannabis plant are called as cannabinoids. There are around 113 cannabinoids that have been isolated from cannabis, among which tetrahydrocannabinol (THC) and cannabidiol (CBD) are the notable compounds exhibiting varied biological activities. The most notable

among the isolated cannabinoids is the tetrahydrocannabinol (THC), the primary psychoactive compound in cannabis. THC has been marketed under the name dronabinol (trade names Marinol and Syndros) and has been approved as an appetite stimulant for AIDS patients and an antiemetic agent for patients receiving chemotherapy. On the other hand, CBD is mainly prescribed for the treatment of epilepsy syndromes in children.

The consumption of cannabis plant has been reported from the Vedic period in Indian traditions. It has been enlisted in the category of abusive drugs. Cannabis has been used in different forms including form of marijuana, kief, hashish, tincture and hash oil.

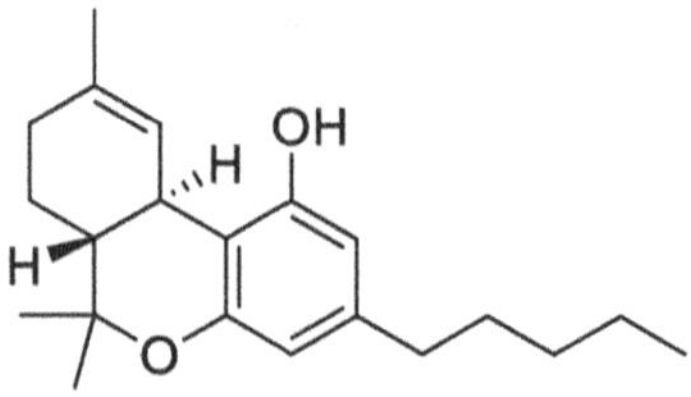

Tetrahydrocannabinol (THC)

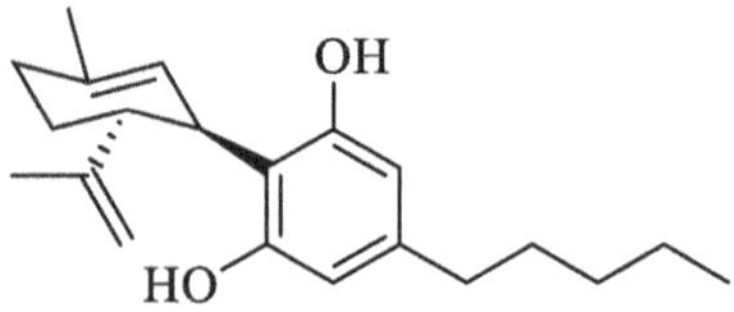

Cannabidiol (CBD)

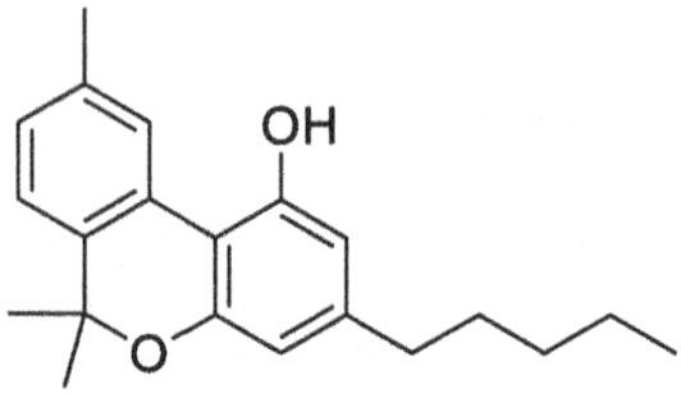

Cannabinol (CBN)

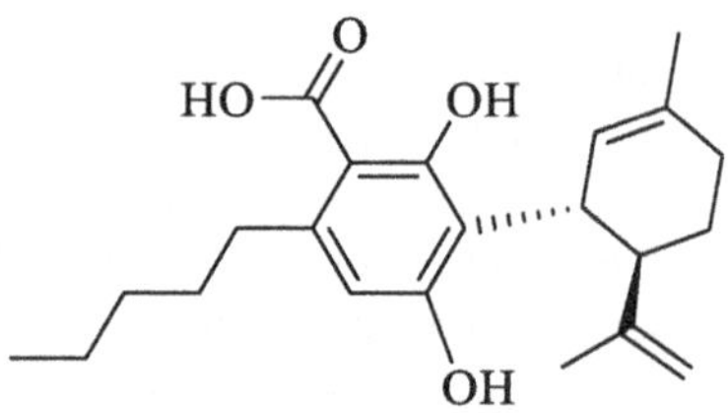

Cannabidiolic Acid

The first cannabinoid that was structurally characterized was cannabinol (CBN) by the British chemist Robert S. Cahn in 1940. CBN has been applied in various ailments including nausea, spasticity and neuropathic pain.

Isolation and Purification: The air-dried (in dark and humid conditions) aerial parts of the plant material were finely ground. The plant material was extracted with ethanol using maceration. The ethanolic extract was processed for purification to obtain pure cannabinoids (Martinenghi et al. 2020).

(31) **Cinnamaldehyde:** Cinnamaldehyde is a naturally occurring organic compound that gives a cinnamon flavour and primarily used as a food flavouring agent. Cinnamaldehyde was first isolated by J. B. Dumas and E. M. Péligot in 1834 from the cinnamon oil and was synthesized by Luigi Chiozza in 1854. Over the years, it has been used to address a variety of disorders including common cold, flu, diarrhoea, cancer, etc.

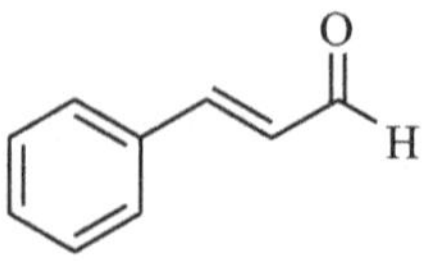

Cinnamaldehyde

Isolation and Purification: The cinnamon essential oil was extracted from the fresh bark by steam distillation in a Clevenger-type apparatus. The distillate was extracted with dichloromethane (DCM) several times using a separating funnel. The combined DCM layer was washed with water, dried on anhydrous sodium sulphate and evaporated with gentle heating to afford the essential oil (Al-Bayati and Mohammed 2009).

(32) **Hispidol:** Hispidol (6,4'-dihydroxyaurone), an antifungal dihydroxyaurone, has been isolated from the fruits of *Poncirus trifoliata* Rafinesque (Xu et al. 2008).

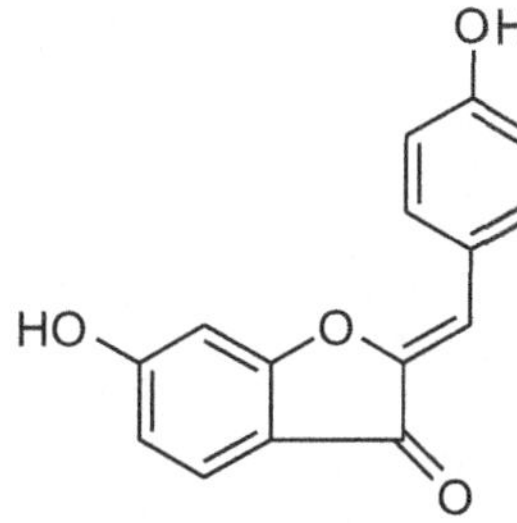

Hispidol

(33) **Vincristine and vinblastine:** Vinblastine and vincristine are the two biologically important alkaloids derived from the periwinkle plant *Catharanthus roseus*. These are the frontline chemotherapeutic drugs to treat different types of cancers. Mechanistically, they found to inhibit tubulin polymerization and prevent formation of the mitotic spindle. Having been practised for centuries in folkloric medicine, rosy periwinkle *Catharanthus roseus* was found to contain more than 120 alkaloids. Many of the alkaloids present in the crude formulations are found to be pharmacologically important; the two most significant compounds are vincristine and vinblastine. Robert Noble and Charles Thomas Beer were the first to isolate vinblastine from the Madagascar periwinkle plant.

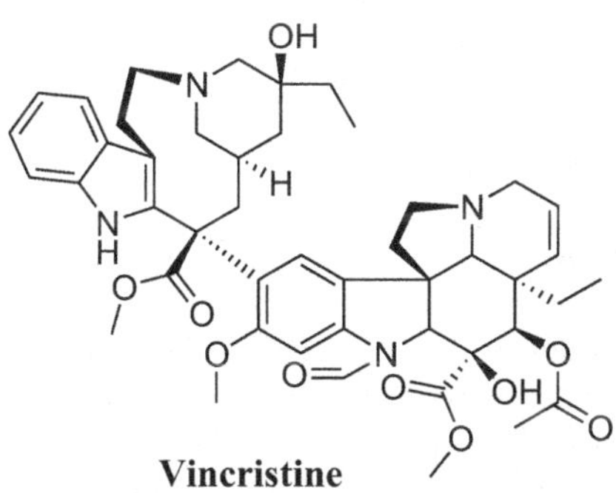

Vincristine

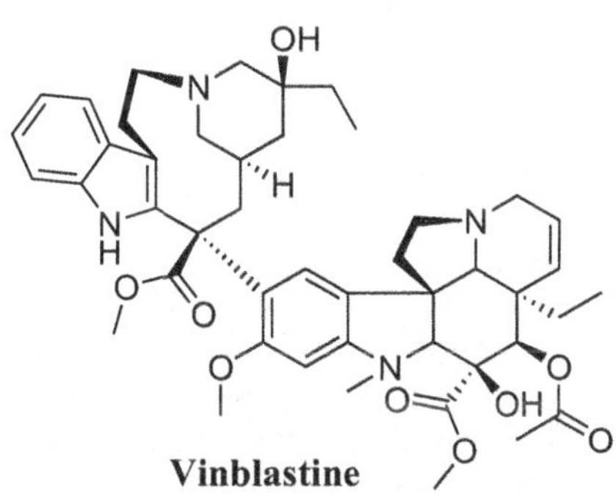

Vinblastine

Isolation and Purification: A two-stage fermentation process is generally used for the isolation of vinblastine and vincristine by using the endophytic fungus *Fusarium oxysporum (*Kumar et al. 2013*)*.

(34) **Withaferin A:** It is a steroidal lactone, isolated from the herbs of Solanaceae family including *Acnistus arborescens*, *Withania somnifera* (Ashwagandha), etc. From the ancient years, it has been part of Indian traditional Ayurvedic system of medicine. Withaferin A has shown several of biological activities including cardioprotective, immuno-modulatory, anti-inflammatory, anti-angiogenesis, anti-metastasis and anti-carcinogenic properties (Vanden Berghe et al. 2012).

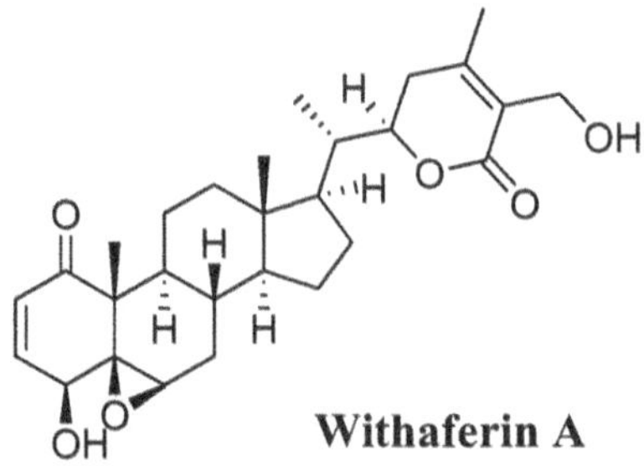

Withaferin A

Isolation and Purification: Finely powdered roots and leaves of Ashwagandha were defatted by refluxing with petroleum ether. The defatted material was then refluxed with methanol for four hours, and then the flask was allowed to attain the ambient temperature and filtered. The collected methanolic extract was transferred into separation funnel and partitioned between water and dichloromethane. The dichloromethane extract was further purified using standard purification techniques (Keesara and Jat 2017).

(35) **Carpaine:** Carpaine, a macrodiolide, is one of the major alkaloid components of papaya leaves, which has a role as a plant metabolite and a cardiovascular drug.

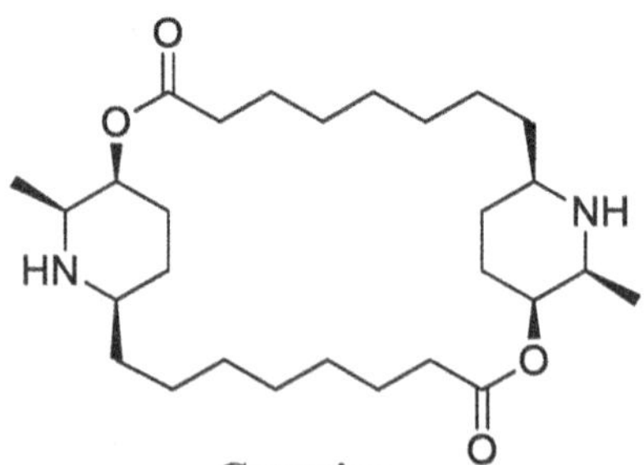

Carpaine

Isolation and Purification: The powdered material of well-powdered *C. papaya* leaves was extracted with water by hot maceration. The aqueous extract of *C. papaya* leaves was partitioned between water and diethyl ether, and the diethyl ether fraction was separated. The aqueous fraction was alkalized with 1N sodium carbonate solution (pH was brought to 11). Again, the aqueous fraction was extracted with diethyl ether. The combined organic fractions were evaporated under vacuum. The diethyl ether fraction was purified through conventional column chromatography to obtain the pure compound (Haldar et al. 2020).

(36) **Taraxerol:** Taraxerol is a naturally occurring pentacyclic triterpenoid present in various plants including *Taraxacum officinale*, *Alnus glutinosa*, *Litsea dealbata*, *Skimmia* spp., *Dorstenia* spp., *Maytenus* spp. and *Alchornea latifolia*.

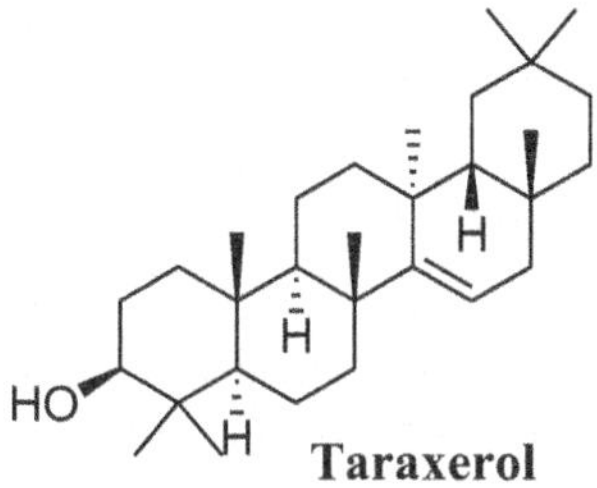

Taraxerol

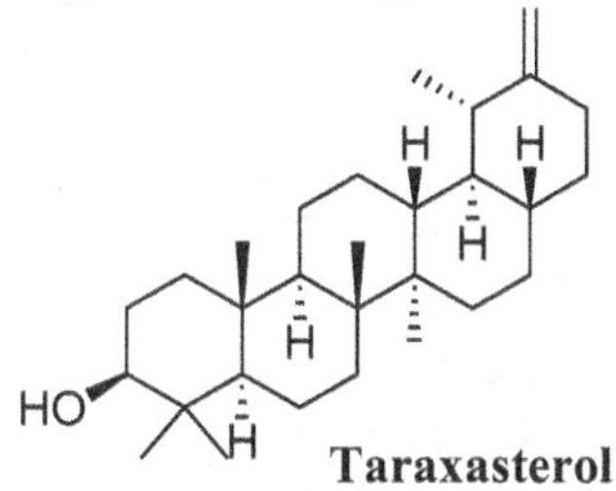

Taraxasterol

Taraxerol and taraxasterol have shown anti-tumour activity. Taraxerol was first isolated in 1923 by Zellner and Röglsperger and named alnulin.

Isolation and Purification: Finely ground plant material was extracted with dichloromethane by maceration for seven days, and this process is repeated three times. After filtration, the combined extracts were evaporated under vacuum to furnish the crude dichloromethane extract, which was further purified on column chromatography to get the pure compounds (Gantait et al. 2010).

(37) **Rosmarinic acid:** Rosmarinic acid is found in a variety of plants. M. L. Scarpati and G. Oriente were the first to isolate and characterized rosmarinic acid in 1958 from rosemary (*Rosmarinus officinalis*). Owing to its antioxidant activity, rosmarinic acid is often used as a natural preservative to increase the shelf life of perishable foods. Rosemary leaves are employed in traditional medicine for their antibacterial and wound healing effects. Compounds in rosemary tea may also have antimicrobial properties, which may help to fight infections.

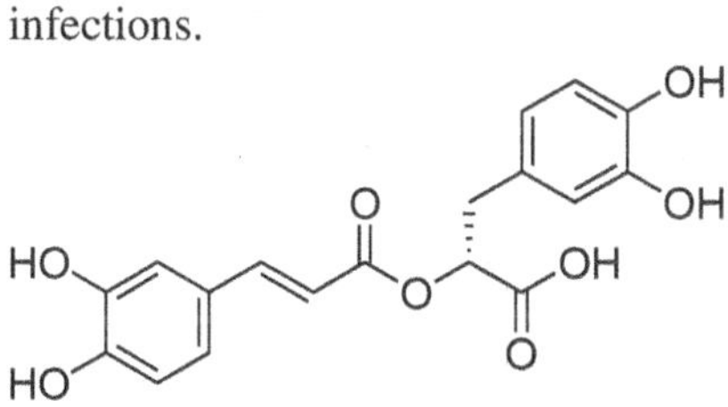

Rosmerinic acid

Isolation: Finely ground plant material (lemon balm leaves) was defatted with n-hexane and extracted with 1:1 water/ethanol solvent system using Soxhlet extraction method. The extract was further processed for purification applying liquid–liquid extraction methods followed by chromatographic purifications (Akoury 2017).

(38) **Sulforaphane:** Sulforaphane is a naturally occurring isothiocyanate present in cruciferous vegetables, viz. broccoli, Brussels sprouts and cabbages.

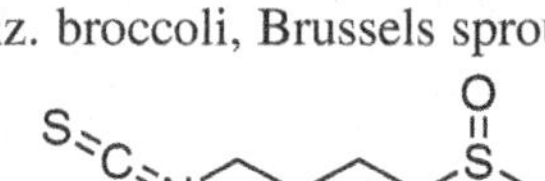

Sulforaphane

It is used for prostate cancer and for the conditions such as autism, asthma and many others.

Isolation and Purification: Crude Extract from Broccoli Seeds. Well-ground broccoli seeds were taken in water (300 mL) and spontaneously autolyse for two hours at ambient temperature. The resulting aqueous solution is repeatedly extracted several times with ethyl acetate. The combined ethyl acetate layer was dried under vacuum to produce ethyl acetate extract. The crude extract was then subjected to liquid–liquid extraction or SPE:

- **Liquid–Liquid Extraction**. The dried crude extract was dissolved in 10% ethanol and treated with hexane to remove nonpolar impurities. The ethanolic phase was fractionated with ethyl acetate. The collective ethyl acetate fractions were evaporated under vacuum to obtain a sulforaphane-rich fraction.
- **Solid-Phase Extraction.** It is an alternate method to the frequently used traditional liquid–liquid extraction method. The crude extract was dissolved in hexane–ethyl acetate (8:2). Activated silica gel (200–300 mesh) was mixed with hexane, and the slurry was packed into an SPE column. The column was then washed with ethyl acetate. Sulforaphane was obtained after eluting the column with ethanol. The eluate was evaporated under vacuum to produce a sulforaphane-rich fraction (Han and Row 2011).

(39) **Resveratrol:** Resveratrol is a natural trans-stilbene and is produced by several plants. Rich sources of resveratrol includes grapes, blueberries, raspberries, mulberries and peanuts (Jasiński et al. 2013). Michio Takaoka has first isolated resveratrol in 1939 from *Veratrum album* and later, in 1963, from the roots of Japanese knotweed (Sales and Resurreccion 2014; Takaoka 1939). Resveratrol is generally used to address the conditions of high cholesterol, cancer, heart disease and many other conditions.

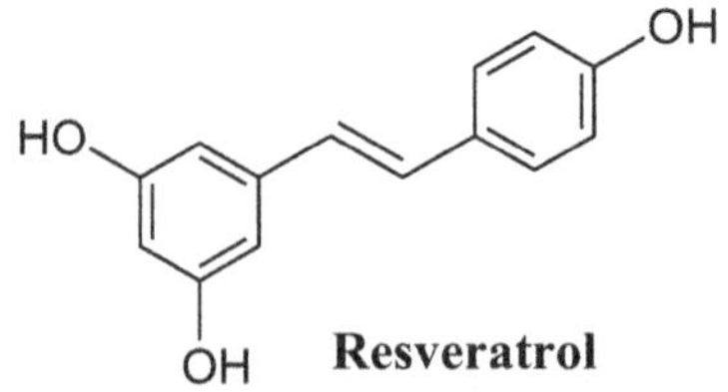

Resveratrol

Isolation and Purification: Finely powdered peanut root (kept in a tube) was homogenized with methanol using a polytron and was heated in a water bath for half an hour. The tubes were centrifuged, and the supernatants from extraction tubes were pooled. The methanolic extract thus obtained was filtered and diluted with water until the solution becomes 20% methanolic. Resveratrol-rich fractions were separated from the methanolic extract by solid-phase extraction (SPE) method. The combined resveratrol-rich fractions was further purified on semi-HPLC (Chen et al. 2002).

(40) **Glycyrrhizin:** Glycyrrhizin is a saponin, practised in traditional folk medicine to alleviate bronchitis, gastritis and jaundice. It is the chief component of *Glycyrrhiza glabra* (liquorice) root.

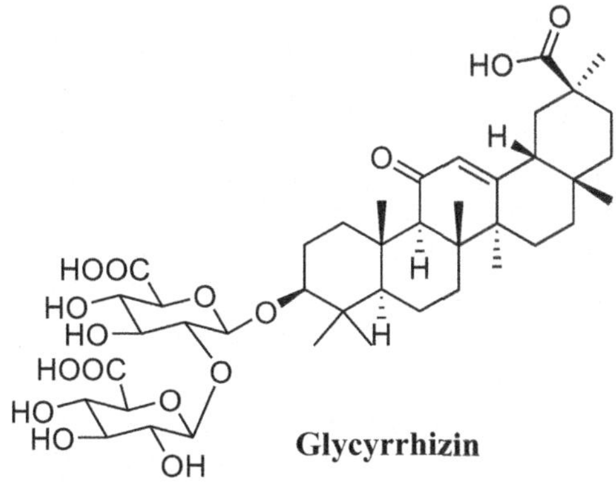

Glycyrrhizin

Isolation and Purification: Powdered plant material of *Glycyrrhiza uralensis* was mixed with NH_4OH solution. The extraction was carried out by a mechanical stirrer at 95–100°C for two hours. The residual solids were removed, and the filtrate was evaporated under vacuum to obtain the impure glycyrrhizin extract, which was further processed for purification (Tianwei et al. 2002).

(41) **Digoxin and Digitoxin: Digoxin** (brand name Lanoxin), a cardiac glycoside, is a medication used to treat various cardiovascular diseases (CVDs) including atrial fibrillation (irregular heartbeat), atrial flutter and heart failure. Digoxin is on the *WHO Model List of Essential Medicines*. Digoxin was first isolated in 1930 from *Digitalis lanata* (*foxglove* plant) (World Health Organization 2014).

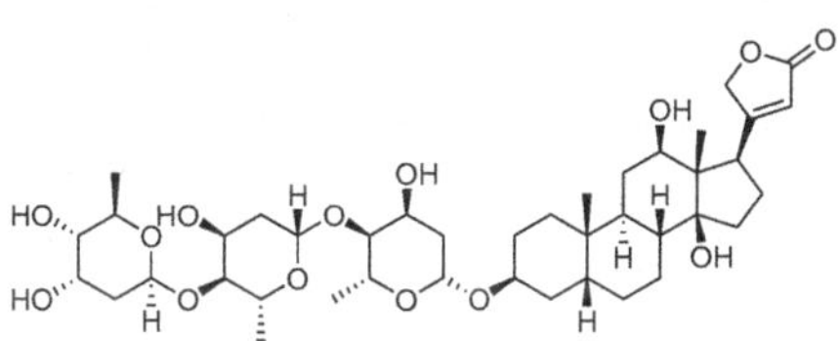

Digoxin

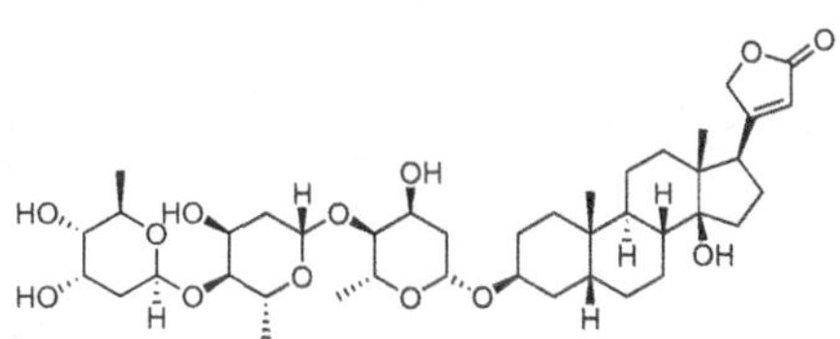

Digitoxin

Digitoxin is a phytosteroidal cardiac glycoside, also isolated from foxglove. Structurally and pharmacologically, it is similar to digoxin. However, digoxin is found to be effective in a proportion of patients treated for heart failure.

(42) **Daidzein:** Daidzein is a natural isoflavonc, mainly present in soybeans and other plants of Leguminosae family. Owing to its potential biological activities, it is under trials for menopausal relief, osteoporosis, blood cholesterol and lowering the risk of some hormone-related cancers and heart diseases.

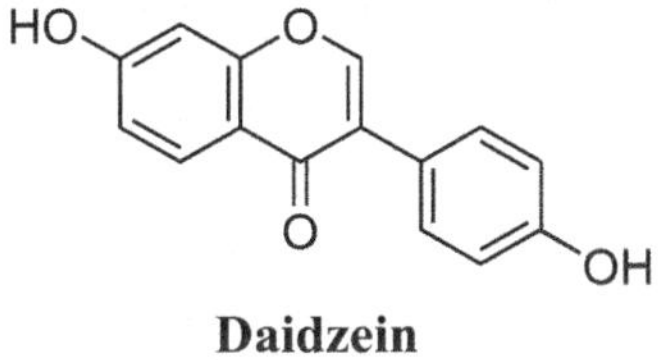

Daidzein

Isolation and Purification: Generally, the isoflavone is isolated from the stem of *P. lobata* (Willd.) Ohwi. Solvents containing different volumes of water and ethanol/n-butanol were used for the extraction. Pure compound was obtained after performing the column chromatography (Xu and He 2007).

(43) **Omacetaxine mepesuccinate** (trade name **Synribo**): Omacetaxine mepesuccinate is formerly known as homoharringtonine (HHT) and is a cytotoxic alkaloid commonly isolated from *Cephalotaxus fortunei* and *Cephalotaxus harringtonia*. US FDA has approved HHT for the treatment of chronic myeloid leukaemia (CML). It was found to be involving in the suppression of SP1/TET1/5hmC/FLT3/MYC signalling pathways in acute myeloid leukaemia (AML).

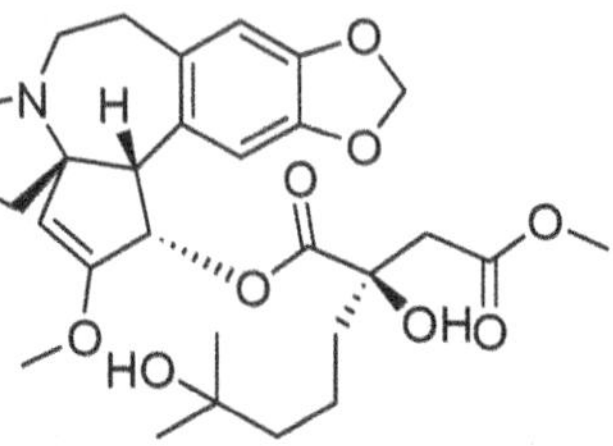

Homoharringtonine (HHT)

Isolation and Purification: Finely ground plant material (bark and needles) was extracted with methanol and further subjected to liquid–liquid extraction with chloroform, and the collective chloroform layer was evaporated under reduced pressure to accomplish the crude chloroform extract. The crude extract was subjected to conventional chromatography to obtain the pure homoharringtonine (Kim and Kim 2008).

(44) **Salicylic acid:** Salicylic (a colourless organic solid), a precursor to aspirin (acetylsalicylic acid), is isolated from the willow tree (genus *Salix*). From the ancient periods, willow (salicin) tree's bark extracts have been used in folkloric medicine to ease pain and to reduce fiver (Topor et al. 2019). Salicylic acid is also used as a *food preservative*, *bactericide*, *antiseptic* and ingredient of anti-*acne* products.

Synthetically it has been used as a precursor for the development of several drugs.

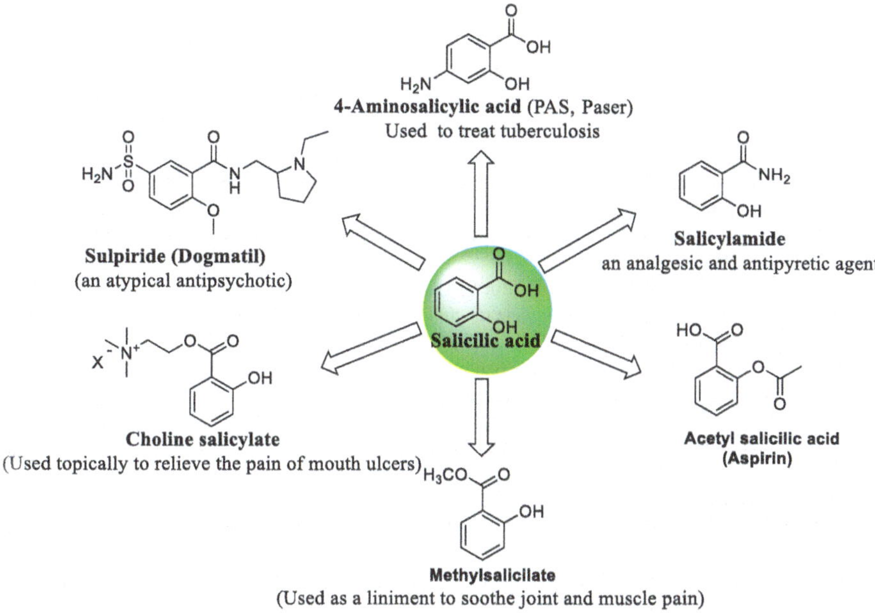

4 Conclusion

Natural products have been a rich source of drug discovery due to their structural diversity and complexity and provide the scope for structural modifications to generate secondary leads. The traditional knowledge of usage of plants/herbs to treat various ailments has greatly influenced and speeded up the drug discovery process. The systematic approach of bioassay-guided fractionation and isolation of bioactive principles has resulted in the discovery of several wonder drugs. Folkloric knowledge of traditional healers led to the discovery of single active constituents and eventually helped for the development of safe and effective drugs. A large number of current marketed drugs are either directly derived from the natural sources or discovered, inspired by the natural scaffolds. Most of them are being used as frontline drugs to treat several diseases including life-threatening diseases. They have a multidimensional chemical structure that can serve as a starting material for the discovery of newer molecules with improved activity and safety profiles. Thus, bioassay-guided fractionation has been proved to be a stepping stone for the modern translational drug discovery.

References

Abe F, Yamauchi T, Nagao T, Kinjo J, Okabe H, Higo H, Akahane H (2002) Ursolic acid as a trypanocidal constituent in rosemary. Biol Pharm Bull **25**(11):1485–1487

Akerele O (1983) Traditional medicine. *World Health*. https://doi.org/10.5363/tits.2.6_25

Akoury E (2017) Isolation and structural elucidation of rosmarinic acid by nuclear magnetic resonance spectroscopy. *Am Res J Chem*. https://doi.org/10.21694/2577-5898.17003

Al-Amin M et al (2012) Neoandrographolide isolated from leaves of *Adhatoda vasica* Nees. *Dhaka Univ J Sci* **60**:1–3

Al-Bayati FA, Mohammed MJ (2009) Isolation, identification, and purification of cinnamaldehyde from Cinnamomum zeylanicum bark oil. An antibacterial study. *Pharm Biol* **47**:61–66

Achor LB, Geiling EMK (1954) Isolation and purification of semimicro quantities of morphine. Anal Chem 26:1061–1062

Arteaga S, Andrade-Cetto A, Cárdenas R (2005) Larrea tridentata (Creosote bush), an abundant plant of Mexican and US-American deserts and its metabolite nordihydroguaiaretic acid. *J Ethnopharmacol* **98**:231–239

Aydin T et al (2014) Insecticidal metabolites from the rhizomes of veratrum album against adults of Colorado potato beetle, Leptinotarsa decemlineata. *Chem Biodivers* **11**:1192–1204

Brachet A, Rudaz S, Mateus L, Christen P, Veuthey JL (2001) Optimisation of accelerated solvent extraction of cocaine and benzoylecgonine from coca leaves. *J Sep Sci* **24**:865–873

Bucar F, Wube A, Schmid M (2013) Natural product isolation-how to get from biological material to pure compounds. *Nat Prod Rep* **30**:525–545

Butler MS (2004) The role of natural product chemistry in drug discovery. *J Nat Prod* **67**:2141–2153

Calatayud J, González Á (2003) History of the development and evolution of local anesthesia since the coca leaf. *Anesthesiology* **98**:1503–1508

Canel C, Dayan FE, Ganzera M, Khan KA, Rimando A, Burandt CL Jr, Moraes RM (2001) High yield of podophyllotoxin from leaves of Podophyllum peltatum by in situ conversion of podophyllotoxin 4-0-D-glucopyranoside. Planta medica 67:97–99

Chandramu C, Manohar RD, Krupadanam DGL, Dashavantha RV (2003) Isolation, characterization and biological activity of betulinic acid and ursolic acid from Vitex negundo L. *Phyther Res* **17**:129–134

Chen RS, Wu PL, Chiou RYY (2002) Peanut roots as a source of resveratrol. *J Agric Food Chem* **50**:1665–1667

Chowdhury AR, Mandal S, Mittra B, Sharma S, Mukhopadhyay S, Majumder HK (2002) Betulinic acid, a potent inhibitor of eukaryotic topoisomerase I: identification of the inhibitory step, the major functional group responsible and development of more potent derivatives. Med Sci Monit 8(7):254–265

Claeson UP, Malmfors T, Wikman G, Bruhn JG (2000) Adhatoda vasica: a critical review of ethnopharmacological and toxicological data. *J Ethnopharmacol* **72**:1–20

Connors NJ, Hoffman RS (2013) Experimental treatments for cocaine toxicity: a difficult transition to the bedside. *J Pharmacol Exp Ther* **347**:251–257

Cragg GM, Newman DJ (2005) Biodiversity: a continuing source of novel drug leads. *Pure Appl Chem* **77**:7–24

Daines A, Payne R, Humphries M, Abell A (2005) The synthesis of naturally occurring vitamin K and vitamin K analogues. *Curr Org Chem* **7**:1625–1634

Das KR, Iwasaki K, Suenaga K, Kato-Noguchi H (2020) Isolation and identification of two phytotoxic compounds from the medicinal plant Cassia alata L. *Weed Biol Manag* **20**:3–11

Denis JN et al (1988) A highly efficient, practical approach to natural taxol. *J Am Chem Soc* **110**:5917–5919

Efferth T et al (2007) Molecular target-guided tumor therapy with natural products derived from traditional chinese medicine. *Curr Med Chem* **14**:2024–2032

Fabricant DS, Farnsworth NR (2001) The value of plants used in traditional medicine for drug discovery. *Environ Health Perspect* **109**:69

Freeman T, Thind A, Stewart M, Brown J, Vingilis E (2009) Research paper. *Healthc Policy | Polit Santé* **5**:e187–e206

Gantait A, Sahu A, Venkatesh P, Dutta PK, Mukherjee PK (2010) Isolation of taraxerol from Coccinia grandis, and its standardization. *J Planar Chromatogr Mod TLC* **23**:323–325

Gao H, Wu L, Kuroyanagi M, Harada K, Kawahara N, Nakane T, Umehara K, Hirasawa A, Nakamura Y (2003) Antitumor-promoting constituents from Chaenomeles sinensis KOEHNE and their activities in JB6 mouse epidermal cells. Chem Pharm Bull **51**(11):1318–1321

Gebhardt R (2005) Choleretic and anticholestatic activities of flavonoids of artichoke (Cynara cardunculus L. subsp. scolymus (L.) Hayek). *Acta Hortic* **681**:429–436

Goldstein RA, DesLauriers C, Burda AM (2009) Cocaine: history, social implications, and toxicity-A review. *Disease-a-Month* **55**:6–38

Graham W, Roberts JB (1953) Intravenous colchicine in the management of gouty arthritis. *Ann Rheum Dis* **12**:16–19

Haefner B (2003) Drugs from the deep: marine natural products as drug candidates. *Drug Discov Today* **8**:536–544

Haldar S, Mohapatra S, Singh R, Katiyar CK (2018) Quantitative evaluation of shatavarin IV by high-performance thin-layer chromatography and its isolation from Asparagus racemosus Willd. *J Planar Chromatogr Mod TLC* **31**:197–201

Haldar S, Mohapatra S, Singh R, Katiyar CK (2020) Isolation and quantification of bioactive Carpaine from Carica papaya L. and its commercial formulation by HPTLC densitometry. *J Liq Chromatogr Relat Technol* **43**:388–393

Halpin CM, Reilly C, Walsh JJ (2010) Nature's anti-Alzheimer's drug: isolation and structure elucidation of galantamine from Leucojum aestivum. *J Chem Educ* **87**:1242–1243

Han D, Row KH (2011) Separation and purification of sulforaphane from broccoli by solid phase extraction. *Int J Mol Sci* **12**:1854–1861

Heinrich M (2004) Snowdrops: the heralds of spring and a modern drug for Alzheimer's disease. *Pharm J* **273**:905–906

Houghton PJ, Raman A (1998) Analysis of crude extracts, fractions and isolated compounds. *Lab Handb Fractionation Nat Extr* 113–138. https://doi.org/10.1007/978-1-4615-5809-5_8

Islam MT et al (2018) Phytol: A review of biomedical activities. *Food Chem Toxicol* **121**:82–94

Ivanova L, Karelson M, Dobchev DA (2018) Identification of natural compounds against neuro-degenerative diseases using in silico techniques. *Molecules* **23**

Jamin E, Naulet N, Martin GJ (1997) Multi-element and multi-site isotopic analysis of nicotine from tobacco leaves. *Plant Cell Environ* **20**:589–599

Jannati N, Honarvar M, Gharachorloo M (2021) Extraction of thymol compound from Thymus vulgaris L. oil. J Med Plant and By-products 1:81–84

Jasiński M, Jasińska L, Ogrodowczyk M (2013) Resveratrol in prostate diseases—a short review. *Cent Eur J Urol* **66**:144–149

Ji ZN, Ye WC, Liu GG, Hsiao WL (2002) 23-Hydroxybetulinic acid-mediated apoptosis is accompanied by decreases in bcl-2 expression and telomerase activity in HL-60 Cells. Life Sci **72**(1):1–9

Keeler RF, Binns W (1964) Chemical compounds of Veratrum californicum related to congenital ovine cyclopian malformations. extraction of active material. *Proc Soc Exp Biol Med* **116**:123–127

Keesara BR, Jat RK (2017) Isolation and characterization of Withaferin-a from the Withania Somnifera (Ashwagandha). *J Drug Deliv Ther* **7**:65–69

Kim BS, Kim JH (2008) Purification of homoharringtonine and removal of residual solvents by spray drying. *Korean J Chem Eng* **25**:108–111

Kinghorn AD, Pan L, Fletcher JN, Chai H (2011) The relevance of higher plants in lead compound discovery programs. *J Nat Prod* **74**:1539–1555

Kumar A, Patil D, Rajamohanan PR, Ahmad A (2013) Isolation, purification and characterization of vinblastine and vincristine from endophytic fungus Fusarium oxysporum isolated from Catharanthus roseus. *PLoS One* **8**

Lattanzio V, Kroon PA, Linsalata V, Cardinali A (2009) Globe artichoke: a functional food and source of nutraceutical ingredients. *J Funct Foods* **1**:131–144

Lilienfeld S (2003) Cholinesterase inhibitors for Alzheimer disease. *JAMA* **289**

Malviya N, Malviya S (2017) Bioassay guided fractionation-an emerging technique influence the isolation, identification and characterization of lead phytomolecules. *Int J Hosp Pharm* 1–6. https://doi.org/10.28933/ijhp-2017-07-0901

Malwade C, Qu H, Rong BG, Christensen LP (2012) *Conceptual process synthesis for isolation and purification of natural products from plants—a case study of artemisinin from Artemisia annua. Comput Aided Chem Eng* 31. Elsevier B.V.

Manda G, Rojo AI, Martínez-Klimova E, Pedraza-Chaverri J, Cuadrado A (2020) Nordihydroguaiaretic acid: from herbal medicine to clinical development for cancer and chronic diseases. *Front Pharmacol* **11**:1–21

Martinenghi LD, Jønsson R, Lund T, Jenssen H (2020) Isolation, purification, and antimicrobial characterization of cannabidiolic acid and cannabidiol from cannabis sativa l. *Biomolecules* **10**:1–16

Mashima R, Okuyama T (2015) The role of lipoxygenases in pathophysiology; new insights and future perspectives. *Redox Biol* **6**:297–310

Maz OF, Foetida PIA (1972) ALKALOIDS. **11**:3529–3531

Mishra BB, Tiwari VK (2011) Natural products: an evolving role in future drug discovery. *Eur J Med Chem* **46**:4769–4807

Müller H, Brackhagen O, Brunne R, Henkel T, Reichel F (2000) Natural products in drug discovery. *Ernst Schering Res Found Workshop* 205–216. https://doi.org/10.1007/978-3-662-04042-3_7

Naman CB, Leber CA, Gerwick WH (2017) Modern natural products drug discovery and its relevance to biodiversity conservation. *Microb Resour From Funct Exist Nat Appl* 103–120. https://doi.org/10.1016/B978-0-12-804765-1.00005-9

Netscher T (2007) Synthesis of vitamin E. *Vitam Horm* **76**:155–202

Newman DJ, Cragg GM (2012) Natural products as sources of new drugs over the 30 years from 1981 to 2010. *J Nat Prod* **75**:311–335

Nitsawang S, Hatti-Kaul R, Kanasawud P (2006) Purification of papain from Carica papaya latex: Aqueous two-phase extraction versus two-step salt precipitation. *Enzyme Microb Technol* **39**:1103–1107

Organización Mundial de la Salud (OMS) (2019) World Health Organization model list of essential medicines. *Ment Holist Heal Some Int Perspect* **21**:119–134

Ottaggio L et al (2008) Taxanes from shells and leaves of Corylus avellana. *J Nat Prod* **71**:58–60

Page JO (1955) Determination of nordihydroguaiaretic acid in creosote bush. *Anal Chem* **27**:1266–1268

Phillipson JD (2001) Phytochemistry and medicinal plants. *Phytochemistry* **56**:237–243

Pomara C et al (2012) Data available on the extent of cocaine use and dependence: biochemistry, pharmacologic effects and global burden of disease of cocaine abusers. *Curr Med Chem* **19**:5647–5657

Pradhan D, Biswasroy P, Suri KA (2013) Isolation of berberine from Berberis vulgaris Linn. and standardization of aqueous extract by RP-HPLC. *Int J Herb Med* **1**:2–7

Pyo SH, Park HB, Song BK, Han BH, Kim JH (2004) A large-scale purification of paclitaxel from cell cultures of Taxus chinensis. *Process Biochem* **39**:1985–1991

Revathy S, Elumalai S, Benny M, Antony B (2011) Isolation, purification and identification of curcuminoids from turmeric (Curcuma longa L.) by column chromatography. *J Exp Sci* **2**:21–25

Rey-Ladino J, Ross AG, Cripps AW, McManus DP, Quinn R (2011) Natural products and the search for novel vaccine adjuvants. *Vaccine* **29**:6464–6471

Robinson WE et al (1996) Dicaffeoylquinic acid inhibitors of human immunodeficiency virus integrase: inhibition of the core catalytic domain of human immunodeficiency virus integrase. *Mol Pharmacol* **50**:846–855

Sales JM, Resurreccion AVA (2014) Resveratrol in peanuts. *Crit Rev Food Sci Nutr* **54**:734–770

Sharifi-Rad M et al (2018) Carvacrol and human health: a comprehensive review. *Phyther Res* **32**:1675–1687

Sharma N, Verma MK, Gupta DK, Satti NK, Khajuria RK (2016) Isolation and quantification of pinitol in Argyrolobium roseum plant, by 1H-NMR. *J Saudi Chem Soc* **20**:81–87

Sharma RA, Gaikar VG (2012) Hydrotropic extraction of reserpine from Rauwolfia vomitoria roots. *Sep Sci Technol* **47**:827–833

Shekelle PG et al (2017) Management of gout: a systematic review in support of an American college of physicians clinical practice guideline. *Ann Intern Med* **166**:37–51

Shingate PN, Dongre PP, Kannur DM (2013) New method development for extraction and isolation of piperine from black pepper. *Int J Pharm Sci Res* **4**:3165

Śramska P, Maciejka A, Topolewska A, Stepnowski P, Haliński ŁP (2017) Isolation of atropine and scopolamine from plant material using liquid-liquid extraction and EXtrelut® columns. *J Chromatogr B Anal Technol Biomed Life Sci* **1043**:202–208

Sreelekshmi U, Sarathchandra G, Vijayarani K, Sp P (2021) Isolation & purification of vasicine from leaves of Adhatoda vasica by modified acid-base extraction method. **10**:171–173

Stierle A, Strobel GSD (1993) Taxol and taxane production by Taxomyces andreanae, an endophytic fungus of Pacific yew. *Science (80-.)* **260**:214–216

Takaoka M (1939) Resveratrol, a new phenolic compound, from Veratrum grandiflorum. *J Chem Soc Japan* 60:1090–1100

Tan Y, Yu R, Pezzuto JM (2003) Betulinic acid-induced programmed cell death in human melanoma cells involves mitogen-activated protein kinase activation. Clin Cancer Res **9**(7): 2866–2875

Tewari D et al (2018) Ethnopharmacological approaches for dementia therapy and significance of natural products and herbal drugs. *Front Aging Neurosci* **10**:1–24

Thakor P et al (2016) Extraction and purification of phytol from: Abutilon indicum: cytotoxic and apoptotic activity. *RSC Adv* **6**:48336–48345

Tianwei T, Qing H, Qiang L (2002) Purification of glycyrrhizin from Glycyrrhiza uralensis Fisch with ethanol/phosphate aqueous two phase system. *Biotechnol Lett* **24**:1417–1420

Topal M et al (2016) Antioxidant, antiradical, and anticholinergic properties of cynarin purified from the Illyrian thistle (Onopordum illyricum L.). *J Enzyme Inhib Med Chem* **31**:266–275

Topor G, Nechita A, Debita M, Ciupilan C, Axente ER (2019) General and particular structural characteristics of acetylsalicylic acid—Aspirine chemical properties. *Rev Chim* **70**:248–253

Van Den Brink DM, Wanders RJA (2006) Phytanic acid: production from phytol, its breakdown and role in human disease. *Cell Mol Life Sci* **63**:1752–1765

Vanden Berghe W, Sabbe L, Kaileh M, Haegeman G, Heyninck K (2012) Molecular insight in the multifunctional activities of Withaferin A. *Biochem Pharmacol* **84**:1282–1291

Vuong QV, Roach PD (2014) Caffeine in green tea: Its removal and isolation. Sep Purif Rev 43:155–174

Wald A (2016) Constipation advances in diagnosis and treatment. *JAMA J Am Med Assoc* **315**:185–191

Wall ME et al (1966) Plant antitumor agents. I. The isolation and structure of Camptothecin, a novel alkaloidal leukemia and tumor inhibitor from Camptotheca acuminata. *J Am Chem Soc* **88**:3888–3890

Wang XH, Huang M, Zhao CK, Li C, Xu L (2019) Design, synthesis, and biological activity evaluation of camptothecin-HAA-Norcantharidin conjugates as antitumor agents in vitro. *Chem Biol Drug Des* **93**:986–992

Wilken R, Veena MS, Wang MB, Srivatsan ES (2011) Curcumin: a review of anti-cancer properties and therapeutic activity in head and neck squamous cell carcinoma. *Mol Cancer* **10**:1–19

World Health Organization (2014) Standard operating protocol assuring medication accuracy at transitions in care. *High5s Proj. – Stand Oper Protoc Medicat Reconcil* 1–36

World Health Organization (WHO) (2002) WHO traditional medicine strategy 2002–2005. *World Heal Organ Geneva* 1–74

Xu GH et al (2008) Terpenoids and coumarins isolated from the fruits of Poncirus trifoliata. *Chem Pharm Bull* **56**:839–842

Xu HN, He CH (2007) Extraction of isoflavones from stem of Pueraria lobata (Willd.) Ohwi using n-butanol/water two-phase solvent system and separation of daidzein. *Sep Purif Technol* **56**:85–89

Zeng XH et al (2013) New and highly efficient column chromatographic extraction and simple purification of camptothecin from Camptotheca acuminata and Nothapodytes pittosporoides. *Phytochem Anal* **24**:623–630

Zimmerman JL (2012) Cocaine intoxication. *Crit Care Clin* **28**:517–526

Zuco V, Supino R, Righetti SC, Cleris L, Marchesi E, Gambacorti-Passerini C, Formelli F (2002) Selective cytotoxicity of betulinic acid on tumor cell lines, but not on normal cells. Cancer Letters. **175**(1):17–25

In Vitro Methodologies for the Safety Assessment of Drugs

Vibha Shukla, Somya Asthana, and Anurag Tripathi

Abstract Since ages, drugs have been used in some form or the other for treatment of several diseases. Safety assessment of drug is an important step required to minimize the drug-associated side effects. Toxicology is an important branch that is required for the safety assessment of drugs. The pharmaceutical companies follow various screening methods during drug development to investigate the toxic effects of effective drugs. The liver, kidney, and heart are the major organs monitored for toxicity evaluation during the drug development. The safety evaluations of drugs are performed at both organ (in vivo) and cellular (in vitro) levels. Considering the increasing demand for alternate to animal models in present times, several cell line-based in vitro models have emerged to assess drug toxicity. Moreover, isolated mitochondria and freshly isolated hepatocytes have also gained importance to understand the mechanism behind the complex processes leading to cellular toxicity. In this chapter, we have discussed the methodologies commonly employed for the safety assessment of drugs.

Keywords Toxicity · Drug · Liver · Kidney · Central nervous system

---

V. Shukla · A. Tripathi (✉)
Food Toxicology Division, Food, Drug and Chemical Toxicology Group, CSIR-Indian Institute of Toxicology Research, Lucknow, India

Academy of Scientific and Innovative Research (AcSIR), Ghaziabad, India
e-mail: anuragtripathi@iitr.res.in

S. Asthana
Food Toxicology Division, Food, Drug and Chemical Toxicology Group, CSIR-Indian Institute of Toxicology Research, Lucknow, India

Department of Biotechnology, Manav Rachna International Institute of Research and Studies (MRIIRS), Faridabad, India

V. S. Rajput, A. Runthala (eds.), *Drugs and a Methodological Compendium*,
https://doi.org/10.1007/978-981-19-7952-1_4

1 Introduction

Humans use drugs for the treatment of several diseases; therefore, evaluation of these drugs must be necessary for their safe administration. For that, toxicologists follow 500-year-old Paracelsus' principle for the safety assessment of drugs (Borzelleca 2000). The concern for the safe administration of drug is due to the side effects caused by the drug compounds. Sometimes these side effects of drugs occurred in such an extent that they caused damage to the whole system to a point of no return (Das 2018). To avoid such type of damage, firstly, we need to identify and then understand the path of disarray to prevent the long-term irreversible destruction of system that might occur due to the unsafe consumption of drug (Das 2018). Different methods have been explored to understand the toxicity of drugs in various ways and up to different extents to decrease the risk factor associated with drugs (Fig. 1). On the other hand, regulatory authorities must be provided with information about the properties of drugs, the results of safety checks, risk assessments, and effective steps to efficiently monitor the risk.

During drug development phases, the pharmaceutical companies use various screening methods in order to minimize the toxic potential of drugs. Withdraw of phase I–III drugs occurred due to the absence of sensitive methods that can detect and determine the human-related toxicity (Li 2004; Dorato and Buckley 2007; Giezen et al. 2008). The required time and cost are 12–15 years and 800 million

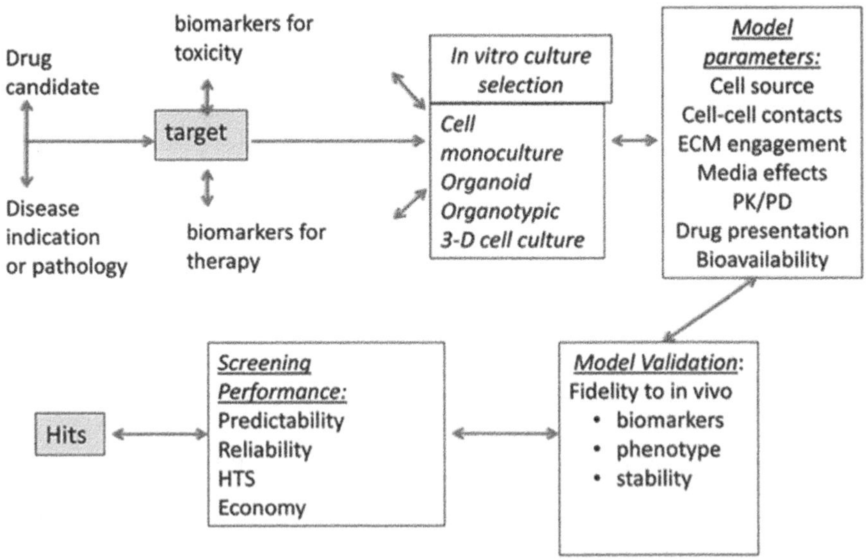

Fig. 1 Strategy for use of in vitro systems for the prediction of in vivo toxicities in preclinical trials (*Reprinted from Pharmacology & Therapeutics, 134/1, Anna Astashkina, Brenda Mann, and David W. Grainger, A critical evaluation of in vitro cell culture models for high-throughput drug screening and toxicity, Pages No.95, Copyright (2012), with permission from Elsevier*)

dollar, respectively, for the successful delivery of drug into the market (DiMasi et al. 2003). The numbers of drugs are withdrawn from the market even though these drugs go through exhaustive exercise that involved huge expenses and attempt (DiMasi et al. 2003). Therefore, to avoid post marketing withdrawal of drugs due to toxicity to humans, sincere test-based approaches have been developed to avoid such circumstances.

In modern toxicology, the key hypotheses are (I) selection of accurate predictive models (other organisms) to evaluate the toxicity of drugs in human (Zbinden 1987), (II) selection of an appropriate model for accurate prediction of potential hazard of drug in man (Gad 1996, 2015), and (III) understanding the strengths and weaknesses of a particular model to assess the cause of hazard and subsequent management of actual risks (Lijinsky 1988). For the collection of descriptive data on human drugs, higher animals have been used to translate methods such as pathology and clinical chemistry (Gad 2017). On the other hand, in order to establish an effective and efficiently applicable product safety assessment program, in vitro as well as other alternative test model systems have been used. The motivation for the interest in in vitro model system comes from philosophical and political ongoing efforts by the diverse groups concerned with animal welfare (Gad 2009). They have come with the concepts of three "Rs" of animal use in research: replacement, reduction, and refinement. Further, in vitro test systems have been used for toxicity evaluation because this system can be used as an alternative definitive test system in place of in vivo test systems (Singer 1975; May et al. 2009; Birnbaum and Stokes 2010). The in vitro model system showed several advantages over in vivo model system during drug toxicity assessment, which include (i) decrease in animal numbers (Boo and Knight 2009), (ii) reduced cost of animal maintenance and care, (iii) small quantity of chemicals required, and (iv) lesser time. In addition, study of chemical metabolism, evaluation of mechanism of toxicity, measurement of enzyme kinetics, and examining dose–response relationships are relatively easy in in vitro system (Soldatow et al. 2013).

In this chapter, we will include in vitro testing methods that can be used for screening of toxicity of drugs in different organs (liver, kidney, and central nervous system) in human.

2 Toxicity Assessment In Vitro

2.1 *In Vitro System for Liver Toxicity Assessment*

The liver is a heterogeneous organ in vertebrates that regulates many biochemical reactions and pharmaceutical drugs' metabolisms and is vulnerable to damage from these chemicals and xenobiotics. The drug-induced liver injury (DILI) occurs due to pharmaceutical constituents, which is one of the critical factors behind the withdrawal of several drugs, such as troglitazone, trovafloxacin, and nefazodone from the market; however some drugs, such as diclofenac and acetaminophen, pose a significant risk but are still in the market (Chen et al. 2015). A huge financial loss for pharmaceutical companies occurs due to the withdrawal of drugs from the market that caused hepatotoxicity. Various liver-derived in vitro model systems have been established in recent years to facilitate the research of possible harmful effects of medicines. For investigation purposes of toxicity of drugs, immortalized cell lines, primary hepatocytes, perfused liver, isolated microsomes, and liver tissue slices have been used (Soldatow et al. 2013). For liver toxicity testing, the most widely used in vitro models are immortalized cell lines and primary isolated liver cells. Apart from these models, various cell lines such as HepG2, H4IIE, HepaRG, and THLE and stem cells (embryonic stem cells and induced pluripotent stem cells) are also used extensively in in vitro test systems for toxicity screening (Yu et al. 2007). These in vitro systems have several advantages and disadvantages that will be discussed later.

Liver Slices

Liver slices can be a valuable model to illustrate the metabolism of xenobiotic, specific cytochrome activity in the liver zone and mechanism toxicity because this model contains all the cell types that are represented in the in vivo model (Lerche-Langrand and Toutain 2000). Liver functions such as albumin production and gluconeogenesis and phase II enzymes remain stable in slices for 20–96 hours after culture procedure (Ghantous et al. 1992; Gokhale and Zurlo 1997; Toutain et al. 1998; Hashemi and Till 1999). In a comparative study, the in vitro liver slices displayed closer correlation to rat liver as compared to cell lines and primary hepatocytes (Boess et al. 2003). In general, culture length for tissue slices ranges from 30 minutes to 5 days (Toutain et al. 1998), and the parameters, which play important role in increasing cell viability and reducing degeneration in tissue slices, include media and supplements, oxygen tension, and culture system (Lerche-Langrand and Toutain 2000). With all these merits, some demerits are also associated with this model, and these are:

(1) After culture, necrosis takes place after 48–72 hours and levels of metabolic enzymes reduce after −72 hours (Fisher et al. 1995; Leeman et al. 1995; Toutain et al. 1998).
(2) In isolated hepatocytes, the rate of drug metabolism and intrinsic clearance are high in comparison to liver slices (Ekins et al. 1995; Worboys et al. 1996).
(3) Kinetic constant (Km) values are mostly high in tissue slices because all the hepatocytes present in the slices are not participated in the metabolism of drug (Worboys et al. 1996).

Primary Hepatocyte Suspensions

Single-cell suspension of hepatocytes is developed by the use of collagenase digestion that disrupts the bond between the cells. For moderately high-throughput toxicity studies, suspension of hepatocytes is the best model, and this method also provides more accurate estimation of internal clearance rate of drugs in comparison to monolayer cultures (Griffin and Houston 2005). Compared to cultivated cells, hepatocyte suspension has a high degree of functioning, allowing for a more accurate correlation with in vivo toxicity (Guillouzo 1998; LeCluyse 2001; O'Brien and Siraki 2005; Hewitt et al. 2007). During isolation of cells, damage occurs at different regions of cells, such as cell surface receptors and antigens, cell membrane, cell junctions, and cytosolic contents (Lerche-Langrand and Toutain 2000; Jauregui et al. 2016). Collagenase digestion causes oxidative stress in hepatocytes during preparation of single-cell suspension, resulting in a decrease in the activity of cytochrome enzymes within 4–8 hours (Duval and Sieg 1995; Wang et al. 1998; Tirmenstein et al. 2000). It was found that hepatocytes in suspension couldn't maintain viability for the time necessary for the development of toxicity, whereas hepatotoxicity is a process that occurs over several hours (Hewitt et al. 2007; Burke et al. 2010).

Primary Hepatocyte Cultures

In in vitro testing, primary hepatocytes act as the best model because their functional activity is maintained for 24–72 hrs. This functional activity can be utilized for the inhibition studies, screening of compounds, and enzyme induction (LeCluyse 2001; Hewitt et al. 2007). An isolated hepatocyte cells mimic like a whole organ, and it is expected that the effect of compounds will be the same in both models (DelRaso 1993). During primary culture of hepatocytes, cells go through the process of de-differentiation in which hepatocytes undergo changes that usually occurred in morphology, structure, gene expression, and liver-specific functions (Nelson et al. 1982; Yarmush 1989; LeCluyse et al. 1996). During isolation, process cells lose their normal microenvironment structure, and responses given by these cells to chemical exposure are different than that occurring in vivo model. Over the time, functionality of cells decreases; that is why hepatocytes are not used in toxicological research (Bader et al. 1992; LeCluyse et al. 1996). In cultured hepatocytes, functionalities such as albumin production and cytochrome P450 expression decline very fast as the differentiation status of cells is lost (LeCluyse 2001).

Immortalized Cell Lines

Liver-derived immortalized cell lines generally do not show the phenotypic features of a liver tissue. These cell lines include HBG, Fa2N-4, PLc/PRFs, HepG2,Huh7, Hep3B, and HepaRG cells (Wooster et al. 1993; MacDonald et al. 1994; Li et al. 2001; Guguen-Guillouzo 2010). The HepG2 cell line expresses many liver-specific genes, but it was found that the expression of genes involved in the phase I and phase II metabolism differs between passages, and as a result of this, outcomes cannot be interpreted accurately (Guguen-Guillouzo 2010). Other than HepG2, HepaRG is a human hepatoma cell line that maintains nuclear receptor, membrane transporters, phase II enzymes, and liver-specific functions (Braiterman and Hubbard 2009; Guguen-Guillouzo 2010). The stable karyotype of HepaRG allows them to grow into hepatocytes or biliary lineages with high proliferation rate. These cell lines have been demonstrated to produce data that is both repeatable and consistent across trials (Guguen-Guillouzo 2010; Lübberstedt et al. 2011). Liver-specific functions are much lower in HepaRG in comparison to primary hepatocytes.

3 Cytotoxicity Endpoints and Hepatic Functional Parameters

One or few endpoints are used for the evaluation of in vitro toxicity. Most parameters are based on the activity of certain enzymes in cells that are earlier exposed to the chemical or estimated by the analysis of cell number, cell viability, and membrane integrity. The release of cytosolic enzymes as a result of membrane alteration and estimation of the functional and metabolic state of the cell are used as cytotoxicity assays. Absorbance, fluorescence, or chemiluminescence measurement methods are used to evaluate the conversion of a substrate to a product. To evaluate the toxicity of drugs, hepatic cells are incubated with different test concentrations of drugs for 24–72 hrs. In these assays, cell viability and general metabolism state are identified, but these methods do not provide a clear picture of the component and the mechanisms that are involved in toxicity. In addition, toxicity is a late-stage indicator of cell viability. Prior to cell death, changes occur in the structure and biochemical or metabolic components that consequently cause serious functional alteration in cells (O'Brien et al. 2006).

Other than cell viability assay, additional parameters should be used for the early evaluation of hepatotoxicity in model organisms (Gómez-Lechón et al. 1988; Guillouzo et al. 1997). Drug-induced liver damage can be identified by using different biochemical and metabolic endpoints that can be helpful for investigating the mechanism of toxicity also. In vitro toxicity in cells can be assessed by using early biomarkers such as GSH levels, generation of ROS, and activation of caspase cascade used in cell injury. Other markers such as gluconeogenesis, glycogen synthesis, ureagenesis, plasma protein synthesis, and synthesis of VLDL are used to evaluate the liver-specific functions. The above point suggests that metabolic parameters are more susceptible than cytotoxicity indicators as well as sublethal concentration of drugs can be used to identify cell injury. It was well established that for mechanistic studies, primary hepatocytes are the best in vitro system in comparison to hepatoma cells because they contain key hepatic functions (Gómez-Lechón et al. 1988; Xu et al. 2004).

Hepatocytes are a good model to study the transport of drugs. To access the possible effects of pharmaceuticals on the hepatic transport process, radiometric analysis, liquid chromatography, and fluorescent dyes can be used (Bi et al. 2006; Noé et al. 2007; Rohacova et al. 2008). In hepatocytes, both detoxification and bioactivation processes occurred so the toxic effects of the drugs can be determined in hepatocytes. During drug–drug interactions and adverse drug reactions, induction of drug metabolism enzymes occurs in the presence of chemical exposure (Gómez-Lechón et al. 2008). The use of hepatocytes in such studies may offer suitable platform to understand species-specific metabolism as well as extrapolation of the outcomes across different useful species (Tuschl et al. 2008).

Instead of functional or mechanistic studies, in hepatic cell lines, cytotoxicity-related screenings of bioactivable hepatotoxins are usually performed. In several steps, in vitro hepatotoxicity studies can be performed. At primary stage, in hepatocytes or liver-derived cell lines, the effect on cell viability is examined by cytotoxicity parameters. Dose viability curve is plotted against 5–6 different concentrations of drug and used to determine the concentration of drug that decreases the cell viability by 50% in comparison to control. At second stage, the investigation of the effect of drugs on three non-cytotoxic concentrations is done to identify the liver-specific functions as well as mechanism. It is suggested that before going more in depth in the mechanism part of the hepatotoxicity, it may be suitable to examine a few important metabolic functions. Liver cell damage induced by hepatotoxins such as paracetamol, valproic acid, paraquat, carbon tetrachloride, and aflatoxin B1 has been demonstrated via using these experimental approaches (Gómez-Lechón et al. 1988; Bort et al. 1998; Ulrich et al. 2001; He et al. 2004; Tong et al. 2005; Tuschl et al. 2008). Thus, the knowledge of liver toxicity mechanism can be increased by using the in vitro studies.

It is observed that in the presence of some drugs such as diclofenac, early stage of oxidative stress (GSH and lipid peroxidation) and changes in calcium homeostatic are not observed in culture cell lines; however, acute toxicity was induced by depletion in the level of ATP and changes in mitochondrial membrane potential (Bort et al. 1998). Gluconeogenesis and plasma protein synthesis are ATP-consuming liver-specific functions, which get inhibited due to dysfunction in the mitochondrial bioenergetic of hepatocytes exposed to diclofenac. *N*,5-dihydroxy-diclofenac or quinone imine structure redox cycling plays an important role in diclofenac hepatotoxicity via decreasing the production of NADPH and increasing the level of ROS in hepatocytes (Bort et al. 1998; Boelsterli 2003). These factors alone or in combination play a role in diclofenac-related toxicity. A better understanding of idiosyncratic hepatotoxicity helps to decrease a patient's specific risk and helps to enhance the certainty of unpredicted drug reaction.

A major limitation of this strategy is that these studies are expressive of the use of cellular models for mechanistic pathways only for one or few drugs. During the drug discovery process, for the screening of a high volume of new chemicals, pharmaceutical companies require faster and more reliable hepatotoxicity methods. Combined use of few parameters in multi-well formats is required to increase the screening of new molecules (Xu et al. 2003; Schoonen et al. 2005a, b; Biagini et al. 2006; O'Brien et al. 2006). For example, mitochondrial activity and total protein synthesis or membrane integrity and lysosomal activity can be assessed sequentially by combined colorimetric assays. In human drugs, induced hepatotoxicity can be accessed via measuring the level of GSH, activity of mitochondria, and synthesis of DNA (O'Brien et al. 2006). Therefore, these different endpoints help to identify the mechanisms that can be involved in toxicity of drugs (Schoonen et al. 2005a, b; Biagini et al. 2006).

The final aim of in vitro hepatotoxicity screening is to generate useful knowledge about drugs that cause toxicity in the human liver. Those drugs that are toxic in nature without metabolism can be identified via simple cytotoxicity assay. However, certain limitations are also associated with in vitro systems. For example, it is challenging to identify the toxicity of drugs, which need a high-order metabolism or longer exposure time to show their toxicity, using in vitro models. Identification of potentially idiosyncratic drugs and adverse effects of these drugs on humans cannot be screened through in vitro models. An accumulation of toxic metabolites and an alteration in cell processes occur when unusual drug metabolism takes place in cells, which finally causes metabolic idiosyncratic toxicity. To overcome such limitations, in vitro human model is a promising tool for the identification of idiosyncratic hepatotoxicity. Other than idiosyncratic hepatotoxicity, immune-mediated hepatotoxicity also occurs due to the high incidences of hypersensitivity reactions and adverse drug reactions (Ulrich et al. 2001; Park et al. 2005). During hypersensitivity reactions, reactive metabolites are formed, but in adverse drug reaction, immunogens (adduct of metabolites of drug–protein) are formed. It is hypothesized that drugs might directly activate T cells by forming covalent bonds (Pichler 2002). Different in vitro models such as recombinant enzymes, genetically engineered cells, liver microsomes, or hepatocytes are used to examine the tendency of a new molecule to undergo bioactivation (Park et al. 2005; Pearce et al. 2005; McDonald and Rettie 2007). For detecting the reactive metabolite formation, covalent binding assays, enzyme inactivation studies, or trapping experiments have been used. Isotope-labeled drugs are required for the determination of covalent binding to macromolecules. Other than radioactive assay, nonradioactive assay is used for the detection of metabolite–cysteine adducts through LC/MS. This assay was potentially used in the case of fipexide in which protein digestion was performed for the identification of metabolite–protein adducts (Sleno et al. 2007).

To increase the reliability of in vitro models, it is necessary to introduce new methods that simultaneously and rapidly evaluate the several hepatotoxicity biomarkers in the same cells (Gomez-Lechon et al. 2010). Multiple parameter analyses can be performed in the same cell preparations to identify the effect of drugs on genome, protein, metabolism, and ultimately on cell viability. The development of high-throughput screening protocols via using these methodologies contributes to our understanding of the mechanisms that are involved in cellular toxicity.

4 In Vitro System for Kidney Toxicity Assessment

The kidney is a vital organ that maintains body homeostasis via regulating water, electrolyte, nitrogen, and acid–base balance. For the treatment or diagnostic purpose, several drugs are used that could lead to poisoning in the kidney, consequently causing damage in tubules, glomerulus, and interstitium and ultimately generating pathology in the kidney. However, in recent medication, 25% of the reported serious adverse effects caused drug-induced nephrotoxicity, of which one third is credited for antimicrobial use (Faria et al. 2019). Such nephrotoxicity induced by drug treatment can be reversed by stopping the treatment of that drug, but sometimes medication causes chronic dysfunction, such as papillary necrosis, prolonged proteinuria, or tubulointerstitial nephritis (Choudhury and Ahmed 2006).

For the screening of drug-induced nephrotoxicity, in vitro models are used. Cultured primary human cells are used as an in vitro model that mimics the physiological state of cells in vivo most closely. However, limitations, such as restricted growth ability and loss of phenotype during successive subculturing, are associated with cultured primary human cells. To study the basic renal cellular functions and the effect of nephrotoxicants, primary renal cells are a suitable model. To overcome the limitations that are associated with primary cells can be defeated by using immortalized cells. Immortalized cells can grow and divide indefinitely. The limitation associated with these cells is that during the procedure of immortalization, the function and characteristic features of the cells might get altered (Bajaj et al. 2018).

4.1 Proximal Tubule

A major target for many nephrotoxicants is the proximal tubule, which is a part of the nephron that expresses many important markers. On basolateral membrane, SLC22A6 (OAT1), SLC22A8 (OAT3), and SLC22A2 (OAT2) are present. On the apical membrane, ABCB1(P-gp), SLC47A1 (MATE1), SLC47A1 (MATE2), ABCC2 (MRP2), ABCC4 (MRP4), ABCG2 (BCRP), and the endocytosis receptors megalin and cubilin are present (Bajaj et al. 2018). HK-2 cell line is immortalized cells that lack the expression of BCRP, MRP2, OCT2, OAT1, and OAT3 and maybe not used to study the nephrotoxicity (Jenkinson et al. 2012).

4.2 Immortalized Cell Lines

For nephrotoxicity screening, two immortalized cell lines have been used extensively. These are RPTEC/TERT1 and ciPTEC. For the preparation of RPTEC/TERT1 cell lines, hTERT was used (Simon-Friedt et al. 2015). In the case of ciPTEC cell lines that were obtained from the kidney, tissues were transfected with SV40T and hTERT. Transfection with these markers allowed cells to grow and mature at 33°C and 37°C, respectively (Wilmer et al. 2010; Jansen et al. 2014). These cell lines are better than HK-2 cell lines because ciPTEC cell lines contained all the relevant markers; however, OAT function was achieved by incorporating lentiviral transduction (Nieskens et al. 2016).

In general, toxicity assays primarily focused on evaluating cell death. However, in the case of renal toxicity, change in the cell polarity, mitochondrial function, and membrane integrity can be manifested. A machine-learning model was developed in which ciPTEC-OAT1 cell line was used for the screening of 62 drugs based on several parameters for the prediction of their nephrotoxicity (Sjögren et al. 2018). A small set of known nephrotoxic and non-nephrotoxic drugs were used to evaluate transepithelial transport in RPTEC/TERT1 cell lines. These cell lines were further used to understand the nephrotoxicity mechanism of drugs via evaluating functional parameters such as transepithelial electronical resistance and reabsorption and secretion of solutes (Secker et al. 2018).

4.3 Three-Dimensional (3D) Structure

To overcome the limitations associated with 2D culture, 3D-organized structures, known as kidney microphysiological systems and organoids, have been utilized to remodel the fluidic characteristics of an in vivo system (Fig. 2). Various sources of stem cells, such as embryonic, adult, and induced pluripotent stem cells, have been used to culture organoids. Human-induced pluripotent stem cells were used to grow kidney organoids, which have cell types from different segments of nephron, but these cells were immature (Takasato et al. 2015). To resolve this problem, chick chorioallantoic membrane was used for the transplantation of kidney organoids to supply a vascularized environment for the maturation of the organoids (Chuva de Sousa and Lopes 2019). Matrigel and Matrigel mixed with collagen I used for 3D culture systems include RPTEC/TERT1 and Nki-2 cells to further mature and form tubular structures. When compared to 2D cultures, both models demonstrated enhanced susceptibility to known nephrotoxins (DesRochers et al. 2013; Secker et al. 2018). To mimic the physiological geometry of the nephron segment, bioengineered geometry was developed. Double coated with (3,4-dihydroxy-1-phenyalanine and collagen IV), polyethersulfone hollow fiber membranes were used to seed ciPTEC-OAT1 cells. Under perfusion condition, elimination of protein-bound uremic toxins and reuptake of albumin can take place via using this system. This condition mimicking like in vivo state and helpful to understand how microbial metabolite levels in the human body can be maintained via remote sensing and signaling pathway (Jansen et al. 2015, 2016). For the study of drug-induced nephrotoxicity, decellularized kidneys can also be a good model. For the decellularization of rat kidney, sodium dodecyl sulfate (SDS) was used, but for recellularization, ci-PTEC-OAT1 was used. This model showed increased sensitivity for cisplatin, tenofovir, and cyclosporine A as compared to 2D model (Fedecostante et al. 2017).

For the maturation of kidney cells, implementation of fluid flow and vasculature is a crucial step in the in vitro model so it can provide the condition that mimics the in vivo model. Another model is organ-on-a-chip, in which microfluidics is applied in cell and tissue cultures. OrganoPlate (3D platform consisting of 96 chips) is used

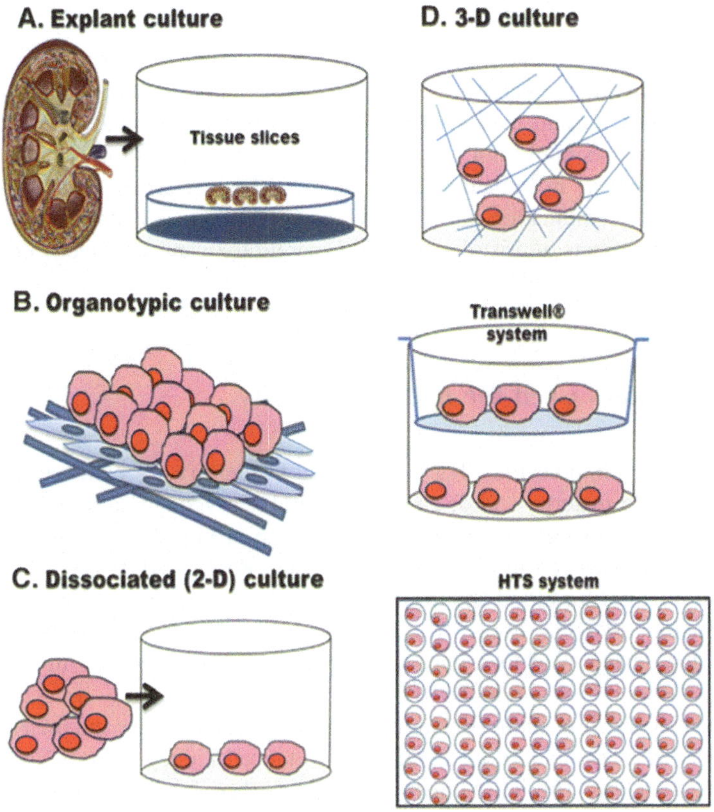

Fig. 2 In vitro cell-based models used for toxicity assessments. (**a**) Explant culture, (**b**) examples of organotypic culture: artificial skin and co-culture on Transwell® inserts, (**c**) dissociated cell culture maintained on 2D surfaces, and (**d**) 3D cell culture in a supporting matrix (*Reprinted from Pharmacology & Therapeutics, 134/1, Anna Astashkina, Brenda Mann, and David W. Grainger, A critical evaluation of in vitro cell culture models for high-throughput drug screening and toxicity, Pages No. 95, Copyright (2012), with permission from Elsevier*)

to culture ci-PTEC-OAT1, and after culturing, it is placed on rocker plate to provide fluid shear stress that is used as a high-throughput screening tool that is in agreement with modern imaging methodologies (Vriend et al. 2018). Adult stem cells were used to generate kidney tubular epithelial organoids or tubuloids, for the development of new microfluidic in vitro system. This system is used to model several diseases such as BK virus, Wilms tumor, and cystic fibrosis because it showed active transepithelial transport function. Due to this feature, this system has a potential to be used in disease modeling and also used for the drug screening (Schutgens et al. 2019).

4.4 *Organ-on-a-Chip*

It is well known that the kidney also works together and responds on remote organs. For the metabolism and excretion of drugs, the liver and kidney work together. Normal 2D cultures could not achieve to predict the biotransformation (liver) and elimination (kidney) in humans (Chang et al. 2016). Organ-on-a-chip provides a platform for the combination of these two complex organs. This platform is used to study the toxic effects of aristolochic acid I (nephrotoxicant) and needs first hepatic bioactivation (Chang et al. 2017).

5 Biomarkers Used for Kidney Toxicity

Biomarkers are biomolecules that can be used to identify the connection between exogenous toxic material and disease. In general, early-stage damages induced by exogenous toxicants can be determined by the use of biomarkers that also suggest the mechanism involved in the toxicity (Finn and Porter 2003). The selection of a suitable biomarker for blood and urine samples after the exposure of nephrotoxicant is a promising task (Shao et al. 2011). Instead of blood sample, urine sample can be used for the detection of toxicity due to noninvasiveness and ease in sampling in significant amounts (Wu et al. 2010).

To identify the drug-induced toxicity in the kidney, creatinine and urea have been recommended as a standard to identify drug-induced kidney toxicity and to sense kidney dysfunction. The limitations with these markers include low sensitivity and specificity, therefore, it is necessary to develop highly specific and sensitivity markers that can detect kidney damage in the early stage of damage induced by drugs/toxicants (Ferguson et al. 2008; Bonventre et al. 2010).

Due to the exposure of nephrotoxic substance, the enzymes present in the tubular epithelial cells of the kidney are secreted into the urine. Such urinary proteins with enzymatic activity can be used as biomarkers; these are alkaline phosphatase, alanine aminopeptidase, γ-glutamyl transpeptidase, α-glutathione-S-transferase, N-acetyl-D-glucosaminidase, and π`-glutathione-S-transferase (Ferguson et al. 2008),

5.1 *Proteinuria*

During filtration process through the glomerulus, high-molecular-weight protein passes from the blood to nephron lumen (Finn and Porter 2003). In some disease condition, the selective infiltration by the glomerulus does not perform properly; as a result, high-molecular-weight proteins can be detected in the urine (Guder and Hofmann 1992). During the early stage of kidney damage (changed in glomerular filtration), high-molecular-weight proteins, such as albumin, are released; therefore, these proteins can be used as a marker for the diagnosis of damage in the kidney (Mogensen 1971). Immunoglobulin G showed damage in the structure of the glomerulus (Tencer et al. 2000), and glomerular damage can be sensitively detected via transferrin (transport iron) (Prinsen et al. 2001). Glomerulus filters and reabsorbs low-molecular-weight proteins that are produced in other organs but not released from the proximal tubule. If the concentration of filtered low-molecular-weight protein increases, it means that absorption of proteins is not adequate in both the glomerulus and proximal tubule. This incident might have occurred due to the overload of proteins or damage in the kidney cells (Bernard et al. 1987). Therefore, the measurement of proteins in urine is used as a marker to detect the early-stage damage in the kidney due to the toxicants. During tubular damage, low-molecular-weight protein such as β_2-microglobulin (Herget-Rosenthal et al. 2004; Emeigh Hart 2005), α_1-microglobulin (Schaub et al. 2005), and retinol-binding protein are secreted and transported retinol from the liver to other organs (Bernard et al. 1987).

5.2 *Kidney Injury Molecule-1 (KIM-1)*

KIM-1 belongs to immunoglobulin mucin family of T cells and is a transmembrane glycoprotein. KIM-1 contains an immunoglobulin-like domain that encodes six unusual cysteine domains and a mucin-like domain at the top of extracellular region. For the identification of damage caused in the kidney via toxic substance (cisplatin) or ischemia, KIM-1 is a more sensitive biomarker than other conventional markers such as serum creatinine, proteinuria, and BUN (Vaidya et al. 2006). It was found that in proximal tubular injury, immune response from nephrotoxicants and renal tubular regeneration are responsible for the expression of KIM-1. In injured kidney, the KIM-1 mRNA and protein level were found to be high. In tubular lumen, KIM-1 is uncovered and matrix metalloproteinases are responsible for cutting their extracellular domain, which is then excreted in the urine (Bailly et al. 2002). Due to nephrotoxicity, the expression of KIM-1 is rapid and can be easily detected in urine sample. As for KIM-1, extracellular domain is very stable and easy to detect; therefore, this marker can be used for the examination of kidney damage.

5.3 *Neutrophil Gelatinase-Associated Lipocalin (NGAL)*

Particularly in neutrophil granulocytes, NGAL (25 kDa) protein binds to gelatinase. During the maturation process of granulocytes (Borregaard et al. 1995). Drug-induced nephrotoxicity or ischemia is involved in the increased expression of NGAL in proximal tubular cells. During infection and inflammation, the concentration of NAGL has been found to be increased in the blood (Xu et al. 1995; Ohlsson et al. 2003). Therefore, for the early diagnosis of acute kidney damage, it can be a sensible biomarker.

5.4 Cytokines

Cytokines in nature are polypeptides that control many important biological functions. They perform as a mediator of immune response and inflammation. They can help in repairing the damaged tissues. During glomerular and tubular damage and repair, several biomarkers such as interferon (Baron et al. 1991), tumor necrosis factor, colony-stimulating factors, and interleukins (Horii et al. 1993; Wada et al. 1994) are used to identify nephrotoxicity (Finn and Porter 2003).

5.5 Clusterin

Clusterin is (sulfated glycoprotein) present at the three different regions in the kidney, in the connecting tubule, at the end of distal convoluted tubule, and in the cytoplasm of proximal convoluted tubule. Patients with acute kidney injury showed the presence of clusterin in the urine sample. Surprisingly, in case of aggravate glomerulonephritis, the level of the cluster gets depleted (Kharasch et al. 2006).

5.6 Osteopontin

This phosphoprotein, osteopontin, is a 44KDa molecule, present in bones. This protein is also known as sialoprotein I, uropontin, secreted phosphoprotein-I, and T-lymphocyte activation-1 (Oldberg et al. 1986; Nomura et al. 1988; Patarca et al. 1989). In the epithelial tissue and bone, the expression level of this protein is high. In the presence of different drugs, such as gentamicin, cisplatin, sevoflurane, and cyclosporin, which may induce kidney damage, an increased expression of osteopontin has been observed; as a consequence, it was also detected in the urine samples of patient (Alchi et al. 2005).

5.7 Type IV Collagen

Damage in the glomerulus increases the level of type IV collagen, which is the main component of basement membrane. During nephrotoxicity, the concentration of collagen (IV) increased in the urine samples and was associated with the change in the structure of the glomerulus. It was due to the structure of collagen to pass through the outer membrane (Donovan et al. 1994; Nerlich et al. 1994).

In addition to acute renal damage, chronic kidney diseases also occur due to the drug abuse. To determine nephrotoxicity, several conventional methods, such as serum creatinine and BUN, are used, but these methods do not have selectivity or sensitivity to identify the early renal damage. Earlier described biomarkers to identify the nephrotoxicity may be useful to provide important information regarding drug abuse. Toxicity assessment of drugs through these biomarkers may be useful for the development of new drugs. In addition to this, it can reduce the economic losses during safety assessment of new drugs. Therefore, the urge of developing selective and sensitive markers to identify the early kidney damage has been enhanced.

6 In Vitro System for the Brain (Central Nervous System) Toxicity Assessment

A major issue related to drug discovery and development is the safety assessment of drugs. Most of the drugs have been withdrawn from the market due to their cardiovascular risk (Sanguinetti and Tristani-Firouzi 2006). Other than that, CNS toxicity also occurs due to the adverse effects of drugs. Failure of nearly one quarter of the whole range of discovery and the development of drugs occurs due to the CNS toxicity (Roberts et al. 2014). In case of cardiovascular and gastrointestinal therapy, toxicity incidences in CNS are very common (Cook et al. 2014). A border set of compounds from different pharmaceutical companies such as Eli Lilly and Company, GlaxoSmithKline, AstraZeneca, and Pfizer have been withdrawn from the market due to the lack of nonclinical toxicological trails that affect drug discovery and development (Waring et al. 2015).

Many registered medicines are cut due to their neurotoxicity, and these medicines carry black box warnings of neurotoxicity. These types of labeling alert consumers and healthcare professionals about the potential health risks of drugs. For example, FDALabel, a web-based application to search labeling documents of FDA-approved drug products, provides a platform to analyze the most common neurotoxicities noted in New Drug Applications (NDAs). The side effects related with these black box warning medicines are sedation, suicidal ideation, abuse liability, seizure, and headache (Fang et al. 2016).

To mimic the key biochemical, morphological, and functional features, isolated neuronal models have been selected as an alternative of the nervous system. Several advantages, such as ease of use, fewer ethical issues, reduced cost, and better control over experiment variable, are associated with in vitro models. For the visualization of individual living cells and examining both electrophysiological and morphological features, primary cultures have been used (Kumaria and Tolias 2008). It is thought that glial and neuronal cells rearrange themselves on the substratum through migration and according to their abilities and functions; these cells differentiate into their respective fate. For the isolation of a single cell of the brain, dissociation procedure is required that can loss the organization of tissue. By using this technique in vivo, like structure cannot be obtained. Over to this lacuna, primary cultures are more available for experimental manipulation than slice cultures, as these cultures remain stable for long periods and are easier to manipulate. From the single cell, biochemical, morphological, molecular, and electrophysiological data can easily be obtained and correlated further .(Banker 1998; Council N.R. 2012).

In a very reproducible manner, large quantities of homogeneous cell populations can be obtained from tumoral origin. The choice of cell lines depends on the properties that are expressed by its normal adult cell counterpart. It must be kept in mind that cell culture systems represent cells that are not an integrated neural network and may show a changed metabolism, appearance, and response to chemicals. The choice of cell lines depends upon the choice of endpoints and experimental design. To predict the neurotoxicity of compounds through in situ methods, in vitro neuronal systems provide such a facility to find out the neurotoxicity of drugs (compounds) (Banker 1998; Council N.R. 2012).

In the later section, we will discuss about the particular advantages and limitations associated with each in vitro neuronal model and its main uniqueness.

6.1 Immortalized Cell Lines

Escaping all the cell-cycle checkpoint regulators, the immortalized cell lines acquire the unique property of uncontrolled proliferation. This feature appears due to the genetic modification that takes place in those cell lines that have a limited life span (Banker 1998). Several advantages are associated with these cell lines, such as:

(i) Along with time, it can retain its main characteristics.
(ii) Large quantity of cells can be easily obtained (Banker 1998; Astashkina et al. 2012).
(iii) Without contamination, the proliferation of culture takes place for a long period.
(iv) Facilitate experiment with similar populations of neuronal cells.
(v) Can be kept in liquid nitrogen forever and utilized as required.
(vi) From a single cell, information regarding biochemical, morphological, molecular, and electrophysiological assays can be extracted and correlated to find out the homeostasis of the cell (Banker 1998).

Although there are cell lines that demonstrate several distinguishing properties of differentiated neurons, such as neurotransmitters, ion channels, and other neuron-specific proteins, yet, they are not ideal models for all neuronal phenotypes. A specific population of differentiated nerve cells shows close resemblance with PC12 cells. These PC12 cells do not form diverse axons and dendrites or form synapse like the sympathetic neurons (Westerink 2008). Neuroblastoma cell lines (e.g., Neuro-2A, SH-SY5Y, or IMR-32 cell lines) (Shastry et al. 2001) are the representative of neurons; however oligodendrocyte (e.g., CG4, HOG, or OLN-93 cell lines), microglia (e.g., BV-2, HAPI, or CHME-5 cell lines), astrocytes (e.g., SFME, C6, or U87MG cell lines), and Schwann cell lines (e.g., IMS32, 33B, or NF1 cell lines) are the representative of glial cells (De Vries and Boullerne 2010). For the neurotoxicity evaluation, information regarding the binding features of the test drugs to the targeted site is an important feature. Thus, the use of correct cell lines with appropriate features should be a matter of concern for the neurotoxicological studies.

6.2 Neuronal Origin's Cell Lines

Neuroblastoma Cell Lines

During the morphogenesis of the adrenal medulla, adrenal neuroblasts are arrested at different levels and correspond to human neuroblastoma cell lines (Shastry et al. 2001). To study the mechanism of neuronal function and differentiation, human neuroblastoma cells are a good model as these cells are representative of human immature neurons. These cell lines synthesize enzymes such as tyrosine hydroxylase (TH), dopamine β-hydroxylase, and choline acetyltransferase that are necessary for the biosynthesis of neurotransmitter (Biedler and Schachner 1978; Påhlman et al. 1983; Sadae et al. 1987; Ciccarone et al. 1989). In chemically defined media, these cell lines proliferate rapidly, form homogenous populations, and extend neurite-like processes. Using different natural and chemical agents, differentiation of neuroblastoma cells may be achieved from immature to mature neuronal cell, in vitro. Therefore, it can help to understand the involvement of several biological processes at different phases of cellular development and differentiation.

In early 1970s, a neuroblastoma cell line, known as SH-SY5Y cell line, was developed using bone marrow biopsy of neuroblastoma patient (with the sympathetic adrenergic ganglia origin). Many biochemical and functional properties of neurons have been attained via these cell lines since then (Biedler and Schachner 1978). These cell lines are used to assess the neurotoxic potential of many compounds (MDMA, MPP+ organic pollutants, and 6-hydroxydopamine) (Presgraves et al. 2003; Tirmenstein et al. 2005; Jung et al. 2009; Tsushima et al. 2012; Chen et al. 2013; Ferreira et al. 2013; Martins et al. 2013). Other than toxicity, different mechanisms, such as mitochondrial dysfunction (Tirmenstein et al. 2005), oxidative stress (Tirmenstein et al. 2005; Tsushima et al. 2012), and disruption in intracellular Ca^{2+} (Hettiarachchi et al. 2012; Barbosa et al. 2014a, b) or disruption in neurite outgrowth (Jung et al. 2009; Chen et al. 2013), can also be accessed. These properties make these cell lines useful in vitro models in the field of neurotoxicity.

Pheochromocytoma Cell Lines

These cells provide the facility to recognize the mechanisms (cellular and molecular) that are involved in drug-induced cellular adaptations and toxicity in an adrenergic system (Eisenhofer et al. 2017). KNA cell line is immortal in nature and derived from sporadic benign human adrenal pheochromocytoma (Pfragner et al. 1998). Another cell line, PC12, is also a pheochromocytoma cell line, useful in neurotoxicology research (Esmaeilzadeh et al. 2013; Im et al. 2013). The process of exocytosis plays an important role during neuronal communication (Eaton and Duplan 2004). These cell lines are also useful as in vitro models for neurosecretory studies (Westerink 2008) as well as to find out the neurotoxin effects of drugs that hinder the release of neurotransmitters (Schubert et al. 1980).

6.3 Cell Lines of Glial Origin

Glioma Cell Lines

The biopsy material can be used to generate stable glioma cell lines via transferring them into tissue culture flasks and successive passaging (You et al. 2007). These cell lines have a regulatory control mechanism and differentiated properties of glial cells. Human U87-MG cell line (Qiu et al. 2012) and rat C6 cell line (Kyeong et al. 2013) have been regularly used for the toxicological studies. In many studies related to toxicity and basic cellular mechanisms, glioma cell lines have been used. Rat C6 glioma cell line is a widely used glioma cell line (Falnoga et al. 2007).

Rat C6 Glioma Cell

Repetitive administration of methylnitrosourea to outbred Wistar rats over a period of 8 months develops C6 glioma cell lines (Benda et al. 1968). These cell lines showed characteristics such as self-renewal and the potential for multilineage differentiations that are present in cancer stem cells during in vitro condition and tumor formation in vivo (Shen et al. 2008). To evaluate the neurotoxicity of compounds such as triorthocresyl phosphate (Liu et al. 2013), methylmercury (Kaur et al. 2010), and manganese (Alaimo et al. 2011), C6 glioma cell line has been used.

6.4 Primary Cultures

Dissociated Primary Culture

In the field of neurotoxicity research, the dissociated primary cultures are the most widely accepted in vitro system. Dissociation of brain tissue is necessary for the preparation of a suspension of individual cells to obtain dissociated primary cultures (Capela et al. 2007, 2013; López-Doménech et al. 2012). The advantages associated with this culture in neurotoxicological studies are (i) accessibility for individual living cells (Banker 1998); (ii) both the morphological and electrophysiological features can be monitored (Banker 1998); (iii) behavior of their growth cones, mode of branching, and growth of axons can be observed; and (iv) single-cell information regarding biochemical, electrophysiological, molecular, and morphological can be correlated (Banker 1998).

Neuronal Primary Culture

Different regions of the brain such as the hippocampus (Capela et al. 2007, 2013), striatum (Oliveira et al. 2006), cerebellum (Jiménez et al. 2004; Giordano et al. 2009), and cortex (Capela et al. 2007; Xu et al. 2012), peripheral nervous system (Söderström and Ebendal 1995) and mouse fetal brain tissues (Oliveira et al. 2006; López-Doménech et al. 2012) are used for the isolation of primary neuronal cultures. Chicken or fetal brain tissue has also been used for the development of primary neuronal culture. For measuring neurotoxicity in vitro, these cultures are useful. To investigate the neurotoxic potential of certain chemicals/drugs, such as MDMA (Capela et al. 2007), cocaine (Cunha-Oliveira et al. 2006), methylmercury (Meacham et al. 2005), organic pollutants, or lead (Hogberg et al. 2009), these primary neuronal cultures have been utilized. Several neurotoxicity outcomes have been evaluated, including cell viability (Capela et al. 2007, 2013; Barbosa et al. 2014a, b), interference with neurite outgrowth (Söderström and Ebendal 1995; Maekawa et al. 2013), and mitochondrial function impairment (Cunha-Oliveira et al. 2006; Hogberg et al. 2009).

Primary Cultures of Cortical Neurons

The heterogeneous neuronal population is found in the cerebral cortex. In the cerebellum cortex, two major classes of cortical neurons are present. Projection neuron is one of them that makes local connections via extending the axon to distal intercortical, subcerebral target, subcortical, and interneurons (Molyneaux et al. 2007). Another key aspect, which makes primary cultures of cortical neurons a valuable system for comprehensive neurobiology and neurotoxicology investigations, is the relatively higher availability of cerebral cortex tissue as compared to other brain parts (Suñol et al. 2008). Some other important features associated with primary cultures of cortical neurons are as follows: (i) functional receptors, such as GABA-A and NMDA, are expressed on these neurons (Suñol et al. 2008), and (ii) they are constituted of both GABAergic (Molyneaux et al. 2007) and cholinergic neurons (Arvidsson et al. 1997). Therefore, these features help to identify the effect of drugs on the cortical neurons via interference with functional receptors as well as the morphology of the neurons.

Primary Cultures of Hippocampus Neurons

The HIP contains a population of neurons with well-defined characteristics that are typical of central nervous system neurons in general. Pyramidal neurons contribute 85 to 95% of the total neuronal population present in the HIP. Some salient features associated with primary culture of hippocampus neuron are as follows: (i) GABA receptors are expressed by hippocampus pyramidal neurons known as glutamatergic excitatory cells (Lambert et al. 1989); (ii) glutamatergic neurons are mainly present in the primary culture of the hippocampus (Lambert et al. 1989); (iii) pyramidal neurons present in high amount, and the percentage of these neurons is (80–90%) of total neuronal population (Fath et al. 2009); and (iv) pyramidal neurons form direct connections with other neurons via the endogenous interneuron population. In the field of neurotoxicology, this primary culture is extensively used to find out the neurotoxic potential of several drugs, thus making it a good model.

Primary Cultures of Cerebellar Granule Neurons

Heterogeneous populations of neuron and non-neuronal cells are present in the cerebellum. Cerebellar granule neurons are the most frequent neuronal population in cerebellum. These neurons control information transfer between cerebellar inputs and outputs and help in motor learning. The primary cultures of these neurons have important features, such as:

(i) They possess NMDA (Babot et al. 2007; Popp et al. 2008) and GABA-A receptor (Vale et al. 1997) and also known as glutamatergic excitatory cells (Smith et al. 2008; Roberts et al. 2010; Hashimoto and Hibi 2012).
(ii) Glutamatergic neurons are mainly present in the primary culture of cerebellar granular cells.
(iii) Granular neurons present in high amount and account for 95% of the total of primary culture of the cerebellum and get more reproducible results (Roberts et al. 2010).
(iv) Purkinje cells and endogenous interneurons such as type II Golgi interneurons form direct connection with cerebellar granule neurons (Kilpatrick et al. 2012).
(v) *In vitro* experimentation of excitotoxicity and oxidative stress-related mechanisms can be performed by using this system (Smith et al. 2008; Suñol et al. 2008; Berntsen et al. 2013).

In the field of neurotoxicology, this primary culture is extensively used to find out the neurotoxic potential of several drugs, thus making it a powerful in vitro system for researching neuroprotective pathways and pharmaceuticals against neurotoxic events.

Primary Cultures of Sensory Neurons

Sensory neurons are present at two locations: (1) spinal ganglia also known as dorsal root ganglia (DRG) and (ii) at cranial nerves V, VII, VIII, IX, and X associated with cranial sensory ganglia. To obtain the primary culture of sensory neurons, dorsal root ganglia (DRG) or rat embryo trigeminal ganglia are used when sensory neurons need to be generated from neonatal animal (E9–16 (Podratz et al. 2011; Yuan et al. 2013) in rats and E10–13 (Lawson and Biscoe 1979; Misko et al. 2010) in mice) and adult animal. These neuronal primary cultures have important features, such as:

(i) They can grow in defined media with vitamin supplement after dissociation from adult animals (Malin et al. 2007).
(ii) They express molecule such as growth-associated protein, galanin, and plasminogen activators that are related to axonal restoration (Woolf et al. 1990).
(iii) During in vivo condition, ion channel's receptor, Ca^{2+} binding proteins, and neuropeptides are expressed on sensory neurons (Malin et al. 2007).
(iv) They respond to different stimuli such as chemical (Jordt et al. 2004), thermal (Reid 2001), and mechanism (McCarter et al. 1999) factors.

Glial Primary Cultures

For the generation of dissociated primary glial cell cultures, many techniques and media conditions are used. Rat or mouse brains between the period of birth and postnatal day 7 are used to generate primary glial cells (Bögler 1997; Weinstein 1997; Weinstein and Wu 1999). Astrocytes, oligodendrocytes, and microglial cells are three phenotypically distinct glial cells that form glial cultures. Cell culture differs severely due to the glial cell composition, which is due to the postnatal period of the mice. In recent study, the behavior of individual cells as well as the interaction between different glial cells has been identified. During in vivo condition, interactions exist between glial cells and neurons to coordinate the functional relationship. During the isolation of these cells, chemicals that are used may cause disruption in the interactions and may be neurotoxic to the cells. It is speculated that the toxicants, which affect neuronal receptors or transduction systems, might affect glial complement. It is also found that enzymes responsible for xenobiotic transformation are present in the glial cells and might metabolize xenobiotic responsible for the neurotoxicity (Philbert et al. 1995; Gradinaru et al. 2012).

Another study showed that these cells might show involvement in the metabolic activation of protoxicants. Astrocytes have monoamine oxidase B, capable of metabolizing MPTP compound into reactive intermediate MPP+, which destroys nigrostriatal dopaminergic neurons (Schildknecht et al. 2009; Bajpai et al. 2013). Schwann cells are a useful model to study the toxicants' toxic effect that induces demyelinating peripheral polyneuropathy (Ydens et al. 2013). To find out the neurotoxicity of several compounds, investigating the interactions of neuronal cells with the glial cells plays a key role. To predict the neurotoxicity potential of drug compounds, glial cells may provide a better platform to understand the effect on the functionality of glial cells.

7 Biomarkers Used to Identify the Neurotoxicity

A number of common drugs and chemicals have been connected to neurotoxicity, but precise methods that can detect such toxicity need to be developed. To fulfill this requirement, more accurate and efficient biomarkers that can help to diagnose and identify the neurotoxicity are required. These biomarkers should be appropriate to animal models and can be translational from nonclinical to clinical data. In addition to tissue, serum, plasma, cerebrospinal fluid (CSF), and urine have been used to identify toxicity via using fluid-based biomarkers.

Fluid-Based Biomarkers: Bodily fluids such as blood (including plasma and serum), CSF, and urine have been used to access neurotoxicity via using fluid-based biomarkers. The future to identify the neurotoxicity, CSF is a precious fluid because it contains most of the biomarkers in it, and it is established that in the target tissue, colocalization of the CSF occurs (Wan et al. 2012). Other than neurotoxicity, CSF biomarkers are also useful in those diseases in which alteration in the CSF composition is a subject of pathology (Tumani et al. 2009). During cell damage, the expression of genes present in neuronal cell is also altered, which can be identified via measuring the level of cellular RNA present in biofluids (Guo et al. 2013; Koh et al. 2014). An ocean of scientific studies provides evidence related to neuronal toxicity that fluid-based biomarkers have a potential to identify toxicity.

Damage in both neuronal and glial cells causes astrogliosis, and to identify such damage, glial fibrillary acid protein (GFAP) is the best biomarker. To identify broad classes of neurotoxic drugs, GFAP ELISA assay is used (O'Callaghan and Sriram 2005). Another biomarker, microtubule-associated protein (MAP-2), is used to identify dendritic injury that occurs during traumatic brain injury (TBI) (Mondello et al. 2012). Oxidative injury that occurs due to the TBI can be recognized by using F_2-isoprostanes. However, these biomarkers did not provide a clear picture about the neurotoxicity (Bayir et al. 2004; Milne and Morrow 2006). With all these markers, the matter of specificity and sensitivity has to be improved. It is very doubtful that one or two biomarkers can provide a required amount of information to distinguish between neurotoxicity and other types of neurotoxicological stress.

Due to drug abuse, severe TBI, ischemia, or hemorrhage may occur that can be detected by using ubiquitin C-terminal hydrolase L1 biomarker (Lewis et al. 2010; Brophy et al. 2011). Neurofilament is a good biomarker to identify injury that occurred in axon rather than in cell body and might be used to identify multiple sclerosis (MS) (Teunissen and Khalil 2012). Other than this biomarker, myelin basic protein is used to identify MS that indicates the disruption in myelin or neural damage. That is why, it is a potentially very useful biomarker (Berger et al. 2006; Belogurov et al. 2008). Translocator protein is a biomarker of activated glial cell, and positron emission tomography is used for imaging of this protein. Neurological and psychiatric disorders and inflammation due to drug abuse could be detected by using translocator protein (Rupprecht et al. 2010; Kreisl et al. 2013). In case of TBI, apoptosis and necrosis occur that cause activation of spectrin degradation products (Berger et al. 2012). Neurodegeneration is detected by spectrin breakdown products (SBDPs). In a recent study, it was found that in the presence of neurotoxic agent, neurodegeneration occurred in rats that could be detected via the presence of SBDF-145 in CSF (Pritt et al. 2014). To track the neurotoxicity, monitoring the production of these biomarkers is necessary. Somewhat these markers cannot be used as a standard to claim the toxicity, so it is necessary to couple additional information, such as body burden of chemicals, patient exposure history, and individual genetics.

8 Conclusion

Humans use drugs for the treatment of several diseases; therefore, evaluation of these drugs must be necessary for their safe administration. The pharmaceutical companies follow various screening methods during drug development to investigate the toxic effects of effective drugs. The liver, kidney, and central nervous system are the major organs monitored for toxicity evaluation during the drug development. The safety evaluations of drugs are performed at both organ (in vivo) and cellular (in vitro) levels. Considering the increasing demand for alternate to animal models in present times, several cell line-based in vitro models have emerged to assess drug toxicity, and several markers are used to achieve that objective. By doing such an exercise, we can reduce the risk of drug-induced toxicity in human.

References

Alaimo A, Gorojod RM, Kotler ML (2011) The extrinsic and intrinsic apoptotic pathways are involved in manganese toxicity in rat astrocytoma C6 cells. Neurochem Int 59:297–308. https://doi.org/10.1016/j.neuint.2011.06.001

Alchi B, Nishi S, Kondo D et al (2005) Osteopontin expression in acute renal allograft rejection. Kidney Int 67:886–896. https://doi.org/10.1111/j.1523-1755.2005.00153.x

Arvidsson U, Riedl M, Elde R, Meister B (1997) Vesicular acetylcholine transporter (VAChT) protein: A novel and unique marker for cholinergic neurons in the central and peripheral nervous systems. J Comp Neurol 378:454–467. https://doi.org/10.1002/(SICI)1096-9861(19970224)378:4<454::AID-CNE2>3.0.CO;2-1

Astashkina A, Mann B, Grainger DW (2012) A critical evaluation of *in-vitro* cell culture models for high-throughput drug screening and toxicity. Pharmacol Ther 134:82–106. https://doi.org/10.1016/j.pharmthera.2012.01.001

Babot Z, Vilaró MT, Suñol C (2007) Long-term exposure to dieldrin reduces γ-aminobutyric acid type A and N-methyl-D-aspartate receptor function in primary cultures of mouse cerebellar granule cells. J Neurosci Res 85:3687–3695

Bader A, Rinkes IH, Closs EI et al (1992) A stable long-term hepatocyte culture system for studies of physiologic processes: cytokine stimulation of the acute phase response in rat and human hepatocytes. Biotechnol Prog 8:219–225

Bailly V, Zhang Z, Meier W et al (2002) Shedding of kidney injury molecule-1, a putative adhesion protein involved in renal regeneration. J Biol Chem 277:39739–39748. https://doi.org/10.1074/jbc.M200562200

Bajaj P, Chowdhury SK, Yucha R et al (2018) Emerging kidney models to investigate metabolism, transport, and toxicity of drugs and xenobiotics. Drug Metab Dispos 46:1692–1702. https://doi.org/10.1124/dmd.118.082958

Bajpai P, Sangar MC, Singh S et al (2013) Metabolism of 1-methyl-4-phenyl-1,2,3/6-tetrahydropyridine by mitochondrion-targeted cytochrome P450 2D6 implications in parkinson disease. J Biol Chem 288:4436–4451. https://doi.org/10.1074/jbc.M112.402123

Banker GGK (1998) Culturing nerve cells, 2nd edn. MIT, Cambridge, MA, pp 11–36

Barbosa DJ, Capela JP, Silva R et al (2014a) The mixture of "ecstasy" and its metabolites is toxic to human SH-SY5Y differentiated cells at *in-vivo* relevant concentrations. Arch Toxicol 88:455–473. https://doi.org/10.1007/s00204-013-1120-7

Barbosa DJ, Capela JP, Silva R et al (2014b) Ecstasy-induced toxicity in SH-SY5Y differentiated cells: Role of hyperthermia and metabolites. Arch Toxicol 88:515–531. https://doi.org/10.1007/s00204-013-1147-9

Baron S, Tyring SK, Fleischmann WR, Coppenhaver DH, Niesel DW, Klimpel GR, Stanton GJHT (1991) The interferons: mechanisms of action and clinical applications. Jama 266:1375–1383

Bayir H, Marion DW, Puccio AM et al (2004) Marked gender effect on lipid peroxidation after severe traumatic brain injury in adult patients. J Neurotrauma 21:1–8. https://doi.org/10.1089/089771504772695896

Belogurov AA, Kurkova IN, Friboulet A et al (2008) Recognition and degradation of myelin basic protein peptides by serum autoantibodies: novel biomarker for multiple sclerosis. J Immunol 180:1258–1267. https://doi.org/10.4049/jimmunol.180.2.1258

Benda P, Lightbody J, Sato G, Levine LSW (1968) Differentiated rat glial cell strain in tissue culture. Science (80-) 161:370–371. https://doi.org/10.1126/science.161.3839.371

Berger RP, Adelson PD, Richichi R, Kochanek PM (2006) Serum biomarkers after traumatic and hypoxemic brain injuries: insight into the biochemical response of the pediatric brain to inflicted brain injury. Dev Neurosci 28:327–335. https://doi.org/10.1159/000094158

Berger RP, Hayes RL, Richichi R et al (2012) Serum concentrations of ubiquitin C-terminal hydrolase-l1 and αiI-spectrin breakdown product 145kDa correlate with outcome after pediatric TBI. J Neurotrauma 29:162–167. https://doi.org/10.1089/neu.2011.1989

Bernard AM, Vyskocil AA, Mahieu P, Lauwerys RR (1987) Assessment of urinary retinol-binding protein as an index of proximal tubular injury. Clin Chem 33:775–779. https://doi.org/10.1093/clinchem/33.6.775

Berntsen HF, Wigestrand MB, Bogen IL et al (2013) Mechanisms of penitrem-induced cerebellar granule neuron death *in-vitro*: Possible involvement of GABAA receptors and oxidative processes. Neurotoxicology 35:129–136. https://doi.org/10.1016/j.neuro.2013.01.004

Bi YA, Kazolias D, Duignan DB (2006) Use of cryopreserved human hepatocytes in sandwich culture to measure hepatobiliary transport. Drug Metab Dispos 34:1658–1665. https://doi.org/10.1124/dmd.105.009118

Biagini CP, Boissel E, Borde F et al (2006) Investigation of the hepatotoxicity profile of chemical entities using Liverbeads® and WIF-B9 *in-vitro* models. Toxicol Vitr 20:1051–1059. https://doi.org/10.1016/j.tiv.2006.01.013

Biedler JL, Schachner M (1978) Multiple neurotransmitter synthesis by human neuroblastoma cell lines and clones. Cancer Res 38:3751–3757

Birnbaum LS, Stokes WS (2010) Safety testing: moving toward alternative methods. Environ Health Perspect 118:12–13. https://doi.org/10.1289/ehp.0901704

Boelsterli UA (2003) Diclofenac-induced liver injury: a paradigm of idiosyncratic drug toxicity. Toxicol Appl Pharmacol 192:307–322. https://doi.org/10.1016/S0041-008X(03)00368-5

Boess F, Kamber M, Romer S et al (2003) Gene expression in two hepatic cell lines, cultured primary hepatocytes, and liver slices compared to the *in-vivo* liver gene expression in rats: possible implications for toxicogenomics use of *in-vitro* systems. Toxicol Sci 73:386–402. https://doi.org/10.1093/toxsci/kfg064

Bögler O (1997) Isolation and purification of primary oligodendrocyte precursors. Curr Protoc Neurosci 00:1–9. https://doi.org/10.1002/0471142301.ns0304s00

Bonventre JV, Vaidya VS, Schmouder R, Feig PDF (2010) Next-generation biomarkers for detecting kidney toxicity. Nat Biotechnol 28:436–440. https://doi.org/10.1038/nbt0510-436. Next-generation

Boo D, Knight A (2009) Replacing, reducing and refining procedures in animal research. J Oral Tissue Eng 6:215–222. https://doi.org/10.11223/jarde.6.215

Borregaard N, Sehested M, Nielsen BS et al (1995) Biosynthesis of granule proteins in normal human bone marrow cells. Gelatinase is a marker of terminal neutrophil differentiation. Blood 85:812–817. https://doi.org/10.1182/blood.v85.3.812.bloodjournal853812

Bort R, Ponsoda X, Jover R, Gómez-Lechón MJCJ (1998) Diclofenac toxicity to hepatocytes: a role for drug metabolism in cell toxicity. J Pharmacol Exp Ther 288:65–72

Borzelleca JF (2000) Paracelsus: Herald of modern toxicology. Toxicol Sci 53:2–4. https://doi.org/10.1093/toxsci/53.1.2

Braiterman LT, Hubbard AL (2009) The liver. John Wiley & Sons, Ltd

Brophy GM, Mondello S, Papa L et al (2011) Biokinetic analysis of ubiquitin C-terminal hydrolase-L1 (UCH-L1) in severe traumatic brain injury patient biofluids. J Neurotrauma 28:861–870. https://doi.org/10.1089/neu.2010.1564

Burke AS, MacMillan-Crow LA, Hinson JA (2010) The hepatocyte suspension assay is superior to the cultured hepatocyte assay for determining mechanisms of acetaminophen hepatotoxicity relevant to *in-vivo* toxicity. Chem Res Toxicol 23:1855–1858. https://doi.org/10.1021/tx1003744

Capela JP, da Costa AS, Costa VM et al (2013) The neurotoxicity of hallucinogenic amphetamines in primary cultures of hippocampal neurons. Neurotoxicology 34:254–263. https://doi.org/10.1016/j.neuro.2012.09.005

Capela JP, Macedo C, Branco PS et al (2007) Neurotoxicity mechanisms of thioether ecstasy metabolites. Neuroscience 146:1743–1757. https://doi.org/10.1016/j.neuroscience.2007.03.028

Chang SY, Weber EJ, Ness KV, Eaton DLKE (2016) Liver and kidney on chips: microphysiological models to understand transporter function. Clin Pharmacol Ther 100:464–478. https://doi.org/10.1002/cpt.436.Liver

Chang SY, Weber EJ, Sidorenko VS et al (2017) Human liver-kidney model elucidates the mechanisms of aristolochic acid nephrotoxicity. JCI Insight 2:1–14. https://doi.org/10.1172/jci.insight.95978

Chen JX, Sun YJ, Wang P et al (2013) Induction of autophagy by TOCP in differentiated human neuroblastoma cells lead to degradation of cytoskeletal components and inhibition of neurite outgrowth. Toxicology 310:92–97. https://doi.org/10.1016/j.tox.2013.05.012

Chen M, Suzuki A, Borlak J et al (2015) Drug-induced liver injury: interactions between drug properties and host factors. J Hepatol 63:503–514. https://doi.org/10.1016/j.jhep.2015.04.016

Choudhury D, Ahmed Z (2006) Drug-associated renal dysfunction and injury. Nat Clin Pract Nephrol 2:80–91. https://doi.org/10.1038/ncpneph0076

Ciccarone V, Spengler BA, Meyers MB et al (1989) Phenotypic diversification in human neuroblastoma cells: expression of distinct neural crest lineages. Cancer Res 49:219–225

Cook D, Brown D, Alexander R et al (2014) Lessons learned from the fate of AstraZeneca's drug pipeline: A five-dimensional framework. Nat Rev Drug Discov 13:419–431. https://doi.org/10.1038/nrd4309

Council N.R. (2012) The future of animals, cells, models, and systems in research, development, education, and testing; Proceedings of a symposium. In: National Academy of Sciences. Washington DC, USA

Cunha-Oliveira T, Rego AC, Cardoso SM et al (2006) Mitochondrial dysfunction and caspase activation in rat cortical neurons treated with cocaine or amphetamine. Brain Res 1089:44–54. https://doi.org/10.1016/j.brainres.2006.03.061

Das S (2018) Extrapolation of *In-vitro* results to predict human toxicity. Academic Press

De Vries GH, Boullerne AI (2010) Glial cell lines: an overview. Neurochem Res 35:1978–2000. https://doi.org/10.1007/s11064-010-0318-9

DelRaso NJ (1993) *In-vitro* methodologies for enhanced toxicity testing. Toxicol Lett 68:91–99

DesRochers TM, Suter L, Roth A, Kaplan DL (2013) Bioengineered 3D human kidney tissue, a platform for the determination of nephrotoxicity. PLoS One 8. https://doi.org/10.1371/journal.pone.0059219

DiMasi JA, Hansen RW, Grabowski HG (2003) The price of innovation: new estimates of drug development costs. J Health Econ 22:151–185. https://doi.org/10.1016/S0167-6296(02)00126-1

Donovan KL, Coles GA, Williams JD (1994) An ELISA for the detection of type IV collagen in human urine—application to patients with glomerulonephritis. Kidney Int 46:1431–1437. https://doi.org/10.1038/ki.1994.415

Dorato MA, Buckley LA (2007) Toxicology testing in drug discovery and development. Curr Protoc Toxicol Chapter 19:1–35. https://doi.org/10.1002/0471141755.tx1901s31

Duval DL, Sieg DJBR (1995) Regulation of hepatic nitric oxide synthase by reactive oxygen intermediates and glutathione. Arch Biochem Biophys 316:699–706

Eaton MJ, Duplan H (2004) Useful cell lines derived from the adrenal medulla. Mol Cell Endocrinol 228:39–52. https://doi.org/10.1016/j.mce.2003.02.001

Eisenhofer G, Klink B, Richter S et al (2017) Metabologenomics of phaeochromocytoma and paraganglioma: an integrated approach for personalised biochemical and genetic testing. Clin Biochem Rev 38:69–100

Ekins S, Murray GI, Burke MD, Williams JA, Marchant NCHG (1995) Quantitative differences in phase I and II metabolism between rat precision-cut liver slices and isolated hepatocytes. Drug Metab Dispos 23:1274–1279

Emeigh Hart SG (2005) Assessment of renal injury *in-vivo*. J Pharmacol Toxicol Methods 52:30–45. https://doi.org/10.1016/j.vascn.2005.04.006

Esmaeilzadeh E, Gardaneh M, Gharib E, Sabouni F (2013) Shikonin protects dopaminergic cell line PC12 against 6-hydroxydopamine-mediated neurotoxicity via both glutathione-dependent and independent pathways and by inhibiting apoptosis. Neurochem Res 38:1590–1604. https://doi.org/10.1007/s11064-013-1061-9

Falnoga I, Šlejkovec Z, Pucer A et al (2007) Arsenic metabolism in multiple myeloma and astrocytoma cells. Biol Trace Elem Res 116:5–28. https://doi.org/10.1007/BF02685915

Fang H, Harris SC, Liu Z et al (2016) FDA drug labeling: rich resources to facilitate precision medicine, drug safety, and regulatory science. Drug Discov Today 21:1566–1570. https://doi.org/10.1016/j.drudis.2016.06.006

Faria J, Ahmed S, Gerritsen KGF et al (2019) Kidney-based *in-vitro* models for drug-induced toxicity testing. Arch Toxicol 93:3397–3418. https://doi.org/10.1007/s00204-019-02598-0

Fath T, Ke YD, Gunning P et al (2009) Primary support cultures of hippocampal and substantia nigra neurons. Nat Protoc 4:78–85. https://doi.org/10.1038/nprot.2008.199

Fedecostante M, Onciu OG, Westphal KGC, Masereeuw R (2017) Towards a bioengineered kidney: recellularization strategies for decellularized native kidney scaffolds. 40:150–158. https://doi.org/10.5301/ijao.5000564

Ferguson MA, Vaidya VS, Bonventre J (2008) Biomarkers of nephrotoxic acute kidney injury. Toxicology 245:182–193. https://doi.org/10.1016/j.tox.2007.12.024.Biomarkers

Ferreira PS, Nogueira TB, Costa VM et al (2013) Neurotoxicity of "ecstasy" and its metabolites in human dopaminergic differentiated SH-SY5Y cells. Toxicol Lett 216:159–170. https://doi.org/10.1016/j.toxlet.2012.11.015

Finn WF, Porter G (2003) Urinary biomarkers and nephrotoxicity. Springer, Dordrecht

Fisher RL, Shaughnessy RP, Jenkins PM et al (1995) Dynamic organ culture is superior to multiwell plate culture for maintaining precision-cut tissue slices: optimization of tissue slice culture, part 1. Toxicol Mech Methods 5:99–113. https://doi.org/10.3109/15376519509045905

Gad SC (1996) ICLAS proceedings: preclinical toxicity testing in the development of new therapeutic agents. Scand J Lab Anim Sci. Supplement (Denmark)

Gad SC (2009) Drug safety evaluation, 2nd edn. John Wiley & Sons, Inc, Hoboken, NJ

Gad SC (2015) Animal models in toxicology, 3rd edn. Marcel Dekker, Marcel Dekker, New York

Gad SC (2017) The application of *in-vitro* techniques in drug safety assessment. Drug Saf Eval 553–582. https://doi.org/10.1002/9781119097440.ch27

Ghantous HN, Fernando J, Gandolfi AJ, Brendel K (1992) Sevoflurane is biotransformed by guinea pig liver slices but causes minimal cytotoxicity. Anesth Analg 75:436–440. https://doi.org/10.1213/00000539-199209000-00021

Giezen TJ, Mantel-Teeuwisse AK, Straus SMJM et al (2008) Safety-related regulatory actions for biologicals approved in the United States and the European Union. JAMA – J Am Med Assoc 300:1887–1896. https://doi.org/10.1001/jama.300.16.1887

Giordano G, Kavanagh TJ, Costa L (2009) Mouse cerebellar astrocytes protect cerebellar granule neurons against toxicity of the polybrominated diphenyl ether (PBDE) mixture DE-71. Neurotoxicology 30:326–329. https://doi.org/10.1016/j.neuro.2008.12.009.Mouse

Gokhale BTMS, Zurlo JYJ (1997) Cytochrome P-450 1A1/1A2 induction, albumin secretion, and histological changes in cultured rat liver slices. In-vitro Toxicol 8:357–368

Gómez-Lechón MJ, Castell JV, Donato MT (2008) An update on metabolism studies using human hepatocytes in primary culture. Expert Opin Drug Metab Toxicol 4:837–854. https://doi.org/10.1517/17425255.4.7.837

Gómez-Lechón MJ, Donato LA, López PTMA, JV C (1988) Liver cells and drugs. Colloque INSERM/John Libbey Ltd.

Gomez-Lechon MJ, Lahoz A, Gombau L et al (2010) *In-vitro* evaluation of potential hepatotoxicity induced by drugs. Curr Pharm Des 16:1963–1977. https://doi.org/10.2174/138161210791208910

Gradinaru D, Minn AL, Artur Y et al (2012) Effect of oxidative stress on UDP-glucuronosyltransferases in rat astrocytes. Toxicol Lett 213:316–324. https://doi.org/10.1016/j.toxlet.2012.07.014

Griffin SJ, Houston JB (2005) Prediction of *in-vitro* intrinsic clearance from hepatocytes: comparison of suspensions and monolayer cultures. Drug Metab Dispos 33:115–120. https://doi.org/10.1124/dmd.33.1.115

Guder WG, Hofmann W (1992) Markers for the diagnosis and monitoring of renal tubular lesions. Clin Nephrol 38:S3–7

Guguen-Guillouzo CGA (2010) General review on *in-vitro* hepatocyte models and their applications. Hepatocytes 640:1–40

Guillouzo A (1998) Liver cell models in *in-vitro* toxicology. Environ Health Perspect 106:511–532. https://doi.org/10.1289/ehp.98106511

Guillouzo A, Morel F, Langouët S et al (1997) Use of hepatocyte cultures for the study of hepatotoxic compounds. J Hepatol Suppl 26:73–80. https://doi.org/10.1016/s0168-8278(97)80499-0

Guo D, Liu J, Wang W et al (2013) Alteration in abundance and compartmentalization of inflammation-related miRNAs in plasma after intracerebral hemorrhage. Stroke 44:1739–1742. https://doi.org/10.1161/STROKEAHA.111.000835

Hashemi E, Till CIC (1999) Stability of Phase II conjugation systems in cultured precision-cut rat hepatic slices. Toxicol Vitr 13:459–466

Hashimoto M, Hibi M (2012) Development and evolution of cerebellar neural circuits. Dev Growth Differ 54:373–389. https://doi.org/10.1111/j.1440-169X.2012.01348.x

He K, Talaat RE, Pool WF et al (2004) Metabolic activation of troglitazone: identification of a reactive metabolite and mechanisms involved. Drug Metab Dispos 32:639–646. https://doi.org/10.1124/dmd.32.6.639

Herget-Rosenthal S, Poppen D, Hüsing J et al (2004) Prognostic value of tubular proteinuria and enzymuria in nonoliguric acute tubular necrosis. Clin Chem 50:552–558. https://doi.org/10.1373/clinchem.2003.027763

Hettiarachchi NT, Boyle JP, Bauer CC, Dallas ML, Pearson HA, Hara S, Gamper NPC (2012) Peroxynitrite mediates disruption of Ca2+ homeostasis by carbon monoxide via Ca2+ ATPase degradation. Antioxid Redox Signal 17:744–755

Hewitt NJ, Lechón MJG, Houston JB et al (2007) Primary hepatocytes: current understanding of the regulation of metabolic enzymes and transporter proteins, and pharmaceutical practice for the use of hepatocytes in metabolism, enzyme induction, transporter, clearance, and hepatotoxicity studies. Drug Metab Rev 39:159–234. https://doi.org/10.1080/03602530601093489

Hogberg HT, Kinsner-Ovaskainen A, Coecke S et al (2009) mRNA expression is a relevant tool to identify developmental neurotoxicants using an *in-vitro* approach. Toxicol Sci 113:95–115. https://doi.org/10.1093/toxsci/kfp175

Horii Y, Iwano M, Hirata E, Shiiki H, Fujii Y, Dohi K IH (1993) Role of interleukin-6 in the progression of mesangial proliferative glomerulonephritis. In: Kidney International Supplement. p 39

Im AR, Kim YH, Uddin MR et al (2013) Betaine protects against rotenone-induced neurotoxicity in PC12 cells. Cell Mol Neurobiol 33:625–635. https://doi.org/10.1007/s10571-013-9921-z

Jansen J, De Napoli IE, Fedecostante M et al (2015) Human proximal tubule epithelial cells cultured on hollow fibers: living membranes that actively transport organic cations. Sci Rep 5:1–12. https://doi.org/10.1038/srep16702

Jansen J, Fedecostante M, Wilmer MJ et al (2016) Bioengineered kidney tubules efficiently excrete uremic toxins. Sci Rep 6:1–12. https://doi.org/10.1038/srep26715

Jansen J, Schophuizen CMS, Wilmer MJ et al (2014) A morphological and functional comparison of proximal tubule cell lines established from human urine and kidney tissue. Exp Cell Res 323: 87–99. https://doi.org/10.1016/j.yexcr.2014.02.011

Jauregui HO, McMillan PN, Hevey KNS (2016) A quantitative analysis of lectin binding to adult rat hepatocyte cell surfaces. Cell Vitr 24:401–412

Jenkinson SE, Chung GW, van Loon E et al (2012) The limitations of renal epithelial cell line HK-2 as a model of drug transporter expression and function in the proximal tubule. Pflugers Arch 464:601–611. https://doi.org/10.1007/s00424-012-1163-2

Jiménez A, Jordà EG, Verdaguer E et al (2004) Neurotoxicity of amphetamine derivatives is mediated by caspase pathway activation in rat cerebellar granule cells. Toxicol Appl Pharmacol 196:223–234. https://doi.org/10.1016/j.taap.2003.12.017

Jordt SE, Bautista DM, Chuang HH et al (2004) Mustard oils and cannabinoids excite sensory nerve fibres through the TRP channel ANKTM1. Nature 427:260–265. https://doi.org/10.1038/nature02282

Jung JE, Moon JY, Ghil SH, Yoo BS (2009) 2,3,7,8-Tetrachlorodibenzo-p-dioxin (TCDD) inhibits neurite outgrowth in differentiating human SH-SY5Y neuroblastoma cells. Toxicol Lett 188: 153–156. https://doi.org/10.1016/j.toxlet.2009.04.004

Kaur P, Evje L, Aschner M, Syversen T (2010) The *in-vitro* effects of Trolox on methylmercury-induced neurotoxicity. Toxicology 276:73–78. https://doi.org/10.1016/j.tox.2010.07.006

Kharasch ED, Schroeder JL, Bammler T et al (2006) Gene expression profiling of nephrotoxicity from the sevoflurane degradation product fluoromethyl-2,2-difluoro-1-(trifluoromethyl)vinyl Ether ("Compound A") in Rats. Toxicol Sci 90:419–431. https://doi.org/10.1093/toxsci/kfj088

Kilpatrick DL, Wang W, Gronostajski RLE (2012) Nuclear factor I and cerebellar granule neuron development: an intrinsic–extrinsic interplay. Cerebellum 11:41–49. https://doi.org/10.1007/s12311-010-0227-0.Nuclear

Koh W, Pan W, Gawad C, Fan HC, Kerchner GA, Wyss-Coray T, Blumenfeld YJ, El-Sayed YYQS (2014) Noninvasive *in-vivo* monitoring of tissue-specific global gene expression in humans. Proc Natl Acad Sci U S A 111:7361–7366. https://doi.org/10.1073/pnas.1411381111

Kreisl WC, Jenko KJ, Hines CS et al (2013) A genetic polymorphism for translocator protein 18 kDa affects both *in-vitro* and *in-vivo* radioligand binding in human brain to this putative biomarker of neuroinflammation. J Cereb Blood Flow Metab 33:53–58. https://doi.org/10.1038/jcbfm.2012.131

Kumaria A, Tolias CM (2008) *In-vitro* models of neurotrauma. Br J Neurosurg 22:200–206. https://doi.org/10.1080/02688690701772413

Kyeong IG, Eum WS, Choi SY, Kang JH (2013) Oxidative modification of neurofilament-L and neuronal cell death induced by the catechol neurotoxin, tetrahydropapaveroline. Toxicol Lett 217:59–66. https://doi.org/10.1016/j.toxlet.2012.11.029

Lambert JDC, Jones RSG, Andreasen M et al (1989) The role of excitatory amino acids in synaptic transmission in the hippocampus. Comp Biochem Physiol -- Part A Physiol 93:195–201. https://doi.org/10.1016/0300-9629(89)90207-7

Lawson SN, Biscoe TJ (1979) Development of mouse dorsal root ganglia: an autoradiographic and quantitative study. J Neurocytol 8:265–274. https://doi.org/10.1007/BF01236122

LeCluyse EL (2001) Human hepatocyte culture systems for the *in-vitro* evaluation of cytochrome P450 expression and regulation. Eur J Pharm Sci 13:343–368. https://doi.org/10.1016/S0928-0987(01)00135-X

LeCluyse EL, Bullock PL, Parkinson A (1996) Strategies for restoration and maintenance of normal hepatic structure and function in long-term cultures of rat hepatocytes. Adv Drug Deliv Rev 22: 133–186. https://doi.org/10.1016/S0169-409X(96)00418-8

Leeman WR, van de Gevel IA, Rutten AAJJL (1995) Cytotoxicity of retinoic acid, menadione and aflatoxin B1 in rat liver slices using Netwell inserts as a new culture system. Toxicol Vitr 9. https://doi.org/10.1016/0887-2333(95)00008-V

Lerche-Langrand C, Toutain HJ (2000) Precision-cut liver slices: characteristics and use for *in-vitro* pharmaco-toxicology. Toxicology 153:221–253. https://doi.org/10.1016/S0300-483X(00)00316-4

Lewis SB, Wolper R, Chi YY et al (2010) Identification and preliminary characterization of ubiquitin C terminal hydrolase 1 (UCHL1) as a biomarker of neuronal loss in aneurysmal subarachnoid hemorrhage. J Neurosci Res 88:1475–1484. https://doi.org/10.1002/jnr.22323

Li AP (2004) Accurate prediction of human drug toxicity: a major challenge in drug development. Chem Biol Interact 150:3–7. https://doi.org/10.1016/j.cbi.2004.09.008

Li JY, Matias J, Scudiero DA, Hite KM et al (2001) P450 enzyme expression patterns in the NCI human tumor cell line panel. Drug Metab Dispos 29:304–312

Lijinsky W (1988) Importance of animal experiments in carcinogenesis research. Environ Mol Mutagen 11:307–314

Liu X, Piao F, Li Y (2013) Protective effect of taurine on the decreased biogenic amine neurotransmitter levels in the brain of mice exposed to arsenic. InTaurine 776:277–287

López-Doménech G, Serrat R, Mirra S et al (2012) The Eutherian Armcx genes regulate mitochondrial trafficking in neurons and interact with Miro and Trak2. Nat Commun 3. https://doi.org/10.1038/ncomms1829

Lübberstedt M, Müller-Vieira U, Mayer M et al (2011) HepaRG human hepatic cell line utility as a surrogate for primary human hepatocytes in drug metabolism assessment *in-vitro*. J Pharmacol Toxicol Methods 63:59–68. https://doi.org/10.1016/j.vascn.2010.04.013

MacDonald C, Vass M, Willett B, Scott AGH (1994) Expression of liver functions in immortalised rat hepatocyte cell lines. Hum Exp Toxicol 13:439–444

Maekawa F, Tsuboi T, Oya M et al (2013) Effects of sodium arsenite on neurite outgrowth and glutamate AMPA receptor expression in mouse cortical neurons. Neurotoxicology 37:197–206. https://doi.org/10.1016/j.neuro.2013.05.006

Malin SA, Davis BM, Molliver DC (2007) Production of dissociated sensory neuron cultures and considerations for their use in studying neuronal function and plasticity. Nat Protoc 2:152–160. https://doi.org/10.1038/nprot.2006.461

Martins JB, Bastos MDL, Carvalho F, Capela JP (2013) Differential effects of methyl-4-phenylpyridinium ion, rotenone, and Paraquat on differentiated SH-SY5Y cells. J Toxicol 2013. https://doi.org/10.1155/2013/347312

May JE, Xu J, Morse HR et al (2009) Toxicity testing: the search for an *in-vitro* alternative to animal testing. Br J Biomed Sci 66:160–165. https://doi.org/10.1080/09674845.2009.11730265

McCarter GC, Reichling DB, Levine JD (1999) Mechanical transduction by rat dorsal root ganglion neurons *in-vitro*. Neurosci Lett 273:179–182. https://doi.org/10.1016/S0304-3940(99)00665-5

McDonald MG, Rettie AE (2007) Sequential metabolism and bioactivation of the hepatotoxin benzbromarone: formation of glutathione adducts from a catechol intermediate. Chem Res Toxicol 20:1833–1842. https://doi.org/10.1021/tx7001228

Meacham CA, Freudenrich TM, Anderson WL et al (2005) Accumulation of methylmercury or polychlorinated biphenyls in *in-vitro* models of rat neuronal tissue. Toxicol Appl Pharmacol 205:177–187. https://doi.org/10.1016/j.taap.2004.08.024

Milne GL, Morrow JD (2006) Isoprostanes and related compounds: update 2006. Antioxidants Redox Signal 8:1379–1384. https://doi.org/10.1089/ars.2006.8.1379

Misko A, Jiang S, Wegorzewska I et al (2010) Mitofusin 2 is necessary for transport of axonal mitochondria and interacts with the Miro/Milton complex. J Neurosci 30:4232–4240. https://doi.org/10.1523/JNEUROSCI.6248-09.2010
Mogensen CE (1971) Urinary albumin excretion in early and long-term juvenile diabetes. Scand J Clin Lab Invest 28:183–193. https://doi.org/10.3109/00365517109086899
Molyneaux BJ, Arlotta P, Menezes JRL, Macklis JD (2007) Neuronal subtype specification in the cerebral cortex. Nat Rev Neurosci 8:427–437. https://doi.org/10.1038/nrn2151
Mondello S, Gabrielli A, Catani S et al (2012) Increased levels of serum MAP-2 at 6-months correlate with improved outcome in survivors of severe traumatic brain injury. Brain Inj 26: 1629–1635. https://doi.org/10.3109/02699052.2012.700083
Nelson KF, Acosta D, Bruckner JV (1982) Long-term maintenance and induction of cytochrome P-450 in primary cultures of rat hepatocytes. Biochem Pharmacol 31:2211–2214. https://doi.org/10.1016/0006-2952(82)90521-4
Nerlich AG, Schleicher ED, Wiest I et al (1994) Immunohistochemical localization of collagen VI in diabetic glomeruli. Kidney Int 45:1648–1656. https://doi.org/10.1038/ki.1994.216
Nieskens TTG, Peters JGP, Schreurs MJ et al (2016) A human renal proximal tubule cell line with stable organic anion transporter 1 and 3 expression predictive for antiviral-induced toxicity. AAPS J 18:465–475. https://doi.org/10.1208/s12248-016-9871-8
Noé J, Portmann R, Brun ME, Funk C (2007) Substrate-dependent drug-drug interactions between gemfibrozil, fluvastatin and other organic anion-transporting peptide (OATP) substrates on OATP1B1, OATP2B1, and OATP1B3. Drug Metab Dispos 35:1308–1314. https://doi.org/10.1124/dmd.106.012930
Nomura S, Wills AJ, Edwards DR et al (1988) Developmental expression of 2ar (osteopontin) and SPARC (Osteonectin) RNA as revealed by in situ hybridization. J Cell Biol 106:441–450. https://doi.org/10.1083/jcb.106.2.441
O'Brien P, Siraki A (2005) Accelerated cytotoxicity mechanism screening using drug metabolising enzyme modulators. Curr Drug Metab 6:101–109. https://doi.org/10.2174/1389200053586082
O'Brien PJ, Irwin W, Diaz D et al (2006) High concordance of drug-induced human hepatotoxicity with *in-vitro* cytotoxicity measured in a novel cell-based model using high content screening. Arch Toxicol 80:580–604. https://doi.org/10.1007/s00204-006-0091-3
O'Callaghan JP, Sriram K (2005) Glial fibrillary acidic protein and related glial proteins as biomarkers of neurotoxicity. Expert Opin Drug Saf 4:433–442. https://doi.org/10.1517/14740338.4.3.433
Ohlsson S, Wieslander J, Segelmark M (2003) Increased circulating levels of proteinase 3 in patients with anti-neutrophilic cytoplasmic autoantibodies-associated systemic vasculitis in remission. Clin Exp Immunol 131:528–535. https://doi.org/10.1046/j.1365-2249.2003.02083.x
Oldberg A, Franzen A, Heinegard D (1986) Cloning and sequence analysis of rat bone sialoprotein (osteopontin) cDNA reveals an Arg-Gly-Asp cell-binding sequence. Proc Natl Acad Sci U S A 83:8819–8823. https://doi.org/10.1073/pnas.83.23.8819
Oliveira JMA, Chen S, Almeida S et al (2006) Mitochondrial-dependent Ca2+ handling in Huntington's disease striatal cells: effect of histone deacetylase inhibitors. J Neurosci 26: 11174–11186. https://doi.org/10.1523/JNEUROSCI.3004-06.2006
Påhlman S, Ruusala AI, Abrahamsson L et al (1983) Kinetics and concentration effects of TPA-induced differentiation of cultured human neuroblastoma cells. Cell Differ 12:165–170. https://doi.org/10.1016/0045-6039(83)90006-4
Park BK, Kitteringham NR, Maggs JL et al (2005) The role of metabolic activation in drug-induced hepatotoxicity. Annu Rev Pharmacol Toxicol 45:177–202. https://doi.org/10.1146/annurev.pharmtox.45.120403.100058
Patarca R, Freeman GJ, Singh RP et al (1989) Structural and functional studies of the early T lymphocyte activation 1 (Eta-1) gene. Definition of a novel T cell-dependent response associated with genetic resistance to bacterial infection. J Exp Med 170:145–161. https://doi.org/10.1084/jem.170.1.145

Pearce RE, Uetrecht JP, Leeder JS (2005) Pathways of carbamazepine bioactivation *in-vitro*: II. The role of human cytochrome P450 enzymes in the formation of 2-hydroxyiminostilbene. Drug Metab Dispos 33:1819–1826. https://doi.org/10.1124/dmd.105.004861
Pfragner R, Behmel A, Smith DP et al (1998) First continuous human pheochromocytoma cell line: KNA biological, cytogenetic and molecular characterization of KNA cells. J Neurocytol 27: 175–186. https://doi.org/10.1023/A:1006959625068
Philbert MA, Beiswanger CM, Manson MM, Green, Novak RF, Primiano T, Reuhl KRLH (1995) Glutathione S-transferases and gamma-glutamyl transpeptidase in the rat nervous systems: a basis for differential susceptibility to neurotoxicants. Neurotoxicology 16:349–362
Pichler WJ (2002) Pharmacological interaction of drugs with antigen-specific immune receptors: the pi concept. Curr Opin Allergy Clin Immunol 2:301–305
Podratz JL, Knight AM, Ta LE et al (2011) Cisplatin induced Mitochondrial DNA damage in dorsal root ganglion neurons. Neurobiol Dis 41:661–668. https://doi.org/10.1016/j.nbd.2010.11.017
Popp RL, Reneau JC, Dertien J (2008) Cerebellar granule cells cultured from adolescent rats express functional NMDA receptors: an *in-vitro* model for studying the developing cerebellum. J Neurochem 106:900–911
Presgraves SP, Ahmed T, Borwege S, Joyce JN (2003) Terminally differentiated SH-SY5Y cells provide a model system for studying neuroprotective effects of dopamine agonists. Neurotox Res 5:579–598. https://doi.org/10.1007/BF03033178
Prinsen BHCMT, Sain-van D, der Velden MGM, Kaysen GA et al (2001) Transferrin synthesis is increased in nephrotic patients insufficiently to replace urinary losses. J Am Soc Nephrol 12: 1017–1025. https://doi.org/10.1681/asn.v1251017
Pritt ML, Hall DG, Jordan WH et al (2014) Initial biological qualification of SBDP-145 as a biomarker of compound-induced neurodegeneration in the rat. Toxicol Sci 141:398–408. https://doi.org/10.1093/toxsci/kfu136
Qiu B, Sun X, Zhang D et al (2012) TRAIL and paclitaxel synergize to kill U87 cells and U87-derived stem-like cells *in-vitro*. Int J Mol Sci 13:9142–9156. https://doi.org/10.3390/ijms13079142
Reid GFM (2001) Cold current in thermoreceptive neurons. Nature 413:480
Roberts RA, Kavanagh SL, Mellor HR, Pollard CE, Robinson SPS (2014) Reducing attrition in drug development: smart loading preclinical safety assessment. Drug Discov Today 19:341–347
Roberts RA, Smith RA, Safe S et al (2010) Toxicological and pathophysiological roles of reactive oxygen and nitrogen species. Toxicology 276:85–94. https://doi.org/10.1016/j.tox.2010.07.009
Rohacova J, Marin ML, Martínez-Romero A et al (2008) Photophysical characterization and flow cytometry applications of cholylamidofluorescein, a fluorescent bile acid scaffold. Photochem Photobiol Sci 7:860–866. https://doi.org/10.1039/b806366d
Rupprecht R, Papadopoulos V, Rammes G et al (2010) Translocator protein (18 kDa) (TSPO) as a therapeutic target for neurological and psychiatric disorders. Nat Rev Drug Discov 9:971–988. https://doi.org/10.1038/nrd3295
Sadae W, Yu VC, Richards ML et al (1987) Expression of neurotransmitter receptors and myc protooncogenes in subclones of a human neuroblastoma cell line. Cancer Res 47:5092–5140
Sanguinetti MC, Tristani-Firouzi M (2006) hERG potassium channels and cardiac arrhythmia. Nature 440:463–469. https://doi.org/10.1038/nature04710
Schaub S, Wilkins JA, Antonovici M, Krokhin O, Weiler T, Rush DNP (2005) Proteomic- based identification of cleaved urinary β2-microglobulin as a potential marker for acute tubular injury in renal allografts. Am J Transplant 5:729–738
Schildknecht S, Pöltl D, Nagel DM et al (2009) Requirement of a dopaminergic neuronal phenotype for toxicity of low concentrations of 1-methyl-4-phenylpyridinium to human cells. Toxicol Appl Pharmacol 241:23–35. https://doi.org/10.1016/j.taap.2009.07.027
Schoonen WG, De Roos JA, Westerink WMDE (2005a) Cytotoxic effects of 110 reference compounds on HepG2 cells and for 60 compounds on HeLa, ECC-1 and CHO cells.: II Mechanistic assays on NAD (P) H, ATP and DNA contents. Toxicol Vitr 19:491–503

Schoonen WG, Westerink WM, de Roos JADE (2005b) Cytotoxic effects of 100 reference compounds on Hep G2 and HeLa cells and of 60 compounds on ECC-1 and CHO cells. I mechanistic assays on ROS, glutathione depletion and calcein uptake. Toxicol Vitr 19:505–516

Schubert D, LaCorbiere M, Klier FG, Steinbach JH (1980) The modulation of neurotransmitter synthesis by steroid hormones and insulin. Brain Res 190:67–79. https://doi.org/10.1016/0006-8993(80)91160-9

Schutgens F, Rookmaaker MB, Margaritis T et al (2019) Tubuloids derived from human adult kidney and urine for personalized disease modeling. Nat Biotechnol 37:303–313. https://doi.org/10.1038/s41587-019-0048-8

Secker PF, Luks L, Schlichenmaier N, Dietrich DR (2018) RPTEC/TERT1 cells form highly differentiated tubules when cultured in a 3D matrix. ALTEX 35:223–234. https://doi.org/10.14573/altex.1710181

Shao C, Li M, Li X et al (2011) A tool for biomarker discovery in the urinary proteome: A manually curated human and animal urine protein biomarker database. Mol Cell Proteomics 10:1–9. https://doi.org/10.1074/mcp.M111.010975

Shastry P, Basu A, Rajadhyaksha M (2001) Neuroblastoma cell lines-A versatile in vitro model in neurobiology. Int J Neurosci 108:109–126

Shen G, Shen F, Shi Z et al (2008) Identification of cancer stem-like cells in the C6 glioma cell line and the limitation of current identification methods. Vitr Cell Dev Biol Anim 44:280–289. https://doi.org/10.1007/s11626-008-9115-z

Simon-Friedt BR, Wilson MJ, Blake DA, Yu H, Eriksson YWJ (2015) The RPTEC/TERT1 cell line as an improved tool for *in-vitro* nephrotoxicity assessments. Biol Trace Elem Res 166:66–71. https://doi.org/10.1007/s12011-015-0339-y.The

Singer P (1975) Animal liberation: a new ethics for our treatment of animals. House, Random, New York

Sjögren AK, Breitholtz K, Ahlberg E et al (2018) A novel multi-parametric high content screening assay in ciPTEC-OAT1 to predict drug-induced nephrotoxicity during drug discovery. Arch Toxicol 92:3175–3190. https://doi.org/10.1007/s00204-018-2284-y

Sleno L, Varesio E, Hopfgartner G (2007) Determining protein adducts of fipexide: mass spectrometry based assay for confirming the involvement of its reactive metabolite in covalent binding. Rapid Commun Mass Spectrom. An Int J Devoted to Rapid Dissem Up-to-the-Minute Res Mass Spectrom 21:4149–4157

Smith AJ, Stone TW, Smith RA (2008) Preconditioning with NMDA protects against toxicity of 3-nitropropionic acid or glutamate in cultured cerebellar granule neurons. Neurosci Lett 440: 294–298. https://doi.org/10.1016/j.neulet.2008.05.066

Söderström S, Ebendal T (1995) *In-vitro* toxicity of methyl mercury: effects on nerve growth factor (NGF)-responsive neurons and on NGF synthesis in fibroblasts. Toxicol Lett 75:133–144. https://doi.org/10.1016/0378-4274(94)03176-8

Soldatow VY, Lecluyse EL, Griffith LG, Rusyn I (2013) *In-vitro* models for liver toxicity testing. Toxicol Res (Camb) 2:23–39. https://doi.org/10.1039/c2tx20051a

de Sousa C, Lopes SM (2019) Accelerating maturation of kidney organoids. Nat Mater 18:303–304. https://doi.org/10.1038/s41563-019-0326-3

Suñol C, Babot Z, Fonfría E, Galofré M, Garcia D, Herrera N, Iraola SVI (2008) Studies with neuronal cells: from basic studies of mechanisms of neurotoxicity to the prediction of chemical toxicity. Toxicol Vitr 22:1350–1355

Takasato M, Er PX, Chiu HS et al (2015) Kidney organoids from human iPS cells contain multiple lineages and model human nephrogenesis. Nature 526:564–568. https://doi.org/10.1038/nature15695

Tencer J, Bakoush O, Torffvit O (2000) Diagnostic and prognostic significance of proteinuria selectivity index in glomerular diseases. Clin Chim Acta 297:73–83. https://doi.org/10.1016/S0009-8981(00)00235-7

Teunissen CE, Khalil M (2012) Neurofilaments as biomarkers in multiple sclerosis. https://doi.org/10.1177/1352458512443092

Tirmenstein MA, Hu CX, Scicchitano MS et al (2005) Effects of 6-hydroxydopamine on mitochondrial function and glutathione status in SH-SY5Y human neuroblastoma cells. Toxicol Vitr 19:471–479. https://doi.org/10.1016/j.tiv.2005.01.006

Tirmenstein MA, Nicholls-Grzemski FA, Schmittgen TD et al (2000) Characterization of nitric oxide production following isolation of rat hepatocytes. Toxicol Sci 53:56–62. https://doi.org/10.1093/toxsci/53.1.56

Tong V, Teng XW, Chang TKH, Abbott FS (2005) Valproic acid II: Effects on oxidative stress, mitochondrial membrane potential, and cytotoxicity in glutathione-depleted rat hepatocytes. Toxicol Sci 86:436–443. https://doi.org/10.1093/toxsci/kfi185

Toutain HJ, Moronvalle-Halley V, Sarsat JP et al (1998) Morphological and functional integrity of precision-cut rat liver slices in rotating organ culture and multiwell plate culture: Effects of oxygen tension. Cell Biol Toxicol 14:175–190. https://doi.org/10.1023/A:1007458408863

Tsushima J, Nishimura K, Tashiro N et al (2012) Protective effect of planarian DJ-1 against 6-hydroxydopamine-induced neurotoxicity. Neurosci Res 74:277–283. https://doi.org/10.1016/j.neures.2012.09.003

Tumani H, Hartung HP, Hemmer B et al (2009) Cerebrospinal fluid biomarkers in multiple sclerosis. Neurobiol Dis 35:117–127. https://doi.org/10.1016/j.nbd.2009.04.010

Tuschl G, Lauer B, Mueller SO (2008) Primary hepatocytes as a model to analyze species-specific toxicity and drug metabolism. Expert Opin Drug Metab Toxicol 4:855–870. https://doi.org/10.1517/17425255.4.7.855

Ulrich RG, Bacon JA, Brass EP, Cramer CT, Petrella DK, Sun EL (2001) Metabolic, idiosyncratic toxicity of drugs: overview of the hepatic toxicity induced by the anxiolytic, Panadiplon. Chem Biol Interact 134:251–270

Vaidya VS, Ramirez V, Ichimura T et al (2006) Urinary kidney injury molecule-1: a sensitive quantitative biomarker for early detection of kidney tubular injury. Am J Physiol Ren Physiol 290:517–529. https://doi.org/10.1152/ajprenal.00291.2005

Vale C, Pomés A, Rodríguez-Farré E, Suñol C (1997) Allosteric interactions between γ-aminobutyric acid, benzodiazepine and picrotoxinin binding sites in primary cultures of cerebellar granule cells. Differential effects induced by γ- and δ-hexachlorocyclohexane. Eur J Pharmacol 319:343–353. https://doi.org/10.1016/S0014-2999(96)00866-7

Vriend J, Nieskens TTG, Vormann MK et al (2018) Screening of drug-transporter interactions in a 3D microfluidic renal proximal tubule on a chip. AAPS J 20. https://doi.org/10.1208/s12248-018-0247-0

Wada T, Yokoyama H, Tomosugi N et al (1994) Detection of urinary interleukin-8 in glomerular diseases. Kidney Int 46:455–460. https://doi.org/10.1038/ki.1994.293

Wan HI, Soares H, Waring J (2012) Use of cerebrospinal fluid biomarkers in clinical trials for schizophrenia and depression. Biomark Med 6:119–129

Wang H, Gao X, Fukumoto S et al (1998) Post-isolation inducible nitric oxide synthase gene expression due to collagenase buffer perfusion and characterization of the gene regulation in primary cultured murine hepatocytes. J Biochem 124:892–899. https://doi.org/10.1093/oxfordjournals.jbchem.a022204

Waring MJ, Arrowsmith J, Leach AR et al (2015) An analysis of the attrition of drug candidates from four major pharmaceutical companies. Nat Rev Drug Discov 14:475–486. https://doi.org/10.1038/nrd4609

Weinstein DE (1997) Isolation and purification of primary rodent astrocytes. Curr Protoc Neurosci 00:1–9. https://doi.org/10.1002/0471142301.ns0305s00

Weinstein DE, Wu R (1999) Isolation and purification of primary Schwann cells. Curr Protoc Neurosci 8:1–9. https://doi.org/10.1002/0471142301.ns0317s08

Westerink RHEA (2008) The PC12 cell as model for neurosecretion. Acta Physiol 192:273–285. https://doi.org/10.1111/j.1748-1716.2007.01805.x.The

Wilmer MJ, Saleem MA, Masereeuw R et al (2010) Novel conditionally immortalized human proximal tubule cell line expressing functional influx and efflux transporters. Cell Tissue Res 339:449–457. https://doi.org/10.1007/s00441-009-0882-y

Woolf CJ, Reynolds ML, Molander C et al (1990) The growth-associated protein gap-43 appears in dorsal root ganglion cells and in the dorsal horn of the rat spinal cord following peripheral nerve injury. Neuroscience 34:465–478. https://doi.org/10.1016/0306-4522(90)90155-W

Wooster R, Ebner T, Sutherland L et al (1993) Drug and xenobiotic glucuronidation catalysed by cloned human liver UDP-Glucuronosyltransferases stably expressed in tissue culture cell lines. Toxicology 82:119–129. https://doi.org/10.1016/0300-483X(93)02607-I

Worboys PD, Bradbury A, Houston JB (1996) Kinetics of drug metabolism in rat liver slices. II. Comparison of clearance by liver slices and freshly isolated hepatocytes. Drug Metab Dispos 24:676–681

Wu Y, Yang L, Su T et al (2010) Pathological significance of a panel of urinary biomarkers in patients with drug-induced tubulointerstitial nephritis. Clin J Am Soc Nephrol 5:1954–1959. https://doi.org/10.2215/CJN.02370310

Xu F, Farkas S, Kortbeek S et al (2012) Mercury-induced toxicity of rat cortical neurons is mediated through N-methyl-DAspartate receptors. Mol Brain 5:1–14

Xu J, Ma M, Purcell WM (2003) Characterisation of some cytotoxic endpoints using rat liver and HepG2 spheroids as *in-vitro* models and their application in hepatotoxicity studies. II. Spheroid cell spreading inhibition as a new cytotoxic marker. Toxicol Appl Pharmacol 189:112–119. https://doi.org/10.1016/S0041-008X(03)00090-5

Xu JJ, Diaz D, O'Brien PJ (2004) Applications of cytotoxicity assays and pre-lethal mechanistic assays for assessment of human hepatotoxicity potential. Chem Biol Interact 150:115–128. https://doi.org/10.1016/j.cbi.2004.09.011

Xu SY, Pauksen K, Venge P (1995) Serum measurements of human neutrophil lipocalin (HNL) discriminate between acute bacterial and viral infections. Scand J Clin Lab Invest 55:125–131. https://doi.org/10.3109/00365519509089604

Yarmush MLKHDJ (1989) Hepatocyte geometry : sandwich function and extracellular culture in a matrix configuration. FASEB J 3:174–177

Ydens E, Lornet G, Smits V et al (2013) The neuroinflammatory role of Schwann cells in disease. Neurobiol Dis 55:95–103. https://doi.org/10.1016/j.nbd.2013.03.005

You F, Osawa Y, Hayaski SI, Nakashima S (2007) Immediate early gene IEX-1 induces astrocytic differentiation of U87-MG human glioma cells. J Cell Biochem 100:256–265. https://doi.org/10.1002/jcb.21082

Yu J, Vodyanik MA, Smuga-Otto K et al (2007) Induced pluripotent stem cell lines derived from human somatic cells. Science (80-) 318:1917–1920. https://doi.org/10.1126/science.1151526

Yuan H, Zhang W, Li H et al (2013) Neuroprotective effects of resveratrol on embryonic dorsal root ganglion neurons with neurotoxicity induced by ethanol. Food Chem Toxicol 55:192–201. https://doi.org/10.1016/j.fct.2012.12.052

Zbinden G (1987) Predictive value of animal studies in toxicology. CMR annual lecture centre for medicines research

In Vivo Models for Evaluation of Drug Efficacy: Demand and Challenges

Somya Asthana, Vibha Shukla, and Anurag Tripathi

Abstract Drug screening for a disease may take more than 5 to 7 years and typically involves scanning of a number of molecules, acting as potential drug candidates. These drugs are tested for pharmacological as well as toxicological evaluations to uncover their therapeutic applications for a particular disease. These tests followed by clinical trials allow studying the safety, efficacy and pharmacology of that molecule. These tests are conducted on a wide range of in vitro and in vivo model systems. Though in vitro investigations are cost-effective and reduce experimental restrictions, these studies do not closely mimic the live scenario of the biological system. Therefore, in vivo testing is still mandated to comprehensively characterise the therapeutic behaviour of the drug. Further, to assess the human risk upon exposure to the drug, in vitro toxicodynamic results are associated with in vivo toxicokinetics using complex mathematical models. Despite significant breakthroughs in the field of in vitro test systems including 3D cell and organ culture techniques, the translational value of these systems is limited, and still mankind needs to traverse a long road ahead in this field. Undoubtedly, the information provided by in vivo models reflects a closer approximation to humans with respect to drug permeability, distribution, metabolism as well as excretion. In view of these facts, this chapter describes several contemporary animal models representing diverse diseases, employed for testing the efficacy of a drug.

S. Asthana
Food Toxicology Division, Food, Drug and Chemical Toxicology Group, CSIR-Indian Institute of Toxicology Research, Lucknow, India

Department of Biotechnology, Manav Rachna International Institute of Research and Studies (MRIIRS), Faridabad, India

V. Shukla · A. Tripathi (✉)
Food Toxicology Division, Food, Drug and Chemical Toxicology Group, CSIR-Indian Institute of Toxicology Research, Lucknow, India

Academy of Scientific and Innovative Research (AcSIR), Ghaziabad, India

V. S. Rajput, A. Runthala (eds.), *Drugs and a Methodological Compendium*,
https://doi.org/10.1007/978-981-19-7952-1_5

1 Drug Efficacy

Drug efficacy is a term used in pharmacology as well as in medicine where it relates to the maximal positive effect shown by a pharmaceutical drug against a disease. It can also be defined as the magnitude for a desirable and sufficient therapeutic effect of a drug. To know the efficacy of a drug, its effect is plotted against the number of doses in a graph, and the dose response curve is analysed (Galandrin et al. 2007).

Lets Assume that the increasing doses are shown on the X axis while the effect, i.e. half as well as maximal response, is shown on the Y axis; then the highest point seen on the curve displays the maximum positive effect or the efficacy of a drug at a particular dose, termed E_{max}.

As shown in Fig. 1, the two drugs show a different dose response effect at similar dose concentration, and it clearly explains that Drug 1 has a better therapeutic effect than Drug 2, with E_{max} greater with Drug 1 at a particular dose concentration.

Similarly in Fig. 2, we can see another example, where the 'increase in dose' for Drug 1 to reach the E_{max} from EC50 is much smaller than the dose increase required for Drug 2, showing the potency (effectiveness) of Drug 1 over Drug 2.

For a drug to show the expected effect, the drug needs to bind to the target (receptor protein), which in turn affects the function of the targeted receptor protein. Efficacy and affinity are two terms that quantify the nature of this interaction. On one hand, affinity is how tightly the drug binds to the receptor, whereas efficacy can be explained as a measure of the action that takes place once the drug binds to the receptor. 'Efficacy' must not be confused with the term 'effectiveness' as the latter defines how the drug works in the real scenario, which can be lower than the efficacy of a drug due to the health conditions of the patient or due to the drug interaction with other medicines (Galandrin et al. 2007).

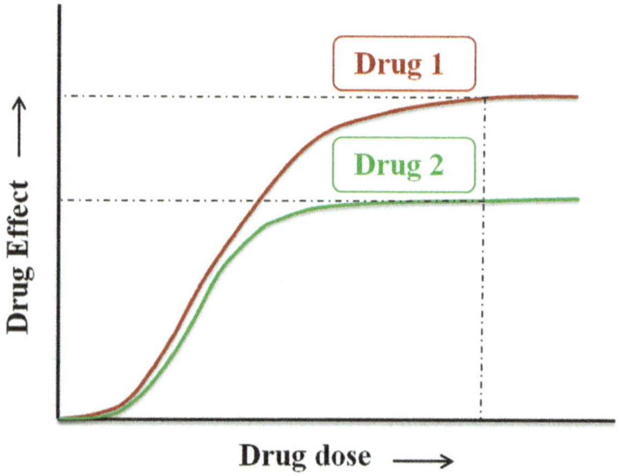

Fig. 1 A graph showing Drug 1 and Drug 2 displaying a different dose response effect at similar dose concentration

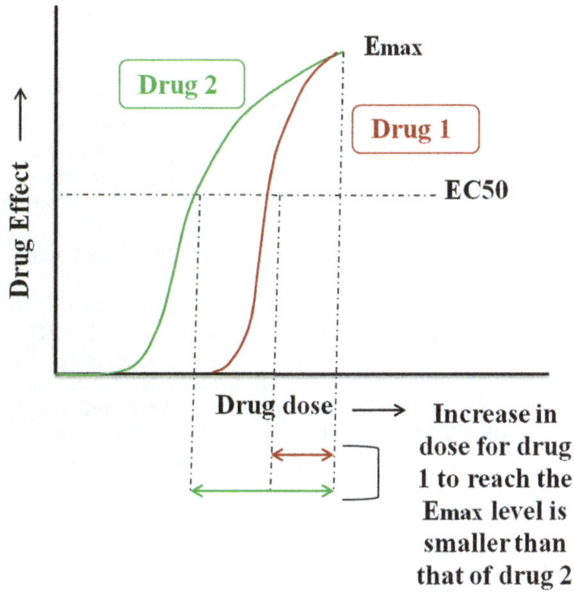

Fig. 2 A graph showing the potency of Drug 1 over Drug 2

2 Need for In Vivo Drug Efficacy Studies

Drug discovery and development involves a screening mechanism that takes more than 5 to 7 years where the drug candidates are screened for their toxicological and pharmacological effects. The new drug is tested for its therapeutic utilisation (Björklund et al. 2002; Lipsky and Sharp 2001; Martini et al. 2001). This pre-clinical step has two categories of studies, viz. in vitro studies and in vivo studies, the outcomes of which are utilised in drug research and development, followed by its clinical trials. These drug efficacy studies are performed on working cells (in vitro) or tissues removed from the living organism (in vivo), i.e. from laboratory animals (Pandey et al. 2010; Thompson 2000). Though the in vitro drug efficacy studies are short term and cost effective, there is still an obvious need for in vivo studies as they enable the evaluation of all the real-time drug characteristics that include biochemical and physiological parameters like drug–drug interactions and the side effects of a drug (Pandey et al. 2010; Thompson 2000).

Successful in vivo drug efficacy studies involve choosing a correct animal model that mimics the human system to the best. Choosing a correct animal model can help the investigators to monitor the bioavailability in the gastrointestinal (GI) tract due to the fasted or the fed state and can mimic the reticuloendothelial system (RES) uptake (Dressman et al. 2007; Shekhawat and Pokharkar 2017). Recent advanced techniques of in vivo drug efficacy studies have also made data collection easy, especially those related to metabolism and drug–drug interaction data before FDA approval (Andrade et al. 2016; Food and Drug Administration et al. 2019) (Fig. 3).

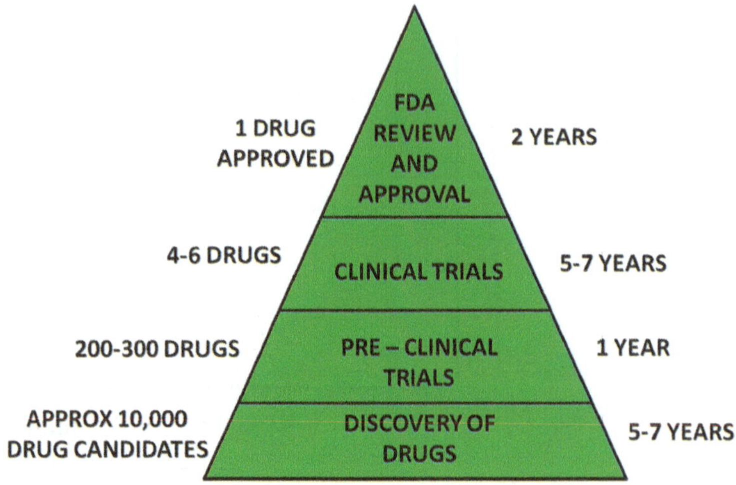

Fig. 3 Timeline from drug discovery and development to drug product approval

The appropriate selection of in vivo animal models helps in interpreting data that yield basic information and knowledge regarding pharmacodynamics of the drug and in improving the drug for future clinical trials. Then In vitro efficacy studies predict primarily the parameters affecting the release of the drug in vivo, while in vivo efficacy studies are important to know about the effects of the drug in a living organism (Hughes et al. 2011).

In vitro efficacy studies give useful information on the drug action mechanisms that play a significant role in the development process of drugs, but until the in vitro toxicity measurements are connected to the in vivo toxicokinetics, the results are irrelevant for human exposure, and thus the risk factor cannot be ignored (Sewell et al. 2017). There are many restrictions on in vitro since drug permeability and its dispersion cannot be easily studied due to its instability to precisely mimic the real and live biological system, where the role of the in vivo efficacy comes (Brake et al. 2017). The data from the in vitro experiments cannot be used exclusively for forecasting the interactions between the organs and their systems with the drug or the real-time interactions of the drug with other drugs (Brake et al. 2017).

Though the in vitro drug dissolution studies have been considered as one of the important steps in the drug efficacy studies, they cannot provide a quantitative clarification on the ADMET, i.e. absorption, distribution, metabolism and excretion tests of a drug, in the animal (pre-clinical) as well as human (clinical) models (Andrade et al. 2016; Singhvi and Singh 2011). Suppose, a drug has poor aqueous solubility, so the solubility factor will be a limitation for drug absorption (Stegemann et al. 2007). The in vitro dissolution study will describe how the medicinal ingredient will squeeze out with time indicating its efficiency of dissolution in vivo, but still it cannot predict any information on the absorption of the medicinal ingredient as well as on the drug–drug interaction within the living system (Bushra et al. 2011; Palleria et al. 2013; Stegemann et al. 2007).

It is still uncertain as to how the in vitro dose concentrations can be translated to the dosage and the pattern in which these drugs are exposed to animals and humans in vivo in natural conditions (Sewell et al. 2017). It is not easy to estimate the exact amount of the chemical or medicine, used in vitro, that reaches the site of action (Sewell et al. 2017). Additionally, all the parameters that influence the drug release and its dissolution cannot be determined through in vitro experiments as the use of cell lines does not permit long-term study of the effects caused due to the presence of a drug candidate (Cardot et al. 2007; Lanao and Fraile 2005).

To summarise, it is due to the difficulty in evaluation of safety, dose effect and its correlation, bioequivalence, side effects and drug–drug interaction in the live scenario through in vitro study translation, in vivo efficacy studies are needed. The in vivo results are multifactorial and explain the detailed effect of drug permeability, absorption, distribution, metabolism and excretion and provide data on pharmacokinetics and toxicological aspects (Andrade et al. 2016). Animal studies are hereby required to estimate the exposure to the drugs and to measure the degree of their toxicity (Palleria et al. 2013; Sewell et al. 2017).

3 Model Animals for Drug Efficacy Studies

The application of animal models in the in vivo research studies has played a major role in the progress of potential drug development, providing health benefits to human being. The animal models are important for the examination of drug absorption, distribution, metabolism and excretion through in vivo testing of the characteristics of the drug (Andrade et al. 2016). Moreover, testing of a drug using an animal model helps in detailing about the drug-to-drug interactions, side effects as well as its probable toxicities apart from studying the pharmacokinetic properties for the safety and efficacy of the drug, prior to human trials (Palleria et al. 2013).

Choosing an animal model entirely depends on the objective of the disease research. For validation of the drug target, the model must restate the phenotype of the disease, administer the similar pathophysiology as that of humans, acknowledge and counteract to the existing human treatments in a similar manner as that of patients (Ducharme et al. 2006). Those models that match the pathophysiology of a disease are very helpful in decoding the safety, toxicity, pharmacokinetics, pharmacodynamics as well as the biomarkers for the upcoming therapeutics (Sim and Kauser 2015). Additionally, such models showing similar characteristics play an important role in anticipating human doses for future clinical trials.

Depending upon the goal, the models may be (Chow et al. 2008; Davidson et al. 1987):

1. *Homologous*: Animal models that entirely resemble humans in pathology, physiology and treatments.
2. *Isomorphic*: Animal models are considered isomorphic when they have the same human disorder that is induced in them by any way.

3. *Predictive*: Animals models are not similar to any human disease; rather, these can be used only for the predictions of any disease or its treatments.

It is important to note that, for choosing an appropriate animal model, the animal ethics, its housing facilities, cost, handling, availabilities and its disease susceptibilities must be considered thoroughly (Chow et al. 2008; Ministry of Environment and Forests 2007).

The animal model used for testing of a drug efficacy can be either a vertebrate or an invertebrate. For traditional drug efficacy studies, the vertebrate animal models used are cows, dogs, guinea pigs, rats and mice. In some cases, baboons as well as macaques are used. On the other hand, an invertebrate animal model is chosen mostly for disorders like neurological, developmental or genetic disorders (Andrade et al. 2016; Chow et al. 2008). For example, for embryological study or for genetic disease study model, zebrafish is commonly used (Lieschke and Currie 2007; Zon and Peterson 2005).

Other parameters that must be focused in choosing an animal disease model is the probability of giving higher number of results and applicability of life cycle (Chow et al. 2008; Kari et al. 2007). For higher number of results, choosing an invertebrate is a better option as it can produce higher number of embryos at one time and at a lower cost (Kari et al. 2007). Moreover, animal models must be chosen on the basis of biochemical and physiological resemblance between the animal model and humans like that of blood pH, blood volume, drug transporter localisation, tissue distribution, residence of metabolising enzymes, etc. (Andrade et al. 2016; Chow et al. 2008; Tang and Prueksaritanont 2010).

4 In Vivo Disease Models

New drug targets are determined using in silico, in vitro and in vivo methodologies. Taking in vivo technologies into consideration, the rodents are majorly chosen for the drug target recognition and its validation. Among the rodents, the rats are generally preferred as a model for the organ transplantation, but rats become the only choice in case of the diseases like Heymann nephritis as the mice lack the antigens that are specific to this disease. On the other hand, mice are considered for the experiments that include gene knock-out. The animal studies rely on many factors like the strain of laboratory animals, age, gender, housing conditions, diet, breeding and handling of the animals. As the characteristics of humans (patients) are important for the clinical trials, it is necessary to know about the characteristics of the animals being used for appropriate population study.

A disease model can be described as the pathological condition that mimics human diseases appearing impromptu or is manipulated so to cause that disease. The disease can be induced either through a surgery, antigen or toxin injections. These disease models depend upon the species or strain, duration of their development, their relevance, technical feasibility and welfare of animals (Anders and Vielhauer 2007).

The occurrence of chronic diseases has been increasing worldwide, especially in the developing countries like India. According to the survey of the WHO, chronic diseases will account for about three-fourths of all the deaths that are occurring worldwide (WHO 1998). There are many diseases that are prevalent all over the world, and to combat the diseases, disease models have been developed through pre-clinical and clinical research trials, in order to find the appropriate medicinal solutions to them. The major and most commonly occurring disease and its in vivo disease models are discussed below in this chapter.

4.1 Gastrointestinal Diseases

For innovation, progress and validation of new remedy for gastrointestinal (GI) disorders, pre-clinical research has been considered very necessary. As discussed before, in vitro assays are important to know about the selectivity and specificity of the drug and receptor–ligand interactions as well as to study the structural activity of new drug compound (Johnson and Greenwood-Van Meerveld 2017). Still, the animal studies are important to elucidate the drug efficacy with the gastrointestinal system. There are many GI disease models that have been developed, though some have been translationally applicable, while others are still under process. It is understood that every animal species has its characteristics that affects its translational value (Johnson and Greenwood-Van Meerveld 2017). For example, rodents like rats, mice, and guinea pigs have almost similar GI tract as well as the nervous system of the GI area, as that of humans. The only difference is that rats do not have the gallbladder. Except that, rodents do not have similar laxative reflex and a definite caecum because of which the GI transport studies become quite problematic. The rodents are also preferred because of their small size as it helps in curtailing the assessment of the GI tract lengthwise. Rabbits like non-rodent species have an advantage over rodents in that these allow manipulations through surgeries, which is not possible in rodents (Johnson and Greenwood-Van Meerveld 2017). Moreover, the progress of disease can be monitored in non-rodents more easily through techniques like endoscopy. The disadvantage of the use of non-rodent species is its availability. Other animal models include dogs, cats or pigs can bear multiple or repeated experimental testings and also provide better regulatory insights (Johnson and Greenwood-Van Meerveld 2017). These animals can also be surgically manipulated so as to induce disease and permit to assess the GI tract regions through endoscopy or fistula. The non-human primates are rarely used as animal models for the GI disorders for reasons being the ethical concerns and their related costs for the GI disease-induced models (Johnson and Greenwood-Van Meerveld 2017). There are many different GI-related disorders, and they can be categorised into two: upper and lower GI disorders, and the major ones are discussed below.

A disease called *gastroparesis* is a condition where the muscles of the stomach get severely affected because of which emptying of stomach becomes difficult. Recent literature suggests the use of rat and mouse models so as to mimic the medical condition. Different agents like dopamine or dexamethasone, etc. are used for healthy mice to induce delay in the emptying of the stomach (Kimura and Sumiyoshi 2012). For some other conditions, type 1 diabetic mice or *knock-in* mice (mutant soluble guanylate cyclase) or TBH4-deficient mice or FORKO mice deficient in follicle-stimulating hormone (FSH) receptor are used (Creedon et al. 2013; Ravella et al. 2013; Welsh et al. 2013). After which, different assays are performed so as to measure the delay like phenol red test or solid food emptying test. The longitudinal variations in gastric emptying are studied through [13C]-octanoic acid breath test. Asano et al. used three different animal models, and deferred emptying was induced by clonidine, dopamine or m-CPBG. For example, in clonidine-induced delayed gastric emptying model, fixed amount of the agent was subcutaneously injected into the mice 5 minutes before administering phenol red. In others, dopamine or m-CPBG was given to mice intraperitoneally, 5–10 min prior to phenol red administration. In all the animal models, the drug to be tested was administered intra-gastrically one hour before phenol red administration (Asano et al. 2016). Phenol red is used to measure the gastric emptying for which the mice are fasted for about 18 h before phenol red was intra-gastrically administered. Few minutes later, mice were sacrificed and their stomachs were isolated (Asano et al. 2016). Treatment with $NaHCO_3$ was followed by the collection and centrifugation of the gastric content. The amount of phenol red in the supernatant is measured through microplate reader by measuring the absorbance at 558 nm. This was followed by some tests like gastric accommodation test, etc. (Asano et al. 2016). Similarly, numerous agents have been used to delay the emptying of the stomach and to mimic the conditions of gastroparesis in rats and mice, as tabulated below (Table 1).

Another serious condition of upper GI disorder includes *peptic ulcer disease (PUD)* where ulcers develop in the inner lining of the stomach. For this, ulcer-inducing models of mice and rats are used so as to evaluate the novel drug compound showing gastro-protective effect. Ulcers can be induced via acetic acid, HCl/ethanol, restraint stress, cold-resistant stress, acetylsalicylic acid, indomethacin, pylorus ligation, piroxicam, etc. (Ali et al. 2014; Balogun et al. 2015; Batista et al. 2015; Chang et al. 2015; Chen et al. 2016; Fatma et al. 2014; Júnior et al. 2014; Nordin et al. 2014; Silva et al. 2015; Wang et al. 2015; Zhao et al. 2015).

Animal was given an oral gavage of 70% to 100% ethanol, which can be given with or without HCl. The ulcer size is measured and compared with the control and the treated group (pre-treated with any drug). This assay had a limitation that the ulcers do not reoccur if healed fully (Balogun et al. 2015; Chang et al. 2015; Fatma et al. 2014; Júnior et al. 2014; Nordin et al. 2014). Few investigators induced ulcer through the treatment of indomethacin. For that, the rats were fasted for 24 h (no food, only water) prior to the administration of the chemical. The control rats were pre-treated with water, while treated groups were pre-treated with drugs. After one hour, all groups except the control group received the dose of indomethacin to induce gastric ulcer. Six hours after induction, the rats were sacrificed, and stomachs were removed for further histopathological tests (Chen et al. 2016).

Table 1 Animal models used to mimic gastroparesis along with the agents and assays performed for delaying the emptying of the stomach

Animal model used	Agents used	Assays performed	References
Healthy mice	Atropine, dopamine, serotonin or the 5-HT3 agonist, 1-(3-chlorophenyl) biguanide	Blue dextran test meal	Johnson and Greenwood-Van Meerveld (2017); Kimura and Sumiyoshi (2012)
Healthy mice	Administration of dexamethasone for three days	Phenol red test meal (direct); measurement of depletion of endogenous nitric oxide production (indirect)	Johnson and Greenwood-Van Meerveld (2017); Reichardt et al. (2014)
Mice (type 1 diabetic)	NOD LtJ	[13C]-octanoic acid breath test (repetitive)	Creedon et al. (2013)
Knock-out mice	Tetrahydrobiopterin (BH4) deficiency	Left-over stomach weight after feeding	Johnson and Greenwood-Van Meerveld (2017); Welsh et al. (2013)
Knock-in mice	Mutant soluble guanylate cyclase	Phenol red meal	Cosyns et al. (2013); Johnson and Greenwood-Van Meerveld (2017)
FORKO mice	Deficiency of FSH receptor	Measurement of decrease in BH4 production	Ravella et al. (2013)
Healthy rats	Acute administration of verapamil	Solid meal emptying	Li and Chen (2014)
Rats with acetic acid-induced gastric hypersensitivity	Acute administration of desvenlafaxine succinate	Solid meal emptying	Dai et al. (2013a)
Rats (Parkinson's disease induced)	6-Hydroxydopamine (6-OHDA) in the substantia nigra	Solid emptying and radiographic analysis of a barium meal	Johnson and Greenwood-Van Meerveld (2017); Song et al. (2014); Zheng et al. (2011)
Rats (Parkinson's disease induced)	6-Hydroxydopamine (6-OHDA) in the substantia nigra	[13C]-octanoic acid breath test	Johnson and Greenwood-Van Meerveld (2017)
Rat (type 1 diabetic)	Streptozotocin (STZ)	Plasma acetaminophen levels post-liquid gavage	Ali et al. (2012); Johnson and Greenwood-Van Meerveld (2017)

Further, restrain stress can also lead to ulcer in the GI tract. For this, animals were kept in a tube for 24 h (at room temperature or cold 4 °C temperature) such that movements of those animals were restrained (immobilised) or the animals were submerged in cold water for about 20 h. The ulcer sizes were measured and again compared with control and treated groups (Agotegaray et al. 2014). Besides, there have been different other approaches through which anti-inflammatory pathways have been studied, including the modification of synthesised compound structure so as to measure drug effectiveness (Agotegaray et al. 2014; Batista et al. 2015; Kudryavtsev et al. 2014; Rathore et al. 2014; Rossato et al. 2015).

Major lower GI disorders include *irritable bowel syndrome (IBS)*, an intestinal disease that causes severe pain in the stomach, diarrhoea, bloating, gas and constipation. Due to the lack of animal models, it is quite difficult to study the mechanism as well as the therapies in detail (Johnson and Greenwood-Van Meerveld 2017). Rodents with few experimental modifications can induce diarrhoea, but there are no other animal models still available to induce constipation or mixed bowel response, as observed in patients with IBS. In most of the researches, male animal models have been preferred over the females, but for the preference of using IBS female animal models, the reason is yet not studied. Besides these restrictions, many other animal models have been studied with modified visceral sensitivity and endow indications for the development of novel therapies for IBS. IBS can be induced through several methods including water avoidance stress, restraint stress, brain or spinal cord manipulation, post-inflammatory hypersensitivity and acetic acid enema (Johnson and Greenwood-Van Meerveld 2017). Colon section of the newborn pups was exposed directly to either chemicals causing inflammation or damage through balloon distension. Distention with high pressure or the use of chemical agent decides the degree of damage; in another method, i.e. in acetic acid enema, the test animal is given dose of dilute acetic acid causing inflammation (Johnson and Greenwood-Van Meerveld 2017). Colonic sensitivity is then studied through distension post administration of enema. There are few variations in different literatures using water avoidance stress, for which the animal is placed on a platform with water all around for one hour for 7–10 days. It induces hypersensitivity of the colon along with changes in the motility and permeability of the GI tract (Johnson and Greenwood-Van Meerveld 2017).

Inflammatory bowel disease affects the digestive tract, constituting two severe medical conditions, i.e. Crohn's disease and ulcerative colitis. For this purpose, animal models with the transmural inflammation are used. A drawback is that in such models, inflammation gets restrained up to the colon and may show patchy inflammation, which may depend upon the inflammatory agent used, as it can vary the markers of inflammation in tissue containing both healthy and inflamed areas. Moreover, gene modification (deletion) and transgene expression have been reported to cause spontaneous colitis in rat and mouse cell lines (HLA-β27). Animals usually develop colitis in adult stage and that too when incubated in dirty cages causing exposure to commensal microbiota. Apart from that, genetically modified rodents have also been used, which may show aggravated or undermined exogenously induced inflammations. In this, colitis is induced through agents like dextran sulphate sodium (DSS), 2,4,6-trinitrobenzene sulphonic acid (TNBS) or the bacterial or parasitic infections (Annahazi et al. 2012; Hosoya et al. 2014; Johnson and Greenwood-Van Meerveld 2017; Lin et al. 2011). Rodents were administered with DSS solution instead of water for about 3–10 days. Though the disease can resolve once the dose is stopped or discontinued, but to develop a chronic model, DSS is administered repeatedly. Initially, it involves only the colonic mucosa, but it can spread to the entire organ. The limitation for this methodology is that DSS can cause mortality if administered in higher percentage and can cause anaphylactic response in few strains of rats (Hosoya et al. 2014). On the other hand, TNBS in ethanol is given to the rodents to disrupt the mucosal lining of the colon (concentrations may vary). Inflammation causes necrosis or ulcer, which depends on the concentration of TNBS or ethanol (Annahazi et al. 2012).

Though animal models have been extensively studied in most of the gastrointestinal disorders, there is a lack of reproducibility in publishing of the pre-clinical studies as acknowledged by different international scientific journals (Freedman et al. 2017; Johnson and Greenwood-Van Meerveld 2017).

4.2 Cardiovascular Diseases

The heart is a complex organ and the cardiac functions involve complicated metabolic, immunologic, haemodynamic and neurological processes interlinked to each other (Sutton and Sharpe 2000). The mechanism of the associated cardiac diseases is not known. To get new therapies and to understand the mechanism behind the diseases and its solutions, the investigators need to mimic the entire complex working process. It is not easy to restate the complex human pathologies experimentally as per the cardiovascular pathophysiology, influenced by genetic and environmental factors, is concerned. For that, appropriate animal model is necessary with minimum translational complications. Literature suggests that lower organism like fishes and flies are capable of providing knowledge about the human heart development and function. It is because of the gene networks that are preserved between these evolutionary species (Olson 2006). Human genome has chromosomal homology with that of a zebrafish (*Danio rerio*) constituting orthologues of 70 percent of the human gene. Zebrafish reproduces numerous offsprings at a time and has external embryonic development, which makes it a suitable model for studying heart development, organogenesis and regeneration (González-Rosa et al. 2017). Due to the lack of translational relevance and for the need to study the processes taking place in a four-chambered mammalian heart, it became necessary to use higher animal models (Hearse and Sutherland 2000).

Small animal models generally include rodent models, the reason being their short gestation period, easier to handle and incubate, low maintenance and easier to manipulate genetically (Milani-Nejad and Janssen 2014). Different mouse models have been listed in Table 2.

The most common among these are myocardial infarction (MI) and cryogenic injury. MI model is a very common and frequently used model so as to mimic the heart attack in humans. To generate this condition, the neonatal mice are sedated via isoflurane induction chamber, after which they are kept over ice water to provide a short cooling period. Animals are bound to the cooling bag through a tape in a lateral

Table 2 Animal models and the methodology used for drug efficacy studies for different cardiovascular diseases

Animal model used	Medical condition mimicked (methodology)	Purpose	References
Mice/rat	Myocardial infarction (ligation induced)	To evaluate cardiac regeneration capability	Camacho et al. (2016)
Mice	Necrosis in the heart (incision in the abdomen via trans-laparotomy/cryogenic injury via thoracotomy)	Regeneration and remodelling investigation	Camacho et al. (2016)
Mice (engineered)	Dilated cardiomyopathy (deletion of actin-associated protein and ca sequestration)	To investigate the progress, progression and reversion of cardiomyopathy	Bullard et al. (2008); Unsöld et al. (2012)
Mice	Hypertrophic cardiomyopathy (knock-out of cMyBP-C gene/ overexpressing human myotrophin gene)	To investigate the progression of hypertrophy, leading to heart failure	de Lange et al. (2013); Sarkar et al. (2004)
Mice	Atrial fibrillation (knock-out of CM-specific liver kinase B1 (LKB1))	To evaluate the bi – Atrial enlargement along with atrial fibrillation and cardiac dysfunction	Ozcan et al. (2015)

position to provide hypothermic anaesthesia during the surgery. The skin of about millimetre length is isolated from the left foreleg, followed by opening up the thorax and ligating the left anterior descending artery. Slight colourlessness appears in the myocardium below the ligation. The thorax and skin are then closed, and the mice are provided heat through heated pad so that they recover from anaesthesia (Camacho et al. 2016; Dayan et al. 2012; Haubner et al. 2012; Porrello et al. 2013).

Cryogenic injury can be induced, either through open thoracotomy on the left ventricle or through abdominal incision on the right ventricle. In the first case, aluminium probe cooled with liquid nitrogen is used straight into the anterior region of the left ventricle for 10 seconds. This is followed by complete thawing for 10–-15 seconds causing the injured area to give deep red colour. More such exposures cause lesions to develop on the midventricular portion of anterolateral wall of the left ventricle (Grisel et al. 2008; Robey and Murry 2008). The second approach is through abdominal incision on the right ventricle. Mice are anesthetised, followed by opening up of the abdomen through transverse laparotomy. The diaphragm is exposed by raising the chest via divaricator. Heart surface is visible through the diaphragm and injury area can be chosen. Von graefe hook with blunt stainless steel tip is used after cooling it by liquid nitrogen to make an injury in the midventricular right ventricle. More such exposures increase the injury to form transmural lesion in the lateroapical portion of the wall of the right ventricle (Camacho et al. 2016; Grisel et al. 2008; Ren et al. 2008).

Apart from mouse models, there are many different rat models as well, like diabetic cardiomyopathy, ligation-induced myocardial infarction, cardiac hypertrophy (induced by overloading) and Duchenne muscular dystrophy (DMD) rat models. Moreover, there are transgenic strains of rats that include hypertensive rats and Goto–Kakizaki rats (type II diabetic rats) (Barretti et al. 2017; Gralinski et al. 2020; Nunes et al. 2017; van den Bos et al. 2005). For example, Duchenne muscular dystrophy (DMD) can be mimicked using rats and can be induced by microinjecting the TALE nuclease mRNA mixture in the zygotes of the rats, generating two generations of DMD rats. The lesions formed in heart and skeletal muscle in the discussed model closely resemble the scenario observed in the DMD patients (Larcher et al. 2014). For the testing efficacy of the anti-hypertensive drugs, a transgenic rat, defined as spontaneously hypertensive rat model, is chosen, or Dahl salt-sensitive rat model is generated by high salt diet administration to rats. There are other hypertension rat models based on stress-induced, pharmacological, renal and environmental hypertension conditions. Rats are preferred over mice due to the following: (1) The myocardium is of large size, which makes post-mortem histological, biological and molecular analysis much easier, and (2) invasive techniques and surgical procedures can be carried out easily (Camacho et al. 2016; McGonigle and Ruggeri 2014).

Atherosclerosis is a condition where high fat diet is administered to the animals to generate models, yet all experimental models like rats and mice are not admissible for the same, as there are variations in their intrinsic genes and they are resistant towards the formation of fat deposits in tissues (atherogenesis). For this, genetically modified models were produced like genetically modified ApoE- and Ldlr-deficient mice have been known to have high levels of plasma cholesterol levels (Badimon 2001). Moreover, the use of transgenic models has been recently reported for the development and deposition of plaque, calcification in the arteries, haemorrhage, ulcer formation, thrombosis, burst of plaques and stenosis (McGonigle and Ruggeri 2014). Therefore, for such experimental objectives, many genetically modified animal models were used like LDL receptor knock-out mice, fatty Zucker rats and cholesteryl ester transfer protein transgenic rats (Liao et al. 2017; Lum-Naihe et al. 2017). Cost being the limitation, these genetically modified and transgenic models have been replaced by adult female Sprague Dawley rats (with ovary removed). These ovariectomised female rats were administered with heated vegetable oil (15% w/w), resulting in the suppression of resistance to atherosclerosis. On the hand, male rats develop atherosclerosis, if similar diet is given for 16–24 weeks continuously (Jaarin et al. 2011; Leong et al. 2008).

Rabbits are considered as medium-sized models, as their myocardia show much resemblance to humans than the smaller rodents, as discussed above. For cardiovascular drug efficacy studies, different types of rabbit models, known as Watanabe heritable hyperlipidaemic myocardial infarction (WHHL-MI) models, have been generated (Fan et al. 2015). These models can be obtained through selective breeding of coronary atherosclerosis-prone WHHL-MI rabbits. As the stenosis at the lumen area of the coronary artery is increased in these rabbits, the occurrence of spontaneous WHHL-MI increases (Shiomi et al. 2003).

Since there are significant divergence in heart rate, consumption of oxygen, architecture of heart, ability of the heart muscle to contract, protein expression, etc. between human and that of rodents, large animals were also employed for cardiovascular drug efficacy studies. Dogs, pigs, etc. show greater resemblance in physiology, structure and function of the heart with humans. There are some examples that include the study of coronary artery micro-embolisation, myocardial infarction and ischemic cardiomyopathy using a canine model (Yue et al. 1999), though there are drawbacks that also exist in the collateral coronary circulation causing variation and discrepancy in myocardial lesions and responses after the injury (Lindsey et al. 2018). Besides, the ethical issues are the other concerns that limit the use of the canine models. Therefore, other animals like porcine animal models have been preferred due to the similar lesion size and anatomical match with humans, especially in the anatomy of arteries and coronary circulations (Montgomery et al. 2017; Mukherjee et al. 2003). Porcine models have been used to induce infarction by inflating a balloon via a catheter in the artery of the femur, for testing balloon blockage in the left anterior descending artery. This model has been found useful in studying the expansion of the infarct and the pharmacological effects associated with it. Moreover, it has also been used to elucidate the effect of transplantation of the stem cells on the contractile function of the left ventricles, distribution of different heart-associated (arteriogenic and angiogenic) growth factors (Lu et al. 2007; Zeng et al. 2007).

There are many animal models with different advantages and disadvantages, as listed (Table 3). Even after the availability of so many animal models for the cardiovascular drug efficacy studies, these yet need special surgical equipments, amenities and expertise. There are several animal models like dog, sheep, pig and

Table 3 Advantages and disadvantages of using small and large animal models for cardiovascular drug efficacy studies

Animal models	Advantages	Disadvantages
Small animal models	Easy to handle and breed	Gross anatomy is different from humans
	Not so costly	Lipoprotein profile is also different from humans
	Distinct genome	Thrombosis as plaque rupture absent
	Genetic modification is easy	Resistant to arthrosclerosis development
	Reproductive cycle is short	
Large animal models	Anatomically much similar to humans	Not cost effective
	Lipoprotein profile is also similar to humans	Difficult to handle and maintain
	Plaque rupture causing rare thrombosis	No genetic modification
	Preferred for translational research studies	Availability issues present
	Structural resemblance in vascular lesions or injury	More ethical problems
	Sensitive to arthrosclerosis development	

non-human primate models for various diseases like myocardial infarction, ischemic heart failure, non-ischemic heart failure, ligation-induced MI, sudden cardiac death, balloon catheter procedures, etc.

For the MI in dogs, general anaesthesia is followed up by opening up of the chest via left, fourth intercostal space. The pericardium is cut open for heart exposure, followed by the ligation of the proximal left anterior descending (LAD) coronary artery. Canine atrial MI is rarely diagnosed and generally goes undetected (Lai et al. 2000; Wei et al. 2007). For ischemic heart failure, chronic infarction is combined with small vessel coronary disease to generate moderate left ventricle dysfunction. In dogs, MI can be induced via ligation of LAD coronary artery. Further, using a coronary catheter, the left main coronary artery was engaged. The artery was recognised and latex beads (sonicated) were gradually infused into it. This procedure was recurring weekly in order to reach left ventricle ejection fraction, ultimately causing chronic ischemic heart failure (Adamson and Vanoli 2001). To mimic non-ischemic heart failure, induction of ventricular dysfunction is done in the right ventricle via the right ventricle tachypacing (TP). Grown-up dogs are fixed with pacemakers that have the leads placed at the apex of the right ventricle. TP is maintained at 180 beats per minute (bpm) for 14 days, followed by 200 bpm for 42 days, then 180 bpm for 60 days and 160 bpm for 180 days, and after that, 120 bpm was maintained, causing arrhythmogenesis during heart failure (Belevych et al. 2011).

There are few limitations in the cardiovascular in vivo drug efficacy studies. Firstly, each species has different demands, which make their cardiovascular system to evolve differently. Secondly, different animal models may have similar range in blood pressure, but the heart rates vary widely among them. Additionally, the orientation of the organ and arrangement of the large vessels of the heart differ in human from those of large animals. The imaging and any delivery of device are difficult, especially in pigs and sheep as the orientation of the heart is along the cranio–caudal axis of their body, close to the sternum. Other major drawbacks are the cost, availability of the animals and fatality issues of animals in some conditions.

4.3 *Neurodegenerative Diseases*

Neurodegenerative diseases are the diseases that involve a gradual deterioration in the structure and function of the central and the peripheral nervous system. These include Alzheimer's disease (AD), Parkinson's disease (PD), amyotrophic lateral sclerosis (ALS) and frontotemporal dementia (FTD). As per literature, no significant genetic reasons have been detected, yet the diseases have been studied through gene mutations. These genetic indications along with the recognition of biochemical proteins (tau, amyloid-β, TDP-43 and α-synuclein) have made it feasible for the investigators to study the pathophysiology of neurodegenerative disorders and to generate appropriate animal model for the same (Dawson et al. 2018).

Though, there is no such model as an ideal animal model present for studying such disease conditions, which could actually mimic the entire condition in humans, yet initial pathology of proteins associated with or the study of early pathological characteristics can be repeated in some models. The animal model studies have provided a better insight into the cellular and molecular sequential events, which ultimately leads to cellular degeneration and dysfunction. Among all animal models, mouse models do not provide reliable prediction for the efficacy of drugs. The available animal models do not have such complex circuits of neurons and glial cells, lack maturation and do not contain such immunological and vascular protein components. For this, there is a need of an engineered animal model that can be modified as per the pathological need and can mimic the neuronal complexity of humans (Liu et al. 2018).

Alzheimer's disease (AD), defined as a neurodegenerative disorder, causes the brain cells to deteriorate and die gradually. It leads to dementia, declination in behavioural and social skills and the ability of a person to think and remember (Schachter and Davis 2000). There has been use of transgenic methodologies in mice, which targets proteins like presenilin, APP and tau or AOPE gene. Besides, vector (*virus*)-based animal models have also been generated in rats and mice. Non-genetic models (*toxic models*) have been generated but not so preferred due to lack of specificity (JNPD 2014). There are few databases like Alzforum that provide entire lists of animal models as per the symptoms and proteins to be targeted (Kinoshita and Clark 2007).

There are basically three major types of pathologies that include neurodegeneration in the cortical or hippocampal region of the brain, neurofibrillary tangles and senile plaques. In neurofibrillary tangles, it is the fibrillar and hyperphosphorylated forms of tau proteins that are found not only in AD but also in FTD (Holtzman et al. 2016). On the other hand, senile plaques are specific to Alzheimer's disease, causing aggregate formation of Aβ fibrillars. AD has early onset with mutations in APP, i.e. an amyloid precursor protein, PSEN1 and PSEN2 or Presenilin 1 and Presenilin 2, respectively. The late-onset period of AD includes APOEe4 allele as the major threat factor (Carmona et al. 2018).

As discussed above, there are more genetic-based models for different pathologies of AD. In amyloid pathology, the deposition of amyloids in senile plaques and in cerebrovascular region has been modelled by the transgenic models of rodents that stimulate the accumulation of the Aβ aggregates. Even the mouse models have similar functions of Alzheimer-associated mutations in APP, PSEN1 and PSEN2 as that of humans. The mutants of APP cause increment in overall Aβ, or it can cause corresponding generation of Aβ42, which is more prone to aggregation (Ashe and Zahs 2010; Golde et al. 2011; LaFerla and Green 2012; Price et al. 1998; Sasaguri et al. 2017). PSEN1 and PSEN2 mutations do not result in the deposition of the amyloids but can cause variation in internal processing of APP of mouse. If the PSEN mutants get co-expressed with that of APP, it tremendously increases the production of Aβ42 (and Aβ43 in some cases), in turn causing hastening of amyloids (Haass and De Strooper 1999).

There are other mouse models, including the knock-in models, that cause generation and accumulation of Aβ without any overexpression of APP. These models show little or no abnormality in their behaviour, which corresponds to pathology of amyloid development and deposition. These models have resemblance with the early/pre-clinical Alzheimer's disease, yet these are not complete AD models (majorly for behavioural dysfunctions), rather engineered to mimic Aβ accumulation (Kim et al. 2013; McGowan et al. 2005; Saito et al. 2014).

As per the tau pathology is concerned, there are transgenic models that are used for the development of neuronal tau inclusions, which are generated via over-expression of mutations, leading to FTD (affecting chromosome 17, FTD–MAPT), though the tauopathy linked to MAPT gene is still in question for its relevance with tauopathy found in FTD. The tau inclusions are indeed very much similar to the Pick bodies, rather than that of NFT pathology (LaFerla and Green 2012).

Parkinson's disease can be explained as the consistent loss of the dopamine in the substantia nigra and as the misfolding of α-synuclein present in Lewy bodies and neurons all over the entire nervous system. There are few genetic-based models that include germline LRRK2 transgenic models that have G2019S over-expression or R1441C/G gene mutation and have different degrees of dopamine abnormalities but still not widely used as they do not have age-related dopamine neurodegeneration (Jankovic and Tan 2020).

Gene LRRK2 and LRRKK1 knock-out results in age-related dopamine neurodegeneration where loss of gene LRRK2 may cause degeneration. It has been observed that when PINK1 gene or parkin gene was deleted from the adult mice, there was strong dopamine degeneration. Moreover, when germline *knock-out* was done for parkin, PINK1 and DJ-1 gene, there was no neurodegeneration in mice. Besides the germline *knock-out* of PINK1 and DJ-1 in rats, cause loss in DA neurons (Creed and Goldberg 2018). Even when c-terminal curtailed human mutant parkin gene was overexpressed, progressive depletion of dopamine was observed in mice. Therefore, it is clear that there are methodologies developed to study the molecular mechanism behind the neurodegeneration in Parkinson's disease through mutation in α-synuclein or that in LRRK2 gene or deletion/knock-out of parkin, PINK1 or DJ-1 gene (Creed and Goldberg 2018).

There are other neuro-diseases like amyotrophic lateral sclerosis (ALS) and frontotemporal dementia (FTD) that are also intensely studied through C9orf72 gene development. The accumulation of the expansion-related transcripts and build-up of dipeptide repeat proteins (DPRs) that are translated from the C9orf72 expansion have been reported to cause increase in the related toxicity. Mouse models specific to this gene have been generated and characterised showing that low levels of C9orf72 may contribute but are insufficient to induce neurodegeneration. Recently, non-rodent models are also being considered, including *Drosophila, C. elegans*, canines and yeast models (Shi et al. 2018; Wen et al. 2017).

4.4 Cancer

As the field of medicine is advancing, the animal research has been increasing in various treatments, including the cancer treatment. Future progress of cancer treatment depends on the in vivo animal experiments; however, no model has been completely satisfactory for predicting all types of human cancers, especially the malignant ones due to the difference in its affecting pathways and mechanisms. Each kind of tumour acts as a separate entity, having its own appearance, behaviour and response to drug (Corbett et al. 2004). The use of transplantable animal tumours has made it easy for the identification of the anticancer agents and has played a very important role in developing therapeutic drugs for the same (Bibby 1999; Corbett et al. 2004; Dykes and Waud 2002). There are three classic in vivo animal models that are used in experiments involving transplantable experimental tumours. These are as follows:

1. Murine P388 leukaemia model (implanted intravenously).
2. Murine B16 melanoma model (implanted subcutaneously).
3. Human tumour xenografts for lung (LX-1) and breast (MX-1) tumours (implanted subcutaneously).

Earlier, the P388 leukaemia model was induced chemically through the application of a carcinogen, methylcholanthrene, on the skin of DBA/2 mouse (Dykes and Waud 2002). Later on, the original leukaemia cells were implanted through the veins in the same host environment of the C2DF1 mice, where the disease initially developed. This initiates the maintenance of the tumours for further studies. The original P388 leukaemia cells stored in liquid nitrogen are thawed rapidly and made into suspensions using diluents, majorly 0.9% NaCl. The DBA/2 female mice were weighed, and the suspensions were implanted into the peritoneal cavity of each mouse. Seven days later when the leukaemia cells propagate, the DBA/2 mice are sacrificed. Using the diluent, few microlitres of cell suspension were isolated from the ascitic fluid and used for further testing and viability. Further, for the chemotherapy experiments, the cell suspensions were taken through the ascitic fluid from donor DBA/2 mice and implanted to the C2DF1 mouse. Moreover, for evaluating the anti-tumour activity of a drug, the increase in the life span and change in the body weight are calculated (Dykes and Waud 2002).

There are some other in vivo animal model which are used, thus including thawing of frozen murine B16 melanoma tumour, followed by its sectioning. Each section is then implanted in the right flank of a female host C57BL/6 mouse (Geran et al. 1972). These mice are considered as the donor and are used to start the experiment. After 14 days, when the B16 tumour grows, the donors are sacrificed for the isolation of another tumour fragment. The fragment is then further implanted into the right flank of another host mouse. The transplantation procedure is repeated, and after five transplantations, the tissue sample is removed from the donor mouse once the implanted tumour reaches its size of 500 mm^3. The sample is then further taken for histopathology or further experimentation like antitumor activity study or to elucidate the assessed effect of test compound on tumour growth, etc. (Langdon et al. 1994).

There are few other examples reported for subcutaneous implantation of human tumour xenografts: one is for lung (LX-1) and the other is for breast (MX-1) tumours. The reason for this model generation was to study the antitumor effect of the drug target or anticancer agent. For this purpose, the homozygous female athymic nude mice were used for the implantation of LX-1 or MX-1 tumours. Similar protocol is followed in this case, as discussed in the B16 melanoma tumours. Here, when the tumour growth reaches the exponential growth phase, the nude mice are sacrificed when the tumour reaches its size of about 10 mm (generally after 3–4 weeks). These steps are followed by multiple transplantations, before it is isolated for the histological analysis or further experimentation (Kruczynski et al. 1998; Langdon et al. 1994).

These are the classical protocols used for the anticancer studies. These protocols form the base of the present modified protocols. There are multiple studies involving modifying protocols. These animal disease models of cancer include a broad spectrum having different types (Ruggeri et al. 2014):

1. Ectopic xenografts (subcutaneous, intramuscular, intraperitoneal or intravenous) of tumour-derived cell lines or tissue explants, which have been implanted in other immunocompromised rodents.
2. Orthotopic models where the tumour explants are implanted within the tissue or particularly organ.
3. Germline transgenic models where the tumour suppressor genes can be modified by regulating its expression patterns.
4. Primary human tumour grafts that maintain the phenotype and the genotype of the primary tumour.
5. Other carcinogenic models that mimic the multiple stages of tumour progression for the study of pathogenesis generated in response to carcinogenic agents.

4.5 Arthritis

Rheumatoid arthritis (RA) and osteoarthritis (OA) are two commonly found diseases, under the category of arthritis. RA is an autoimmune disease with chronic inflammatory effects on the joints of the hands and the legs, specifically the feet. Though it is generally found in the hands and legs, but it can affect other joints, thus causing systemic complications (Smolen et al. 2016). RA can be characterised by inflammation in the synovial membrane along with increase in the cellularity (hyperplasia) in synovial membrane, erosion in bones and cartilage, etc. (Scott et al. 2010). This kind of severe inflammation and gradual destruction in tissues lead to structural deformities that are irreversible and cause possible functional loss in the affected joint (Alarcon et al. 2015).

Extensive studies have been done in this field along with the study of animal models for detailed RA pathogenesis for the development of therapies. The most commonly used animal species used are mice and rat. There are spontaneous and induced types of arthritis models. The spontaneous models of RA include KxB/N, SKG, TNF-tg, IL-1-tg, IL-1Ra −/− models. On the other hand, the induced models include collagen-, adjuvant-, oil-, pristane-, proteoglycan-, antigen- and streptococcal cell wall-induced arthritis models. Besides them, collagen antibody transfer model, KRN serum transfer model and ova TCR transfer models are also present in the induced arthritis category (Vincent et al. 2012). For human RA, collagen-induced arthritis (CIA) in mice and adjuvant-induced arthritis (AIA) in rats are preferred. It is commonly used and can be initiated in DBA/I mice through intradermal injections of emulsions made up of bovine collagen type II in complete Freund's adjuvant (CFA). This is followed by a booster dose of collagen after 21 days and is administered intraperitoneally, thus developing chronic arthritis majorly affecting the hind paws of the mouse. CIA of mice resembles that of human RA in many clinical, immunological and histological features. Moreover, the production of the rheumatoid factor has been reported (Courtenay et al. 1980). Later, a protocol was also developed for inducing CIA in C57BL/6 mice. Two intradermal injections of emulsion of chicken type II collagen in CFA were administered on 0th day and 21st day, causing the development of arthritis. The disadvantages are that it is less severe and has less swelling in paws as compared to that observed in DBA/I mice (Campbell et al. 2000).

In rats, three collagen-based proteins have been known that can initiate arthritis, which are collagen type II, collagen type XI and cartilage oligomeric matrix protein. Collagen type I and collagen III induce immune reactions but do not cause the development of CIA in rats as well as mice (Stuart et al. 1984). The most common model type is collage type II in which an emulsion of collagen is prepared in Freud's incomplete adjuvant. It develops two weeks post immunisation, leading to severe polyarthritis in DA and Lewis rats. Arthritis in rats can be characterised by swollen paws (both forelimbs and hind limbs), which can stay for weeks and decrease slowly. It can cause chronic arthritis by reappearing with ultimate destruction of the bones (Duris et al. 1994).

Besides, the adjuvant-induced arthritis (AIA) has been studied in Lewis rats where CFA is injected at the base of the tail of the rats (intradermally) using *Mycobacterium tuberculosis* H37 RA. The rats were daily evaluated for the swelling in paws using plethysmometer and through the signs of inflammation in the right ankle joint after the intra-articular injections. The rats were anaesthetised using methoxyflurane, followed by 31-gauge needle insertion anterolaterally into the right ankle joint. Paw swelling starts around day 10. The treated rats were sacrificed after seven days through carbon dioxide necrosis (Tak et al. 2001). In this, knowledge of genetic environment of the rats is necessary as the Major Histocompatibility Complex (MHC) and non-MHC of both types of genes take part in causing susceptibility to AIA (Kim and Moudgil 2009).

Apart from spontaneous and induced model, arthritis can also be developed through genetic modification in mice, which is transgenic mouse models (Asquith et al. 2009; Kannan et al. 2005). There are two well-identified transgenic models, K/BxN and human TNF-transgenic mouse models. In the former model, K/BxN mice is generated after crossing KRN-C57BL/6 T-cell receptor-transgenic mice with NOD mice. In this model, arthritis is developed through immunisation of mice with recombinant human glucose 6-phosphate isomerase (rhu G6PI). The antigen was mixed with CFA in 1:1 ratio and emulsified. The emulsion was injected subcutaneously at the base of the tail on each side. Monitoring of the clinical index was done at regular time interval for both paws (Schubert et al. 2004). This induces T-cell-dependant arthritis in genetically unaltered mice. For the latter, human TNF-transgenic mouse models, Tg197 mice express increased levels of soluble transmembrane human TNFα and develop arthritis spontaneously. Symptoms like synovial hyperplasia and inflammations are visible from weeks 3–4, and the mice develop the disease completely by the age of 10 weeks. All the symptoms of RA are observed like pannus formation, cartilage destruction, bone resorption, etc. (Keffer et al. 1991; Probert et al. 1996). In this model, a targeting vector containing the genomic fragment with entire human TNFα gene encoded in it was used. The fragment contains AU rich element (ARE) containing 3'UTR that was replaced with 3'UTR isolated from β-globin gene (Li and Schwarz 2003). Thus, therapies that are efficient in models of arthritis may do not succeed in RA. Still, these models remain a very vital means for pre-clinical testing of new therapeutic drugs. Deep concern as to which animal model should be used is needed. It ought to be understood that therapeutic drugs that are stopped due to unsatisfactory outcomes in animal models of arthritis might still be effective in decreasing inflammation and bone damage in patients with RA.

4.6 *Infectious Disease*

There are many infectious diseases caused by different pathogens that include bacteria, virus, parasite or a fungus. Microbial infections are one of the important threats that arise due to the increase in the microbial resistance to the already existing anti-microbial drugs. This is the major reason for better drug efficacy testing methodologies and for development of new anti-microbial medicines. There has been an increase in interest for the breakthrough and progress of new anti-microbial drugs as all the antibiotics lose its efficacy due to the microbial resistance.

There are some traditional in vitro methods or antimicrobial bioassays like well diffusion, agar dilution, disc diffusion method, time kill test, TLC bioautography or flow cytometry using plating agar medium with different modifications for testing the drug's efficacy (Balouiri et al. 2016; Barry 1999; Choma and Grzelak 2011; Favre-Godal et al. 2013; Hamburger and Cordell 1987; Klepser et al. 1998; Magaldi et al. 2004; Valgas 2007). One of the most common methods for antifungal assays is the poisoned food technique (Li et al. 2016).

To test the drug's efficacy, in vivo traditional methods were used to localise as well as to quantify the pathogens in the animal models by sacrificing the animal, taking out the tissue, followed by counting the number of colony-forming units. Now, after the discovery of the light-generating enzymes, called as luciferase, a methodology called as bioluminescence imaging had made it very easy to study the infections and its probable recovery after the application of drugs. The principle behind the process includes the model animal that is infected with the microbial strain expressing the bacterial luciferase system, where the light generation is proportional to the concentration of the microbes (Dai et al. 2011; Huang et al. 2014; Witcomb et al. 2017). The light can be measured using spectrophotometer

(490 nm for *P. luminescens* variant), and the bioluminescence signal forms infections in small animals, or the animal models can be visualised using highly sensitive CCD camera. Moreover, by sacrificing the model organism and removing the tissue, followed by weighing and homogenising the samples so as to calculate colony-forming units, can help in correlating the bioluminescence imaging with the CFU values (Dai et al. 2011; Huang et al. 2014; Witcomb et al. 2017). Using bioluminescence imaging (BLI), various infections can be studied, which can be summarised below.

For external wound infections, animal models generally used are male or female Balb/c mice. Different bioluminescent microorganisms are used. For dermal abrasions, photodynamic therapy was done using phenothiazinium salts and red light for which the superficial skin was scrapped using a scalpel blade and the effected red injured area is inoculated with the bioluminescent *C. albicans* (Dai et al. 2011). Male Balb/c mice are used as dermal excision wound model along with *E. coli* as the bioluminescent microorganism where complete transdermal excisional wounds are made on the dorsal surface of mice followed by photodynamic therapy with pLce6 conjugate and red light (Hamblin et al. 2003). For the burn wounds, male and female Balb/c mice are used using different microorganisms like *Acinetobacter baumannii*, *S. aureus*, *E. coli*, *Pseudomonas aeruginosa* or *C. albicans* for all the dermal or even the third-degree burn wounds that were created on the dorsal surface of mice or infected with *S. aureus*. These methods were used to test the effectiveness of UVC light treatment (against *C. albicans* infection), blue light treatment (against *P. aeruginosa*) or antimicrobial phytodynamic theory mediated by meso-mono-phenyl-tri (N-methyl-4 pyridyl)-porphyrin, monitored by BLI to treat burn wounds (Dai et al. 2013b; Dai et al. 2011; Golberg et al. 2014; Huang et al. 2014; Lambrechts et al. 2005). There are some other bacterial infections that include osteomyelitis, UTI, mycobacterial Buruli ulcer, gastrointestinal and endodontic infections, bacterial pneumonia, meningitis and lung infection. These use different animal models and can also be observed using BLI.

For osteomyelitis, generally mouse models are preferred where *S. aureus* or *Listeria monocytogenes* are taken as the microorganism. Investigators infected the mouse model with *S. aureus* in the femur monitored using BLI. This model was found to be very useful for testing the pathophysiology of acute as well as chronic osteomyelitis and to evaluate the efficacy of new antibiotic drugs (Hamblin and Hasan 2004). Others developed *Listeria* bone infections, which showed that bacteria can grow in various locations causing localised listeriosis (Hamblin et al. 2003; Hardy et al. 2009). For gastrointestinal infections, C57BL/6 mice were orally inoculated by *E. coli* (Enteropathogenic Escherichia coli (EPEC) and Enterohemorrhagic Escherichia coli (EHEC)), which showed that EPEC stays longer if mice are given ketamine or xylazine through anaesthesia (Rhee et al. 2011). Another 2-day-old Wistar infant rat was fed with *E. coli K1* A192PP-lux2 bacteria through a dropper where *E. coli K1* enters into the systemic circulation through the oesophagus (Witcomb et al. 2015). The site of infection includes the tongue, oesophagus, stomach and intestine, followed by the brain and whole body.

For *Mycobacterium tuberculosis*, the intranasal inoculation of the bacteria is done in CB-17 SCID mice, and the efficacy of the isoniazid therapy was elucidated (Andreu et al. 2013). On the other hand, for bacterial pneumonia affecting the lungs, female Balb/c mice were inoculated by *S. pneumoniae* (A66.1-encapsulated strain) intranasally by some investigators and found improvement in therapy for severe pneumonia (Ghoneim and McCullers 2014). As per other lung infections are concerned, cystic fibrosis animal model C3H/HeN mice were infected by *P. aeruginosa* TBCF10839 and D8A6 mutant through intra-tracheal instillation, and it was found that the former pathogenic strain caused 50% mortality while attenuated latter strain allows monitoring of infection (Munder et al. 2014). Infections like meningitis mainly use transgenic mice CD46 inoculated intravenously by *Neisseria meningitides* affecting the whole body, spinal cord and brain (Sjolinder and Jonsson 2007). There are literatures that suggest that *S. pneumoniae* A 66.1 inoculation into cistern magna of immunocompetent hairless mice also causes meningitis, while dexamethasone along with daptomycin or vancomycin gave very good and recovering results (Mook-Kanamori et al. 2009). Besides the bacterial infections, there are few fungal infections that use in vivo drug efficacy studies using BLI method. There are literatures regarding fungal infections like candidiasis where CD1 female mice are immunosuppressed with cyclophosphamide and infected with *C. albicans*. This disease model was used to study systemic infections in the kidneys (Enjalbert et al. 2009). An another candidiasis disease model was prepared when oral inoculation of *C. albicans* to C57BL/5 mice treated with subcutaneous cortisone acetate every 2 days. This infection spreads to the oesophagus and stomach (Mosci et al. 2013). Moreover, there are some parasitic infections that have been known to be monitored using BLI technique. This includes malaria where the in vivo animal model used is Swiss and C57BL/6 mice inoculated intravenously. This model has been useful to study the withdrawal pattern of the Schizont stage within 1–2 days after infection (Franke-Fayard et al. 2006). For infections like leishmaniasis, C57BL/6 mice were inoculated via *Leishmania major* in the ear pinna dermis, and depending on the inocula, immune response can be generated clinically with small population of the pathogen (Thalhofer et al. 2010).

For viral infection smallpox (*Orthopoxvirus*), female Balb/c mice were inoculated by the virus (vaccinia WR strain) intra-peritoneally. The investigators suggested that small pox virus called Dryvax was used to immunise, and mice pre-treated via human IV vaccinia immunoglobulin (VIGIV) were protected (Henriques et al. 2012). Similarly, cowpox and monkeypox virus is inoculated in Balb/c and CAST/Ei mice intranasally and found that CAST/Ei mice were 100X more susceptible than Balb/c (Americo et al. 2014). Hepatitis B and C can be studied using in vivo disease models Balb/c and female C57BL/6 mice, respectively. Both were inoculated by the virus intravenously. It was found that enhancers have no effect in hepatitis B, while in hepatitis C, signal was found to be sensitive to N53/4A protease, and it is reduced by N53/4A-specific shRNA and IFN-alpha (Wang et al. 2010). Influenza used ferrets that were inoculated by H1N1 virus intranasally and were found to monitor intra-host dissemination and inter-host transmission (Karlsson et al. 2015).

SARS-CoV-2: Recent Pandemic Infection

Wuhan City, Hubei, China experienced a pandemic outbreak in December 2019, reason being later identified as coronavirus or SARS-CoV-2 virus by CDC of China (Zhu et al. 2020). As the pandemic progresses, the search for new drug continues. The investigators are hastening to find appropriate animal model to study the same. This virus has been a global threat, and the WHO named this syndrome as COVID-19 (Gralinski and Menachery 2020). Coronavirus belong to Coronaviridae family and Coronavirinae subfamily, which can infect a broad host range, resulting in acute respiratory infection. Like other coronaviruses, COVID-19 virus transmits among humans through direct contact, droplets, any physical contact or aerosols (WHO 2020). As this virus causes high rate of mortality, morbidity and that of transmission, there is a urgent need of remedial drug and strategy to treat this deadly diseases (Sarma et al. 2020). This leads to the urge for an appropriate animal model to investigate the pathogenicity and counteractions for the disease. There were various animal models that were used for the viral strain examination. Yu et al. (2020) infected the rhesus monkeys of 3–5 and 15 years of age through intra-tracheal route. Viral replication and histopathological signs were analysed, which was found more in the lungs of old monkeys when compared with the young ones. Old monkeys were found to have severe pneumonia (Yu et al. 2020). Similarly, Lu et al. (2020a, 2020b) infected rhesus monkey with SARS-CoV-2 virus through throat and nasal route. Increased body temperature, pulmonary infiltration and high level of viral genome RNA showed abnormal chest radiograph results. These experiments were suitable for vaccine and therapeutic studies. The investigators evaluated the effect of SARS-CoV-2 virus when infected to other animal models like *Macaca fascicularis* and common marmoset through throat and nasal route. This mimics the pathogenesis closer to the real scenario of disease clinically, but lower level of RNA, high cost and limited size availability limit its application (Lu et al. 2020a, 2020b). Woolsey et al. developed an in vivo disease model that was considered as the gold standard model for infectious disease pathogens, by infecting the African green monkey by SARS-CoV-2 virus and showed increased level of SARS-CoV-2 replication along with acute respiratory illness more than other non-human primate models like cynomolgus and rhesus macaques (Rockx et al. 2020; Woolsey et al. 2020, 2021).

Apart from non-human primate models, mouse models including the transgenic mouse models were also used. The Balb/c mice inoculated with mouse-adapted SARS-CoV-2 strain affected all the age groups of the mice. Moderate pneumonia along with inflammation has been the common symptom observed in mice. Increase in lung viral load and change in histopathological images of control and infected mice clearly showed inflammation in the alveoli, tracheal denaturation and viral antigen presence in the trachea, bronchiole and pneumocytes (type II). Like those of the clinical symptoms, increases in the cytokines and chemokines were observed in the serum and macrophages of the lungs (Gu et al. 2020a, 2020b; Zhang et al. 2021). Bao et al. used transgenic human ACE2 (hACE2) mice infected with the 10^5 $TCID_{50}$ strain of SARS-CoV-2 virus through nasal route. The infected mice were observed to have weight loss and increment in the levels of the virus in the lungs. Moreover, the histopathological data revealed clear presence of interstitial pneumonia. This model proved to be an effective disease model to study the pathology of COVID-19 (Bao et al. 2020). Another experiment performed by Dinnon et al. mimicked the COVID-19 infection by constructing a recombinant virus called as SARS-CoV-2 MA, which had the capability to replicate in both the upper and the lower airways in all age groups of mice and observed severity of the infection in aged mice more than those in young mice (Dinnon et al. 2020). There were some other mouse model-based experiments performed by different investigators that included transgenic hACE2 mice and used different techniques like CRISPR/Cas9 *knock-in* technique (compared with the wild-type C57BL/6 mouse model) or used lung ciliated epithelial cell-specific HFH4/FOXJ1 promoter 81 (Hassan et al. 2020; Jiang et al. 2020; Sun et al. 2020a, 2020b; Yao et al. 2021).

Besides, few investigators studied the pathogenesis of COVID-19 using hamster and ferret models and observed promising results. The clinical symptoms observed in COVID-19 were very much similar to those observed in SARS-CoV-2 hamster model. The infection was observed from nasal up to the pulmonary alveoli along with the inflammatory changes, cellular viral protein expression and high viral load in the first week of COVID-19 infection. Decrease in the weight and activity and increase in the rate of respiration were observed (Huang et al. 2020).

Similar study was performed by other investigators, thus elucidating the pathogenesis and the transmission of the infection in hamster model (golden Syrian hamster model). Similar weight loss and replication of virus in lungs were observed. High viral load was observed, which increased later, but quick clearance was also seen due to the presence of T lymphocytes (CD3+) (Sia et al. 2020). In ferrets, Kim et al. (2020) studied the COVID-19 infection and its transmission. Ferrets were found to be more sensitive as per infection when transmission from one ferret to another is concerned. Transmission to the new healthy ferrets can be through direct or indirect contact and showed high viral replication rate in the upper respiratory tract. This disease model was considered to show very good results and mimicked human conditions very well, thus ensuring it to be an important model to study COVID-19 therapeutic measures (Kim et al. 2020; Park et al. 2020).

5 Conclusion

These in vivo studies have been proved to be very essential for the development of a new drug along with the evaluation of different pharmacological, pharmacodynamic or pharmacokinetic characteristics of a new drug target. As discussed in the chapter, the in vivo studies help so many investigators in studying the pathogenesis of different diseases. Though the animal models are important, these studies lack crucial data with respect to the rationale for the study design.

References

Adamson PB, Vanoli E (2001) Early autonomic and repolarization abnormalities contribute to lethal arrhythmias in chronic ischemic heart failure: characteristics of a novel heart failure model in dogs with postmyocardial infarction left ventricular dysfunction. J Am Coll Cardiol 37:1741–1748

Agotegaray M, Gumilar F, Boeris M, Toso R, Minetti A (2014) Enhanced analgesic properties and reduced ulcerogenic effect of a mononuclear copper (II) complex with fenoprofen in comparison to the parent drug: promising insights in the treatment of chronic inflammatory diseases. BioMed Res Int 2014:505987

Alarcon RT, da Rocha Corrêa Fernandes A, Laurindo IM, Bértolo MB, Pinheiro GC, Andrade LE (2015) Characterization of cumulative joint damage patterns in patients with rheumatoid arthritis: a clinical, serological, and gene polymorphism perspective. J Rheumatol 42:405–412

Ali MS, Tiscareno-Grejada I, Locovei S, Smiley R, Collins T, Sarosiek J, McCallum R (2012) Gender and estradiol as major factors in the expression and dimerization of nNOSα in rats with experimental diabetic gastroparesis. Dig Dis Sci 57:2814–2825

Ali Y, Alam MS, Hamid H, Husain A, Kharbanda C, Bano S, Nazreen S, Haider S (2014) Attenuation of inflammatory mediators, oxidative stress and toxic risk evaluation of Aporosa lindleyana Baill bark extract. J Ethnopharmacol 155:1513–1521

Americo JL, Sood CL, Cotter CA, Vogel JL, Kristie TM, Moss B, Earl PL (2014) Susceptibility of the wild-derived inbred CAST/Ei mouse to infection by orthopoxviruses analyzed by live bioluminescence imaging. Virology 449:120–132

Anders H-J, Vielhauer V (2007) Identifying and validating novel targets with in vivo disease models: guidelines for study design. Drug Discov Today 12:446–451

Andrade E, Bento A, Cavalli J, Oliveira S, Schwanke R, Siqueira J, Freitas C, Marcon R, Calixto J (2016) Non-clinical studies in the process of new drug development-part II: good laboratory practice, metabolism, pharmacokinetics, safety and dose translation to clinical studies. Braz J Med Biol Res 49:e5646

Andreu N, Zelmer A, Sampson SL, Ikeh M, Bancroft GJ, Schaible UE, Wiles S, Robertson BD (2013) Rapid in vivo assessment of drug efficacy against Mycobacterium tuberculosis using an improved firefly luciferase. J Antimicrob Chemother 68:2118–2127

Annahazi A, Dabek M, Gecse K, Salvador-Cartier C, Polizzi A, Rosztoczy A, Roka R, Theodorou V, Wittmann T, Bueno L (2012) Proteinase-activated receptor-4 evoked colorectal analgesia in mice: an endogenously activated feed-back loop in visceral inflammatory pain. Neurogastroenterol Motil 24:76–e13

Asano T, Aida S, Suemasu S, Mizushima T (2016) Anethole restores delayed gastric emptying and impaired gastric accommodation in rodents. Biochem Biophys Res Commun 472:125–130

Ashe KH, Zahs KR (2010) Probing the biology of Alzheimer's disease in mice. Neuron 66:631–645

Asquith DL, Miller AM, McInnes IB, Liew FY (2009) Animal models of rheumatoid arthritis. Eur J Immunol 39:2040–2044

Badimon L (2001) Atherosclerosis and thrombosis: lessons from animal models. Thromb Haemost 86:356–365

Balogun SO, Damazo AS, de Oliveira Martins DT (2015) Helicteres sacarolha A. St.-Hil. et al.: gastroprotective and possible mechanism of actions in experimental animals. J Ethnopharmacol 166:176–184

Balouiri M, Sadiki M, Ibnsouda SK (2016) Methods for in vitro evaluating antimicrobial activity: a review. J Pharm Anal 6:71–79

Bao L, Deng W, Huang B, Gao H, Liu J, Ren L, Wei Q, Yu P, Xu Y, Qi F (2020) The pathogenicity of SARS-CoV-2 in hACE2 transgenic mice. Nature 583:830–833

Barretti D, Melo S, Oliveira E, Barauna V (2017) Resistance training attenuates salt overload-induced cardiac remodeling and diastolic dysfunction in normotensive rats. Braz J Med Biol Res 50

Barry AL (1999) Methods for determining bactericidal activity of antimicrobial agents: approved guidelines. National committee for clinical laboratory standards, Wayne, PA

Batista LM, Lima GRDM, De Almeida ABA, Magri LDP, Calvo TR, Ferreira AL, Pellizzon CH, Hiruma-Lima CA, Vilegas W, Sano PT (2015) Ulcer healing and mechanism (s) of action involved in the gastroprotective activity of fractions obtained from Syngonanthus arthrotrichus and Syngonanthus bisulcatus. BMC Complement Altern Med 15:1–9

Belevych AE, Terentyev D, Terentyeva R, Nishijima Y, Sridhar A, Hamlin RL, Carnes CA, Györke S (2011) The relationship between arrhythmogenesis and impaired contractility in heart failure: role of altered ryanodine receptor function. Cardiovasc Res 90:493–502

Bibby M (1999) Making the most of rodent tumour systems in cancer drug discovery. Br J Cancer 79:1633–1640

Björklund LM, Sánchez-Pernaute R, Chung S, Andersson T, Chen IYC, McNaught KSP, Brownell A-L, Jenkins BG, Wahlestedt C, Kim K-S (2002) Embryonic stem cells develop into functional dopaminergic neurons after transplantation in a Parkinson rat model. Proc Natl Acad Sci 99: 2344–2349

van den Bos EJ, Mees BM, de Waard MC, de Crom R, Duncker DJ (2005) A novel model of cryoinjury-induced myocardial infarction in the mouse: a comparison with coronary artery ligation. Am J Phys Heart Circ Phys 289:H1291–H1300

Brake K, Gumireddy A, Tiwari A, Chauhan H, Kumari D (2017) In vivo studies for drug development via oral delivery: challenges, animal models and techniques. Pharm Anal Acta 8:560. https://doi.org/10.4172/2153-2435.1000560

Bullard TA, Protack TL, Aguilar F, Bagwe S, Massey HT, Blaxall BC (2008) Identification of Nogo as a novel indicator of heart failure. Physiol Genomics 32:182–189

Bushra R, Aslam N, Khan AY (2011) Food-drug interactions. Oman Med J 26:77

Camacho P, Fan H, Liu Z, He J-Q (2016) Small mammalian animal models of heart disease. Am J Cardiovasc Dis 6:70

Campbell IK, Hamilton JA, Wicks IP (2000) Collagen-induced arthritis in C57BL/6 (H-2b) mice: new insights into an important disease model of rheumatoid arthritis. Eur J Immunol 30:1568–1575

Cardot J, Beyssac E, Alric M (2007) In vitro-in vivo correlation: importance of dissolution in IVIVC. Dissolut Technol 14:15

Carmona S, Hardy J, Guerreiro R (2018) The genetic landscape of Alzheimer disease. Handb Clin Neurol 148:395–408

Chang X, Luo F, Jiang W, Zhu L, Gao J, He H, Wei T, Gong S, Yan T (2015) Protective activity of salidroside against ethanol-induced gastric ulcer via the MAPK/NF-κB pathway in vivo and in vitro. Int Immunopharmacol 28:604–615

Chen X-Y, Chen H-M, Liu Y-H, Zhang Z-B, Zheng Y-F, Su Z-Q, Zhang X, Xie J-H, Liang Y-Z, Fu L-D (2016) The gastroprotective effect of pogostone from Pogostemonis Herba against indomethacin-induced gastric ulcer in rats. Exp Biol Med 241:193–204

Choma IM, Grzelak EM (2011) Bioautography detection in thin-layer chromatography. J Chromatogr A 1218:2684–2691

Chow PK, Ng RT, Ogden BE (2008) Using animal models in biomedical research: a primer for the investigator. World Scientific

Corbett T, Polin L, LoRusso P, Valeriote F, Panchapor C, Pugh S, White K, Knight J, Demchik L, Jones J (2004) In vivo methods for screening and preclinical testing. In: Anticancer drug development guide. Springer, pp 99–123

Cosyns S, Dhaese I, Thoonen R, Buys E, Vral A, Brouckaert P, Lefebvre R (2013) Heme deficiency of soluble guanylate cyclase induces gastroparesis. Neurogastroenterol Motil 25:e339–e352

Courtenay J, Dallman MJ, Dayan A, Martin A, Mosedale B (1980) Immunisation against heterologous type II collagen induces arthritis in mice. Nature 283:666–668

Creed RB, Goldberg MS (2018) New developments in genetic rat models of Parkinson's disease. Mov Disord 33:717–729

Creedon CT, Verhulst P-J, Choi KM, Mason JE, Linden DR, Szurszewski JH, Gibbons SJ, Farrugia G (2013) Assessment of gastric emptying in non-obese diabetic mice using a [13C]-octanoic acid breath test. J Vis Exp 73:e50301. https://doi.org/10.3791/50301

Dai F, Lei Y, Li S, Song G, Chen JD (2013a) Desvenlafaxine succinate ameliorates visceral hypersensitivity but delays solid gastric emptying in rats. Am J Physiol Gastrointest Liver Physiol 305:G333–G339

Dai T, Gupta A, Huang Y-Y, Yin R, Murray CK, Vrahas MS, Sherwood ME, Tegos GP, Hamblin MR (2013b) Blue light rescues mice from potentially fatal Pseudomonas aeruginosa burn infection: efficacy, safety, and mechanism of action. Antimicrob Agents Chemother 57:1238–1245

Dai T, Kharkwal GB, Zhao J, Denis TGS, Wu Q, Xia Y, Huang L, Sharma SK, d'Enfert C, Hamblin MR (2011) Ultraviolet-C light for treatment of Candida albicans burn infection in mice. Photochem Photobiol 87:342–349

Davidson M, Lindsey J, Davis J (1987) Requirements and selection of an animal model. Isr J Med Sci 23:551–555

Dawson TM, Golde TE, Lagier-Tourenne C (2018) Animal models of neurodegenerative diseases. Nat Neurosci 21:1370–1379

Dayan V, Yannarelli G, Filomeno P, Keating A (2012) Human mesenchymal stromal cells improve scar thickness without enhancing cardiac function in a chronic ischaemic heart failure model. Interact Cardiovasc Thorac Surg 14:516–520

Dinnon KH, Leist SR, Schäfer A, Edwards CE, Martinez DR, Montgomery SA, West A, Yount BL, Hou YJ, Adams LE (2020) A mouse-adapted model of SARS-CoV-2 to test COVID-19 countermeasures. Nature 586:560–566

Dressman J, Vertzoni M, Goumas K, Reppas C (2007) Estimating drug solubility in the gastrointestinal tract. Adv Drug Deliv Rev 59:591–602

Ducharme J, Dudley A, Thompson R (2006) Pharmacokinetic issue in drug discovery. In: Drug discovery and development. Churchill Livingstone Elsevier, Philadelphia, pp 141–161

Duris FH, Fava RA, Noelle RJ (1994) Collagen-induced arthritis as a model of rheumatoid arthritis. Clin Immunol Immunopathol 73:11–18

Dykes DJ, Waud WR (2002) Murine L1210 and P388 leukemias. In: Tumor models in cancer research. Humana Press, Totowa, pp 23–40

Enjalbert B, Rachini A, Vediyappan G, Pietrella D, Spaccapelo R, Vecchiarelli A, Brown AJ, d'Enfert C (2009) A multifunctional, synthetic Gaussia princeps luciferase reporter for live imaging of Candida albicans infections. Infect Immun 77:4847–4858

Fan J, Kitajima S, Watanabe T, Xu J, Zhang J, Liu E, Chen YE (2015) Rabbit models for the study of human atherosclerosis: from pathophysiological mechanisms to translational medicine. Pharmacol Ther 146:104–119

Fatma G, Mouna BF, Mondher M, Ahmed L (2014) In-vitro assessment of antioxidant and antimicrobial activities of methanol extracts and essential oil of Thymus hirtus sp. algeriensis. Lipids Health Dis 13:1–12

Favre-Godal Q, Queiroz EF, Wolfender J-L (2013) Latest developments in assessing antifungal activity using TLC-bioautography: a review. J AOAC Int 96:1175–1188

Food and Drug Administration, CDER, CBER, US Department of Human Health Services (2019) M4E(R2): the CTD—efficacy guidance for industry, Revision 1. http://www.fda.gov/downloads/Drugs/GuidanceComplianceRegulatoryInformation/Guidances/ucm072101.pdf

Franke-Fayard B, Waters AP, Janse CJ (2006) Real-time in vivo imaging of transgenic bioluminescent blood stages of rodent malaria parasites in mice. Nat Protoc 1:476–485

Freedman LP, Venugopalan G, Wisman R (2017) Reproducibility2020: progress and priorities. F1000Research 6:604

Galandrin S, Oligny-Longpré G, Bouvier M (2007) The evasive nature of drug efficacy: implications for drug discovery. Trends Pharmacol Sci 28:423–430

Geran RI, Greenberg NH, MacDonald MM, Schumacher AM (1972) Protocols for screening chemical agents and natural products against animal tumors and other biological systems. Cancer Chemother Rep 3:1–103

Ghoneim HE, McCullers JA (2014) Adjunctive corticosteroid therapy improves lung immunopathology and survival during severe secondary pneumococcal pneumonia in mice. J Infect Dis 209:1459–1468

Golberg A, Broelsch GF, Vecchio D, Khan S, Hamblin MR, Austen WG Jr, Sheridan RL, Yarmush ML (2014) Eradication of multidrug-resistant A. baumannii in burn wounds by antiseptic pulsed electric field. Technology 2:153–160

Golde TE, Schneider LS, Koo EH (2011) Anti-Aβ therapeutics in Alzheimer's disease: the need for a paradigm shift. Neuron 69:203–213

González-Rosa JM, Burns CE, Burns CG (2017) Zebrafish heart regeneration: 15 years of discoveries. Regeneration 4:105–123

Gralinski LE, Menachery VD (2020) Return of the coronavirus: 2019-nCoV. Viruses 12:135

Gralinski M, Neves LA, Tiniakova O (2020) Methods to induce cardiac hypertrophy and insufficiency. In: Drug discovery and evaluation: pharmacological assays. Springer, Berlin, p 287

Grisel P, Meinhardt A, Lehr H-A, Kappenberger L, Barrandon Y, Vassalli G (2008) The MRL mouse repairs both cryogenic and ischemic myocardial infarcts with scar. Cardiovasc Pathol 17: 14–22

Gu H, Chen Q, Yang G, He L, Fan H, Deng Y-Q, Wang Y, Teng Y, Zhao Z, Cui Y (2020a) Adaptation of SARS-CoV-2 in BALB/c mice for testing vaccine efficacy. Science 369:1603–1607

Gu H, Chen Q, Yang G, He L, Fan H, Deng Y-Q, Wang Y, Teng Y, Zhao Z, Cui Y (2020b) Rapid adaptation of SARS-CoV-2 in BALB/c mice: novel mouse model for vaccine efficacy. BioRxiv. https://doi.org/10.1101/2020.05.02.073411

Haass C, De Strooper B (1999) The presenilins in Alzheimer's disease—proteolysis holds the key. Science 286:916–919

Hamblin MR, Hasan T (2004) Photodynamic therapy: a new antimicrobial approach to infectious disease? Photochem Photobiol Sci 3:436–450

Hamblin MR, Zahra T, Contag CH, McManus AT, Hasan T (2003) Optical monitoring and treatment of potentially lethal wound infections in vivo. J Infect Dis 187:1717–1726

Hamburger MO, Cordell GA (1987) A direct bioautographic TLC assay for compounds possessing antibacterial activity. J Nat Prod 50:19–22

Hardy J, Chu P, Contag CH (2009) Foci of Listeria monocytogenes persist in the bone marrow. Dis Model Mech 2:39–46

Hassan AO, Case JB, Winkler ES, Thackray LB, Kafai NM, Bailey AL, McCune BT, Fox JM, Chen RE, Alsoussi WB (2020) A SARS-CoV-2 infection model in mice demonstrates protection by neutralizing antibodies. Cell 182:744–753.e4

Haubner BJ, Adamowicz-Brice M, Khadayate S, Tiefenthaler V, Metzler B, Aitman T, Penninger JM (2012) Complete cardiac regeneration in a mouse model of myocardial infarction. Aging (Albany NY) 4:966

Hearse DJ, Sutherland FJ (2000) Experimental models for the study of cardiovascular function and disease. Pharmacol Res 41:597–603

Henriques C, Castro DP, Gomes LH, Garcia ES, de Souza W (2012) Bioluminescent imaging of Trypanosoma cruzi infection in Rhodnius prolixus. Parasit Vectors 5:1–15

Holtzman DM, Carrillo MC, Hendrix JA, Bain LJ, Catafau AM, Gault LM, Goedert M, Mandelkow E, Mandelkow E-M, Miller DS (2016) Tau: from research to clinical development. Alzheimers Dement 12:1033–1039

Hosoya T, Matsumoto K, Tashima K, Nakamura H, Fujino H, Murayama T, Horie S (2014) TRPM 8 has a key role in experimental colitis-induced visceral hyperalgesia in mice. Neurogastroenterol Motil 26:1112–1121

Huang C, Wang Y, Li X, Ren L, Zhao J, Hu Y, Zhang L, Fan G, Xu J, Gu X (2020) Clinical features of patients infected with 2019 novel coronavirus in Wuhan, China. Lancet 395:497–506
Huang L, Wang M, Dai T, Sperandio FF, Huang Y-Y, Xuan Y, Chiang LY, Hamblin MR (2014) Antimicrobial photodynamic therapy with decacationic monoadducts and bisadducts of [70] fullerene: in vitro and in vivo studies. Nanomedicine 9:253–266
Hughes JP, Rees S, Kalindjian SB, Philpott KL (2011) Principles of early drug discovery. Br J Pharmacol 162:1239–1249
Jaarin K, Mustafa MR, Leong X-F (2011) The effects of heated vegetable oils on blood pressure in rats. Clinics 66:2125–2132
Jankovic J, Tan EK (2020) Parkinson's disease: etiopathogenesis and treatment. J Neurol Neurosurg Psychiatry 91:795–808
Jiang R-D, Liu M-Q, Chen Y, Shan C, Zhou Y-W, Shen X-R, Li Q, Zhang L, Zhu Y, Si H-R (2020) Pathogenesis of SARS-CoV-2 in transgenic mice expressing human angiotensin-converting enzyme 2. Cell 182:50–58.e8
EU Joint Programme – Neurodegenerative Disease Research (JPND) (2014) Experimental models for neurodegenerative diseases. JNPD Research: EU Joint programme – Neurodegenerative disease research, pp 5–42. http://www.extranet-jpnd.eu
Johnson AC, Greenwood-Van Meerveld B (2017) Critical evaluation of animal models of gastrointestinal disorders. Handb Exp Pharmacol 239:289–317
Júnior FE, de Oliveira DR, Boligon AA, Athayde ML, Kamdem JP, Macedo GE, da Silva GF, de Menezes IR, Costa JG, Coutinho HDM (2014) Protective effects of Croton campestris A. St-Hill in different ulcer models in rodents: evidence for the involvement of nitric oxide and prostaglandins. J Ethnopharmacol 153:469–477
Kannan K, Ortmann RA, Kimpel D (2005) Animal models of rheumatoid arthritis and their relevance to human disease. Pathophysiology 12:167–181
Kari G, Rodeck U, Dicker AP (2007) Zebrafish: an emerging model system for human disease and drug discovery. Clin Pharmacol Ther 82:70–80
Karlsson EA, Meliopoulos VA, Savage C, Livingston B, Mehle A, Schultz-Cherry S (2015) Visualizing real-time influenza virus infection, transmission and protection in ferrets. Nat Commun 6:1–10
Keffer J, Probert L, Cazlaris H, Georgopoulos S, Kaslaris E, Kioussis D, Kollias G (1991) Transgenic mice expressing human tumour necrosis factor: a predictive genetic model of arthritis. EMBO J 10:4025–4031
Kim EY, Moudgil KD (2009) The determinants of susceptibility/resistance to adjuvant arthritis in rats. Arthritis Res Ther 11:1–9
Kim J, Chakrabarty P, Hanna A, March A, Dickson DW, Borchelt DR, Golde T, Janus C (2013) Normal cognition in transgenic BRI2-Aβ mice. Mol Neurodegener 8:1–12
Kim YI, Kim SG, Kim SM, Kim EH, Park SJ, Yu KM, Chang JH, Kim EJ, Lee S, Casel MAB, Um J, Song MS, Jeong HW, Lai VD, Kim Y, Chin BS, Park JS, Chung KH, Foo SS, Poo H, Mo IP, Lee OJ, Webby RJ, Jung JU, Choi YK (2020) Infection and rapid transmission of SARS-CoV-2 in ferrets. Cell Host Microbe 27:704
Kimura Y, Sumiyoshi M (2012) Effects of an Atractylodes lancea rhizome extract and a volatile component β-eudesmol on gastrointestinal motility in mice. J Ethnopharmacol 141:530–536
Kinoshita J, Clark T (2007) Alzforum. Methods Mol Biol 401:365–381
Klepser ME, Ernst EJ, Lewis RE, Ernst ME, Pfaller MA (1998) Influence of test conditions on antifungal time-kill curve results: proposal for standardized methods. Antimicrob Agents Chemother 42:1207–1212
Kruczynski A, Colpaert F, Tarayre J-P, Mouillard P, Fahy J, Hill BT (1998) Preclinical in vivo antitumor activity of vinflunine, a novel fluorinated Vinca alkaloid. Cancer Chemother Pharmacol 41:437–447
Kudryavtsev KV, Markevich AO, Virchenko OV, Falalyeyeva TM, Beregova TV, Ostapchenko LI, Zabolotnev DV, Zefirov NS (2014) Pharmacological correction of stress-induced gastric

ulceration by novel small-molecule agents with antioxidant profile. ScientificWorldJournal 2014:217039
LaFerla FM, Green KN (2012) Animal models of Alzheimer disease. Cold Spring Harb Perspect Med 2:a006320
Lai AC, Wallner K, Cao JM, Chen LS, Karagueuzian HS, Fishbein MC, Chen PS, Sharifi BG (2000) Colocalization of tenascin and sympathetic nerves in a canine model of nerve sprouting and sudden cardiac death. J Cardiovasc Electrophysiol 11:1345–1351
Lambrechts SA, Demidova TN, Aalders MC, Hasan T, Hamblin MR (2005) Photodynamic therapy for Staphylococcus aureus infected burn wounds in mice. Photochem Photobiol Sci 4:503–509
Lanao J, Fraile M (2005) Drug tissue distribution: study methods and therapeutic implications. Curr Pharm Des 11:3829–3845
Langdon S, Hendriks H, Braakhuis B-J, Pratesi G, Berger D, Fodstad Ø, Fiebig H, Boven E (1994) Preclinical phase II studies in human tumor xenografts: a European multicenter follow-up study. Ann Oncol 5:415–422
de Lange WJ, Grimes AC, Hegge LF, Ralphe JC (2013) Ablation of cardiac myosin–binding protein-C accelerates contractile kinetics in engineered cardiac tissue. J Gen Physiol 141:73–84
Larcher T, Lafoux A, Tesson L, Remy S, Thepenier V, François V, Le Guiner C, Goubin H, Dutilleul M, Guigand L (2014) Characterization of dystrophin deficient rats: a new model for Duchenne muscular dystrophy. PLoS One 9:e110371
Leong XF, Aishah A, Aini UN, Das S, Jaarin K (2008) Heated palm oil causes rise in blood pressure and cardiac changes in heart muscle in experimental rats. Arch Med Res 39:567–572
Li P, Schwarz EM (2003) The TNF-α transgenic mouse model of inflammatory arthritis. Springer Semin Immunopathol 25:19–33
Li S, Chen J (2014) Decreased L-type calcium current in antral smooth muscle cells of STZ-induced diabetic rats. Neurogastroenterol Motil 26:971–979
Li W-R, Shi Q-S, Dai H-Q, Liang Q, Xie X-B, Huang X-M, Zhao G-Z, Zhang L-X (2016) Antifungal activity, kinetics and molecular mechanism of action of garlic oil against Candida albicans. Sci Rep 6:1–9
Liao J, Huang W, Liu G (2017) Animal models of coronary heart disease. J Biomed Res 31:3
Lieschke GJ, Currie PD (2007) Animal models of human disease: zebrafish swim into view. Nat Rev Genet 8:353–367
Lin Y, Roman K, Foust K, Kaspar B, Bailey M, Stephens R (2011) Glutamate transporter GLT-1 upregulation attenuates visceral nociception and hyperalgesia via spinal mechanisms not related to anti-inflammatory or probiotic effects. Pain Res Treat 2011:507029
Lindsey ML, Bolli R, Canty JM, Du X-J, Frangogiannis NG, Frantz S, Gourdie RG, Holmes JW, Jones SP, Kloner RA et al (2018) Guidelines for experimental models of myocardial ischemia and infarction. Am J Physiol Heart Circ Physiol 314:H812–H838
Lipsky MS, Sharp LK (2001) From idea to market: the drug approval process. J Am Board Fam Pract 14:362–367
Liu C, Oikonomopoulos A, Sayed N, Wu JC (2018) Modeling human diseases with induced pluripotent stem cells: from 2D to 3D and beyond. Development 145:dev156166
Lu H, Xu X, Zhang M, Cao R, Bråkenhielm E, Li C, Lin H, Yao G, Sun H, Qi L (2007) Combinatorial protein therapy of angiogenic and arteriogenic factors remarkably improves collaterogenesis and cardiac function in pigs. Proc Natl Acad Sci 104:12140–12145
Lu S, Zhao Y, Yu W, Yang Y, Gao J, Wang J, Kuang D, Yang M, Yang J, Ma C (2020a) Comparison of nonhuman primates identified the suitable model for COVID-19. Signal Transduct Target Ther 5:1–9
Lu S, Zhao Y, Yu W, Yang Y, Gao J, Wang J, Kuang D, Yang M, Yang J, Ma C (2020b) Comparison of SARS-CoV-2 infections among 3 species of non-human primates. BioRxiv:1–27. https://doi.org/10.1101/2020.04.08.031807
Lum-Naihe K, Toedebusch R, Mahmood A, Bajwa J, Carmack T, Kumar SA, Ardhanari S, DeMarco VG, Emter CA, Pulakat L (2017) Cardiovascular disease progression in female Zucker Diabetic Fatty rats occurs via unique mechanisms compared to males. Sci Rep 7:1–16

Magaldi S, Mata-Essayag S, De Capriles CH, Pérez C, Colella M, Olaizola C, Ontiveros Y (2004) Well diffusion for antifungal susceptibility testing. Int J Infect Dis 8:39–45
Martini L, Fini M, Giavaresi G, Giardino R (2001) Sheep model in orthopedic research: a literature review. Comp Med 51:292–299
McGonigle P, Ruggeri B (2014) Animal models of human disease: challenges in enabling translation. Biochem Pharmacol 87:162–171
McGowan E, Pickford F, Kim J, Onstead L, Eriksen J, Yu C, Skipper L, Murphy MP, Beard J, Das P (2005) Aβ42 is essential for parenchymal and vascular amyloid deposition in mice. Neuron 47:191–199
Milani-Nejad N, Janssen PM (2014) Small and large animal models in cardiac contraction research: advantages and disadvantages. Pharmacol Ther 141:235–249
Ministry of Environment and Forests, Govt. of India (2007) Guidelines on the regulation of scientific experiments on animals, pp 1–144
Montgomery M, Ahadian S, Huyer LD, Rito ML, Civitarese RA, Vanderlaan RD, Wu J, Reis LA, Momen A, Akbari S (2017) Flexible shape-memory scaffold for minimally invasive delivery of functional tissues. Nat Mater 16:1038–1046
Mook-Kanamori BB, Rouse MS, Kang C-I, van de Beek D, Steckelberg JM, Patel R (2009) Daptomycin in experimental murine pneumococcal meningitis. BMC Infect Dis 9:1–9
Mosci P, Pericolini E, Gabrielli E, Kenno S, Perito S, Bistoni F, d'Enfert C, Vecchiarelli A (2013) A novel bioluminescence mouse model for monitoring oropharyngeal candidiasis in mice. Virulence 4:250–254
Mukherjee R, Brinsa TA, Dowdy KB, Scott AA, Baskin JM, Deschamps AM, Lowry AS, Escobar GP, Lucas DG, Yarbrough WM (2003) Myocardial infarct expansion and matrix metalloproteinase inhibition. Circulation 107:618–625
Munder A, Wölbeling F, Klockgether J, Wiehlmann L, Tümmler B (2014) In vivo imaging of bioluminescent Pseudomonas aeruginosa in an acute murine airway infection model. Pathog Dis 72:74–77
Nordin N, Salama SM, Golbabapour S, Hajrezaie M, Hassandarvish P, Kamalidehghan B, Majid NA, Hashim NM, Omar H, Fadaienasab M (2014) Anti-ulcerogenic effect of methanolic extracts from Enicosanthellum pulchrum (King) Heusden against ethanol-induced acute gastric lesion in animal models. PLoS One 9:e111925
Nunes S, Rolo AP, Palmeira CM, Reis F (2017) Diabetic cardiomyopathy: focus on oxidative stress, mitochondrial dysfunction and inflammation. In: Cardiomyopathies—types and treatments. Intech, pp 235–257
Olson EN (2006) Gene regulatory networks in the evolution and development of the heart. Science 313:1922–1927
Ozcan C, Battaglia E, Young R, Suzuki G (2015) LKB1 knockout mouse develops spontaneous atrial fibrillation and provides mechanistic insights into human disease process. J Am Heart Assoc 4:e001733
Palleria C, Di Paolo A, Giofrè C, Caglioti C, Leuzzi G, Siniscalchi A, De Sarro G, Gallelli L (2013) Pharmacokinetic drug-drug interaction and their implication in clinical management. J Res Med Sci 18:601
Pandey S, Pandey P, Tiwari G, Tiwari R (2010) Bioanalysis in drug discovery and development. Pharm Methods 1:14–24
Park S-J, Yu K-M, Kim Y-I, Kim S-M, Kim E-H, Kim S-G, Kim EJ, Casel MAB, Rollon R, Jang S-G (2020) Antiviral efficacies of FDA-approved drugs against SARS-CoV-2 infection in ferrets. MBio 11:e01114–e01120
Porrello ER, Mahmoud AI, Simpson E, Johnson BA, Grinsfelder D, Canseco D, Mammen PP, Rothermel BA, Olson EN, Sadek HA (2013) Regulation of neonatal and adult mammalian heart regeneration by the miR-15 family. Proc Natl Acad Sci 110:187–192
Price DL, Tanzi RE, Borchelt DR, Sisodia SS (1998) Alzheimer's disease: genetic studies and transgenic models. Annu Rev Genet 32:461–493

Probert L, Akassoglou K, Alexopoulou L, Douni E, Haralambous S, Hill S, Kassiotis G, Kontoyiannis D, Pasparakis M, Plows D (1996) Dissection of the pathologies induced by transmembrane and wild-type tumor necrosis factor in transgenic mice. J Leukoc Biol 59: 518–525

Rathore A, Rahman MU, Siddiqui AA, Ali A, Shaharyar M (2014) Design and synthesis of benzimidazole analogs endowed with oxadiazole as selective COX-2 inhibitor. Arch Pharm 347:923–935

Ravella K, Al-Hendy A, Sharan C, Hale A, Channon K, Srinivasan S, Gangula P (2013) Chronic estrogen deficiency causes gastroparesis by altering neuronal nitric oxide synthase function. Dig Dis Sci 58:1507–1515

Reichardt SD, Weinhage T, Rotte A, Foeller M, Oppermann M, Lühder F, Tuckermann JP, Lang F, van den Brandt J, Reichardt HM (2014) Glucocorticoids induce gastroparesis in mice through depletion of l-arginine. Endocrinology 155:3899–3908

Ren C, Wang F, Li G, Jiao Q, Bai J, Yu D, Hao W, Wang R, Cao J-M (2008) Nerve sprouting suppresses myocardial Ito and IK1 channels and increases severity to ventricular fibrillation in rat. Auton Neurosci 144:22–29

Rhee K-J, Cheng H, Harris A, Morin C, Kaper JB, Hecht G (2011) Determination of spatial and temporal colonization of enteropathogenic E. coli and enterohemorrhagic E. coli in mice using bioluminescent in vivo imaging. Gut Microbes 2:34–41

Robey TE, Murry CE (2008) Absence of regeneration in the MRL/MpJ mouse heart following infarction or cryoinjury. Cardiovasc Pathol 17:6–13

Rockx B, Kuiken T, Herfst S, Bestebroer T, Lamers MM, Munnink BBO, De Meulder D, Van Amerongen G, Van Den Brand J, Okba NM (2020) Comparative pathogenesis of COVID-19, MERS, and SARS in a nonhuman primate model. Science 368:1012–1015

Rossato MF, Oliveira SM, Trevisan G, Rotta M, Machado P, Martins MA, Ferreira J (2015) Structural improvement of compounds with analgesic activity: AC-MPF4, a compound with mixed anti-inflammatory and antinociceptive activity via opioid receptor. Pharmacol Biochem Behav 129:72–78

Ruggeri BA, Camp F, Miknyoczki S (2014) Animal models of disease: pre-clinical animal models of cancer and their applications and utility in drug discovery. Biochem Pharmacol 87:150–161

Saito T, Matsuba Y, Mihira N, Takano J, Nilsson P, Itohara S, Iwata N, Saido TC (2014) Single App knock-in mouse models of Alzheimer's disease. Nat Neurosci 17:661–663

Sarkar S, Chawla-Sarkar M, Young D, Nishiyama K, Rayborn ME, Hollyfield JG, Sen S (2004) Myocardial cell death and regeneration during progression of cardiac hypertrophy to heart failure. J Biol Chem 279:52630–52642

Sarma P, Kaur H, Kumar H, Mahendru D, Avti P, Bhattacharyya A, Prajapat M, Shekhar N, Kumar S, Singh R (2020) Virological and clinical cure in COVID-19 patients treated with hydroxychloroquine: a systematic review and meta-analysis. J Med Virol 92:776–785

Sasaguri H, Nilsson P, Hashimoto S, Nagata K, Saito T, De Strooper B, Hardy J, Vassar R, Winblad B, Saido TC (2017) APP mouse models for Alzheimer's disease preclinical studies. EMBO J 36:2473–2487

Schachter AS, Davis KL (2000) Alzheimer's disease. Dialogues Clin Neurosci 2:91–100

Schubert D, Maier B, Morawietz L, Krenn V, Kamradt T (2004) Immunization with glucose-6-phosphate isomerase induces T cell-dependent peripheral polyarthritis in genetically unaltered mice. J Immunol 172:4503–4509

Scott DL, Wolfe F, Huizinga TW (2010) Rheumatoid arthritis. Lancet 376:1094–1108

Sewell F, Aggarwal M, Bachler G, Broadmeadow A, Gellatly N, Moore E, Robinson S, Rooseboom M, Stevens A, Terry C (2017) The current status of exposure-driven approaches for chemical safety assessment: a cross-sector perspective. Toxicology 389:109–117

Shekhawat PB, Pokharkar VB (2017) Understanding peroral absorption: regulatory aspects and contemporary approaches to tackling solubility and permeability hurdles. Acta Pharm Sin B 7: 260–280

Shi Y, Lin S, Staats KA, Li Y, Chang W-H, Hung S-T, Hendricks E, Linares GR, Wang Y, Son EY (2018) Haploinsufficiency leads to neurodegeneration in C9ORF72 ALS/FTD human induced motor neurons. Nat Med 24:313–325

Shiomi M, Ito T, Yamada S, Kawashima S, Fan J (2003) Development of an animal model for spontaneous myocardial infarction (WHHLMI rabbit). Arterioscler Thromb Vasc Biol 23:1239–1244

Sia SF, Yan LM, Chin AWH, Fung K, Choy K-T, Wong AYL, Kaewpreedee P, Perera RAPM, Poon LLM, Nicholls JM, Peiris M, Yen H-L (2020) Pathogenesis and transmission of SARS-CoV-2 in golden hamsters. Nature 583:834–838

Silva LP, de Angelis CD, Bonamin F, Kushima H, Mininel FJ, Dos Santos LC, Delella FK, Felisbino SL, Vilegas W, da Rocha LRM (2015) Terminalia catappa L.: a medicinal plant from the Caribbean pharmacopeia with anti-Helicobacter pylori and antiulcer action in experimental rodent models. J Ethnopharmacol 159:285–295

Sim DS, Kauser K (2015) In vivo target validation using biological molecules in drug development. In: New approaches to drug discovery. Springer, Cham, pp 59–70

Singhvi G, Singh M (2011) In-vitro drug release characterization models. Int J Pharm Stud Res 2: 77–84

Sjolinder H, Jonsson AB (2007) Imaging of disease dynamics during meningococcal sepsis. PLoS One 2:e241

Smolen JS, Aletaha D, McInnes IB (2016) Rheumatoid arthritis. Lancet 388:2023–2038

Song J, Zheng L, Zhang X, Feng X, Fan R, Sun L, Hong F, Zhang Y, Zhu J (2014) Upregulation of β1-adrenoceptors is involved in the formation of gastric dysmotility in the 6-hydroxydopamine rat model of Parkinson's disease. Transl Res 164:22–31

Stegemann S, Leveiller F, Franchi D, De Jong H, Lindén H (2007) When poor solubility becomes an issue: from early stage to proof of concept. Eur J Pharm Sci 31:249–261

Stuart JM, Townes AS, Kang AH (1984) Collagen autoimmune arthritis. Annu Rev Immunol 2: 199–218

Sun J, Zhuang Z, Zheng J, Li K, Wong RL-Y, Liu D, Huang J, He J, Zhu A, Zhao J (2020a) Generation of a broadly useful model for COVID-19 pathogenesis, vaccination, and treatment. Cell 182:734–743.e5

Sun S-H, Chen Q, Gu H-J, Yang G, Wang Y-X, Huang X-Y, Liu S-S, Zhang N-N, Li X-F, Xiong R (2020b) A mouse model of SARS-CoV-2 infection and pathogenesis. Cell Host Microbe 28: 124–133.e4

Sutton MGSJ, Sharpe N (2000) Left ventricular remodeling after myocardial infarction: pathophysiology and therapy. Circulation 101:2981–2988

Tak PP, Gerlag DM, Aupperle KR, Van De Geest DA, Overbeek M, Bennett BL, Boyle DL, Manning AM, Firestein GS (2001) Inhibitor of nuclear factor κB kinase β is a key regulator of synovial inflammation. Arthritis Rheum 44:1897–1907

Tang C, Prueksaritanont T (2010) Use of in vivo animal models to assess pharmacokinetic drug-drug interactions. Pharm Res 27:1772–1787

Thalhofer CJ, Graff JW, Love-Homan L, Hickerson SM, Craft N, Beverley SM, Wilson ME (2010) In vivo imaging of transgenic Leishmania parasites in a live host. J Vis Exp 41:1980. https://doi.org/10.3791/1980

Thompson TN (2000) Early ADME in support of drug discovery: the role of metabolic stability studies. Curr Drug Metab 1:215–241

Unsöld B, Schotola H, Jacobshagen C, Seidler T, Sossalla S, Emons J, Klede S, Knöll R, Guan K, El-Armouche A (2012) Age-dependent changes in contractile function and passive elastic properties of myocardium from mice lacking muscle LIM protein (MLP). Eur J Heart Fail 14: 430–437

Valgas C, de Souza SM, Smânia EF, Smânia A Jr (2007) Screening methods to determine antibacterial activity of natural products. Braz J Microbiol 38:369–380

Vincent TL, Williams RO, Maciewicz R, Silman A, Garside P (2012) Mapping pathogenesis of arthritis through small animal models. Rheumatology (Oxford) 51:1931–1941

Wang J, Zhang T, Zhu L, Ma C, Wang S (2015) Anti-ulcerogenic effect of Zuojin Pill against ethanol-induced acute gastric lesion in animal models. J Ethnopharmacol 173:459–467
Wang L, Fu Q, Dong Y, Zhou Y, Jia S, Du J, Zhao F, Wang Y, Wang X, Peng J, Yang S et al (2010) Bioluminescence imaging of Hepatitis C virus NS3/4A serine protease activity in cells and living animals. Antivir Res 8:50–56
Wei Y-j, Tang Y, Li J, Cui C-j, Zhang H, Zhang X-l, Zhang H, Hu S-s (2007) Cloning and expression pattern of dog SDF-1 and the implications of altered expression of SDF-1 in ischemic myocardium. Cytokine 40:52–59
Welsh C, Enomoto M, Pan J, Shifrin Y, Belik J (2013) Tetrahydrobiopterin deficiency induces gastroparesis in newborn mice. Am J Physiol Gastrointest Liver Physiol 305:G47–G57
Wen X, Westergard T, Pasinelli P, Trotti D (2017) Pathogenic determinants and mechanisms of ALS/FTD linked to hexanucleotide repeat expansions in the C9orf72 gene. Neurosci Lett 636: 16–26
WHO (1998) The World Health Report 1998: life in the 21st century a vision for all. The world health report 1998: life in the 21st century—a vision for all, p 241
WHO (2020) Coronavirus disease 2019 (COVID-19) situation report. https://www.who.int/emergencies/diseases/novel-coronavirus-2019/situation-reports
Witcomb LA, Collins JW, McCarthy AJ, Frankel G, Taylor PW (2015) Bioluminescent imaging reveals novel patterns of colonization and invasion in systemic Escherichia coli K1 experimental infection in the neonatal rat. Infect Immun 83:4528–4540
Witcomb LA, Czupryna J, Francis KP, Frankel G, Taylor PW (2017) Non-invasive three-dimensional imaging of Escherichia coli K1 infection using diffuse light imaging tomography combined with micro-computed tomography. Methods 127:62–68
Woolsey C, Borisevich V, Prasad AN, Agans KN, Deer DJ, Dobias NS, Heymann JC, Foster SL, Levine CB, Medina L (2020) Establishment of an African green monkey model for COVID-19. BioRxiv. 10.1101/2020.05.17.100289. Update in: Nat Immunol. 2020 Nov 24;: PMID: 32511377; PMCID: PMC7263506
Woolsey C, Borisevich V, Prasad AN, Agans KN, Deer DJ, Dobias NS, Heymann JC, Foster SL, Levine CB, Medina L (2021) Establishment of an African green monkey model for COVID-19 and protection against re-infection. Nat Immunol 22:86–98
Yao Y-F, Wang Z-J, Jiang R-D, Hu X, Zhang H-J, Zhou Y-W, Gao G, Chen Y, Peng Y, Liu M-Q (2021) Protective efficacy of inactivated vaccine against SARS-CoV-2 infection in mice and non-human primates. Virol Sin 36:879–889. https://doi.org/10.1007/s12250-021-00376-w
Yu P, Qi F, Xu Y, Li F, Liu P, Liu J, Bao L, Deng W, Gao H, Xiang Z (2020) Age related rhesus macaque models of COVID-19. Animal Model Exp Med 3:93–97
Yue L, Melnyk P, Gaspo R, Wang Z, Nattel S (1999) Molecular mechanisms underlying ionic remodeling in a dog model of atrial fibrillation. Circ Res 84:776–784
Zeng L, Hu Q, Wang X, Mansoor A, Lee J, Feygin J, Zhang G, Suntharalingam P, Boozer S, Mhashilkar A (2007) Bioenergetic and functional consequences of bone marrow–derived multipotent progenitor cell transplantation in hearts with postinfarction left ventricular remodeling. Circulation 115:1866–1875
Zhang Y, Huang K, Wang T, Deng F, Gong W, Hui X, Zhao Y, He X, Li C, Zhang Q (2021) SARS-CoV-2 rapidly adapts in aged BALB/c mice and induces typical pneumonia. J Virol 95:e02477–e02420
Zhao Z, Gong S, Wang S, Ma C (2015) Effect and mechanism of evodiamine against ethanol-induced gastric ulcer in mice by suppressing Rho/NF-κB pathway. Int Immunopharmacol 28: 588–595
Zheng L-F, Wang Z-Y, Li X-f, Song J, Hong F, Lian H, Wang Q, Feng X-Y, Tang Y-y, Zhang Y (2011) Reduced expression of choline acetyltransferase in vagal motoneurons and gastric motor dysfunction in a 6-OHDA rat model of Parkinson's disease. Brain Res 1420:59–67
Zhu N, Zhang D, Wang W, Li X, Yang B, Song J, Zhao X, Huang B, Shi W, Lu R (2020) A novel coronavirus from patients with pneumonia in China, 2019. N Engl J Med 382:727
Zon LI, Peterson RT (2005) In vivo drug discovery in the zebrafish. Nat Rev Drug Discov 4:35–44

Preclinical In Vivo Drug Development Studies: Limitations, Model Organisms, and Techniques

Seema Negi, Sanjay Kumar, and Ajeet Singh

Abstract The process of drug discovery and development encompasses target identification, validation, assay development, identification of hits, lead optimization, preclinical evaluation, and finally human clinical trials. Once a new chemical entity (NCE) is discovered, it progresses toward the development process that includes preclinical and clinical pharmacology. Preclinical research includes in silico, in vitro, ex vivo, and in vivo studies using cell lines, tissues, and animal models for predicting pharmacokinetic and pharmacodynamic properties of lead candidates. In vivo studies are critical in the drug development process because these investigations are useful to assess the properties of drugs and physiological and biochemical processes like adverse drug reactions and drug-drug interactions, which are difficult to examine in vitro. This chapter provides the detailed insight on in vivo studies that includes animal models and toxicology testing methodologies to identify a safe, potent, and efficacious drug. This chapter also highlights the importance of predictive and validated animal models for absorption, distribution, metabolism, and excretion (ADME) studies, along with the disease-based animal models for understanding disease pathophysiology that ultimately helps in making decisions that lead to human clinical trials for a drug candidate.

Keywords Drug discovery · Drug development · In vivo studies · Animal models · Pharmacokinetic · Toxicology · Techniques

---

S. Negi (✉)
Central Research Station, Subharti Medical College, Swami Vivekanand Subharti University, Meerut, U.P., India

S. Kumar
Glocal College of Paramedical Science and Research Centre, The Glocal University, Mirzapur Pole, Saharanpur, U.P., India

School of Life and Allied Health Sciences, The Glocal University, Mirzapur Pole, Saharanpur, U.P., India

A. Singh
Department of Pharmaceutical Sciences, J. S. University, Shikohabad, Firozabad, U.P., India

V. S. Rajput, A. Runthala (eds.), *Drugs and a Methodological Compendium*,
https://doi.org/10.1007/978-981-19-7952-1_6

1 Introduction

The process of discovering possible new medicines is known as drug discovery and development. It involves a broad range of scientific disciplines, including biology to molecular biology, chemistry to computational chemistry, and pharmacology to molecular pharmacology and takes an average of 10 to 15 years to bring a single drug into market (Csermely et al. 2013; Hughes et al. 2011). The first steps in this process are carried out largely by basic studies, and their findings facilitate the identification of potential new targets for drug discovery. The whole procedure of drug discovery and development follows a defined process and is guided by regulatory requirements, with the goal of avoiding excessive costs by eliminating unlikely drug candidates early on (Haber and Spaventi 2017). A schematic diagram of the overall drug discovery and development process is depicted in Fig. 1. The process is divided into the following five primary steps: drug discovery, preclinical research, clinical research/trial, Food and Drug Administration (FDA) review, and FDA post-market safety monitoring with three subdivisions under clinical research (Nys and Fillet 2018). Thousands of compounds are assessed before moving on to the preclinical step of the drug development process, which takes an average of 6 years. Target identification and lead discovery are the first steps in drug development, which can then advance to the preclinical stage for determining the drug's efficacy and safety. The new drug is biologically evaluated in preclinical studies for

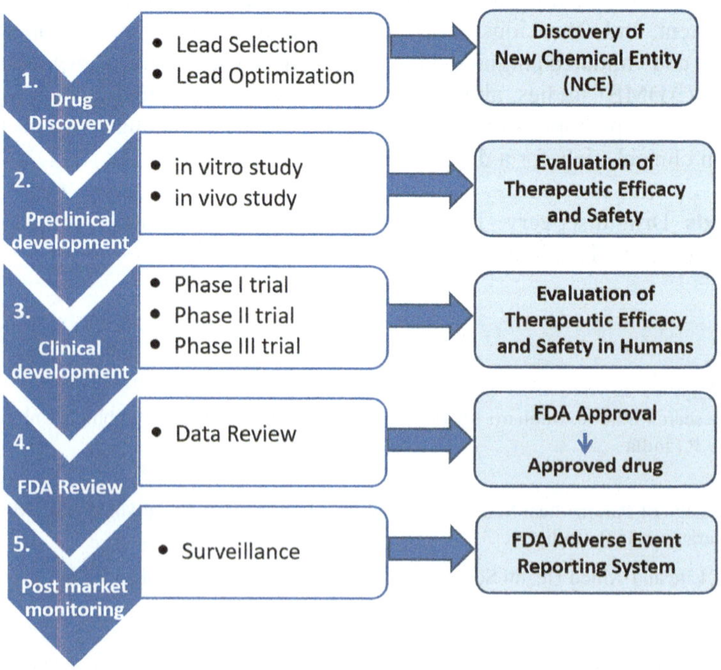

Fig. 1 An illustration of the stages involved in the drug discovery and development process with major strategies and aims at different phases

pharmacological and toxicological effects, as well as potential therapeutic applications. In vitro and in vivo studies are used in the preclinical stage to develop a safe and effective drug that can then be assessed in clinical trials. For assessing the safety, efficacy, and pharmacology of a drug in humans, clinical trials are further divided into three phases (Bjorklund et al. 2002; Lipsky and Sharp 2001; Martini et al. 2001). The procedure then moves from clinical trials to FDA approval, with the FDA either approving or rejecting the drug following its evaluation. If the application is denied, the applicant is given an explanation for the rejection of the application as well as the information to enhance the claim (Lipsky and Sharp 2001).

Validation procedures used in preclinical investigations range from in vitro (studies performed in cell lines and tissues separated from living organisms) to in vivo (studies performed on laboratory animals). In vivo studies are critical to determine the safety, bioequivalence, dosing regimen, adverse drug reactions, and drug-drug interactions in a living system, as well as to monitor and observe the drug's long-term effects. In vitro-in vivo correlation (IVIVC) data is utilized to choose suitable excipients and optimize the formulation process for quality control, leading in lower total costs (Nainar et al. 2012; Segovia-Zafra et al. 2021). Fig. 2

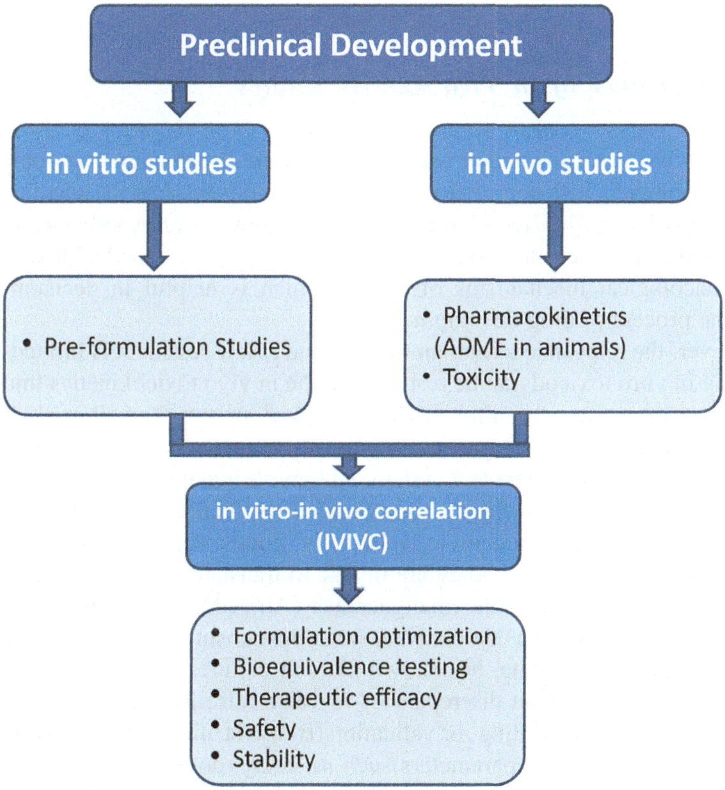

Fig. 2 A model representing preclinical development including in vitro and in vivo studies to predict in vitro-in vivo correlation (IVIVC)

depicts the role of in vitro and in vivo studies for predicting IVIVC at the preclinical phase of drug discovery and development. In vitro studies not only reduce the cost of a drug testing but also reduce ethical conflicts and experimental restriction. In vivo studies are important for drug development because these studies are useful for assessing a drug's characteristics such as therapeutic effects, side effects, drug metabolism, and drug-drug interactions that are difficult to detect in vitro. This chapter's aim is to highpoint the importance of in vivo studies in drug development, discussing in detail various diseases including cancer and metabolic, genetic, cardiovascular, and neurodegenerative diseases based on in vivo animal models.

2 Preclinical In Vivo Studies of ADME in Drug Development

Preclinical studies are used for establishing a starting, safe dose for first-in-human studies and to analyze the molecule's potential toxicity, which usually includes prescription drugs as well as diagnostics and new medical devices.

2.1 Importance of In Vivo ADME Studies

Routine in vivo experimentation was mostly used to screen at an early point in the drug development phase before target-directed methods became the standard. Many essential drugs (e.g., thiazide diuretics, benzodiazepines, cyclosporin) were discovered on the basis of their in vivo effects. In vitro assays provide valuable data related to pharmacological mechanisms of action, which is helpful in decision-making during the process of drug development.

However, the relevance to human toxicity and risk assessment is limited without corelating in vitro toxicodynamic results with the in vivo toxicokinetics findings, as in vivo systems accurately mimic a live biological system (Sewell et al. 2017). In vitro studies can anticipate organ and organ system interactions with drugs, as well as drug-drug interactions inside a system, and give a quantitative data of ADME in animal and human models (Singhvi and Singh 2011; Pellegatti 2012).

In vitro studies are unable to accurately mimic the system's micro- and macroenvironment; therefore, they are unable to translate in vivo at the preclinical stage in the case of metabolic malignancies (Amoedo et al. 2017). Anticancer activity of benzimidazole derivatives, amidino-substituted benzimidazole and benzimidazo[1,2-a]quinoline, has shown 2D cell cultures were comparable to 3D cell cultures, but significant discrepancies revealed false-positive findings that ultimately require in vivo profiling for validation (Brajsa et al. 2015). In vivo research is essential to assess various parameters such as safety, dosage schedule, bioequivalence, effects of the drug, side effects, and drug interactions to develop a safe and effective drug (Pellegatti 2012). In vivo findings are multifactorial, combining the effects of permeability, distribution, metabolism, and excretion to produce a valuable data related to pharmacokinetic parameters and toxicological endpoints. Although in vitro assays are useful to determine various parameters, animal studies are essential to analyze the therapeutic effect and potential toxicities associated with the drugs (Sewell et al. 2017). There are a wide variety of animals used in preclinical in vivo studies. Rodents are commonly used in animal testing, particularly mice and rats. Since they are low cost and only need a little quantity of test chemical, they are the first animal species utilized to assess drug exposure. Laboratory rats and mice provide ideal animal models for drug development because the anatomy and physiology of rats and mice are more similar to humans. Similarly, rats, mice, and humans each contain about 30,000 genes, with 95% of them being shared by all three species (Waterston et al. 2002; Bryda 2013). In vivo rat investigations can highlight ADME issues with a novel chemical series, for example, whether there is a low absorption level or high level of clearance, resulting in unacceptable pharmacokinetics (PK).

2.2 *Challenges to Design In Vivo Studies for Drug Discovery and Development*

Drugs, chemical drugs, or biologics such as antibodies, vaccines, and peptides can be administered into the body through different routes of administration such as mouth (gastrointestinal lining), upper respiratory airway (pulmonary epithelium), and intravenous (vascular endothelium). Intravenous route is particularly used for tumor vasculature and blood-brain barrier targeting. Biological barriers typically occur during the delivery of lead drugs to target areas, and these barriers have a considerable impact on drug bioavailability and potential therapeutic action. To reach the blood compartment, the lead molecule(s) must pass through epithelia of the lung or gastrointestinal (GI) tract, tumoral vascular endothelial lining, or the blood-brain barrier (BBB) (Sjogren et al. 2014). The pharmacokinetic parameters are influenced by the in vivo effect of the drug and the biological barriers related to drug delivery.

2.3 *Limitations of In Vivo Studies*

In vivo studies provide many detailed information in the drug development process, but there are few limitations that warrant attention. About 75% of drugs flop in phase II or phase III human clinical trials owing to lack of efficacy or safety data (Van Norman 2019). Dependence on non-human animal models in preclinical investigations remains a major factor in this failure. It is difficult to anticipate the efficacy and safety of a drug in small animals like mice because of fundamental biological differences (Pound and Ritskes-Hoitinga 2018; Weaver and Valentin 2019; Khalil et al. 2020; Ferreira et al. 2020). The shape, size, and regenerating capability of tissues and organs, along with physiological variations in immunology, metabolism, and drug transportation, all affect drug development in humans and small animals (Pound and Ritskes-Hoitinga 2018; Weaver and Valentin 2019; Khalil et al. 2020). Large animal models, for instance, dogs, pigs, and non-human primates, are more identical to human anatomy and physiology and therefore can ameliorate the predictive worth of in vivo models (Tsang et al. 2016; Ziegler et al. 2016; Khalil et al. 2020). Nevertheless, large animal models increase cost, time, and more ethical considerations significantly. Additionally, there is a remarkable difference between humans and animals at molecular, genetic, cellular, anatomical, and physiological levels (Pound and Ritskes-Hoitinga 2018; Weaver and Valentin 2019; Khalil et al. 2020). Therefore, there is requirement of biological models based on human tissue for better representation of human biology (Khalil et al. 2020; Pound 2020). Novel in vitro and in vivo preclinical models that mimic human tissues are needed to address this constraint.

3 Preclinical Animal Models Used for ADME Parameter Optimization in Drug Discovery and Development

Animal models are required for bridging the preclinical-clinical research gap. Preclinical research in fit-for-purpose animal models may improve success rates of drugs during clinical development (Pound and Ritskes-Hoitinga 2020). In vivo experiments can be designed to determine efficacy in a specific biological model based on early findings from in vitro and ex vivo research, in addition to information regarding therapeutic target, clinical symptoms, and pharmacokinetic profile of the drug candidate. Furthermore, the scientifically relevant in vivo studies will be selected on a case-by-case manner. There are three types of disease models to choose from: physiological, pharmacological, and genetically engineered animal models (genetic) (Andrade et al. 2016). All of these models are intended to develop abnormalities that are comparable to those seen in the disease under investigation. Furthermore, depending on the duration of the disease, the in vivo models may be classified as acute or chronic (Andrade et al. 2016).

When evaluating the efficacy of a new chemical entity in preclinical in vivo studies, it is also crucial to establish the therapeutic target/protein. In the case of *Mus musculus* and *Rattus norvegicus*, proof-of-principle assays (proof-of-concept testing) are usually done, and if no association with the target is detected, the animal studies will not give significant results. Using several animal species during the drug development process is one of the primary reasons of failure because of the variances between species and the complications in translating the findings to humans (Oreff et al. 2021). Indeed, the pathophysiology of a variety of diseases varies significantly between species (Mestas and Hughes 2004; Wang and Urban 2004). Furthermore, the ADME profile in animals and humans is frequently different, which might cause variations in the duration of the test substance action, influencing both pharmacology and toxicology and ultimately leading to ambiguous findings (Martignoni et al. 2006).

3.1 The Role of Animal Models

Animal models are frequently used in drug absorption, metabolism, distribution, and excretion investigations, and the scientist's ability to increase and improve human and animal well-being is wholly dependent on breakthroughs in research employing animal models for evaluating pharmacological properties in vivo (Landskroner et al. 2011). Although animal models can provide helpful information regarding a substance's nonclinical efficacy, they are not able to replicate all of the indications and symptoms of human disease pathology, and ultimate efficacy confirmation can only be validated when phase II clinical trials are completed. Before a medicine is investigated in humans, it must first be tested in an animal model to record toxicity, side effect, and drug interactions, among other pharmacokinetic parameters. Furthermore, animal studies are required to assist development in the early stages, especially when deciding whether to precede with the human research studies (go/no-go choice).

Human disease-based animal models are only considered significant if these models aid in the improvement of intervention and therapy techniques by recapitulating disease pathophysiology. A model must be able to precisely represent the morphological and biochemical components of the pathophysiology and also able to imitate the typical physiology and anatomy of human organs as well as tissues of interest.

Scientists use models to create an artificial condition in a lab animal that mimics the etiology of human disease. There are several animal models, both vertebrates and invertebrates that may be used to study disease pathogenesis that affects both humans and animals. Invertebrate animal models including zebrafish, *Caenorhabditis elegans*, and *Drosophila melanogaster* have been widely used in neuroscience, genetics, and metabolic and cancer research. A variety of vertebrate animal including mice, rats, guinea pigs, rabbits, cats, dogs, and primates are essential for translational research in biomedical sciences (Harman et al. 2020).

3.2 Choosing an Animal Model

Animal models are of three main types: homologous, isomorphic, or predictive, which is largely determined by the study objective (Davidson et al. 1987; Brake et al. 2017). An appropriate model for preclinical in vivo studies would be the disease-based animal model that shares the same pathophysiology as humans and recapitulates the disease phenotype and responds to human treatments in a way that is analogous to humans. In every way, human physiology, pathology, and treatment are replicated in homologous animal models (Davidson et al. 1987; Brake et al. 2017). Isomorphic models have identical symptoms to humans, although they are not represented by the same events (Davidson et al. 1987; Brake et al. 2017). Predictive animal models are not similar to human disorders; yet they do allow for some comparisons or predictions of human disease, treatment, and treatment effect (Davidson et al. 1987; Brake et al. 2017). Mechanisms of actions, pharmacokinetics, and pharmacodynamics along with biomarkers and safety and toxicity of prospective therapies must also be determined using animal models that correctly mimic disease pathophysiology (Franklin et al. 2022). In therapeutic studies, such animal models might potentially help in assessing human dose (Sim and Kauser 2015). Availability, cost-effectiveness, ethical concern, ease of handling, housing requirements, and disease vulnerability are all factors to consider while choosing an animal model (Brake et al. 2017).

Various animal models including invertebrate and vertebrate are being used in preclinical drug testing. In pharmacological research involving neurological, genetic, and developmental problems, invertebrate animal models are frequently employed (Wilson-Sanders 2011). The zebrafish is one such creature that is frequently utilized (Zon and Peterson 2005; Takaki et al. 2018). This model is particularly useful when researchers are looking for a disease model that is both embryologically and genetically docile (Lieschke and Currie 2007). Traditionally, mice, rats, guinea pigs, rabbits, dogs, baboons, cows, and macaques have been used as vertebrate models. In translational research, these models may be the most useful (Hickman et al. 2017).

When selecting an animal model for preclinical studies, general principles such as a large number of results and a related life cycle must be considered. If a large number of findings are required, an invertebrate would be an excellent candidate; nevertheless, relevancy of life cycle and ability of the biological sample must be considered. A pair of zebrafish may produce a large number of embryos every week, leading to the generation of huge results and findings at low cost (Kari et al. 2007). Although the zebrafish is a more popular animal model, genetically engineered mice, rats, dogs, and non-human primates are commonly used animal models in drug testing (Hickman et al. 2017; Khan et al. 2018; Sobczuk et al. 2020). Furthermore, biochemical and physiological resemblance between animal models and humans, as well as the underlying mechanism of drug ADME in animals, should be considered while choosing the suitable animals (Tang and Prueksaritanont 2010). There are several instances of well-established animal models that have been utilized to study particular diseases (Khan et al. 2018). In addition, when compared to human physiological and biochemical parameters, including blood pH, blood volume, organ blood flow, tissue distribution, localization of metabolizing enzymes, and drug transporters, are used for selecting an appropriate animal model (Tang and Prueksaritanont 2010).

4 Human Disease Models for the Preclinical In Vivo Studies

Human disease animal models are extremely useful for the advancement of innovative and effective diagnostic and therapeutic approaches. These models are the valuable tool to understand the disease pathology and also helpful for assessing the safety of novel drugs. Animal models have provided new insights and an extensive knowledge of the onset and diagnosis of human disease. They have been used to evaluate new chemical entity and other biologics, like vaccines, hormones, and antibodies (Gong et al. 2020).

Methodological advancements in recombinant DNA technology now allow for precise manipulation of laboratory animals, such as the introduction or the deletion of a gene. These advances have resulted in the production of transgenic and knockout (KO) mice, which are useful tools for understanding the molecular basis of various human diseases and also for the development of novel medication and therapeutic procedure for the treatment of diseases. Current trends in animal models indicate the inclusion of advance technologies like genetic manipulations and stem cell technology, which may be even more potent in the development of successful drugs, vaccines, and medical devices (Gong et al. 2020; Gong et al. 2021). The mostly used models are listed below.

4.1 Mouse Model

The drug discovery and development process has been transformed due to advancement in genetic engineering. As a result, genetic engineered mouse models have emerged as precious tools for modeling of human disease and drug development. Transgenic mouse models with knock-in and knockout technologies have proven effective in basic and applied research to find answers to fundamental questions (Table 1). Furthermore, more advanced mouse models are essential for cutting-edge research.

LDLR$^{2/2}$ Mouse Model

The LDLR$^{2/2}$ mice have been used to study familial hypocholesterolemia (low-density lipoprotein (LDL) receptor). The plasma lipoprotein profile of these mice is comparable to that of humans because of the mutation in LDLR. On a typical chow diet, the genetic abnormality causes a delay in the disposal of very low-density lipoprotein (VLDL) and LDL from the plasma, resulting in an elevated plasma level of cholesterol (Bentzon and Falk 2010). A high-cholesterol and high-fat diet worsens lesions associated with atherosclerosis and hypercholesterolemia in LDLR$^{2/2}$ mouse model (Knowles and Maeda 2000).

Table 1 Disease-based mouse models for preclinical in vivo studies (modified from Khan et al. 2018)

Diseases	Mouse models	References
Atherogenesis	$ApoE^{2/2}$ mice	Plump et al. (1992)
Hypercholesterolemia	Calcium chloride–induced abdominal aortic aneurysm (AAA)	Freestone et al. (1997)
Aneurysm	Spontaneous mutant mouse strain	Brophy et al. (1998)
Hyperlipidemia and atherosclerosis	Mutant E3L, ApoE mice	Leppanen et al. (1998)
Colon cancer	Human colorectal cancer (CRC) cell lines in mice	Rashidi and Gamagami (2000)
Familial hypocholesterolemia	$LDLR^{2/2}$ mice	Jawien et al. (2004)
Diabetic cardiomyopathy and atherosclerosis	$LDLR^{2/2}$ mice and $ApoE^{2/2}$ mice	Hayek et al. (2005)
Colon cancer	C57BL/6 mice applying murine colon adenocarcinoma (MCA) cells	de Jong and Aarts (2009)
Liver diseases	Fatty liver disease-associated mouse model	Chung et al. (2010)

On a normal chow diet, $LDLR^{2/2}$ and ApoE double-deficient mice ($LDLR^{2/2}ApoE^{2/2}$) could indeed develop severe atherosclerosis and hyperlipidemia. As a result, these models make it easier to study diseases without having to worry about feeding atherogenic diets to the mice (Jawien et al. 2004).

$ApoE^{2/2}$ Mouse Model

In 1992, two different embryonic stem cell research groups employed the homogeneous recombination technique to produce ApoE mice (Zhang et al. 1992; Plump et al. 1992; Zhang et al. 1992). A homogeneous loss in the ApoE gene causes plasma levels of VLDL and LDL to rise, resulting in the inability of the LDL receptor and associated proteins to function. It was the first mouse model to display a wide range of atherogenesis lesions, making it the first mouse model to resemble human-like lesions (Plump et al. 1992).

Transgenic Mouse Model

The use of transgenic mice in the research of hyperlipidemia and atherosclerosis is common, and the mutant ApoE3 Leiden (E3L) and ApoE (Arg 112-Cys-142) are often utilized transgenic mice in such studies. These transgenic mice have a lipoprotein profile that is analogous to the profile of people having dysbetalipoproteinemia (Hofker et al. 1998). The E3L mice exhibit the features of human vasculopathy in mild, moderate, and severe atherosclerotic plaques (Leppanen et al. 1998).

Diabetes-Associated Atherosclerosis Model

One of the primary causes of cardiovascular disease is diabetes. The LDLR$^{2/2}$ and ApoE$^{2/2}$ mouse models are frequently used to examine diabetes-related cardiomyopathy and atherosclerosis. Injecting the models with viral injections or streptozotocin causes them to develop type 1 diabetes (Shen and Bornfeldt 2007). Streptozotocin injections cause calcification in the proximal aorta as well as atherosclerosis inside aortic sinus, abdominal aorta, and carotid artery in mice (Khan et al. 2018).

Abdominal Aortic Aneurysm (AAA) Calcium Chloride-Induced Model

This model was created initially in rabbits and subsequently in mice. Calcium chloride was injected intravenously into the region between the iliac bifurcation and the renal artery during the model's development. The aorta dilates significantly after 14 days, resulting in the formation of an aneurysm. Calcium chloride and thioglycolate can be used to augment the severity. The animals can also be fed a high-cholesterol diet to get similar outcomes (Freestone et al. 1997).

Spontaneous Mutant Mouse Model

In the X chromosome, a spontaneous mutation was done to create the blotchy mouse model, which results in an abnormal shift in the rate of intestinal copper absorption. This mutant model develops aneurysms in the thoracic aorta, aortic arch, and abdominal aorta because of inadequate cross-linkage within collagen and weaker elastin tissues. However, as mutation leads to several effects, besides aneurysm, it becomes difficult to interpret the results drawn from such models (Brophy et al. 1998).

Liver Metastasis Mouse Model

In roughly 50% to 60% of patients, liver metastasis develops in the colorectal area. Better treatment options are urgently needed to extend the life span of patients suffering from this condition. A system for animal trials on rodents was devised for this purpose (de Jong and Aarts 2009). Immunocompetent rodents were used in this study because they have an advantage in that their immune systems are similar to those of patients with colorectal cancer that develop metastases. Therefore, to induce liver metastases, first, the mice were examined for immunotherapy effects, and then the human colorectal cancer (CRC) cells were inoculated in five different locations of the animal, including the colonic wall, subcutaneous, intraportal, intrasplenic, and intrahepatic (Kobaek-Larsen and Thorup 2000). The advantage of employing this model is believed that they exhibit pathologic behavior that is quite comparable to human pathological behavior.

Colon Cancer Mouse Model

To establish a hepatic tumor model, scientists used orthotopic injection of tumors into the cecal walls. The intrasplenic or intrahepatic injection of tumor cells is similar to the hematogenous spread of tumor cells in the liver. Moreover, these models are useful to create macroscopic metastases about in all cells within the entire body of the animal. The C57BL/6 mouse model with MCA cells and Wistar, WAG/Rij, or BDIX rat models having N-methyl-N-nitrosoguanidine-induced adenocarcinoma cells, CC531, or DHDK12/TR colon cancerous cells are the most useful animal models of hepatic tumor (Burtin et al. 2020). Most of the desired qualities are covered by injecting heterotopic syngeneic tumor cells into immunocompetent animals (de Jong and Aarts 2009; Ben-david et al. 2019; Guerin et al. 2020).

Fatty Liver Disease Mouse Model

The complication of the metabolic syndrome is non-alcoholic steatohepatitis (Zivkovic et al. 2007). Choline- and methionine-deficient diet is provided to the non-alcoholic steatohepatitis mouse models. The particular diet causes an increase in liver triglycerides and total bilirubin levels in the blood, fibrosis, and hepatic steatosis. Ultimately, mice not only had dramatically reduced overall weight but also liver weight and total protein concentration. Non-alcoholic steatohepatitis has been developed in these mice without showing any other indications of metabolic syndrome (Chung et al. 2010).

Neurodegenerative Disease Mouse Model

Parkinson's Disease

In the study of neurodegenerative illnesses, mouse models have shown to be invaluable. They have been shown to be a good model organism for Parkinson's disease (PD). PD is a degenerative neurological condition characterized by a deficiency of dopaminergic neurons (DNs) within the substantia nigra as well as extensive buildup of the protein α-synuclein, which results in motor deficits and eventually cognitive dysfunction (Youssef et al. 2019; Shadrina and Slominsky 2021).

Alzheimer's Disease Model

A new transgenic mouse model, APPPS1, has been developed with strain C57 black 6/Jackson (C57BL/6 J) genetic background. The transgenic mouse model has been co-expressed with KM670/671NL-mutated amyloid precursor protein (APP) and

Table 2 Transgenic mouse models of neurodegenerative diseases (modified from Khan et al. 2018)

Disease	Name of the model	Target gene	References
Alzheimer's disease	APPPS1	Co-expression of KM670/671NL-mutated APP and L166P-mutated presenilin 1	Francis et al. (2009)
Parkinson's disease	KO mice	Overexpression of α-synuclein with mutations in familial A53T or A30P	Janus and Welzl (2010)

L166P-mutated presenilin 1 controlled by a neuron-specific Thy1 promoter element. The APPPS1 mouse models are suitable tools for Alzheimer's disease research due to the early development of amyloid plaques, known genetic background, and ease of breeding (Francis et al. 2009). In Table 2, transgenic mouse models of Alzheimer's and Parkinson's disease are summarized.

Heart Failure Models

The ligation of the left coronary artery is a way of producing myocardial injury in rats and mice that permanently occludes arteries. Partially obstructed arteries have been found in recent investigations to generate comparable effects (Michael et al. 1995). As this method has proven to be effective, cryoinjuries are currently being used to cause cardiac injury in rat and mouse models (Ryu et al. 2010).

4.2 *Rat Models*

Rat models have speeded preclinical in vivo cardiovascular disease research. To generate myocardial injury in the rat heart, three methods are typically used: surgical, electrical, and pharmacological. Myocardial damage is caused in the rat by ligating the left coronary artery (Pfeffer et al. 1979). Isoproterenol, an agonist of the β-1 adrenergic receptor, was first used to inflict pharmacological damage in the heart tissue in 1963. Isoproterenol has a cardioprotective effect when given before ischemia, but when given at a proper dose, it produces myocyte necrosis, severe hypertrophy, and left ventricular dilatation. This method has been used to investigate the fundamental mechanisms of heart attacks (Zbinden and Bagdon 1963) as well as to better understand the role of potential heart attack prevention drugs.

In order to cause electrically induced myocardial damage in rats, an electric shock is delivered to the left ventricle of the heart. Though this is a highly validated method for causing cardiac injury, its results are not shown to be consistently reproducible (Adler et al. 1976).

Celiac Disease Rat Model

Celiac disease is classified as "immune-mediated small intestinal enteropathy." It is caused by the intake of gluten in the diet. Gluten causes this reaction exclusively in people who are genetically prone to the condition. The condition is diagnosed by looking for serum antibodies produced by the body's reaction to the enzyme tissue transglutaminase 2. Gluten-dependent enteropathies are studied in vivo using gluten-sensitized rat models. There are two types of rat models: HLA independent and HLA dependent. An HLA-independent model was developed based on the T-cell transfer colitis model that was used to investigate chronic inflammatory bowel disease of the colon in a rat (Freitag and Rietdijk 2009). In RAG1 mice, expansion of crypt hyperplasia and villous atrophy was induced by giving gluten orally and transferring in vitro gliadin primer. When Wistar rats were given gluten orally plus INF-γ intraperitoneal injection, they exhibited lower villus height, higher TNF levels, and cellular infiltrates within the small intestinal lamina propria (Laparra and Olivares 2012). As a result, the progression of disease-based animal models provided us a plethora of novel therapeutic targets and numerous pathways for testing that could ultimately lead to prevention of celiac disease and support to the discovery leading to the chain of events accountable for the disease (Laparra and Olivares 2012; Costes and Meresse 2015).

Nile Rat

Nile rat (*Arvicanthis niloticus*), also branded as African grass rat, has been used as an animal model for obesity and diabetes studies. Metabolic disease develops in these rats when a high-fat diet is given to them, but wild type rats do not develop diabetes (Noda et al. 2010). These rats show signs of dyslipidemia and hyperglycemia at the age of 1 year. Other symptoms, including abdominal fat deposition, hypertension, hyperinsulinemia, and liver steatosis, have also been shown in these animal models. They provide significant results in case of metabolic diseases when a regular diet is given to them, in contrast to people who are fed a high-fat, high-carbohydrate diet (Noda et al. 2010; Chaabo et al. 2010).

4.3 Porcine Model

Pigs are particularly valuable model organisms in the preclinical stage of drug development, especially for research on neurobiology, as anatomical and physiological properties of pigs are similar to humans. Different technologies have been utilized to generate genetically modified pigs, including DNA microinjection into pronuclei for zygote collected from super-ovulated women, lentivirus and retrovirus gene translation into swine oocytes, sperm-mediated gene transfer, and nuclear

Table 3 Transgenic pig models for Alzheimer's disease, Huntington's disease, Parkinson's disease, and amyotrophic lateral sclerosis (modified from Khan et al. 2018)

Disease	Name of the model	Mutation in genes	References
Alzheimer's disease	Gottingen minipig model for AD	Mutation in amyloid precursor protein (APP) gene	Swindle et al. (2012)
Parkinson's disease	Minipig models for PD	Homolog of FBXO7 gene	Swindle et al. (2012)
Huntington's disease	Transgenic HD (TgHD) minipig model for HD	Mutation in N-terminal HTT (huntingtin) fragment (208 amino acids and 105 Q)	Bassols et al. (2014)
Amyotrophic lateral sclerosis (ALS)	Transgenic pig model for ALS	Mutant G93A hSOD1 gene expression	Yang et al. (2014)

Table 4 Zebrafish models used in neurodegenerative diseases (modified from Khan et al. 2018)

Disease	Name of the model	Genes involved	References
Alzheimer's disease	Zebrafish model for AD	Two homologs of APP	Newman et al. (2007)
Parkinson's disease	Zebrafish model for PD	*DJ-1* gene expression (DJ-1 knockdown zebrafish)	Best and Alderton (2008)
Epilepsy	Zebra fish models for epilepsy	Expression of early proto-oncogenes, e.g., *c-fos* *Mind-bomb* mutant zebrafish, zebrafish Nav1.1 mutants	Hortopan and Baraban (2011), Kalueff et al. (2013)

transfer and cloning (Dolezalova et al. 2014). Pigs are also a perfect model for research on accelerated atherosclerosis in the presence of diabetes and hypercholesterolemia because they resemble the instability of human plaques (Gerrity et al. 2001). The porcine models for coronary atherosclerosis make it easier to study vascular remodeling, adventitial neovascularization, and the makeup of atherosclerotic plaque (Alviar et al. 2010). Table 3 summarizes transgenic pig models utilized in in vivo research.

4.4 Zebrafish Model

Zebrafish has become quite prominent for neurological research. Brain cell processes in both normal and diseased conditions have been studied using adult and larval zebrafish as models. The commonly used zebrafish models for preclinical in vivo research are tabulated in Table 4.

4.5 *Rabbit Models*

Rabbit models have largely been utilized in cardiovascular disease research to see how statins or diet affects cholesterol levels and plaque formation. These findings increased our understanding of pathways involved in atheroma inflammatory processes including accumulation of macrophage and lipid reduction, as mentioned further below (Khan et al. 2018).

Inflammation-Associated Atherosclerosis Rabbit Models

In magnetic resonance imaging (MRI) quantification studies, rabbits are employed as animal models to determine and image the atherosclerotic aortic component (Helft et al. 2001). Although aortic arteries of rabbits are lesser in diameter than human carotid arteries, they are extensively used in developing endovascular therapies. Furthermore, rabbits have numerous benefits as models for cardiovascular disease research, the most notable of which is the high degree of resemblance between the appearance of aneurysm in rabbits and the incidence of aneurysm in humans. Because they can be readily checked in the femoral artery, rabbit aneurysms are useful models for researching endovascular treatments (Dai et al. 2008).

Myocardial Damage Rabbit Model

Rabbits are useful models for studying myocardial damage because their sarcomere protein composition is comparable to that of humans. The rabbit strain WHHLMI serves as a non-surgical model of spontaneous myocardial infarction. The strain was created via selective breeding of WHHL rabbits with coronary atherosclerosis. A fundamental flaw in this model is the deficiency in plaque formation, conflicting with true myocardial infarction and related with coronary plaque rupture and intravascular thrombosis (Kuge et al. 2010; Shin et al. 2021).

5 In Vivo Research Techniques

As mentioned earlier, before a drug can be approved, its metabolism and drug interactions must be properly studied. For analyzing specific in vivo properties of a drug, a variety of methodologies and sampling protocols are available. Many approaches including equilibrium dialysis, microdialysis, isolated lung perfusion, and imaging techniques are widely employed for determining the distribution of a drug of interest. Advanced techniques, for example, microdialysis, positron emission tomography (PET), and magnetic resonance spectroscopy (MRS), provide a number of advantages over traditional approaches such as saliva sampling, tissue biopsy, and skin blister fluid sampling, to name a few. These methods have a number of advantages, including a semi-invasive method, direct concentration measurement, multiple location measurement, continuous monitoring, low technological complexity, and low cost (Brunner and Langer 2006). These techniques are briefly exemplified for their functional roles.

5.1 *Equilibrium Dialysis*

Equilibrium dialysis is utilized for determining how much ligand is bound to a macromolecule (Lanao and Fraile 2005). Despite the fact that there is no standard for measuring in vitro protein binding, equilibrium dialysis remains routinely employed to determine therapeutic protein binding characteristics (Zeitlinger et al. 2011).

5.2 *Isolated Organ Perfusion*

By using a single pass or recirculation with the medium, the isolated organ perfusion technique can keep an organ alive. Distribution studies use a single pass, whereas metabolism and excretion investigations benefit from recirculation. This technique is often employed in distribution investigations involving various organs, including kidney, lung, and brain (Lanao and Fraile 2005). Chemotherapy is given to the target organ without disrupting the functionality of other organs using these isolated organ perfusion techniques.

5.3 *Microdialysis*

Microdialysis remains a preferable technique for assessing the pharmacokinetics of a drug as it is extremely valuable to determine in vivo protein bonding (Zeitlinger et al. 2011). It is a powerful semi-invasive sampling technique especially effective for explaining drug distribution and receptor phase pharmacokinetics (Brunner and Langer 2006). Microdialysis enables simultaneous monitoring of a number of physiological parameters including locomotor activity, convulsive activity, and blood pressure, making it an appropriate tool for drug pharmacokinetic-pharmacodynamic studies. The reverse microdialysis approach (Hocht et al. 2004; Rudin and Weissleder 2003) is a strong and effective tool for studying local drug effects in diverse tissues, particularly liver, brain nuclei, and skeletal muscle.

5.4 *Imaging Techniques*

Non-invasive imaging techniques such as autoradiography, magnetic resonance imaging (MRI), magnetic resonance spectroscopy (MRS), and positron emission tomography (PET) are routinely utilized in in vivo drug distribution research (Brunner and Langer 2006). Because imaging technologies are non-invasive, they may be used to conduct longitudinal investigations on a single animal, and that statistically enhances the significance of a study (Rudin and Weissleder 2003).

Neuroimaging techniques give precise anatomical, functional, and metabolic details of the human or animal brain in real time, which helps researchers better understand drug impacts on brain systems. MRI and PET can be used to explore disease pathogenesis in vivo, diagnose patients, and offer quantitative markers for disease status assessment (Wise and Tracey 2006; Gustafsson et al. 2017). Early biomarkers associated with neurological diseases, for example, epilepsy, brain tumors, PD, schizophrenia and Alzheimer's disease (AD), are commonly identified using these techniques (Wise and Tracey 2006; McGuire et al. 2008; Bertoglio et al. 2017; Zhao et al. 2017).

6 Conclusion

Preclinical in vivo studies are essential for assessing the pharmacokinetic and pharmacodynamic properties of drugs during development. These studies are necessary since in vitro research cannot provide quantitative data on absorption, distribution, metabolism, and excretion in animal and human models. The animal models are crucial in the process of drug discovery and development, and they have played a critical role in elucidating the critical processes behind many deadly human diseases. Animal models that are more similar to the human genome have shown to be very useful in drug development and discovery. These animals were chosen for their physiological and biochemical parallels to humans, as well as their underlying drug absorption, distribution, metabolism, and excretion systems. Transgenic models can modify the genetic composition of animal models, which is beneficial to examine the molecular mechanisms of human genome-related activities and develop new medications and testing procedures. Many modern techniques, such as MRI and microdialysis, have gradually superseded traditional approaches, such as skin blistering, in in vivo investigations. Undeniably, in vivo studies are the required stage in the drug discovery and development process; however, considerable effort remains to be done in order to make animal study results more comparable to human clinical trials.

References

Adler N, Camin LL, Shulkin P (1976) Rat model for acute myocardial infarction: application to technetium labeled glucoheptonate, tetracycline, and polyphosphate. J Nucl Med 17(3): 203–207

Alviar CL, Tellez A, Wallace-Bradley D et al (2010) Impact of adventitial neovascularisation on atherosclerotic plaque composition and vascular remodelling in a porcine model of coronary atherosclerosis. EuroIntervention 5(8):981–988

Amoedo N, Obre E, Rossignol R (2017) Drug discovery strategies in the field of tumor energy metabolism: limitations by metabolic flexibility and metabolic resistance to chemotherapy. Biochimica et Biophysica Acta (BBA)-Bioenergetics 1858:674–685

Andrade EL, Bento AF, Cavalli J et al (2016) Non-clinical studies required for new drug development - part I: early *in silico* and *in vitro* studies, new target discovery and validation, proof of principles and robustness of animal studies. Braz J Med Biol Res 49(11):e5644

Bassols A, Costa C, Eckersall PD et al (2014) The pig as an animal model for human pathologies: a proteomics perspective. Proteomics Clin Appl 8(9–10):715–731

Ben-david U, Beroukhim R, Golub TR (2019) Genomic evolution of cancer models: perils and opportunities. Nat Rev Cancer 19:97–109

Bentzon JF, Falk E (2010) Atherosclerotic lesions in mouse and man: is it the same disease? Curr Opin Lipid 21(5):434–440

Bertoglio D, Verhaeghe J, Dedeurwaerdere S et al (2017) Neuroimaging in animal models of epilepsy neuroscience. Neuroscience 358:277–299

Best J, Alderton WK (2008) Zebrafish: an *in vivo* model for the study of neurological diseases. Neuropsychiatr Dis Treat 4(3):567

Bjorklund LM, Sanchez-Pernaute R, Chung S et al (2002) Embryonic stem cells develop into functional dopaminergic neurons after transplantation in a Parkinson rat model. Proc Natl Acad Sci USA 99:2344–2349

Brajsa K, Vujasinovic I, Jelic D, Trzun M, Zlatar I et al (2015) Antitumor activity of amidino-substituted benzimidazole and benzimidazo [1, 2-a] quinoline derivatives tested in 2D and 3D cell culture systems. J Enzyme Inhib Med Chem 31:1139–1145

Brake K, Gumireddy A, Tiwari A et al (2017) *In vivo* studies for drug development via Oral delivery: challenges, animal models and techniques. Pharm Anal Acta 8(9):1–11

Brophy CM, Netzer D, Forster D (1998) Detonation studies of JP-10 with oxygen and air for pulse detonation engine development. AIAA Paper 98:4003

Brunner M, Langer O (2006) Microdialysis versus other techniques for the clinical assessment of *in vivo* tissue drug distribution. AAPS J 8:E263–E271

Bryda EC (2013) The mighty mouse: the impact of rodents on advances in biomedical research. Mo Med 110(3):207–211

Burtin F, Mullins CS, Linnebacher M (2020) Mouse models of colorectal cancer: past, present and future perspectives. World J Gastroenterol 26(13):1394–1426

Chaabo F, Pronczuk A, Maslova E et al (2010) Nutritional correlates and dynamics of diabetes in the Nile rat (Arvicanthis niloticus): a novel model for diet-induced type 2 diabetes and the metabolic syndrome. Nutr Metab 7:29

Chung S, Yao H, Caito S et al (2010) Regulation of SIRT1 in cellular functions: role of polyphenols. Arch Biochem Biophys 501(1):79–90

Costes LM, Meresse B (2015) The role of animal models in unravelling therapeutic targets in coeliac disease. Best Pract Res Clin Gastroenterol 29(3):437–450

Csermely P, Korcsmáros T, Kiss HJ et al (2013) Structure and dynamics of molecular networks: a novel paradigm of drug discovery: a comprehensive review. Pharmacol Ther 138:333–408

Dai D, Ding YH, Danielson MA et al (2008) Endovascular treatment of experimental aneurysms with use of fibroblast transfected with replication-deficient adenovirus containing bone morphogenetic protein-13 gene. Am J Neuroradiol 29(4):739–744

Davidson M, Lindsey J, Davis J (1987) Requirements and selection of an animal model. Isr J Med Sci 23:551–555

Dolezalova D, Hruska-Plochan M, Bjarkam CR et al (2014) Pig models of neurodegenerative disorders: utilization in cell replacement-based preclinical safety and efficacy studies. J Comp Neurol 522(12):2784–2801

Ferreira GS, Veening-Griffioen DH, Boon WPC, Moors EHM, Peter JK, Van Meer (2020) Levelling the translational gap for animal to human efficacy data. Animals (Basel) 10(7):1199

Francis YI, Fa M, Ashraf H et al (2009) Dysregulation of histone acetylation in the APP/PS1 mouse model of Alzheimer's disease. J Alzheimer Dis 18(1):131–139

Franklin MR, Platero S, Saini KS et al (2022) Immuno-oncology trends: preclinical models, biomarkers, and clinical development. J Immunother Cancer 10(1):e003231. https://doi.org/10.1136/jitc-2021-003231

Freestone T, Turner RJ, Higman DJ et al (1997) Influence of hypercholesterolemia and adventitial inflammation on the development of aortic aneurysm in rabbits. Arterioscler Thromb Vasc Biol 17(1):1017

Freitag TL, Rietdijk S (2009) Gliadin-primed CD4þCD45RBlowCD25-T cells drive gluten-dependent small intestinal damage after adoptive transfer into lymphopenic mice. Gut 58(12): 1597–1605

Gerrity RG, Natarajan R, Nadler JL et al (2001) Diabetes-induced accelerated atherosclerosis in swine. Diabetes 50(7):1654–1665

Gong W, Liang Y, Mi J, Jia Z, Xue Y, Wang J, Wang L, Zhou Y, Sun S, Wu X (2021) Peptides-based vaccine MP3RT induced protective immunity against mycobacterium tuberculosis infection in a humanized mouse model. Front Immunol 12:666290

Gong W, Liang Y, Wu X (2020) Animal models of tuberculosis vaccine research: an important component in the fight against tuberculosis. Biomed Res Int 2020:4263079

Guerin MV, Finisguerra V, Van Den Eynde BJ (2020) Preclinical murine tumor models: a structural and functional perspective. elife 9:1–24

Gustafsson S, Eriksson J, Syvanen S et al (2017) Combined PET and microdialysis for *in vivo* estimation of drug blood-brain barrier transport and brain unbound concentrations. NeuroImage 155:177–186

Haber VE, Spaventi R (2017) Discovery and development of novel drugs. Blue Biotechnol J 55:91–104

Harman NL, Sanz-Moreno A, Papoutsopoulou S, Lloyd KA, Ameen-Ali KE, Macleod M, Williamson PR (2020) Can harmonisation of outcomes bridge the translation gap for pre-clinical research? A systematic review of outcomes measured in mouse models of type 2 diabetes. J Transl Med 18:468

Hayek T, Hussein K, Aviram M et al (2005) Macrophage foam cell formation in streptozotocin-induced diabetic mice: stimulatory effect of glucose. Atherosclerosis 183(1):2533

Helft G, Worthley SG, Fuster V et al (2001) Atherosclerotic aortic component quantification by noninvasive magnetic resonance imaging: an *in vivo* study in rabbits. J Am Coll Cardiol 37(4): 1149–1154

Hickman DL, Johnson J, Vemulapalli TH et al (2017) Commonly used animal models. Principles of animal research for graduate and undergraduate students. Academic Press, pp 1–175

Hocht C, Opezzo JA, Taira CA (2004) Microdialysis in drug discovery. Curr Drug Discov Technol 1:269–285

Hofker MH, Van Vlijmen BJM, Havekes LM (1998) Transgenic mouse models to study the role of APOE in hyperlipidemia and atherosclerosis. Atherosclerosis 137(1):1–11

Hortopan GA, Baraban SC (2011) Aberrant expression of genes necessary for neuronal development and notch signaling in an epileptic mind bomb zebrafish. Dev Dynam 240(8):1964–1976

Hughes JP, Rees S, Kalindjian SB et al (2011) Principles of early drug discovery. Br J Pharmacol 162:1239–1249

Janus C, Welzl H (2010) Mouse models of neurodegenerative diseases: criteria and general methodology. Methods Mol Biol 602:323–345

Jawien J, Nastałek P, Korbut R (2004) Mouse models of experimental atherosclerosis. J Physiol Pharmacol 55(3):503–517
de Jong GM, Aarts F (2009) Animal models for liver metastases of colorectal cancer: research review of preclinical studies in rodents. J Surg Res 29:167–176
Kalueff AV, Gebhardt M, Stewart AM et al (2013) Towards a comprehensive catalog of zebrafish behavior 1.0 and beyond. Zebrafish 10(1):70–86
Kari G, Rodeck U, Dicker AP (2007) Zebrafish: an emerging model system for human disease and drug discovery. Clin Pharm Therap 82:70–80
Khalil AS, Jaenisch R, Mooney DJ (2020) Engineered tissues and strategies to overcome challenges in drug development. Adv Drug Deliv Rev 158:116–139
Khan A, Waqar K, Shafique A, Irfan R, Gul A (2018) In vitro and in vivo animal models: the engineering towards understanding human diseases and therapeutic interventions. Omics Technologies and Bio-Engineering Towards Improving Quality of Life 1:431–448
Knowles JW, Maeda N (2000) Genetic modifiers of atherosclerosis in mice. Arterioscler Thromb Vasc Biol 20(11):2336–2345
Kobaek-Larsen M, Thorup I (2000) Review of colorectal cancer and its metastases in rodent models: comparative aspects with those in humans. Comp Med 50(1):16
Kuge Y, Takai N, Ogawa Y et al (2010) Imaging with radiolabelled anti-membrane type 1 matrix metalloproteinase (MT1-MMP) antibody: potentials for characterizing atherosclerotic plaques. Eur J Nucl Med Mol Imaging 37:2093–2104
Lanao J, Fraile M (2005) Drug tissue distribution: study methods and therapeutic implications. Curr Pharm Des 11:3829–3845
Landskroner KA, Hess P, Treiber A (2011) Surgical and pharmacological animal models used in drug metabolism and pharmacokinetics. Xenobiotica 41(8):687–700
Laparra JM, Olivares M (2012) Bifidobacterium longum CECT 7347 modulates immune responses in a gliadin-induced enteropathy animal model. PLoS One 7(2):e30744
Leppanen P, Luoma JS, Hofker MH et al (1998) Characterization of atherosclerotic lesions in apo E3-leiden transgenic mice. Atherosclerosis 136(1):147–152
Lieschke GJ, Currie PD (2007) Animal models of human disease: zebrafish swim into view. Nat Rev Genet 8:353–367
Lipsky MS, Sharp LK (2001) From idea to market: the drug approval process. J Am Board Fam Pract 14:362–367
Martignoni M, Groothuis GM, de KR. (2006) Species differences between mouse, rat, dog, monkey and human CYP-mediated drug metabolism, inhibition and induction. Expert Opin Drug Metab Toxicol 2:875–894
Martini L, Fini M, Giavaresi G et al (2001) Sheep model in orthopedic research. A Literature Review Comp Med 51:292–299
McGuire P, Howes OD, Stone J et al (2008) Functional neuroimaging in schizophrenia: diagnosis and drug discovery. Trends Pharmacol Sci 29:91–98
Mestas J, Hughes CC (2004) Of mice and not men: differences between mouse and human immunology. J Immunol 172:2731–2738
Michael LH, Entman ML, Hartley CJ et al (1995) Myocardial ischemia and reperfusion: a murine model. Am J Phys 269(6):H2147–H2154
Nainar S, Rajiah K, Angamuthu S et al (2012) Biopharmaceutical classification system in *in vitro-in vivo* correlation: concept and development strategies in drug delivery tropical. J Pharm Res 11:319–329
Newman M, Musgrave F, Lardelli M (2007) Alzheimer disease: amyloidogenesis, the presenilins and animal models. Biochim Biophys Acta 1772(3):285–297
Noda K, Melhorn MI, Zandi S et al (2010) An animal model of spontaneous metabolic syndrome: nile grass rat. FASEB J 24(7):2443–2453
Nys G, Fillet M (2018) Microfluidics contribution to pharmaceutical sciences: from drug discovery to post marketing product management. J Pharm Biomed Anal 159:348–362

Oreff GL, Fenu M, Vogl C, Ribitsch I, Jenner F (2021) Species variations in tenocytes' response to inflammation require careful selection of animal models for tendon research. Sci Rep 11:12451
Pellegatti M (2012) Preclinical *in vivo* ADME studies in drug development: a critical review. Expert Opin Drug Metab Toxicol 8(2):161–172
Pfeffer MA, Pfeffer JM, Fishbein MC (1979) Myocardial infarct size and ventricular function in rats. Circ Res 44(4):503–512
Plump AS, Smith JD, Hayek T et al (1992) Severe hypercholesterolemia and atherosclerosis in apolipoprotein E- deficient mice created by homologous recombination in ES cells. Cell 71(2): 343–353
Pound P (2020) Are animal models needed to discover, develop and test pharmaceutical drugs for humans in the 21st century? Animals (Basel) 10(12):2455
Pound P, Ritskes-Hoitinga M (2018) Is it possible to overcome issues of external validity in preclinical animal research? Why most animal models are bound to fail. J Transl Med 16:1–8
Pound P, Ritskes-Hoitinga M (2020) Can prospective systematic reviews of animal studies improve clinical translation? J Transl Med 18:15
Rashidi B, Gamagami R (2000) An orthotopic mouse. Clin Cancer Res 6(6):2556–2561
Rudin M, Weissleder R (2003) Molecular imaging in drug discovery and development. Nat Rev Drug Discov 2:123–131
Ryu JUH, Kim ILK, Cho SW et al (2010) Implantation of bone marrow mononuclear cells using injectable fibrin matrix enhances neovascularization in infarcted myocardium. Biomaterials 26(3):319–326
Segovia-Zafra A, Daniel E, Di Zeo-Sánchez, López-Gómez C et al (2021) Preclinical models of idiosyncratic drug-induced liver injury (iDILI): moving towards prediction. Acta Pharm Sin B 11(12):3685–3726
Sewell F, Aggarwal M, Bachler G et al (2017) The current status of exposure-driven approaches for chemical safety assessment: a cross-sector perspective. Toxicology 389:109–117
Shadrina M, Slominsky P (2021) Modeling Parkinson's disease: not only rodents? Front Aging Neurosci 13:695718
Shen X, Bornfeldt KE (2007) Mouse models for studies of cardiovascular complications of type 1 diabetes. Ann N Y Acad Sci 1103:202–217
Shin HS, Shin HH, Shudo Y (2021) Current status and limitations of myocardial infarction large animal models in cardiovascular translational research. Front Bioeng Biotechnol 9:673683
Sim DS, Kauser K (2015) *In vivo* target validation using biological molecules in drug development. Hand book of Exp Pharmacol 232:59–70
Singhvi G, Singh M (2011) Review: *in vitro* drug release characterization models. Int J Pharm Stud Res 2:77–84
Sjogren E, Abrahamsson B, Augustijns P et al (2014) *In vivo* methods for drug absorption: comparative physiologies, model selection, correlations with *in vitro* methods (IVIVC), and applications for formulation/ API/excipient characterization including food effects. Eur J Pharm Sci 57:99–151
Sobczuk P, Brodziak A, Khan MI et al (2020) Choosing the right animal model for renal cancer research. Transl Oncol 13(3):100745
Swindle M, Makin A, Herron A et al (2012) Swine as models in biomedical research and toxicology testing. Vet Pathol 49(2):344–356
Takaki K, Ramakrishnan L, Basu S (2018) A zebrafish model for ocular tuberculosis. PLoS One 13(3):e0194982
Tang C, Prueksaritanont T (2010) Use of *in vivo* animal models to assess pharmacokinetic drug-drug interactions. Pharm Res 27:1772–1787
Tsang HG, Rashdan NA, Whitelaw CBA et al (2016) Large animal models of cardiovascular disease. Cell Biochem Funct 34(3):113–132
Van Norman G (2019) Limitations of animal studies for predicting toxicity in clinical trials: is it time to rethink our current approach? J Am Coll Cardiol Basic Trans Science 4:845–854

Wang J, Urban L (2004) The impact of early ADME profiling on drug discovery and development strategy. Drug Discov World Fall 5:73–86
Waterston RH, Lindblad-Toh K, Birney E et al (2002) Initial sequencing and comparative analysis of the mouse genome. Nature 420(6915):520–562
Weaver RJ, Valentin J (2019) Today's challenges to de-risk and predict drug safety in human "mind-the-gap". Soc Toxicol 167:307–321
Wilson-Sanders SE (2011) Invertebrate models for biomedical research, testing, and education. ILAR J 52(2):126–152
Wise RG, Tracey I (2006) The role of fMRI in drug discovery. J Magn Reson Imaging 23:862–876
Yang H, Wang G, Sun H et al (2014) Species-dependent neuropathology in transgenic SOD1 pigs. Cell Res 24(4):464–481
Youssef K, Tandon A, Rezai P (2019) Studying Parkinson's disease using Caenorhabditis elegans models in microfluidic devices. Integr Biol (Camb) 11(5):186–207
Zbinden G, Bagdon RE (1963) Isoproterenol-induced heart necrosis, an experimental model for the study of angina pectoris and myocardial infarct. Rev Can Biol 22:257–263
Zeitlinger MA, Derendorf H, Mouton JW et al (2011) Protein binding: do we ever learn? Antimicrob Agents Chemother 55:3067–3074
Zhang SH, Reddick RL, Piedrahita JA et al (1992) Spontaneous hypercholesterolemia and arterial lesions in mice lacking apolipoprotein E. Science 258(5081):468–471
Zhao Y, Raichle ME, Wen J et al (2017) *In vivo* detection of microstructural correlates of brain pathology in preclinical and early Alzheimer disease with magnetic resonance imaging. Neuro Image 148:296–304
Ziegler A, Gonzalez L, Blikslager A (2016) Large animal models: the key to translational discovery. Cell Mol Gastroenterol Hepatol 2:716–724
Zivkovic M, German JB, Sanyal AJ (2007) Comparative review of diets for the metabolic syndrome: implications for nonalcoholic fatty liver disease. Am J Clin Nutr 86(2):285–300
Zon LI, Peterson RT (2005) *In vivo* drug discovery in the zebrafish. Nat Rev Drug Discov 4:35–44

Peptide-Based Therapeutics and Drug Delivery Systems

Aman Kumar Mahto, Shalini Kumari, Saleem Akbar, Shweta Paroha, Pravat Kumar Sahoo, Ajay Kumar, and Rikeshwer Prasad Dewangan

Abstract Peptides are functional biomolecules that hold great potential for the discovery of new therapeutics and drug delivery systems. Structurally, these are short molecules, but they possess similar functionality of large proteins. Peptides can overcome the physiological barriers posed by the diseases due to their intrinsic properties. In recent years, bioactive peptides such as antimicrobial, anticancer, antithrombotic, antidiabetic, and anti-Alzheimer effects have been successfully identified, and many of them are in clinical trials. Apart from that, peptides are also reported for the drug delivery applications as targeting molecules or as self-assembling soft nano-materials. This chapter offers a brief introduction to peptides as therapeutics and drug delivery agents and their application in the management of different diseases through targeted therapy. Here, we discuss the recent development on peptides as carriers to penetrate different physiological barriers such as gastrointestinal (GI) tract and blood-brain barriers.

Keywords Peptides · Therapeutic peptides · Drug delivery systems · Self-assembly · Nanocarrier

---

A. K. Mahto · S. Akbar · R. P. Dewangan (✉)
Department of Pharmaceutical Chemistry, School of Pharmaceutical Education and Research, Jamia Hamdard (Deemed to be University), New Delhi, India
e-mail: rpdewangan@jamiahamdard.ac.in

S. Kumari
CSIR-Institute of Genomics and Integrative Biology, Mathura Road Sukhdev Vihar, New Delhi, India

S. Paroha · P. K. Sahoo
Department of Pharmaceutics, Delhi Institute of Pharmaceutical Sciences and Research (DIPSAR), Delhi Pharmaceutical Sciences and Research University, New Delhi, India

A. Kumar
Government Pharmacy College, BRD Medical College Campus, Gorakhpur, India

V. S. Rajput, A. Runthala (eds.), *Drugs and a Methodological Compendium*,
https://doi.org/10.1007/978-981-19-7952-1_7

1 Introduction

The first synthetic peptide Gly-Gly was prepared by Emil Fischer a century ago (Fischer and Fourneau 1906), who later won the Nobel Prize for the same (Goodman et al. 2003). After the Second World War, Vigneaud's group in the USA and the Swiss industrial groups of Robert Schwyzer (Ciba) and Huguenin (Sandoz) successfully prepared pure peptides that led to the use of synthetic peptides in clinical applications (Loffet 2002). Earlier peptide synthesis was a time-consuming and challenging task. A single peptide took a couple of years to produce using the standard method; thus, peptides were often perceived as the "drug of the future." Later in 1963, Bruce Merrifield discovered an easy and fast technique for synthesis of peptides based on solid-phase synthesis, and he was awarded the Nobel Prize in chemistry in 1984 (Merrifield 1985, 1986, 1997).

Peptides are a chain of amino acid residues linked together in a specific manner by amide bonds. Also, peptides are a stretch or short fragment of proteins, which are essential biomolecules that play various physiological roles in biological systems (Joseph et al. 2017b). Bioactive peptides are oligopeptides (typically containing 2–50 amino acids) that exhibit various biological properties through interactions with enzymes and receptors. The discovery of new peptide molecules is becoming increasingly important due to small size and nontoxic nature, similar to that of drug molecules. Furthermore, advances in conventional chemical synthesis and recombinant DNA techniques have increased the production of these peptides and peptide-based therapeutics. Peptides, both natural and synthetic, exhibit a wide range of biological responses, making them useful as therapeutic and diagnostic agents in biotechnology (Kaspar and Reichert 2013; Joseph et al. 2017b).

Human diseases and disorders are associated within a localized area of an organ (e.g., solid tumors, wounds, arthritis, etc.) or in the whole body through systemic circulations (e.g., sarcoidosis, multiple sclerosis, Crohn's disease, etc.). Localized lesions, wounds, arthritis, or any organ-specific diseases can be treated by the local applications of the drugs in the lesions or by oral administration. However, the systemically administered drugs get distributed throughout the body, including the non-disease parts, which might cause some toxicity to the normal tissues or organs. On the other hand, local application may not penetrate the drugs to the affected tissues or cells to achieve the optimal therapeutic concentration. Hence, the targeted drug delivery vehicles have been adopted for site-specific delivery in a controlled manner. Peptide-based delivery vehicles are preferred for targeted and controlled delivery of many drugs and bio-therapeutics by entrapment or covalent conjugations. The smart approach of peptide-based systems is widely accepted for their diversified structures, environmental responsive behavior, selective receptor targeting, cell-penetrating, and scalable chemical synthesis (Tesauro et al. 2019a). Apart from DDS, there are more than 50 peptide-based active pharmaceutical ingredients available in the pharmaceutical market, and over 100 peptides are in several clinical phases (Zompra et al. 2009). The conventional advantages of peptides (e.g., high bioloigcal activity and lower toxicity) and current trends of peptide research encouraged pharmaceutical industries for development of new peptide-based therapeutics. At present, peptides have achieved sound success in therapeutics design and smart drug delivery system (DDS), which are notable challenges in drug delivery technology from last decades. In this chapter, we have discussed the therapeutic and drug delivery aspects of peptides with various examples.

2 Classifications of Bioactive Peptides

On the basis of the application, the bioactive peptides can be classified in the following classes:

- Therapeutic peptides.
- Stimulus-responsive peptides.
- Targeting peptides.
- Cell-penetrating peptides.

2.1 Therapeutic Peptides

Peptides represent a certain category of pharmaceutical drugs that has similarity between small heterocyclic drug molecules and proteins in terms of molecular size but different in therapeutic and biochemical aspects from both. Throughout the period, significant development on peptide therapeutics included it as an important class of therapeutic molecules, which leads to change in the treatment paradigm (Lau and Dunn 2018). In the early twentieth century, peptides extracted from natural sources were used to create life-saving drugs (e.g., insulin and adrenocorticotropic hormone (ACTH)). When peptide sequence elucidation and chemical synthesis became possible, several synthetic peptides such as oxytocin and vasopressin entered into clinical application. Afterward, the isolation of bioactive peptides from exotic sources (such as venoms from arthropods and cephalopods) became a widespread trend for identifying novel potential therapeutics (Lau and Dunn 2018). However, protease instability and poor pharmacokinetic properties were major obstacles for peptide therapeutics. The presence of serum proteases and metabolizing enzymes that inactivate and eliminate peptides from the body results in short half-life of many peptide hormones (Mentlein 2004). Researchers have emphasized on making peptides similar to drug-like molecules by utilizing various pharmaceutical approaches by improving the protease stability and receptor specificity (Lombardino and Lowe 2004).

Consequently, improved hormonal peptide analogues have entered into clinical use. Chemically, peptides represent diversified structures, demonstrating that size and complex structure do not restrict their role as therapeutic agents. For example, 60-residue peptide ecallantide (Bernstein and Qazi 2010) and 31-residue peptide liraglutide (Iepsen et al. 2015) are larger peptides, whereas telavancin (lipoglycopeptide) (Smith and Drew 2009), romidepsin (a natural bicyclic depsipeptide) (Kitagaki et al. 2015), boceprevir, and telaprevir (Wilby et al. 2012) are shorter peptides without any natural amino acid. Icatibant is a synthetic peptide peptidomimetic containing both natural and non-natural amino acid residues used for the treatment of hereditary angioedema (Table 1) (Vlieghe et al. 2010). In the following paragraph, we will discuss emerging therapeutic categories of peptides with numerous examples.

Antimicrobial Peptides

All the living organisms ubiquitously produce many antimicrobial peptides (e.g., defensins and cathelicidins) as a part of their innate immunity against pathogens. The sequences of antimicrobial peptides are diversified in which most of them are short, cationic, and acquire amphipathic secondary structures in non-polar environments. These peptides exhibits broad-spectrum antibacterial activity with lower tendency to select for resistance compared to conventional antibiotics. In contrast to cell selectivity, these peptides are neutral toward the mammalian cells, whereas antimicrobial peptides tend to disrupt the anionic bacterial cell membranes through electrostatic interactions (Lien and Lowman 2003; Marr et al. 2006).

Anticancer Peptides

Anticancer peptides (ACPs) are short oligopeptides that usually contain less than 100 amino acid residues. They can inhibit growth and migration of tumor cell proliferations and reduce the risk of drug resistance. Because ACPs are highly specific for cancer cells and therefore these are utilizing as an excellent tools for a targeted drug delivery system in cancer chemotherapy. According to the National Cancer Institute (NCI), these peptides have emerged as one of the most promising therapeutic ligands in the field of nanotechnology to fight cancer, and several anticancer drugs are transported by these nanoparticles (Xie et al. 2020). Moreover, peptides can self-assemble in various nanostructures, e.g., nanospheres, nanotubes, and nanofibers, etc. Anticancer drugs such as curcumin, paclitaxel, and doxorubicin can be entrapped into these self-assembled peptides (as peptide-drug conjugates), several of which have already been studied in preclinical and clinical studies. Peptide drug conjugates have become popular due to easy procedure, low cost, and versatility in targeted administration (Debnath et al. 2021).

Antidiabetic Peptides

Higher organisms like mammals secrete peptide hormones in response to certain stimuli, such as insulin, to control the blood sugar level. These peptide hormones are

Table 1 Examples of different therapeutic peptides (Vlieghe et al. 2010)

Peptide class	Peptide name (INNs)	Sequence	Companies	Application	Reference
Anticancer peptides [GnRH and analogues (agonists)]	Buserelin	Peptidomimetics	Sanofi-Aventis	Prostate cancer	Kuhn et al. (1989)
	Gonadorelin or GnRH or LHRH	Pyr-H-W-S-Y-G-L-R-P-G-NH_2	Sanofi-Aventis, Wyeth Pharmaceuticals, Baxter healthcare, Ferring Pharmaceuticals	Gonadotropin secretion stimulator during fertility disturbances. Diagnosis and monitoring of the anterior pituitary gonadotrophs	Sansone et al. (2021)
	Goserelin	Peptidomimetics	AstraZeneca Pharmaceuticals	Prostate cancer of advanced stage and breast cancer	Perry and Brogden (1996)
	Histrelin	Peptidomimetics	Endo Pharmaceuticals, Roberts pharma, Shire	Prostate cancer and breast cancer	Hou and Flaig (2012)
	Leuprolide or leuprorelin	Peptidomimetics	Abbott, Bayer, Genzyme, Johnson & Johnson, Sanofi-Aventis, Takeda, Teva, Wyeth, etc.	Breast and prostate cancer and central precocious puberty (a condition that causes early sexual development in kids)	Marberger et al. (2010)
Antidiabetic peptides	Exenatide	H-G-E-G-T-F-T-S-D-L-S-K-Q-M-E-E-E-A-V-R-L-F-I-E-W-L-K-N-G-G-P-S-S-G-A-P-P-P-S-NH_2	Amylin Pharmaceuticals, Eli Lilly	Type 2 diabetes mellitus	Kolterman et al. (2005)
	Liraglutide (GLP-1 receptor agonist)	H-A-E-G-T-F-T-S-D-V-S-S-Y-L-E-G-Q-A-A-N6-[N-(1-oxohexadecyl)-L-g-E]-K-E-F-I-A-W-L-V-R-G-R-G-OH	Novo Nordisk	Chronic weight management and type 2 diabetes	Iepsen et al. (2015)
	Pramlintide acetate	K-c[C-N-T-A-T-C]-A-T-Q-R-L-A-N-F-L-V-H-S-S-N-N-F-G-P-I-L-P-P-T-N-V-G-S-N-T-Y-NH_2, acetate	Amylin pharma	Both type 1 and type 2 diabetes	Singh-Franco et al. (2007)

(continued)

Table 1 (continued)

Peptide class	Peptide name (INNs)	Sequence	Companies	Application	Reference
Anti-HIV peptides	Enfuvirtide	Ac-Y-T-S-L-I-H-S-L-I-E-E-S-Q-N-Q-Q-E-K-N-E-Q-E-L-L-E-L-D-K-W-A-S-L-W-N-W-F-NH_2	Roche	AIDS/HIV-1 infections	Patel et al. (2005)
Cardiovascular peptides	Bivalirudin	(D)F-P-R-P-G-G-G-G-N-G-D-F-E-E-I-P-E-E-Y-L-OH	Nycomed pharma	Anticoagulant, antianginal	Sangeetha et al. (2019)
	Ularitide	T-A-P-R-S-L-R-R-S-S-C(1)-F-G-G-R-M-D-R-I-G-A-Q-S-G-L-G-C(1)-N-S-F-R-Y-OH	Cardiorentis, Switzerland	Treatment of acute heart and kidney failure	Emani et al. (2015)
	Eptifibatide	c[Mpa-homo(R)-G-D-W-P-C]-NH_2	GSK, Schering-Plough	Coronary artery syndrome and unstable angina undergoing PCI	Giugliano et al. (2005)
	Icatibant acetate	(D)R-R-P-Hyp-G-T-S-(D)-tic-Oic-R-OH, acetate	Jerini AG	Hereditary angioedema	Gras (2009)
Antimicrobial peptide	Daptomycin	N-(decanoyl)-W-(D)-N-D-T-G-Orn-D-l-(D)A-D-G-(D)S-T-(3-methyl)-glutamyl-(3-anthraniloyl)-A-ε1-lactone	Eagle Pharmaceuticals, Phosphagenics ltd., Xellia Pharmaceuticals, Cubist Pharmaceuticals	Gram-positive infection	Tedesco and Rybak (2004)
Peptides acting on CNS	Cilengitide	c[R-G-D-(D)F-(N-me)-V]	Merck Serono	Glioblastoma multiforme	Lassen et al. (2014)
	Taltirelin hydrate	Peptidomimetics	Tanabe Seiyaku	Spinocerebellar	Brown (2001)
	Ziconotide acetate	Peptidomimetics	Elan pharmaceuticals	Severe chronic pain	Atanassoff et al. (2000); Smith and Deer (2009)
		Peg-[H-G-E-G-T-F-T-S-D-L-S-K-Q-M-E-E-E-A-V-R-L-F-I-E-	Neuraly Inc.	Neurodegenerative disorders	Yun et al. (2018)

	PEGylated exendin-4 (NLY01)	W-L-K-N-G-G-P-S-S-G-A-P-P-P-S]			
Angiotensin II receptor blocker	Saralasin	S-R-V-Y-V-H-P-A-OH	Procter & gamble	Hypertension	Phillips et al. (1977)
ACE inhibitors	Enalapril maleate	(S)-1-[N-[1-(ethoxycarbonyl)-3-phenylproline	Merck Sharp & Dohme, Biovail Pharmaceuticals, Sandoz-Novartis, Apothecon, Genpharm, Mylan Pharmaceuticals, Ivax Pharmaceuticals, sun pharma, Taro, Teva, Wockhardt	Hypertension	Kubo and Cody (1985)
	Lisinopril	(S)-1-[N2-(1-carboxy-3-phenylpropyl)-K]-P-OH	Merck Sharp & Dohme, Apotex, Ivax pharms, LEK Pharmaceuticals, Lupin, Mylan, Par Pharmaceutical, Ranbaxy, Sandoz-Novartis Pharma, Teva Pharmaceuticals, Watson labs, west Ward, Wockhardt	Hypertension and congestive heart failure	Thomson et al. (1989)

responsible for normal homeostasis and regulate a wide range of cellular functions when released into the bloodstream. It was not long enough before diabetes was recognized as a leading cause of death during the early twentieth century, with no effective treatment insight. A lack of pancreatic hormone was suggested as the cause of diabetes at that time. In 1921, Banting and Best discovered insulin by successfully isolating the insulin from the pancreatic islets, which later proved to be a life-saving drug (Wetzler and Hamilton 2018). Glucose-dependent insulinotropic polypeptide (GIP) and glucagon-like peptide 1 (GLP-1) are endogenous peptides that control the blood glucose level in response to certain stimuli (Wetzler and Hamilton 2018).

Analgesic Peptides

Venoms from marine cone snails contain conopeptide, which attracted many researchers as it targets a wide range of receptors, transporters, and ion channels. Considering that only a small fragment of conopeptides have been investigated so far, making these peptides is a potentially untouched resource for drug discovery (Lewis 2012; Schroeder and Craik 2012). Efforts have been made toward the modifications of conopeptides to improve their biological activities for accelerating their clinical application. The in vivo study showed that several conopeptides exhibit antinociceptive activity, and χ-conopeptides (Xen2174) undergoing clinical trials and ω-conotoxin MVIIA (Prialt®) are in the market to treat severe pain. Conopeptides are used as probes to determine the role of many essential membrane peptides and proteins in normal and disease physiology (Vetter and Lewis 2012; Lewis 2012). Some endogenous analgesic peptides like endorphins have a very high selectivity toward μ (mu), opioid receptors in mammals, and are considered a potential lead for the discovery of analgesic peptides (Gu et al. 2017).

Tree frog skin contained novel analgesic peptides named Analgesin-HJ (sequence: Phe-Trp-Pro-Val-Ile-NH_2) and Analgesin-HJ(I5T) (sequence: Phe-Trp-Pro-Val-Thr-NH_2). It consists of 171 amino acid residues, which encodes Analgesin-HJs (Zhu et al. 2014).

Anti-Alzheimer Peptides

Alzheimer is a neurodegenerative central nervous system (CNS) disorder that results in a continuous process of degeneration of neuronal cells, which increasingly deteriorate over time. To date, there has not been any single potential therapeutic agent available that can be proven beneficial for patients suffering from these neurodegenerative disorders. Therefore, novel drugs or early diagnosis tools are urgently required, explicitly identifying and selectively targeting the origin of the disease. Exendin-4 is an antidiabetic peptide having agonistic activity like glucagon peptide receptor agonist that promotes insulin secretion. NLY01 is a PEGylated exendin-4 peptide that has been shown to protect dopaminergic neuronal loss by suppressing microglia (Yun et al. 2018). NLY01 was found well tolerated in early clinical trials, with a longer half-life than non-PEGylated GLP-1R agonists. NLY01 has been recently approved investigational new drug against Alzheimer's and Parkinson's diseases (Sterling et al. 2020).

Anti-HIV Peptides

AIDS is one of the global health issues caused by human immunodeficiency virus type 1 (HIV-1), and it is essential to discover novel therapeutics with clinical efficacy. Enfuvirtide is a peptide-derived antiviral drug that acts as a fusion inhibitor and prevents the penetration of HIV-1 in mammalian cells. It is efficacious in the early stages of HIV infection along with other anti-HIV drugs (Kapić et al. 2005).

Cardiovascular Peptides

Cardiovascular disorders are among the major healthcare problems responsible for an estimated 17.9 million death toll every year. The role of peptides in cardiovascular functions has been investigated as an alternative treatment option. Angiotensin II and its analogues, associated with the renin-angiotensin-aldosterone system, are a classic example of it. Angiotensin-converting enzyme inhibitors (ACE inhibitors) and angiotensin receptor antagonists have been highly prescribed clinical drugs for many cardiovascular diseases. Some other examples of peptide drugs used in different cardiovascular problems are as follows:

- Ularitide is a 20-mer cysteine knot peptide that completed phase III clinical trials for the treatment of heart and kidney failure introduced by Cardiorentis biopharmaceutical in 2018 (Emani et al. 2015).
- Eptifibatide is a cyclic heptapeptide, derived from a disintegrin protein, used as an antiplatelet drug (O'Shea and Tcheng 2002).
- Bivalirudin (Hirulog®, Angiomax®), a peptide drug, which, unlike heparin, is a direct inhibitor of thrombin and is utilized in indications such as percutaneous coronary intervention (thrombosis). It is cleared by both proteolytic cleavage and renal mechanisms, mainly through glomerular filtration (Shammas 2005; Berlioz and Sanghavi 2020).
- Icatibant, a peptidomimetic drug consisting of ten amino acids, is a selective and specific antagonist of bradykinin receptors. The marketed brand name of this drug is used for the symptomatic relief of hereditary angioedema in adults (Gras 2009).

Miscellaneous Peptides

Some therapeutic peptides targeting different diseases do not belong to a specific major class and have miscellaneous bioactivity. Hepcidin (LJPC- 401) is a synthetic peptide analogue used for resolving conditions of hemochromatosis (a condition of iron accumulation in the liver and heart) (Billesbølle et al. 2020). Oxeia Biopharmaceuticals developed OXE-103 as a synthetic analogue of human endogenous hormone "ghrelin" to treat patients with concussions or traumatic brain injury. OXE-103 easily penetrates the blood brain barrier (BBB) and aids in the stabilization of metabolic and energy brain dysfunction after a concussion (Cabri et al. 2021). Similarly, TAK-639 was developed by Takeda, which is a synthetic linear peptide derived from cornea-permeable C-type natriuretic peptide that enhances the cellular level of cyclic guanosine monophosphate which can lower intraocular pressure. Recently phase I clinical trial of TAK-639 has been completed for treatment of glaucoma (Millar et al. 2019; Martin et al. 2020).

2.2 *Stimulus-Responsive Peptides*

External, physical, and physiological stimuli primarily affect the secondary structure of some peptide sequences, which can significantly change their biological interactions. These stimuli include the change in pH of the environment, presence of salts, light, reactive oxygen species (ROS), or enzymes present in cellular or tissue environments. By using peptides where cellular environments can be different from the normal cells, the stimulus-responsive behavior of peptides is beneficial in a site-specific drug delivery (Chockalingam et al. 2007). For example, tumors have an acidic environment compared to normal cells, so the pH-responsive peptides can be used to transport very specific anticancer agents (Chockalingam et al. 2007). Several peptide sequences have been reported that can modulate their characteristics in change in environmental stimulus, e.g., pH difference, temperature change, or enzyme overexpression (Fig. 1 and Table 2).

pH-Responsive Peptides

The peptide contains both basic and acidic amino acid residues essential for ionization of peptides at specific pH, and the number of these specific amino acid residues decides the total charge of the peptide. The charge on peptides decides the pKa value, which affects the conformations of the peptides. For instance, pH (low) insertion peptide named pHLIP displays a random coil shape at a pH of 7.4, while at acidic pH, it changes conformation into α-helix with C-terminal inside and N-terminal outside the cell. Thus, pHLIP and its derivatives are considered promising candidates for cytoplasmic delivery of therapeutic agents (Wyatt et al. 2017, 2018). Poly-histidine is also an example of the pH-responsive peptide that represents the hydrophobic character at physiological pH, whereas at acidic pH (specific to tumor cells), it exhibits the hydrophilic character. This happens due to the increase in the extent of ionization of the imidazole group present in histidine. These characteristic features of poly-histidine can be utilized for optimization of cellular uptake or controlled release of therapeutic agents to the targeted sites (Zhao et al. 2012, 2016).

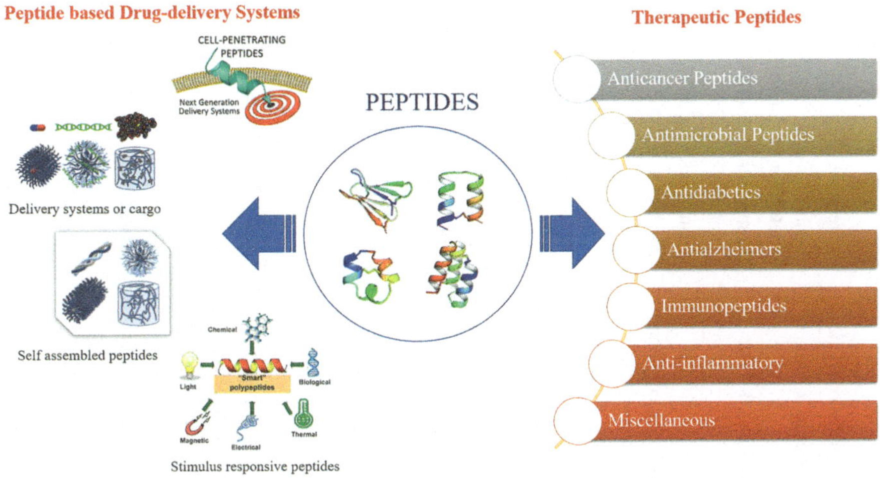

Fig. 1 Classifications of bioactive peptides

Thermo-Responsive Peptides

Hydrophobicity is an important characteristic for stability of conformations of the peptides, which are also modulated by change in temperature. The fluctuation in the hydrophobic hydration energy of the peptide is affected by the size of non-polar solutes present in peptide; if it is <1 nm, the hydration energy becomes positive, and if it is >1 nm, the hydration energy is negative. These results drew a theoretical correlation for understanding the thermo-responsive transitions in the molecular assembly of peptides (Li et al. 2005, 2016, 2020; Davis et al. 2013a).

For instance, elastin-like polypeptides (ELPs) are highly explored thermally active biopolymers containing repetitive units of VPGXG (X = hydrophobic amino acid residue except proline) (Zhang et al. 2020). ELPs show an assisted hydrophobic aggregation when heated up above their transition temperature (Lyons et al. 2013). Usually, the transition temperature (Tt) of these ELPs is somewhat maintained around 38–42 °C (exceeding the Tt), which is specific to tumor regions. The amphiphilic domain of ELPs transitions into a micelle, where the hydrophilic part acts as a linker exposed to the targeting moieties, which tends aggregation and accumulation of the conjugated drugs within tumor cells (Helawe Betre et al. 2002, Zhang et al. 2020).

Enzyme-Responsive Peptides

In diseased or tumor tissue, a range of enzymes are overexpressed, which distribute in extracellular matrix, cell membranes on cancer-associated fibroblasts, or inside of cells. Enzyme-responsive peptide motifs in DDS can be obtained from peptide

Table 2 List of some stimulus-responsive peptides (Chockalingam et al. 2007)

Application	Stimulus	Peptide sequence	Conformational change	Reference
Amyloid study	Temperature and pH	ADADADADARARARAR	β-Sheet to α-helix	Schneider et al. (2002), Kretsinger et al. (2005)
	Temperature, salt	SIRELEARIRELELRIG	α-Helix to β-sheet	Kammerer et al. (2004)
	Temperature	YGCVAALETKIAALETKKAALETKI-AALC	–	Ciani et al. (2002)
Amyloid disease	Zn ion	DAEFRHDSGYEVHHQK	Poorly folded alpha helix to 3_{10} helix	Zirah et al. (2006)
Biomaterials	Calcium	GGXGXDXX (here X = L/F/I)	Disordered structure to β-roll	Ringler and Schulz (2003)
	Porphyrin	IQQLKNQIKQLLKQ	Random coil to α-helix	Kovaric et al. (2006)
Biomineralization	pH	CCCCGGGSRGD	–	Hartgerink et al. (2001)
Bioremediation	Temperature, pH, salt	VPGXG	Random coil to β-turn	Kostal et al. (2003), Prabhukumar et al. (2004)
Conformational studies	Salt addition	GIGAVLKVLTTGLPALISWIKRKRQQ	Disordered structure to α-helix	Raghuraman and Chattopadhyay (2006)
	pH	(a) EAALEAALELAAELAA. (b) KAALKAALKLAAKLAA. (c) KAALEAALKLAAELAA. (d) EAALKAALELAAKLAA.	β-Sheet to α-helix	Mutter et al. (1991)
	Redox state	CGGEIRALKYEIARLKQAAQAKIRALEQKIAALEGGC	Dimeric coiled coil to monomeric α-helix	Pandya et al. (2004)
	Zinc	YIHALHRKAFAKIARLERHIRALEH-AA	Coiled coil to α-helix	Cerasoli et al. (2005)
	Redox state	YLKAMLEAMAKLMAKLMA	α-Helix to β-sheet	Dado and Gellman (1993)
	Light	EACARV $_A$ AACEAAARQ [a]	Disordered structure to α-helix	Kumita et al. (2000)
	Solvent polarity	ELALKAKAELELKAG	β-Sheet to α-helix	(Dado and Gellman 1993)

DNA purification	pH, salt, and temperature	VPGXG (here X = any hydrophobic amino acid except proline)	Disordered structure to β-turn	Kostal et al. (2004)
Tumor-targeted drug delivery	Temperature	VPGXG (here X = any hydrophobic amino acid except proline)	–	Raucher and Chilkoti (2001)
Change in protein binding	Temperature	VPGXG (here X = any hydrophobic amino acid except proline)	Disordered structure to β-turn	Reiersen and Rees (1999), Megeed et al. (2006)
Hydrogels, tissue engineering	pH	VKVKVKVKVPPTKVKVKVKV	Disordered structure to β-hairpin	Schneider et al. (2002), Kretsinger et al. (2005)

residues obtained from enzyme-cleavable sites; for example, the PLGLAG peptide sequence can be cleaved by matrix metalloproteinase, and it has been used in nanoparticles for selective retention of nanoparticles in infarcted tissues of angina pectoris (Zhang et al. 2012). Another example of an enzyme-responsive peptide is GPA-X, where X can be any amino acid residue. GPA-X sequences have been used for developing cancer-associated fibroblasts targeting DDSs for cancer therapy, and they can be cleaved by FAP-α (Li et al. 2020). Also, the tetrapeptide DEVD from caspase-3 has been functionalized into many intracellular trigger-responsive nanoparticles (Tang et al. 2013). Enzyme-responsive peptides have been used within non-degradable PEG hydrogels along with RGD peptide to facilitate the enzyme-based release of entrapped vascular endothelial growth factor (VEGF) (Aimetti 2011; Wanakule 2012).

Reactive Oxygen Species (ROS)–Responsive Peptides

Excessive ROS generation causes pathological stress in cells and tissues, resulting in the breakdown of crucial protein, lipid peroxidation, and DNA damage, among other things (Darley-Usmar and Halliwell 1996). The thiol group of cysteine has favorable oxidative targets by ROS (e.g., hydrogen peroxide), which can cause the breakdown of disulfide linkage to subsequently result in the significant alteration in protein's tertiary and quaternary structures. As a result of this phenomenon, redox-responsive peptide assemblies have been developed based on cysteine/cystine derivatives as ROS sensors. For example, cholesterol-conjugated poly-cysteine peptide adopts an antiparallel β-sheet in solution and forms micelles (Liu et al. 2018). After treatment with acidic 10% hydrogen peroxide, the side chains of poly-cysteine are oxidized.

2.3 *Targeting Peptides*

Selective delivery of any therapeutics or gene to the specific target is the biggest challenge to ameliorate the toxicity associated with off-targeting (Wang et al. 2020). The intracellular and extracellular environments surrounding the disease tissues are biologically complex, and hence, selective targeting is inferred by off-targeting mechanisms. Selective targeting can be obtained by using the specific ligands of the proteins or receptors found specific to the disease tissues. These ligands can be small molecules or metabolites associated with the host; for example, folic acid targets the folate receptor or any synthesized small heterocyclic molecules (Chen et al. 2013). However, the inclusions of these molecules may cause the generation of some other pharmacological actions or toxic effects to the cells. Peptides are biocompatible molecules, and due to their smaller sizes, they have been used as potential tools for developing targeted DDS. There are numerous methods for the identification of peptides for targeted therapy. Targeting function of ligand can be selected out from phage display library, in silico approach through protein-peptide interactions, identifications of a specific fragment from proteins, or de novo design of peptide by using physicochemical properties, important for interactions. Integrin receptor is overexpressed in most cancer cells and tumors, and RGD is identified as the smallest peptide motif to bind with these receptors. RGD peptide is the most utilized sequence for the development of cancer targeting DDS (Pasqualini et al. 1997; Ruoslahti 2017; Alipour and Baneshi 2019) (Table 3).

2.4 *Cell-Penetrating Peptides (CPPs)*

CPPs are usually lysine- and arginine-rich cationic peptides that can easily penetrate the plasma membrane of most mammalian cells. Their ability to transport large biomolecules like proteins/oligonucleotides across the cell membrane makes them a desirable candidate for the drug delivery vehicle (Snyder and Dowdy 2004). CPPs are a diverse class of peptides due to several factors; for instance, when it comes to the source of CPPs, some are derived from natural origin (e.g., TAT from HIV), and some are generated artificially (e.g., polyarginine). Chimeric peptides are a hybrid class of CPPs, which is formed by joining two very distinct proteins. For example, transportan is a chimeric peptide formed by conjugating "mastoparan" (from wasp venom) and "galanin" (human neuropeptide) (Pooga et al. 1998). CPPs can also be grouped broadly on the basis of their physicochemical parameters like (1) charge, (2) hydrophobicity, and (3) amphipathicity (Table 4).

Moreover, CPPs can be conjugated with small molecules to enhance the in vivo delivery to the specific sites; for example, topical application of cyclosporin A against skin disease fails to penetrate the skin, but upon conjugation with hepta-arginine (R7), the penetration of cyclosporin A through the epidermis and dermis is enhanced without disturbing the functionality of cyclosporin A for contact dermatitis in a mouse model (Rothbard et al. 2000). Similarly, doxorubicin is rapidly excreted out from cancer cells via P-glycoprotein efflux pump. To bypass this efflux system, doxorubicin has been conjugated to the Antp peptide, and the conjugated Antp-dox displayed enhanced absorption in the brain perfusion model (Rousselle et al. 2000). This enhanced uptake of the conjugated drug was not observed in the P-glycoprotein-deficient model, confirming that the effect was due to the P-glycoprotein bypass (Mazel et al. 2001). Moreover, the conjugates showed cytotoxicity, higher than free doxorubicin, against cultured multidrug resistance (MDR) tumor cells, which demonstrates that cellular absorption of CPPs is not controlled by the P-glycoprotein pump (Mazel et al. 2001). So peptide-drug conjugates are promising strategies for targeted drug delivery to the specific tissues where the activity of P-glycoprotein is rate limiting. Recently, stimuli-responsive CPPs have been developed that respond toward pH changes, enzymatic activity, or oxidative stress. CPPs containing amino acid sequences that can alter the net charge depending on the pH are considered responsive molecules (e.g., histidine can tune the net charge of a peptide) (Tesauro et al. 2019b). Zhang et al. reported an α-helical pH-responsive histidine-rich CPP showed neutral charge at physiological pH, while it became cationic under acidic milieu, which triggered its cell penetration capability (Zhang et al. 2013). PEGylated liposomes functionalized with this

Table 3 List of some targeting peptides with examples

S. no.	Target/receptor	Targeting peptide sequence	Disease	Reference
1	Integrin receptors	CRGDC, CDCRGDCFC, c (RGDXi)n	Glioblastoma, melanoma, lung cancer, ovarian cancer, colorectal cancer, breast cancer, fibrosarcoma, neuroblastoma	Pasqualini et al. (1997), Koivunen et al. (1993)
2	CD13/aminopeptidase N	CNGRC, CRNGRGPDC	Lymphoma, melanoma, sarcoma, breast cancer, glioblastomas	Arap et al. (1998)
3	FABP3/MDGI	CGLSGLGVA	Glioblastoma	Hyvönen et al. (2014)
4	Transferrin receptor	THRPPMWSPVWP, HAIYPRH	Glioblastoma, colorectal cancer, brain targeting	Lee et al. (2001)
5	Low-density lipoprotein receptor	TFFYGGSRGKRNNFKTEEY	Brain tumor/brain metastasis	Demeule et al. (2008)
6	EGFR	YHWYGYTPQNVI	Hepatocellular carcinoma	Li et al. (2005)
7	Melanocortin 1 receptor (MC1-R)	SYSMEHFRWGKPV	Melanoma	Eberle and Froidevaux (2003)
8	MMP2/MMP9	CTTHWGFTLC	Kaposi's sarcoma	Koivunen et al. (1999), Wang et al. (2009)
9	FXIIIa	GNQEQVSPLTLLKC	Arthritis	Schultz (2019)
10	Lysosome	YQRLC	Gold nanoparticle for lysosomal delivery	Dekiwadia et al. (2012)
11	Mitochondrion	(KLAKLAK)2	Anticancer pro-apoptotic	Ellerby et al. (1999)
12	Nucleus	PKKKKRKV, TSFAEYWNLLSP	Gene delivery	Bremner et al. (2004), Qu et al. (2013), Liang et al. (2020)

particular peptide sequence have shown more effective internalization into C26 colon tumors with high apoptotic levels with an increased rate of tumor cell inhibition (Tesauro et al. 2019b).

Table 4 Classifications and examples of CPPs

CPP class	Common name	Peptide sequence	Source	Application	Reference
Cationic	TAT (48–60), TAT	^{48}GRKKRRQRRRPPQ^{60} YGRKKRRQRRR	Natural (HIV-1 protein) HIV-TAT domain	Intracellular delivery of drugs and nanocarrier systems	Green and Loewenstein (1988), Frankel and Pabo (1988)
	Polyarginine	R_n (n = 8–12)	Designer peptides	Imaging and drug delivery systems	Tünnemann et al. (2008)
	Penetratin	RQIKIWFQNRRMKWKK	Natural (Antennapedia *Drosophila melanogaster*)	Directing toward human fibroblasts	Derossi et al. (1998)
	Human calcitonin (hCT)–derived peptide	LGTYTQDFNKFHTFPQTAIGVGAP	Hormone secreted from thyroid	Management of diseases like hypercalcemia or osteoporosis	
Amphipathic	Pep-1	KETWWETWWTEWSQPKKKRKV	Natural (SV40 T antigen NLS)	Gene delivery and peptides	Morris et al. (2001)
	MAP	KLALKLALKALKAALKLA	Designer peptides	Antimicrobial peptides	Oehlke et al. (1998)
	Transportan	GWTLNSAGYLLGKINLKALAALA-KKIL	Synthetic	Antimicrobial peptides	Pooga et al. (1998)
	VP22	NAATATRGRSAASRPTQRP-RAPARSASRPRRPVQ	Herpes virus	Nuclear transport	Elliott and O'Hare (1997)
	FGF	PIEVCMYREP	Viral proteins	Antiapoptotic	Nakayama et al. (2011)
Hydrophobic	Pep-7	SDLWEMMMVSLACQY	CHL8	Peptide and gene delivery	Gao et al. (2002)
	PFV	PFVYLI	C105Y	Intracellular delivery of peptides	Rhee and Davis (2006)

3 Peptide-Based Drug Delivery System

Peptide-based DDSs are acquiring more and more limelight in the field of medicine, as their role in pathophysiology is well understood, and all this is due to the rapid development in the field of biotechnology. Also, the progress in the field of conventional chemical synthesis and DNA-recombinant techniques has scaled up the production of these peptides and peptide-based therapeutics. Peptides, both natural and synthetic, have a wide spectrum of biological responses, making them useful in biotechnological applications such as treatments and diagnostics.

However, their capacity is often overwhelmed by the inherent limitation of natural peptides, i.e., the inability to reach their target site intact (in vivo) due to the poor absorption from the mucous membrane and rapid degradation by proteolytic enzymes. Generally, peptides are effective and selective, and therefore, a lower concentration is required to act on their target site, but the metabolism of peptides is significant to limit their action compared to other small molecules.

Fortunately, there are many alternative methods to improve the drug delivery of peptides with enhanced biological activity such as chemical modification of the peptide backbone, substitution with unnatural or non-gene encoded amino acids, and cyclization. Other methods include the modification/alteration of peptide bonds, terminal modifications, and polymer conjugation. For example, the D- and L-peptide derivatives of "C-X-C" type 4 chemokine receptor were compared for their stability profiles, where the L-enantiomer of the peptide degraded within 24 hours, while the D-enantiomer of the peptide demonstrated dramatically improved stability and showed no sign of degradation even after 72 hours. Peptides can also be used as carriers for enhanced transport; for example, cellular uptake of the peptide "dalargin" by the brain was augmented when conjugated to chimeric peptide vectors such as SynB1 and SynB3 (Zou et al. 2013).

3.1 Peptide-Conjugated Drugs

Peptide-conjugated drugs (PCDs) represent therapeutic agents where a drug molecule is conjugated with a peptide via a cleavable or biodegradable linker. Conjugating a drug molecule with peptides generally applies to the prodrug strategy, which enhances the pharmacokinetic profile of drugs while minimizing the toxicity. The engineering on peptide sequence allows adjusting of the net hydrophobic and charge character of the conjugate; both characteristics play an important role on bioavailability of PCDs (Wang et al. 2017). For example, doxorubicin conjugated with RGD peptide in a free amino group showed improved antitumor activity in comparison to free drug in in vivo study. Interestingly, ex vivo histopathological assays demonstrated that prodrug was less toxic to the liver and heart than free drug (Wang et al. 2017).

In chemotherapy, selective delivery of drugs in tumor sites without affecting the healthy tissues is highly desired. Cancer specific over expressing receptors, e.g., integrin receptors, on the malignant tissues are good target for selective interactions of PCDs because the primary interaction between any therapeutic drug and cell surface via overexpressing receptors on cell membrane (Wang et al. 2017). Therefore, a number of target groups can be incorporated onto the PCD design for selective target. Another example of PCDs such as doxorubicin (DOX) was conjugated with FRRG peptide sequence (cathepsin B susceptible) acting as a prodrug that was self-assembled into nanoparticles (Shim et al. 2019). These nanoparticles (FRRG-DOX) showed rapid cellular uptake indicating the successful cleavage of the FRRG-DOX by the enzymatic cleavage by cathepsin B enzyme (overexpressed in tumor cells). The cleaved part G-DOX induces cancer cell tumor apoptosis specifically, without affecting non-cancerous cells due to the lower level of cathepsin B (Shim et al. 2019).

Similarly, Kim et al. conjugated resveratrol (RSV) with leucine- and lysine-rich cell penetrating α-helical peptide (LKKLLKLLKKLLKLAG) for the treatment of chronic rhinosinusitis via nasal route. The resultant prodrug (RSV-peptide conjugate) showed improved uptake with high potency and efficacy and with reduction in the frequency of administration than free RSV. This idea of developing a new delivery system utilizing cell-penetrating peptides (CPPs) provided a different strategy for curing localized nasal disease (Kim et al. 2020).

3.2 *Peptide-Modified Nanocarriers*

In recent years, nanotechnology has advanced to the forefront of drug delivery technologies. Like other DDS, nanoparticles/nanocarriers can be derived from natural sources like bovine serum albumin or synthetic polymers. Polymeric nanocarriers have been explored extensively for the design and development of smart DDS. These nanocarriers offer various benefits as drug carriers, such as improved biocompatibility with enhanced bioavailability, minimized toxicity, and improved drug solubility. Polymeric nanocarriers are simply colloidal nanoparticles (size ranging from 1 to 1000 nm) and are collectively named spheres and capsules (Balaji and Parimala Devi 2010; Gundersen 2016). They are considered outstanding carriers for peptide delivery because they are more efficiently taken up by cells due to their smaller size than larger macromolecules and have been reported for their site-specific, controlled, and improved biological application. For example, co-administration of peptides like P3 and QBP1 has been reported to significantly reduce the polyglutamine toxicity, but due to their enzyme susceptibility and inadequate delivery system, they face problems in the clinical administration and cannot also penetrate the cell membrane for cellular uptake (Kim et al. 2019). Moreover, these peptides can be encapsulated into the polymeric nanocarriers and can be co-delivered to meet the need of the combination therapy for the treatment of neurodegenerative disorders (Kim et al. 2019).

Therapeutic applications like the delivery of genes or drugs, pathogenic bio-detection, biological fluorescent material, tumor destruction by heating, and tissue engineering can be augmented by nanocarriers. The nanoparticle surfaces can be modified by specific ligands to achieve active targeting. The small size allows nanocarriers to have a longer systemic circulation as well as increased cell penetration (Bhatia 2016).

A broad class of medicinal substances can be encapsulated in nanoparticles, regardless of their physiological properties, such as size, hydrophilicity, or hydrophobicity (Joseph et al. 2017a). The cytosine-phosphate-guanosine (HBV CpG) nanoparticles of the hepatitis B virus (HBV) can promote therapeutic immunity against HBV infection. In HBV mouse models, this delivery strategy elicited an "anti-hepatitis B" surface antigen response and had a substantial immunostimulatory effect on lymphocytes (Lv et al. 2014). Spaliviero utilized Raman spectroscopy for analysis of nanoparticles to study the metastatic condition of lymph nodes. In mice with prostate cancer, nanoparticles were administered intravenously and were found in larger concentrations in normal lymph nodes than in metastasized lymph nodes of mice using surface-enhanced resonance raman spectroscopy (SERRS). This technique can be used to distinguish between healthy and metastasized lymph nodes (Spaliviero et al. 2016).

A theranostic approach was adopted to deliver indocyanine green and doxorubicin simultaneously using poly lactic-co-glycolic acid (PLGA) nanoparticles. The in vivo and ex vivo investigations revealed that nanoparticles accumulated in solid tumors, with some particles permeating into tumor tissues in a mice model (Joseph et al. 2017a). Several different delivery systems have been reported so far by utilizing polymeric nanoparticles, e.g., hydrogels, micelles, liposomes, dendrimers, polymer-drug conjugates, peptide-metal complexes, etc. The biomedical applications of all the above-mentioned polymeric nanocarriers have been narrowed down to discuss only the applications that involve the delivery of peptides to targeted sites or peptide functionalization to these nanocarriers for therapeutic and diagnostic purposes (Joseph et al. 2017a).

Hydrogels

Hydrogels are 3D polymeric networks that can retain a large amount of water and/or aqueous fluids (Hoare and Kohane 2008). Polymeric hydrogels form through chemical crosslinking and have a robust 3D network to entrap the solvent molecule, and therefore, these are stable in physiological environments. Similarly, peptide hydrogels self-assembled into a 3D network associated with secondary structures such as α-helices, β-sheets, vesicles, micelles, fibers, ribbons, tapes, tubes, and coils (Bennet and Kim 2014). Peptide hydrogels can be modified to exhibit stimuli-responsive behavior, which means they can respond to external physical stimulus, e.g., pH change, temperature change, ionic strength, enzymes, stress, light, etc. (Qiu and Park 2001; Mart et al. 2006). More importantly, characteristic features like non-toxicity and stability under varying biological conditions make them the ideal

Table 5 Examples of self-assembling peptide hydrogel for drug delivery

Peptide category	Peptide sequence	Application	Reference
Ionic self-complementary	Ac-RADARADARADARADA-$CONH_2$	Sustained release action and insulin delivery	Koutsopoulos et al. (2009), Nishimura et al. (2012)
Ionic self-complementary	1. Ac-RADARADARADARADA-$CONH_2$. 2. Ac-RADARADARADARADA-GGDGEA-$CONH_2$. 3. Ac-RADARADARADARADA-GCPFSSTKT-$CONH_2$.	Regenerative medicine	Gelain et al. (2010)
HBPA	C16-AAAAGGGLRKKLGKA	Islet transplantation	Chow et al. (2010)
β-Hairpin	MAX8 (VKVKVKVKV^{D}PPTKVEVKVKV-NH_2)	Treatment of spinal cord injuries	Lindsey et al. (2015)
DCH	$K_{180}L_{20}$ and $E_{180}L_{20}$	Delivery of protein CNS	Song et al. (2012)

candidate for the targeted delivery systems. Also, these hydrogels can be fine-tuned depending on their gelation properties. The loading capacity, release pattern, and gelation effects of these peptide hydrogels can be engineered by selection of amino acid sequence and size used in the peptide modifying their hydrophobicity and hydrophilicity (Mondal et al. 2020). Fmoc-RGD is a short but effective peptide hydrogelator capable of producing rigid and porous hydrogels with exceptional stability at a concentration of 10% w/v. Drug molecules were easily loaded into the hydrogel by dissolving them in an aqueous medium before or after the self-assembly process. The presence of the RGD integrin-binding motif, which promotes cell adhesion, also opened up various possibilities for the biological applications of these hydrogels (Iglesias and Marchesan 2017).

Lui et al. conjugated RGD with polyethylene glycol (PEG) through triazole linkage by using click chemistry which forms a polymeric network that augments gelation within 2–30 minutes. Peptide-diazide interlinked hydrogels exhibited high loading capacity and shorter spacers between azide groups (Liu et al. 2009) (Table 5).

Hydrogels can be administered in different routes; for example, they can be inserted surgically, injected locally, or infused systemically via intravenous infusion. For a given application, the choice of a delivery method is based on optimizing efficacy and patient compliance. A family of biodegradable PEG hydrogels with half-life up to 17 days has been developed using ester bonds that can hydrolyze slowly. Also, anionic alginate hydrogels have been utilized to deliver growth factors such as VEGF via electrostatic interactions to facilitate tissue regeneration (Silva and

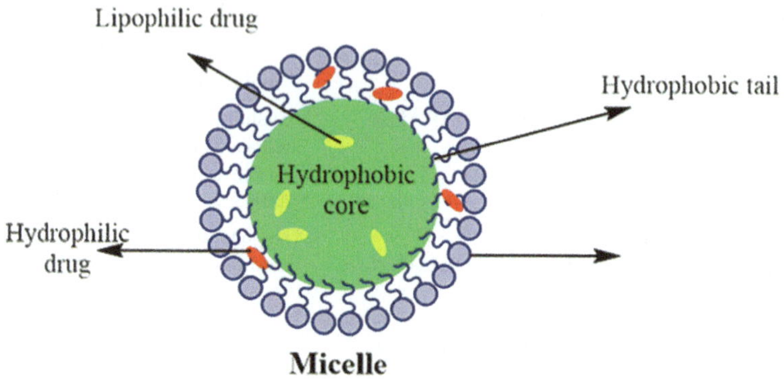

Fig. 2 Encapsulation of drug/peptide into a micelle

Mooney 2010; Kolambkar et al. 2011). A more comprehended version of polymer and peptide hydrogels was reviewed by Mondal et al. (Mondal et al. 2020).

Micelles

In general, micelles are widely utilized for delivery of poor water-soluble drugs; however, micelles are amphoteric in nature, and therefore, they have the capacity to deliver both hydrophilic and hydrophobic drugs (Wennerström and Lindman 1979). The lipophilic core provides a microenvironment for the loading of hydrophobic drugs which increases the drug solubility and loading efficiency of the hydrophobic drugs including peptides (Fig. 2). Simultaneously, the hydrophilic chain increases colloidal stability while preventing unwanted interactions with other components (Lombardo et al. 2019). Micelles can also deliver macromolecules, providing controlled and sustained release, chemical and physical stability, improved pharmacokinetics, and desired tissue distribution and improved therapeutic bioavailability of the encapsulated drug. Generally, micelles are spherical, with dimensions ranging from 2 to 20 nanometers depending on the composition. Oil-in-water emulsion, solid dispersion, solvent evaporation, and dialysis procedures are some of the most prevalent methods for micelle formation (Keskin and Tezcaner 2017).

Peptides self-assemble with micelles and adopt the α-helical conformation, which favors the receptor interaction and improves the stability of peptides against protease degradation (Banerjee and Onyuksel 2012). The trans-activating transcriptional protein (TAT) metabolizes rapidly by the protease enzyme in human plasma or a medium containing trypsin. TAT modified with PEG-PE offers high stability against proteolysis and hold great potential for developing a multifunctional smart drug delivery system (Grunwald et al. 2009). TAT-conjugated polymers can be employed for generation of non-invasive and effective delivery tools for brain targeting. Kanazawa and coworkers evaluated the distribution of coumarin in the brain and observed the effect of TAT conjugation. TAT was conjugated with polymer MPEG-PCL and loaded coumarin within formed nanoparticles that increased the cellular uptake of coumarin. Interestingly, it was notable that the distribution of coumarin

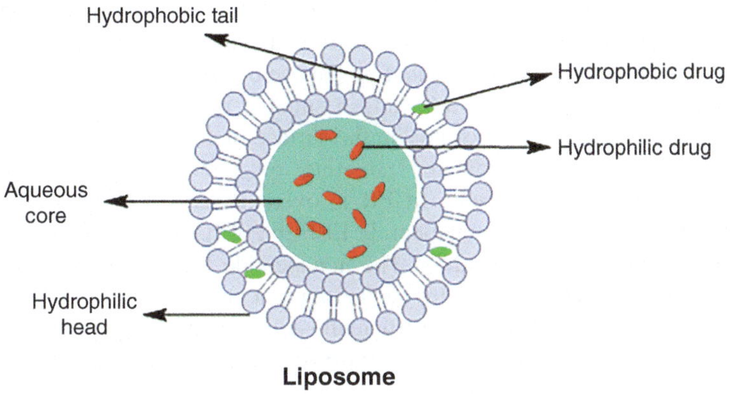

Fig. 3 General structure of liposomes

was lower in non-targeted organs after MPEG-PCL-TAT administration (Kanazawa et al. 2011).

Liposomes

Liposomes are vesicles of phospholipid bilayers (lamellas) that help to encapsulate the medicinal substances within their lipid bilayers and hydrophilic cores (Fig. 3). They are sensitive to environmental factors like pH and temperature, and therefore, these can be modified depending on the nature of phospholipids. These unique properties make liposomes ideal for drug delivery carriers (Kraft et al. 2014; Pudlarz and Szemraj 2018). To improve the efficiency of nanoliposomes, it is beneficial to conjugate an amphiphilic molecule letting the liposomes to reach the desired site. This was achieved by a strategies of developing pH-sensitive nanoliposomes to target cancer cells/tumors whose cytoplasmic pH is lower than normal cells. When internalizing into the cancer cell, pH-sensitive liposomes destabilize and disintegrate, thereby releasing the active ingredient or drug inside the cell (Kim and Huang 2012). Liposomes can be divided into three types: large unilamellar vesicles (LUVs), small unilamellar vesicles (SUVs), and multilamellar vesicles. The methods used to prepare these liposomes decide their types, size, and shape. Liposomal amphotericin and liposomal doxorubicin are two of the clinically approved drugs used for the treatment of mucormycosis and cancer, respectively. Liposomes have been shown to protect encapsulated drugs and peptides from oxidation and deamidation. Furthermore, drug- or therapeutic peptide-encapsulated liposomes are known to be absorbed by the lymphatic tissues that directly transfer these liposomes into the systemic circulation. Hence, this mode of delivery bypasses the first-pass metabolism effect (Lalanne et al. 2007; Porter and Charman 1997). Liposomes can deliver multiple biomolecules simultaneously, such as DNA and proteins. For example, a potential anticancer peptide, "stoppin," competitively inhibits the p53-MDM2/MDMX complex, limiting the binding of MDM2 and MDMX complex to p53 protein, which triggers the apoptosis in cancer cells. In vitro studies on liposomes containing peptides and nucleic acids against A549 cells (lung cancer cell line) demonstrated enhanced anticancer activity augmented by the use of liposomes (Gao et al. 2015).

Another possible strategy is attaching a suitable molecule on liposomes that directs liposomes toward the receptor on the surface of cells. PEGylated liposomes were attached with anisamide for targeting of sigma receptors, which are overexpressed in the tumor cells. EGFR is overexpressed in many cancer cells (Kim and Huang 2012). The nonapeptide EEEEpYFELV (EV) was developed to inhibit EGFR downstream signaling. The nonapeptide EV was attached to Lipid-polymer hybrid (LPH) nanoparticles, which have a core structure with a surface-grafted PEG. The resulting conjugate efficiently delivered the EV peptide into cancer cells, where EV inhibited STAT5b phosphorylation, stopped cell proliferation, and induced major apoptosis (Kim and Huang 2012).

Dendrimers

Dendrimers are branched, three-dimensional, and mono-dispersed macromolecules composed of two layers (exterior and interior layers) (Fig. 4). The external layers are made up of functional groups for bio-conjugation of the ligands or targeting molecules. In contrast, the internal layers have voids appropriate for encapsulation of drugs (Kaga et al. 2016).

Dendrimers differ from the self-assembled systems discussed thus far in that they are generated by utilizing a complex synthetic procedure that controls molecular design parameters, resulting in formation of highly monodisperse nanostructures (Lombardo et al. 2019). The factors affecting the structure of dendrimers in solutions include synthesis method, spacer length, surface functionalization, pH and ionic strength of a solution, and temperature (Ballauff and Likos 2004). Interactions between dendrimers and lipid bilayers deciphered by different mechanisms include adsorption, pore formation, and disruption of vesicles. The size and charge of dendrimers are deciding factors for interaction mechanisms. Functional groups on the external layer of dendrimers allow the binding of other targeting moieties (such as folate and antibodies) and enhance the drug delivery efficiency (Liu et al. 2012).

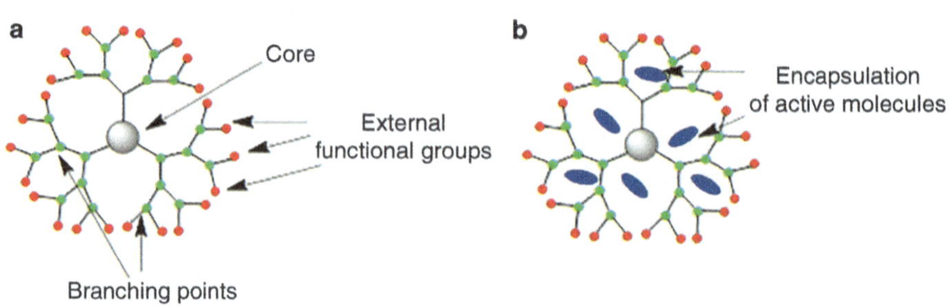

Fig. 4 (**a**) Dendrimer; (**b**) encapsulation of ligands/peptides

Peptide-grafted dendrimers have been used as vectors for selective and targeted gene delivery. For example, arginine functionalized peptide dendrimer was reported as a vector where the plasmid DNA (pDNA) was condensed for delivery as well as to protect the pDNA from nuclease digestion (Luo et al. 2012). Similarly, Pandita et al. reported RGD-grafted G5 and G6 PAMAM dendrimers, and the resulting conjugates form complexes with pDNA through electrostatic interactions. The resulting systems delivered pDNA to bone marrow–derived mesenchymal stem cells (Pandita et al. 2011).

Polymer-Drug Conjugates

Polymeric nanocarriers offer a range of options for structural modification and functionalization that can be used to conjugate targeting ligands to enhance biodistribution (Couvreur and Puisieux 1994). Modification on physicochemical properties of nanocarrier systems can also enhance bioactivity of the drug or peptides and simultaneously reduces the side effects caused by the drug when administered alone (Mottaghitalab et al. 2015). Polymer-drug conjugates are another impressive aspect of polymeric nanocarrier, designed by incorporating the bioactive molecules into polymers through specific functional moieties (such as -COOH, -OH, and -NH_2) (Larson and Ghandehari 2012).

The basic idea behind functionalizing nanocarriers was to improve performance and expanded broad application areas in biomedicine, such as diagnostic and therapeutic purposes. Both linear and branched polymeric carriers can be used for the production of polymer-drug conjugates. Polyvinylpyrrolidone (PVP), poly-aspartamides, polyvinyl alcohol (PVA), polymalic acid (PMA), and polyethylene glycol (PEG) are some examples of linear polymers. Similarly, poly-ethyleneimine dendrimers, PAMAM, and polymeric micelles are examples of branched polymers (Larson and Ghandehari 2012). The first polymeric-drug conjugate was reported in 1975 by Helmut Ringsdorf (Ringsdorf 1975). In general, a polymer-drug conjugate has a biodegradable polymeric backbone and is comprised of three basic units: (1) a hydrophilic unit for solubility, (2) a bioactive molecule (drug or peptide) conjugated with a specific linker or spacer, and (3) a targeting moiety for site-specific delivery (Van et al. 2010; Pasut and Veronese 2007).

Peptide-Metal Complex

Peptide self-assembly allows the generation of a diverse range of nanoarchitectures alone or in combination with a variety of materials, and they hold great potential for the designing of smart DDS. The introduction of a peptide-metal complex guides the self-assembly process so that it produces various nanostructures with different morphologies (Das et al. 2018). The peptide-metal complex, such as histidine-copper coordination complex in the self-assembled supramolecular hydrogel, can be utilized as a tool for understanding the development of a spontaneous and straightforward delivery system (Shao et al. 2020). The use of the peptide-metal complex was reported by Das and coworkers as an optical-based drug displacement system where they showed the efficiency of this system utilizing doxorubicin. The system has a turn-on emission response and delivers the drug via a fluorescence-based strategy. Histidine displaces the drug bound to the peptide-metal complex and releases the drug in a controlled fashion (Das et al. 2018).

Rajchakit and coworkers showed in their study that the conjugation of antimicrobial peptides (AMPs) with gold nanoparticles (AuNPs) enhanced their serum stability and significantly enhanced antimicrobial activity against drug-resistant bacterial pathogens (Rajchakit and Sarojini 2017). Furthermore, Mukaya et al. reported poly-aspartamide conjugated bisphosphonate and platinum (II) complexes for antimalarial activity. The initial studies showed that the conjugates of bisphosphonate and platinum compounds exhibited increased antimalarial activity in comparison to the standard antimalarial drug chloroquine (Mukaya et al. 2017).

Peptide Nanofiber

Mazza et al. (2013) reported utilized an amphiphilic peptide for the brain delivery of an opioid receptor agonist named dalargin. The amphiphilic peptide self-assembled in nanofibers while conjugating the dalargin peptide wrapped in the nanofiber core and successfully delivered into the brain (Mazza et al. 2013). Ultrashort peptides like Fmoc-FF-COOH peptide have been explored for various biotechnological applications. The monomers of Fmoc-FF arranges in the nanofibers which forms a hydrogel in aquoeus system after cross linking. This hydrogel was further explored for various applications in drug delivery, antibacterial application, and 3D cell culture. Zhou et al. reported a hydrogel of Fmoc-RGD as biomimetic nanofiber-based hydrogels which was further explored as 3D scaffold for tissue and organ regeneration purposes. The RGD sequence plays a dual role in imparting the structural component and acting as biological ligand for cell adhesion. Formation of the nanofibrils from these ultrashort peptides was inspired from the peptide sequence of amyloid peptides and posing a promising candidates for future biomaterials (Zhou et al. 2009; Seroski and Hudalla 2018).

3.3 *Self-Assembled Peptide Nanomaterials*

One of the most popular ways for producing nanomaterials is through molecular self-assembly. Since its discovery, molecular self-assembly has attracted a lot of attention as a fresh "bottom-up" technique that complements classic "top-down" strategies. Natural biomolecules like lipid, protein, peptide, and DNA and RNA are usually imparted in self-assembly for biological structure and functions of the cells and act as building units for preparation of self-assembling materials as models (Zhang et al. 2018).

Peptides are a highly explored building block for self-assembling material because a diversified and specific nanostructure can be designed that can be modulated by various physicochemical factors like pH, temperature, and ionic solutes (Kopeček and Yang 2009). Self-assembled peptides can generate a variety of well-defined nanostructures for various biomedical applications (Lee et al. 2019). For a range of biomedical purposes, self-assembled peptides can build a variety of well-defined nanostructures. Self-assembled nanomaterials can be engineered with peptide sequences, developing them as smart and responsive to stimuli, which could be important for the creation of next-generation biomaterials and efficient systems in the future (Lee et al. 2019). The advantages offered by self-assembled peptides over other drug delivery methods include structural diversity, biodegradability and biocompatibility, the high loading capacity of the drugs, and their tendency to target specific receptors and enzymes.

One such successful example is the self-assembly of phloretic acid conjugated tetrapeptides into uniform nanofibrils to form a stiff hydrogel at physiological pH. The hydrogel exhibits properties like thixotropy and injectability and can be used for the entrapment of drugs for targeted and sustained release action (Dewangan et al. 2020). Self-assembled peptide hydrogels can also be combined with other biomaterials, like proteins, drugs, or cells for the treatment of metabolic disorders and comorbidities such as diabetes, cardiovascular diseases, and stroke (Castillo-Díaz et al. 2020). Hydrogels were also discussed in the "Peptide-Modified Nanocarriers" Sect. 3.2, where a polymer was modified with peptide molecules which forms hydrogel. Self-assembled peptides hydrogels have advantageous over polymeric hydrogels that they can self-assemble into hydrogel structures without addition of any polymeric scaffold.

4 Peptides in Clinical Use

Peptides are known to be selective and effective solutions with safe and well-tolerable biomolecules. Peptides have a broad range of pharmaceutical and biotechnological applications. Therapeutic peptides can be used against various diseases that include bone disorders, diabetes, cancer, HIV infection, multiple sclerosis, inflammations, chronic pain, etc. (Muttenthaler et al. 2021). Peptide drug discovery has been extended, and it is not limited only up to endogenous human peptides. The broader perspectives are being explored due to structural diversity identified in various natural sources or chemistry in designing aspects. Therefore, an increase in the interest in peptides has been seen by the pharmaceutical research and development sector in the past decades. In the last five years, the United States Food and Drug Administration (US FDA) has approved 208 new drugs in which 150 were new chemical entities and 58 were biologics. Within this list, 15 were peptides and related molecules (as shown in Table 6), accounting for 7% of the total approved drugs.

Mostly peptide drugs can be administered via parenteral route or as injectables because peptides are susceptible to proteases. However, according to the respective advancements, alternative routes of administration are increasing for oral, intranasal, and transdermal delivery systems. A combination of gold nanoparticles and PharmFilmTMAQ71 (MonoSol Rx) was reported as trans-buccal delivery system.

Table 6 FDA-approved peptide drugs from 2015 to 2020

S. no.	Approved drug and year of approval	Synthetic peptide analogue	Application
1	Tresiba (2015)	Insulin degludec	Diabetes
2	Ninlaro (2015)	Ixazomib	Multiple myeloma
3	Adlyxin (2016)	Lixisenatide	Improves glycemic control (blood sugar level)
4	Trulance (2017)	Plecanatide	Chronic idiopathic constipation in adults
5	Parsabiv (2017)	Etelcalcetide	Hyperparathyroidism and renal diseases
6	Tymlos (2017)	Abaloparatide	Osteoporosis
7	Hemlibra (2017)	Semaglutide	Type 2 diabetes mellitus
8	Fu Laimei Polyethylene Glycol Loxenatide Injection (2019)	PEG-loxenatide	Type 2 diabetes
9	Vyleesi (2019)	Bremelanotide	Treatment of hypoactive sexual desire in women
10	Scenesse (2019)	Afamelanotide	To prevent skin damage of patients with erythropoietic protoporphyria who have previously experienced phototoxic responses
11	Imcivree (2020)	Setmelanotide	To treat obesity and hunger control in people who have proopiomelanocortin deficiency, a rare condition that produces severe obesity from a young age

This system is an example of alternative administration routes and currently this is under pre-clinical phase. MidaSol Therapeutics is currently under clinical trials for a trans-buccal delivery system for insulin through gold nanoparticles.

Another example is ActoGeniX's TopAct™ technology for oral delivery of peptides directly expressed in the gastrointestinal tract (GIT) (Wetzler and Hamilton 2018). Peptide-based drugs like Lupron™ (Abbott Laboratories) for prostate cancer therapy had global sales of more than 2.3 billion dollars in 2011, while Lantus™ (Sanofi) had global sales of 7.9 billion dollars in 2013. Glucagon-like peptide 1 (GLP-1) receptor agonists for the remedy of type 2 diabetes mellitus (T2DM) generated over 2.6 billion dollars in sales in 2013 (Fosgerau and Hoffmann 2015).

5 Challenges and Future of Peptide-Based Therapeutics and Delivery Systems

Peptides are good alternatives for the design of new therapeutics and modulation of drug delivery systems. However, bulk production of peptides is challenging due to high cost over the conventional drug molecules. At present, peptides are synthesized by chemical methods [solid-phase peptide synthesis (SPPS) or liquid-phase peptide synthesis] or biological methods (recombinant DNA technology). SPPS is an advanced and the most adopted technique for synthesis of therapeutic peptides. However, SPPS involves the use of hazardous solvents and intensive chromatographic processes for synthesis of pure grade of peptides. Hence, this field requires further advancement in synthetic techniques of peptides for low cost and green synthesis methods.

In biological aspects, peptides are better than small molecules based on target specificity and potency, but they have poor in vivo stability and membrane impermeability because of the inherent constraints of amino acids. Development of effective drug delivery systems can be a solution to these problems, but it is almost impossible for a single delivery system to deliver a therapeutic peptide to every part/tissue of the mammalian body. Anatomically our body parts are made from variety of tissues and threfore it have different physiological barriers and environments for entry of the drug molecules. Improving the efficiency and pharmacokinetics of a therapeutic peptide is the prime objective of a delivery system. The design of a better carrier or delivery system should contain a broad range of biological response modifiers, and in silico methods can be adopted for precise selection of amino acids. The development of therapeutically viable delivery systems and therapeutics would be a multidisciplinary task due to the complex properties and functions of regulatory peptides. Therefore, it is essential to comprise information from different disciplines (such as biochemistry that involves the action of enzymes, physiology that explains about the membrane permeability, immunology for immunogenicity, physical chemistry for molecular characteristics, solubility, and stability) for the development of smart-delivery systems (Davis et al. 2013b). Additionally, with the continuing advancement of fields like synthetic biology and bio-nanotechnology, the development of potential medical innovations has augmented. Development of the new delivery systems with the help of peptides and their conjugates can be a better and smart delivery systems. Meanwhile, the rational design and discovery of novel peptides and proteins with distinct biological functions should be utilized to further broaden these exciting efforts.

6 Conclusions

Peptides have common features of heterocyclic drug molecules and biologics depending on their molecular size. Because of these characteristics, the number of peptide discoveries has sped up in the past decades especially due to technical advancement in the synthesis of peptides and modification strategies for protease stability. New peptide-based lead discovery takes shorter time as compared to the conventional heterocyclic molecules. Similarly, in the development of drug delivery systems, the use of peptides accelerated in the development of effective and selective targeting nanocarriers as well as biotherapeutics. Developments on self-assembled peptide nanomaterials are an emerging field for the preparation of cost-effective and biodegradable nanomaterials for biotechnological applications. Overall, peptides have a lot of potential both in therapeutics and drug delivery, and their use in the pharmaceutical industry will continue to grow due to their high receptor specificity and binding affinity.

Acknowledgments The corresponding author, RPD, acknowledged the Indian Council of Medical Research (ICMR), New Delhi, Government of India, for providing financial assistance through the ICMR extramural ad hoc project grant (File No. OMI/14/2020-ECD-I). AKM is thankful to this project for Junior Research Fellowship.

References

Aimetti A (2011) Synthetic peptide design for functionalized hydrogels: development of cellularly responsive drug delivery platforms and cyclic, multivalent peptide derivatives using radical-mediated thiol-ene/thiol-yne chemistries. Dissertation, University of Colorado

Alipour M, Baneshi M, A SH (2019) Recent progress in biomedical applications of RGD-based ligand: from precise cancer theranostics to biomaterial engineering: a systematic review. Wiley Online Libr 108:839–850. https://doi.org/10.1002/jbm.a.36862

Arap W, Pasqualini R, Ruoslahti E (1998) Cancer treatment by targeted drug delivery to tumor vasculature in a mouse model. Science 279(5349):377–380

Atanassoff PG, Hartmannsgruber MWB, Thrasher J et al (2000) Ziconotide, a new N-type calcium channel blocker, administered intrathecally for acute postoperative pain. Regional Anesthesia Pain Med 25(3):274–278

Balaji S, Parimala Devi B (2010) Nanotechnology and cancer-an overview. Int J Pharm Bio Sci 1(4):186–201

Ballauff M, Likos CN (2004) Dendrimers in solution: insight from theory and simulation. Angew Chem Int Ed 43(23):2998–3020

Banerjee A, Onyuksel H (2012) Peptide delivery using phospholipid micelles. WIREs Nanomedicine and Nanobiotechnology 4(5):562–574. https://doi.org/10.1002/wnan.1185

Bennet D, Kim S (2014) Polymer nanoparticles for smart drug delivery. In: Application of nanotechnology in drug delivery, pp 257–310. https://www.intechopen.com/chapters/46807. https://doi.org/10.5772/58422

Berlioz BE, Sanghavi D (2020) Bivalirudin. In: StatPearls [internet]. https://www.ncbi.nlm.nih.gov/books/NBK557823/

Bernstein JA, Qazi M (2010) Ecallantide: its pharmacology, pharmacokinetics, clinical efficacy and tolerability. Expert Rev Clin Immunol 6(1):29–39. https://doi.org/10.1586/ECI.09.60

Betre H, Setton LA, Meyer DE, Chilkoti A (2002) Characterization of a genetically engineered elastin-like polypeptide for cartilaginous tissue repair. Biomacromolecules 3(5):910–916. https://doi.org/10.1021/BM0255037

Bhatia S (2016) Nanoparticles types, classification, characterization, fabrication methods and drug delivery applications. In: Natural polymer drug delivery systems. Springer, Cham, pp 33–93. https://doi.org/10.1007/978-3-319-41129-3_2

Billesbølle CB, Azumaya CM, Kretsch RC et al (2020) Structure of hepcidin-bound ferroportin reveals iron homeostatic mechanisms. Nature 586(7831):807–811. https://doi.org/10.1038/s41586-020-2668-z

Bremner KH, Seymour LW, Logan A, Read ML (2004) Factors influencing the ability of nuclear localization sequence peptides to enhance nonviral gene delivery. Bioconjug Chem 15(1):152–161

Brown WM (2001) Taltirelin (Tanabe Seiyaku). IDrugs 4(12):1389–1400

Cabri W, Cantelmi P, Corbisiero D et al (2021) Therapeutic peptides targeting PPI in clinical development: overview, mechanism of action and perspectives. Front Mol Biosci 8:1–21. https://doi.org/10.3389/fmolb.2021.697586
Castillo-Díaz LA, Ruiz-Pacheco JA, Elsawy MA et al (2020) Self-assembling peptides as an emerging platform for the treatment of metabolic syndrome. Int J Nanomedicine 15:10349–10370. https://doi.org/10.2147/IJN.S278189
Cerasoli E, Sharpe BK, Woolfson DN (2005) ZiCo: a peptide designed to switch folded state upon binding zinc. J Am Chem Soc 127:15008–15009
Chen C, Ke J, Zhou XE et al (2013) Structural basis for molecular recognition of folic acid by folate receptors. Nature 500:486. https://doi.org/10.1038/NATURE12327
Chockalingam K, Blenner M, Banta S (2007) Design and application of stimulus-responsive peptide systems. Protein Eng Des Sel 20(4):155–161. https://doi.org/10.1093/protein/gzm008
Chow LW, Wang L, Kaufman DB, Stupp SI (2010) Self-assembling nanostructures to deliver angiogenic factors to pancreatic islets. Biomaterials 31(24):6154–6161
Ciani B, Gail Hutchinson E, Sessions RB, Woolfson DN (2002) A designed system for assessing how sequence affects α to β conformational transitions in proteins *. J Biol Chem 277(12): 10150–10155. https://doi.org/10.1074/JBC.M107663200
Couvreur P, Puisieux F (1994) Nanoparticles for the delivery of peptides and proteins. Targeting Drugs 4:153–159
Dado GP, Gellman SH (1993) Redox control of secondary structure in a designed peptide. J Am Chem Soc 115(26):12609–12610. https://doi.org/10.1021/ja00079a060
Darley-Usmar V, Halliwell B (1996) Blood radicals: reactive nitrogen species, reactive oxygen species, transition metal ions, and the vascular system. Pharm Res 13(5):649–662. https://doi.org/10.1023/A:1016079012214
Das P, Pan I, Cohen E, Reches M (2018) Self-assembly of a metallo-peptide into a drug delivery system using a "switch on" displacement strategy. J Mat Chem B 6(48):8228–8237. https://doi.org/10.1039/C8TB01483C
Davis JG, Rankin BM, Gierszal KP, Ben-Amotz D (2013a) On the cooperative formation of non-hydrogen-bonded water at molecular hydrophobic interfaces. Nat Chem 5(9):796–802. https://doi.org/10.1038/NCHEM.1716
Davis SS, Illum L, Tomlinson E (2013b) Delivery systems for peptide drugs. Springer Science & Business Media
Debnath S, Gogoi B, Swetha D (2021) Peptide-drug conjugates in targeted drug delivery to cancer. In: Multifunctional theranostic nanomedicines cancer, pp 147–161. https://doi.org/10.1016/b978-0-12-821712-2.00001-3
Dekiwadia CD, Lawrie AC, Fecondo JV (2012) Peptide-mediated cell penetration and targeted delivery of gold nanoparticles into lysosomes. J Pept Sci 18(8):527–534
Demeule M, Currie J, Bertrand Y et al (2008) Involvement of the low-density lipoprotein receptor-related protein in the transcytosis of the brain delivery vector Angiopep-2. J Neurochem 106(4): 1534–1544
Derossi D, Chassaing G, Prochiantz A (1998) Trojan peptides: the penetratin system for intracellular delivery. Trends Cell Biol 8(2):84–87
Dewangan RP, Kumari S, Kumar Mahto A et al (2020) Self assembly and hydrogelation of N-terminal modified tetrapeptide for sustained release and synergistic action of antibacterial drugs against methicillin resistant S. aureus. Bioorg Chem 102:104052. https://doi.org/10.1016/j.bioorg.2020.104052
Eberle AN, Froidevaux S (2003) Radiolabeled α-melanocyte-stimulating hormone analogs for receptor-mediated targeting of melanoma: from tritium to indium. J Mol Recognit 16(5): 248–2544
Ellerby HM, Arap W, Ellerby LM et al (1999) Anti-cancer activity of targeted pro-apoptotic peptides. Nat Med 5(9):1032–1038
Elliott G, O'Hare P (1997) Intercellular trafficking and protein delivery by a herpesvirus structural protein. Cell 88(2):223–233

Emani S, Meyer M, Palm D et al (2015) Ularitide: a natriuretic peptide candidate for the treatment of acutely decompensated heart failure. Futur Cardiol 11(5):531–546. https://doi.org/10.2217/FCA.15.53

Fischer E, Fourneau E (1906) Übereinige Derivate des Glykocolls. In: Untersuchungenüber Amin Polypeptide und Proteïne, pp 279–289. https://doi.org/10.1007/978-3-642-99499-9_21

Fosgerau K, Hoffmann T (2015) Peptide therapeutics: current status and future directions. Drug Discov Today 20(1):122–128. https://doi.org/10.1016/j.drudis.2014.10.003

Frankel AD, Pabo CO (1988) Cellular uptake of the tat protein from human immunodeficiency virus. Cell 55(6):1189–1193

Gao C, Mao S, Ditzel HJ et al (2002) A cell-penetrating peptide from a novel pVII–pIX phage-displayed random peptide library. Bioorg Med Chem 10(12):4057–4065. https://doi.org/10.3892/mmr.2015.4024

Gao Y, Wei X, Ye X et al (2015) Anticancer activity of stoppin based on a novel peptide delivery system. Mol Med Rep 12(4):5437–5442

Gelain F, Unsworth LD, Zhang S (2010) Slow and sustained release of active cytokines from self-assembling peptide scaffolds. J Control Release 145(3):231–239

Giugliano RP, Newby LK, Harrington RA et al (2005) The early glycoprotein IIb/IIIa inhibition in non–ST-segment elevation acute coronary syndrome (EARLY ACS) trial: a randomized placebo-controlled trial evaluating the clinical benefits of early front-loaded eptifibatide in the treatment of patients with n. Am Heart J 149(6):994–1002

Goodman M, Cai W, Smith ND (2003) The bold legacy of Emil Fischer. J Pept Sci 9:594–603. https://doi.org/10.1002/PSC.476

Gras J (2009) Icatibant for hereditary angioedema. Drugs Today (Barc) 45(12):855–864

Green M, Loewenstein PM (1988) Autonomous functional domains of chemically synthesized human immunodeficiency virus tat trans-activator protein. Cell 55(6):1179–1188

Grunwald J, Rejtar T, Sawant R et al (2009) TAT peptide and its conjugates: proteolytic stability. Bioconjug Chem 20(8):1531–1537

Gu Z-H, Wang B, Kou Z-Z et al (2017) Endomorphins: promising endogenous opioid peptides for the development of novel analgesics. Neurosignals 25(1):98–116. https://doi.org/10.1159/000484909

Gundersen ET (2016) The production and characterization of drug-loaded liposomal and PLGA nanocarriers for targeted treatment of acute myeloid leukemia. Dissertation, The University of Bergen

Hartgerink JD, Beniash E, Stupp SI (2001) Self-assembly and mineralization of peptide-amphiphile nanofibers. Science 294(5547):1684–1688

Hoare TR, Kohane DS (2008) Hydrogels in drug delivery: Progress and challenges. Polymer (Guildf) 49(8):1993–2007

Hou X, Flaig TW (2012) Redefining hormone sensitive disease in advanced prostate cancer. Adv Urol 2012:1–6. https://doi.org/10.1155/2012/978531

Hyvönen M, Enbäck J, Huhtala T et al (2014) Novel target for peptide-based imaging and treatment of brain tumors. Mol Cancer Ther 13(4):996–1007. https://doi.org/10.1158/1535-7163.MCT-13-0684

Iepsen EW, Torekov SS, Holst JJ (2015) Liraglutide for type 2 diabetes and obesity: a 2015 update. Expert Rev Cardiovasc Ther 13(7):753–767

Iglesias D, Marchesan S (2017) Short peptide self-assembled nanostructures for therapeutics innovative delivery. Nanostruct Novel Therapy, In, pp 227–250. https://doi.org/10.1016/B978-0-323-46142-9.00009-8

Joseph M, Trinh HM, Mitra AK (2017a) Peptide and protein-based therapeutic agents. In: emerging nanotechnologies for diagnostics. Drug Deliv Med Dev, Elsevier Inc, pp 145–167. https://doi.org/10.1016/B978-0-323-42978-8.00007-3

Joseph M, Trinh HM, Mitra AK (2017b) Chapter 7 - peptide and protein-based therapeutic agents∗. In: Mitra AK, Cholkar K (eds) Mandal drug delivery and medical devices ABT-EN for D (eds) micro and Nano Technologies. Elsevier, Boston, pp 145–167. https://doi.org/10.1016/B978-0-323-42978-8.00007-3

Kaga S, Arslan M, Sanyal R, Sanyal A (2016) Dendrimers and Dendrons as versatile building blocks for the fabrication of functional hydrogels. Molecules 21(4):497. https://doi.org/10.3390/molecules21040497

Kammerer RA, Kostrewa D, Zurdo J et al (2004) Exploring amyloid formation by a de novo design. Proc Natl Acad Sci U S A 101(13):4435–4440

Kanazawa T, Taki H, Tanaka K et al (2011) Cell-penetrating peptide-modified block copolymer micelles promote direct brain delivery via intranasal administration. Pharm Res 28(9): 2130–2139. https://doi.org/10.1007/s11095-011-0440-7

Kapić E, Becić F, Zvizdić S (2005) Enfuvirtide, mechanism of action and pharmacological properties. MedicinskiArhiv 59(5):313–316

Kaspar AA, Reichert JM (2013) Future directions for peptide therapeutics development. Drug Discov Today 18(17):807–817. https://doi.org/10.1016/j.drudis.2013.05.011

Keskin D, Tezcaner A (2017) Micelles as delivery system for cancer treatment. Curr Pharm Des 23(35):5230–5241

Kim MR, Feng T, Zhang Q et al (2019) Co-encapsulation and co-delivery of peptide drugs via polymeric nanoparticles. Polymers (Basel) 11(2):288

Kim SK, Huang L (2012) Nanoparticle delivery of a peptide targeting EGFR signaling. J Control Release 157(2):279–286

Kim Y, Hwang S, Khalmuratova R et al (2020) α-Helical cell-penetrating peptide-mediated nasal delivery of resveratrol for inhibition of epithelial-to-mesenchymal transition. J Control Release 317:181–194. https://doi.org/10.1016/j.jconrel.2019.11.034

Kitagaki J, Shi G, Miyauchi S, Murakami S, Yang Y et al (2015) Cyclic depsipeptides as potential cancer therapeutics. Anti-Cancer Drugs 26(3):259–271. https://doi.org/10.1097/CAD.0000000000000183

Koivunen E, Arap W, Valtanen H et al (1999) Tumor targeting with a selective gelatinase inhibitor. Nat Biotechnol 17:768–774

Koivunen E, Gay DA, Ruoslahti E (1993) Selection of peptides binding to the alpha 5 beta 1 integrin from phage display library. J Biol Chem 268(27):20205–20210

Kolambkar YM, Dupont KM, Boerckel JD et al (2011) An alginate-based hybrid system for growth factor delivery in the functional repair of large bone defects. Biomaterials 32(1):65–74

Kolterman OG, Kim DD, Shen L et al (2005) Pharmacokinetics, pharmacodynamics, and safety of exenatide in patients with type 2 diabetes mellitus. Am J Health Syst Pharm 62(2):173–181

Kopeček J, Yang J (2009) Peptide-directed self-assembly of hydrogels. Acta Biomater 5(3): 805–816. https://doi.org/10.1016/J.ACTBIO.2008.10.001

Kostal J, Mulchandani A, Chen W (2004) Affinity purification of plasmid DNA by temperature-triggered precipitation. BiotechnolBioeng 85(3):293–297. https://doi.org/10.1002/bit.10890

Kostal J, Mulchandani A, Gropp KE, Chen W (2003) A temperature responsive biopolymer for mercury remediation. Environ Sci Technol 37(19):4457–4462

Koutsopoulos S, Unsworth LD, Nagai Y, Zhang S (2009) Controlled release of functional proteins through designer self-assembling peptide nanofiber hydrogel scaffold. Proc Natl Acad Sci 106(12):4623–4628

Kovaric BC, Kokona B, Schwab AD et al (2006) Self-assembly of peptide porphyrin complexes: toward the development of smart biomaterials. J Am Chem Soc 128(13):4166–4167

Kraft JC, Freeling JP, Wang Z, Ho RJY (2014) Emerging research and clinical development trends of liposome and lipid nanoparticle drug delivery systems. J Pharm Sci 103(1):29–52

Kretsinger JK, Haines LA, Ozbas B et al (2005) Cytocompatibility of self-assembled β-hairpin peptide hydrogel surfaces. Biomaterials 26(25):5177–5186. https://doi.org/10.1016/j.biomaterials.2005.01.029

Kubo SH, Cody RJ (1985) Clinical pharmacokinetics of the angiotensin converting enzyme inhibitors. Clin Pharmacokinet 10(5):377–391

Kuhn J-M, Billebaud T, Navratil H et al (1989) Prevention of the transient adverse effects of a gonadotropin-releasing hormone analogue (buserelin) in metastatic prostatic carcinoma by administration of an antiandrogen (nilutamide). New England J Med 321(7):413–418

Kumita JR, Smart OS, Woolley GA (2000) Photo-control of helix content in a short peptide. Proc Natl Acad Sci 97(8):3803–3808

Lalanne M, Andrieux K, Paci A et al (2007) Liposomal formulation of a glycerolipidic prodrug for lymphatic delivery of didanosine via oral route. Int J Pharm 344:62–70

Larson N, Ghandehari H (2012) Polymeric conjugates for drug delivery. Chem Mater 24(5): 840–853

Lassen U, Mau-Sørensen M, Poulsen HS (2014) Orphan drugs in glioblastoma multiforme: a review. Orphan Drugs: Res Rev 4:83–91. https://doi.org/10.2147/ODRR.S46018

Lau JL, Dunn MK (2018) Therapeutic peptides: historical perspectives, current development trends, and future directions. Bioorg Med Chem 26(10):2700–2707

Lee JH, Engler JA, Collawn JF, Moore BA (2001) Receptor mediated uptake of peptides that bind the human transferrin receptor. Eur J Biochem 268(7):2004–2012

Lee S, Trinh THT, Yoo M et al (2019) Self-assembling peptides and their application in the treatment of diseases. Int J Mol Sci 20(23):5850. https://doi.org/10.3390/ijms20235850

Lewis RJ (2012) Discovery and development of the χ-conopeptide class of analgesic peptides. Toxicon 59(4):524–528. https://doi.org/10.1016/J.TOXICON.2011.07.012

Li K, Liu CJ, Zhang XZ (2020) Multifunctional peptides for tumor therapy. Adv Drug Deliv Rev 160:36–51. https://doi.org/10.1016/J.ADDR.2020.10.009

Li Y, Wang F, Cui H (2016) Peptide-based supramolecular hydrogels for delivery of biologics. Bioeng Transl Med 1(3):306–322. https://doi.org/10.1002/btm2.10041

Li Z, Zhao R, Wu X et al (2005) Identification and characterization of a novel peptide ligand of epidermal growth factor receptor for targeted delivery of therapeutics. FASEB J 19(14): 1978–1985

Liang C, Yan X, Zhang R et al (2020) Enhanced cellular uptake and nuclear accumulation of drug-peptide nanomedicines prepared by enzyme-instructed self-assembly. J Control Release 317: 109–117

Lien S, Lowman HB (2003) Therapeutic peptides. Trends Biotechnol 21(12):556–562

Lindsey S, Piatt JH, Worthington P et al (2015) Beta hairpin peptide hydrogels as an injectable solid vehicle for neurotrophic growth factor delivery. Biomacromolecules 16(9):2672–2683

Liu H, Wang R, Wei J et al (2018) Conformation-directed micelle-to-vesicle transition of cholesterol-decorated polypeptide triggered by oxidation. J Am Chem Soc 140(21): 6604–6610. https://doi.org/10.1021/JACS.8B01873

Liu J, Gray WD, Davis ME, Luo Y (2012) Peptide-and saccharide-conjugated dendrimers for targeted drug delivery: a concise review. Interface Focus 2(3):307–324

Liu SQ, Rachel Ee PL, Ke CY et al (2009) Biodegradable poly(ethylene glycol)–peptide hydrogels with well-defined structure and properties for cell delivery. Biomaterials 30(8):1453–1461. https://doi.org/10.1016/j.biomaterials.2008.11.023

Loffet A (2002) Peptides as drugs: is there a market? J Peptide Sci 8(1):1–7

Lombardino JG, Lowe JA (2004) The role of the medicinal chemist in drug discovery — then and now. Nat Rev Drug Discov 23(10):853–862. https://doi.org/10.1038/nrd1523

Lombardo D, Kiselev MA, Caccamo MT (2019) Smart nanoparticles for drug delivery application: development of versatile Nanocarrier platforms in biotechnology and nanomedicine. J Nanomater 2019:3702518. https://doi.org/10.1155/2019/3702518

Luo K, Li C, Li L et al (2012) Arginine functionalized peptide dendrimers as potential gene delivery vehicles. Biomaterials 33(19):4917–4927. https://doi.org/10.1016/j.biomaterials.2012.03.030

Lv S, Wang J, Dou S et al (2014) Nanoparticles encapsulating hepatitis B virus cytosine-phosphate-guanosine induce therapeutic immunity against HBV infection. Hepatology 59(2):385–394. https://doi.org/10.1002/HEP.26654

Lyons DF, Le V, Bidwell GL et al (2013) Structural and hydrodynamic analysis of a novel drug delivery vector: ELP[V5G3A2-150]. Biophys J 104(9):2009–2021. https://doi.org/10.1016/J.BPJ.2013.03.040

Marberger M, Kaisary AV, Shore ND et al (2010) Effectiveness, pharmacokinetics, and safety of a new sustained-release leuprolide acetate 3.75-mg depot formulation for testosterone suppression in patients with prostate cancer: a phase III, open-label, international multicenter study. Clin Ther 32(4):744–757

Marr AK, Gooderham WJ, Hancock REW (2006) Antibacterial peptides for therapeutic use: obstacles and realistic outlook. Curr Opin Pharmacol 6:468–472. https://doi.org/10.1016/j.coph.2006.04.006

Mart RJ, Osborne RD, Stevens MM, Ulijn RV (2006) Peptide-based stimuli-responsive biomaterials. Soft Matter 2(10):822–835

Martin P, Cohen A, Uddin S et al (2020) Randomized, double-masked, placebo-controlled dose escalation study of TAK-639 topical ophthalmic solution in subjects with ocular hypertension or primary open-angle glaucoma. Clin Ophthalmol 14:885–896. https://doi.org/10.2147/OPTH.S242932

Mazel M, Clair P, Rousselle C et al (2001) Doxorubicin-peptide conjugates overcome multidrug resistance. Anti-Cancer Drugs 12(2):107–116

Mazza M, Notman R, Anwar J, Rodger A, Hicks M, Parkinson G, McCarthy D, Daviter T, Moger J, Garrett N, Mead T, Briggs M, Andreas G, Schätzlein AG, Uchegbu IF (2013) Nanofiber-based delivery of therapeutic peptides to the brain. ACS Nano 7(2):1016–1026. https://pubs.acs.org/doi/abs/10.1021/nn305193d

Megeed Z, Winters RM, Yarmush ML (2006) Modulation of single-chain antibody affinity with temperature-responsive elastin-like polypeptide linkers. Biomacromolecules 7(4):999–1004. https://doi.org/10.1021/bm0507002

Mentlein R (2004) Cell-Surface Peptidases. Int Rev Cytol 235:165. https://doi.org/10.1016/S0074-7696(04)35004-7

Merrifield B (1997) Concept and early development of solid-phase peptide synthesis. Methods Enzymol 289:3–13. https://doi.org/10.1016/S0076-6879(97)89040-4

Merrifield RB (1985) Solid phase synthesis (Nobel lecture). AngewandteChemie International Edition in English 24(10):799–810. https://doi.org/10.1002/anie.198507993

Merrifield RB (1986) Solid phase synthesis. Science 232(4748):341–348

Millar JC, Savinainen A, Josiah S, Pang IH (2019) Effects of TAK-639, a novel topical C-type natriuretic peptide analog, on intraocular pressure and aqueous humor dynamics in mice. Exp Eye Res 188:107763. https://doi.org/10.1016/J.EXER.2019.107763

Mondal S, Das S, Nandi AK (2020) A review on recent advances in polymer and peptide hydrogels. Soft Matter 16:1404–1454. https://doi.org/10.1039/C9SM02127B

Morris MC, Depollier J, Mery J et al (2001) A peptide carrier for the delivery of biologically active proteins into mammalian cells. Nat Biotechnol 19(12):1173–1176

Mottaghitalab F, Farokhi M, Shokrgozar MA et al (2015) Silk fibroin nanoparticle as a novel drug delivery system. J Control Release 206:161–176

Mukaya HE, Van Zyl RL, Van Vuuren NJ, Mbianda XY (2017) Synthesis and characterization of water-soluble polyaspartamides containing platinum (II) complex and bisphosphonate as potential antimalarial drug. Polym Bull 74(8):3161–3178

Muttenthaler M, King GF, Adams DJ, Alewood PF (2021) Trends in peptide drug discovery. Nat Rev Drug Discov 20(4):309–325. https://doi.org/10.1038/s41573-020-00135-8

Mutter M, Gassmann R, Buttkus U, Altmann K (1991) Switch peptides: pH-induced α-helix to β-sheet transitions of bis-amphiphilic oligopeptides. Angew Chem Int Ed Engl 30(11): 1514–1516

Nakayama F, Yasuda T, Umeda S et al (2011) Fibroblast growth factor-12 (FGF12) translocation into intestinal epithelial cells is dependent on a novel cell-penetrating peptide domain: involvement of internalization in the in vivo role of exogenous FGF12. J Biol Chem 286(29): 25823–25834

Nishimura A, Hayakawa T, Yamamoto Y et al (2012) Controlled release of insulin from self-assembling nanofiber hydrogel, PuraMatrix™: application for the subcutaneous injection in rats. Eur J Pharm Sci 45:1–7

O'Shea JC, Tcheng JE (2002) Eptifibatide: a potent inhibitor of the platelet receptor integrin glycoprotein IIb/IIIa. Expert Opin Pharmacother 3(8):1199–1210. https://doi.org/10.1517/14656566.3.8.1199

Oehlke J, Scheller A, Wiesner B et al (1998) Cellular uptake of an α-helical amphipathic model peptide with the potential to deliver polar compounds into the cell interior non-endocytically. Biochim et Biophysica Acta (BBA)-Biomembranes 1414:127–139

Pandita D, Santos JL, Rodrigues J et al (2011) Gene delivery into mesenchymal stem cells: a biomimetic approach using RGD nanoclusters based on poly (amidoamine) dendrimers. Biomacromolecules 12(2):472–481

Pandya MJ, Cerasoli E, Joseph A et al (2004) Sequence and structural duality: designing peptides to adopt two stable conformations. J Am Chem Soc 126(51):17016–17024

Pasqualini R, Koivunen E, Ruoslahti E (1997) αv integrins as receptors for tumor targeting by circulating ligands. Nat Biotechnol 15(6):542–546

Pasut G, Veronese FM (2007) Polymer–drug conjugation, recent achievements and general strategies. Prog Polym Sci 32:933–961

Patel IH, Zhang X, Nieforth K et al (2005) Pharmacokinetics, pharmacodynamics and drug interaction potential of enfuvirtide. Clin Pharmacokinet 44(2):175–186

Perry CM, Brogden RN (1996) Goserelin Drugs 51(2):319–346. https://doi.org/10.2165/00003495-199651020-00009

Phillips MI, Mann JFE, Haebara H et al (1977) Lowering of hypertension by central saralasin in the absence of plasma renin. Nature 270(5636):445–447

Pooga M, Hällbrink M, Zorko M, Langel Ü (1998) Cell penetration by transportan. FASEB J 12(1):67–77

Porter CJH, Charman WN (1997) Uptake of drugs into the intestinal lymphatics after oral administration. Adv Drug Deliv Rev 25(1):71–89

Prabhukumar G, Matsumoto M, Mulchandani A, Chen W (2004) Cadmium removal from contaminated soil by tunable biopolymers. Environ Sci Technol 38(11):3148–3152

Pudlarz A, Szemraj J (2018) Nanoparticles as carriers of proteins, peptides and other therapeutic molecules. Open Life Sciences 13(1):285–298. https://doi.org/10.1515/biol-2018-0035

Qiu Y, Park K (2001) Environment-sensitive hydrogels for drug delivery. Adv Drug Deliv Rev 53(3):321–339. https://doi.org/10.1016/S0169-409X(01)00203-4

Qu W, Qin SY, Ren S et al (2013) Peptide-based vector of VEGF plasmid for efficient gene delivery in vitro and vessel formation in vivo. Bioconjug Chem 24(6):960–967. https://doi.org/10.1021/bc300677n

Raghuraman H, Chattopadhyay A (2006) Effect of ionic strength on folding and aggregation of the hemolytic peptide melittin in solution. Biopolymers 83:111–121. https://doi.org/10.1002/bip.20536

Rajchakit U, Sarojini V (2017) Recent developments in antimicrobial-peptide-conjugated gold nanoparticles. Bioconjug Chem 28:2673–2686. https://doi.org/10.1021/acs.bioconjchem.7b00368

Raucher D, Chilkoti A (2001) Enhanced uptake of a thermally responsive polypeptide by tumor cells in response to its hyperthermia-mediated phase transition. Cancer Res 61:7163–7170

Reiersen H, Rees AR (1999) An engineered Minidomain containing an elastin turn exhibits a reversible temperature-induced IgG binding. Biochemistry 38:14897–14905. https://doi.org/10.1021/bi991243a

Rhee M, Davis P (2006) Mechanism of uptake of C105Y, a novel cell-penetrating peptide. J Biol Chem 281:1233–1240

Ringler P, Schulz GE (2003) Self-assembly of proteins into designed networks. Science 302:106–109. https://doi.org/10.1126/science.1088074

Ringsdorf H (1975) Structure and properties of pharmacologically active polymers. J Polym Sci: Polym Symp:135–153
Rothbard JB, Garlington S, Lin Q et al (2000) Conjugation of arginine oligomers to cyclosporin a facilitates topical delivery and inhibition of inflammation. Nat Med 6:1253–1257
Rousselle C, Clair P, Lefauconnier J-M et al (2000) New advances in the transport of doxorubicin through the blood-brain barrier by a peptide vector-mediated strategy. Mol Pharmacol 57:679–686
Ruoslahti E (2017) Tumor penetrating peptides for improved drug delivery. Adv Drug Deliv Rev 110–111:3–12. https://doi.org/10.1016/J.ADDR.2016.03.008
Sangeetha N, Selvamani P, Latha S (2019) Emerging trends in therapeutic peptide pharmaceuticals: prospects and perspectives. J Drug Deliv Therapeut 9:606–610
Sansone A, Schubert M, Tüttelmann F et al (2021) Pituitary response to GnRH stimulation tests in different FSHB-211 G/T genotypes. Hum Reprod 36:1376–1382
Schneider JP, Pochan DJ, Ozbas B et al (2002) Responsive hydrogels from the intramolecular folding and self-assembly of a designed peptide. J Am Chem Soc 124:15030–15037. https://doi.org/10.1021/ja027993g
Schroeder CI, Craik DJ (2012) Therapeutic potential of conopeptides. Future Med Chem 4:1243–1255. https://doi.org/10.4155/FMC.12.70
Schultz C (2019) Targeting the extracellular matrix for delivery of bioactive molecules to sites of arthritis. Br J Pharmacol 176:26–37
Seroski DT, Hudalla GA (2018) Self-assembled peptide and protein nanofibers for biomedical applications. In: Biomedical applications of functionalized nanomaterials: concepts, development and clinical translation. Elsevier, pp 569–598. https://doi.org/10.1016/b978-0-323-50878-0.00019-7
Shammas NW (2005) Bivalirudin: pharmacology and clinical applications. Cardiovasc Drug Rev 23:345–360. https://doi.org/10.1111/J.1527-3466.2005.TB00177.X
Shao T, Falcone N, Kraatz H-B (2020) Supramolecular peptide gels: influencing properties by metal ion coordination and their wide-ranging applications. ACS Omega 5:1312–1317. https://doi.org/10.1021/acsomega.9b03939
Shim MK, Park J, Yoon HY et al (2019) Carrier-free nanoparticles of cathepsin B-cleavable peptide-conjugated doxorubicin prodrug for cancer targeting therapy. J Control Release 294:376–389. https://doi.org/10.1016/j.jconrel.2018.11.032
Silva EA, Mooney DJ (2010) Effects of VEGF temporal and spatial presentation on angiogenesis. Biomaterials 31:1235–1241
Singh-Franco D, Robles G, Gazze D (2007) Pramlintide acetate injection for the treatment of type 1 and type 2 diabetes mellitus. Clin Ther 29:535–562
Smith HS, Deer TR (2009) Safety and efficacy of intrathecal ziconotide in the management of severe chronic pain. Ther Clin Risk Manag 5:521
Smith WJ, Drew RH (2009) Telavancin: a new lipoglycopeptide for gram-positive infections. Drugs Today 45:159–173. https://doi.org/10.1358/DOT.2009.45.3.1343792
Snyder EL, Dowdy SF (2004) Cell penetrating peptides in drug delivery. Pharm Res 21:389–393
Song B, Song J, Zhang S et al (2012) Sustained local delivery of bioactive nerve growth factor in the central nervous system via tunable deblock copolypeptide hydrogel depots. Biomaterials 33:9105–9116
Spaliviero M, Harmsen S, Huang R et al (2016) Detection of lymph node metastases with SERRS nanoparticles. Mol Imaging Biol 18:677–685
Sterling JK, Adetunji MO, Guttha S et al (2020) GLP-1 receptor agonist NLY01 reduces retinal inflammation and neuron death secondary to ocular hypertension. Cell Rep 33:10827. https://doi.org/10.1016/J.CELREP.2020.108271
Tang A-M, Wang W-J, Mei B et al (2013) DEVD-based Hydrogelator minimizes cellular apoptosis induction. Sci Rep 3:1–7. https://doi.org/10.1038/srep01848
Tedesco KL, Rybak MJ (2004) Daptomycin. Pharmacotherapy: the journal of human pharmacology and drug. Therapy 24:41–57

Tesauro D, Accardo A, Diaferia C et al (2019a) Peptide-based drug-delivery systems in biotechnological applications: recent advances and perspectives. Molecules 24:1–27. https://doi.org/10.3390/molecules24020351
Tesauro D, Accardo A, Diaferia C et al (2019b) Peptide-based drug-delivery systems in biotechnological applications: recent advances and perspectives. Molecules 24:351
Thomson AH, Kelly JG, Whiting B (1989) Lisinopril population pharmacokinetics in elderly and renal disease patients with hypertension. Br J Clin Pharmacol 27:57–65
Tünnemann G, Ter-Avetisyan G, Martin RM et al (2008) Live-cell analysis of cell penetration ability and toxicity of oligo-arginines. J Peptide Sci 14:469–476
Van S, Das SK, Wang X et al (2010) Synthesis, characterization, and biological evaluation of poly (L-γ-glutamyl-glutamine)-paclitaxel nanoconjugate. Int J Nanomedicine 5:825
Vetter I, Lewis RJ (2012) Therapeutic potential of cone snail venom peptides (conopeptides). Curr Top Med Chem 12(14):1546–1552. https://doi.org/10.2174/156802612802652457
Vlieghe P, Lisowski V, Martinez J, Khrestchatisky M (2010) Synthetic therapeutic peptides: science and market. Drug Discov Today 15:40–56. https://doi.org/10.1016/j.drudis.2009.10.009
Wanakule P (2012) Development and evaluation of enzymatically-degradable hydrogel microparticles for pulmonary delivery of nanoparticles and biologics. Dissertation, University of Texas. http://repositories.lib.utexas.edu/handle/2152/23398
Wang D, Wang K, Cai Y (2020) An overview of development in gene therapeutics in China. Gene Ther 27:1. https://doi.org/10.1038/S41434-020-0163-7
Wang W, Shao R, Wu Q et al (2009) Targeting gelatinases with a near-infrared fluorescent cyclic his-try-Gly-Phe peptide. Mol Imaging Biol 11:424–433
Wang Y, Cheetham AG, Angacian G et al (2017) Peptide–drug conjugates as effective prodrug strategies for targeted delivery. Adv Drug Deliv Rev 110–111:112–126. https://doi.org/10.1016/j.addr.2016.06.015
Wennerström H, Lindman B (1979) Micelles. Physical chemistry of surfactant association. Phys Rep 52:1–86. https://doi.org/10.1016/0370-1573(79)90087-5
Wetzler M, Hamilton P (2018) Peptides as therapeutics. In: Peptide applications in biomedicine, biotechnology and bioengineering. Elsevier Ltd, pp 215–230. https://doi.org/10.1016/B978-0-08-100736-5.00008-9
Wilby KJ, Partovi N, Ford JA et al (2012) Review of boceprevir and telaprevir for the treatment of chronic hepatitis C. Can J Gastroenterol 26(4):205–210. https://doi.org/10.1155/2012/751057
Wyatt LC, Lewis JS, Andreev OA et al (2017) Applications of pHLIP Technology for Cancer Imaging and Therapy. Trends Biotechnol 35:653–664. https://doi.org/10.1016/J.TIBTECH.2017.03.014
Wyatt LC, Moshnikova A, Crawford T et al (2018) Peptides of pHLIP family for targeted intracellular and extracellular delivery of cargo molecules to tumors. Proc Natl Acad Sci 115:E2811–E2818. https://doi.org/10.1073/PNAS.1715350115
Xie M, Liu D, Yang Y (2020) Anti-cancer peptides: classification, mechanism of action, reconstruction and modification. Open Biol 10:200004
Yun SP, Kam T-I, Panicker N et al (2018) Block of A1 astrocyte conversion by microglia is neuroprotective in models of Parkinson's disease. Nat Med 24:931–938. https://doi.org/10.1038/s41591-018-0051-5
Zhang Q, Tang J, Fu L et al (2013) A pH-responsive α-helical cell penetrating peptide-mediated liposomal delivery system. Biomaterials 34:7980–7993
Zhang W, Yu L, Ji T, Wang C (2020) Tumor microenvironment–responsive peptide-based supramolecular drug delivery system. Front Chem 0:549. https://doi.org/10.3389/FCHEM.2020.00549
Zhang W, Yu X, Li Y et al (2018) Protein-mimetic peptide nanofibers: motif design, self-assembly synthesis, and sequence-specific biomedical applications. Prog Polym Sci 80:94–124. https://doi.org/10.1016/J.PROGPOLYMSCI.2017.12.001

Zhang XX, Eden HS, Chen X (2012) Peptides in cancer nanomedicine: drug carriers, targeting ligands and protease substrates. J Control Release 159:2–13. https://doi.org/10.1016/J.JCONREL.2011.10.023

Zhao B-X, Zhao Y, Huang Y et al (2012) The efficiency of tumor-specific pH-responsive peptide-modified polymeric micelles containing paclitaxel. Biomaterials 33:2508–2520

Zhao Y, Ren W, Zhong T et al (2016) Tumor-specific pH-responsive peptide-modified pH-sensitive liposomes containing doxorubicin for enhancing glioma targeting and anti-tumor activity. J Control Release 222:56–66

Zhou M, Smith AM, Das AK et al (2009) Self-assembled peptide-based hydrogels as scaffolds for anchorage-dependent cells. Biomaterials 30:2523–2530. https://doi.org/10.1016/j.biomaterials.2009.01.010

Zhu Y, Li Z, Liu H et al (2014) Novel analgesic peptides from the tree frog of Hyla japonica. Biochimie 99:38–43. https://doi.org/10.1016/J.BIOCHI.2013.10.017

Zirah S, Kozin SA, Mazur AK et al (2006) Structural changes of region 1-16 of the Alzheimer disease amyloid β-peptide upon zinc binding and in vitro aging. J Biol Chem 281:2151–2161

Zompra AA, Galanis AS, Werbitzky O, Albericio F (2009) Manufacturing peptides as active pharmaceutical ingredients. Future Med Chem 1:361–377. https://doi.org/10.4155/fmc.09.23

Zou L-L, Ma J-L, Wang T et al (2013) Cell-penetrating peptide-mediated therapeutic molecule delivery into the central nervous system. Curr Neuropharmacol 11(2):197. https://doi.org/10.2174/1570159X11311020006

Basics of Clinical Drug Development: Clinical Trial and Drug Development

Parul Gupta and Ajay Kumar Verma

Abstract The development of new drugs typically goes through various complications and requires the involvement of distinct scientific streams. Extensive preclinical studies that provide the initial information about the efficacy, toxicity, safety, and pharmacokinetics of the drug decide whether it goes for the clinical trial. Clinical trial of a drug is executed in four different phases (phase I, II, III, and IV). Different types of trials are based on observational (non-experimental) and experimental designs. In the observational study, there is no intervention or manipulation in a group of people. Experimental design (randomized control trial) proves to be the most convincing and is thought to be the most dependable way to study the effectiveness of treatment. Alongside these experimental designs, descriptive research is also very useful in the case of different studies, i.e., epidemiological studies. Besides all these suitable application of statistics for calculation of sample size, data collection, compiling, and analysis improve the outcomes of the clinical research. In this chapter, we have discussed the different methodologies used in clinical research and clinical trial.

Keywords Clinical trials · Epidemiological study · Experimental and non-experimental research

1 Introduction

Drug development is a time-consuming, intricate, and costly process with no guarantee that a drug will be successful and reach the market (Taylor 2015). The expense of pharmaceutical research has steadily increased since the 1950s. In fact, R&D costs for new licensed treatments increased linearly over time on a log scale, and costs doubled every 9 years (Yildirim et al. 2016). On average, it takes

P. Gupta · A. K. Verma (✉)
Department of Respiratory Medicine, King George's Medical University, Lucknow, Uttar Pradesh, India

V. S. Rajput, A. Runthala (eds.), *Drugs and a Methodological Compendium*,
https://doi.org/10.1007/978-981-19-7952-1_8

10–15 years for a new drug to enter the market and costs $1.5–2 billion. Clinical phases take approximately half of this time and money that was used in the drug development process. The remaining half of the R&D investment is spent on preclinical testing of new drugs as well as regulatory processes (Harrer et al. 2019). Despite increased R&D investment, attrition rates remain high and, worse, are increasing. Only around 10% of medications that go through clinical trials (CTs) are authorized by regulatory organizations (Akhondzadeh 2016). This poses a serious threat to the pharmaceutical industry and must be addressed.

The drug development process may be split into two parts: the preclinical (non-clinical) phase, which involves drug discovery, and the clinical phase, which comprises drug development (Strovel et al. 2016a). We shall concentrate on clinical research, with a particular emphasis on clinical trials, because the preclinical phase is covered elsewhere in this book. Clinical trials (CTs) are critical to the drug development process because they ensure the efficacy and safety of new drugs. They are also the foundation for bringing new and improved treatments to the market (Gal et al. 2018). Since the first randomized clinical trial in the 1940s, the drug development process has been modified and improved to address crucial concerns linked to drug safety and minimize the attrition rate (Preziosi 2007). However, the fundamental ideas that underpin the clinical evaluation process have remained considerably intact.

Clinical trials for drugs and medical devices have a lot of potential failures. There are many culprits for the failure including lack of efficacy, issues with safety, or a lack of funding to complete a trial, as well as other factors such as failing to maintain good manufacturing protocols, failing to follow FDA guidance, or problems with patient recruitment, enrolment, and retention (Fogel 2018). However, the majority of the time failures occur as a result of poor planning or a misunderstanding of important biology and/or pharmacological development concepts. Most errors and technical failures in clinical trials arise due to the improper selection of clinical trial designs. The clinical trial design selection will have a strong impact on the cost and time associated with clinical trials. The expected outcomes shall be thought out and carefully addressed while designing a clinical trial. Therefore, it will be critical to adopt an appropriate research design to give the needed information.

The clinical study design is the formulation of experiments, trials, and studies in medical, clinical, and other types of research involving human beings. Clinical studies are classified into two types: experimental studies, in which the investigator intervenes to prevent or treat a condition, and observational studies, in which the investigator makes no intervention and the patients are assigned treatment based on clinical choices (Grimes and Schulz 2002). The selection of clinical trial designs varies with the phase of the trial, stage of drug development, expected outcome, patient population factors, disease type, and many other such factors.

In this chapter, we will discuss key concepts involved in the conduct of clinical trials. We will also discuss various kinds of study designs and methodologies that are involved in the clinical development process of a drug.

2 Overview of Clinical Research Designs

Clinical research, or studies involving human beings, is broadly divided into two categories: experimental and observational research. In experimental studies, researchers administer interventions to the individuals or population in order to assess the effectiveness of any therapy. The most prevalent type of experimental clinical research is the randomized controlled trial (RCT), which is regarded as the gold standard (Hariton and Locascio 2018). Observational research makes no attempt to intervene in the study sample for the sole purpose of examination. It seems immoral and unethical to conduct randomized controlled trials that intentionally expose participants to potentially dangerous conditions (Deaton and Cartwright 2018). In this case, a case-control study may be the most effective way to find a possible cause of a rare disease because it is started with an existing case. If nothing is known about how the problem progresses over time, a cohort study may be the optimal design (Song and Chung 2010). Here, we'll go through some of the most frequent study designs for these two sorts of studies. Figure 1 illustrates the diagrammatic representation of different clinical research designs.

3 Experimental Study Design

Experimental research is described as a study in which the researcher intervenes and controls one or more variables and then evaluates the resulting variance in other variables. In the study of interventions, experimental designs provide a structure for evaluating the causal link between a set of independent and dependent variables. They allow researchers to draw important conclusions about the observed

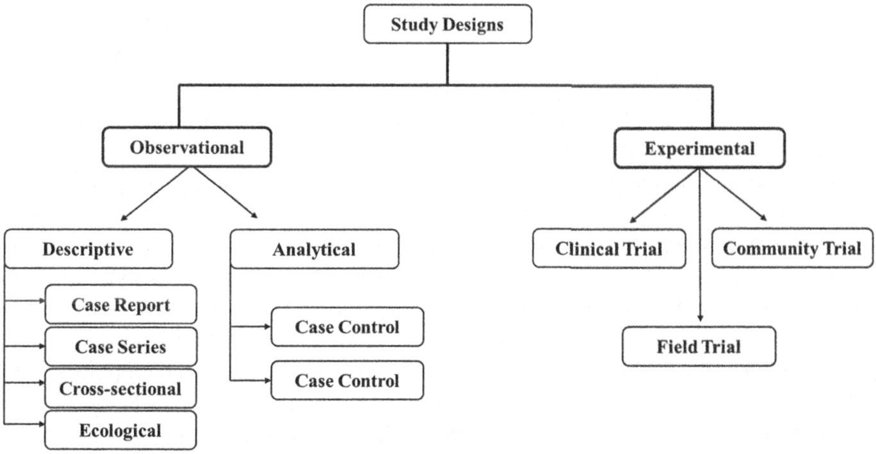

Fig. 1 Diagramatic illustration of clinical research design

differences by controlling or taking into account the effects of external variables (Song and Chung 2010; Chiang et al. 2015; Chidambaram and Josephson 2019; Nair 2019).

Experimental studies are broadly divided into two groups: those that have controls (controlled) and those that do not (uncontrolled). In an uncontrolled trial, the outcome is directly assigned to the therapy received in one group without comparison to another treatment or control. As a result, these studies tend to have less information about treatment than controlled studies (Caparrotta et al. 2019). This does not establish whether the outcome was caused by the intervention or by coincidence. Uncontrolled studies, on the other hand, cannot be ruled out in the process of assessing novel medications, especially in the early phases of clinical development when they are utilized to identify pharmacokinetic characteristics and tolerated dosage. Controlled trials, on the other hand, have at least one study group that is compared to a control group. Unless historical data is used as a control, the control group can be given a placebo or another beneficial therapy. In phase III drug development, these studies are the most prevalent. Controlled trials distinguish a patient's result from one induced by other factors such as the disease's natural history or the patient's or investigator's expectations.

Experimental designs may be split into three major groups from an epidemiological standpoint: clinical trial, field trial, and community trial.

3.1 Clinical Trial

A clinical trial is a human subject-based prospective controlled study that is used to assess the effectiveness of a potential treatment, diagnostic method, or healthcare technology (Evans 2010). Clinical trials are usually planned on a large scale, with participants coming from a variety of locations or treatment institutions. Due to the fact that a clinical trial is prospective rather than retrospective, study participants must be tracked over time. Study participants must always be observed from a certain point in time, which serves as the study's time zero or baseline for the individual participant (Trials, Charles H. Evans and Ildstad 2001). James Lind is credited as being the first modern-day physician to perform a controlled clinical study (Torre and Shahriari 2017). Clinical trials are divided into two types: therapeutic and preventative. Therapeutic trials look at how a therapy or intervention affects a certain condition. Preventative trials determine if a surgery or drug lowers the risk of a disease (Schwartz and Lellouch 2015). Clinical trials unfold in different phases to give information regarding the therapy in terms of dose, safety, and efficacy, with increasing complexity in establishing the intervention's efficacy and tolerability (Karlberg and Speers 2010). During the early phases, the emphasis is on reducing subject risk and ensuring that resources are handled wisely. The purpose of the studies is to first show efficacy and then to show efficiency under actual situations, with the main aim of implementing the treatment as a new norm. Before getting into clinical trial designs, we must first grasp the many phases of clinical trials.

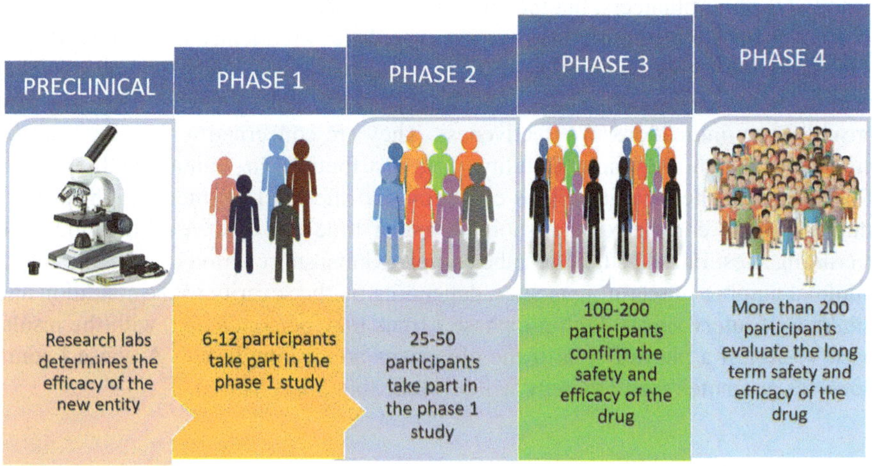

Fig. 2 The four phases of clinical trial

Clinical trials have traditionally been divided into four phases: I, II, III, and IV (Fig. 2). However, the FDA recently launched phase 0 trials in 2006 in the hopes of streamlining the drug development process (Kummar et al. 2008; Burt et al. 2020).

Clinical Phase 0

Phase I trials are traditionally the first time a medicine is examined in humans. In order to accelerate the discovery and development of novel molecular entities, the FDA released exploratory Investigational New Drug (IND) guidelines in 2006 to aid clinical evaluation prior to the dose escalation, safety, and tolerance studies associated with a conventional IND. These guidelines resulted in phase 0 clinical studies or micro-dosing studies (Marchetti and Schellens 2007; Kummar et al. 2008; Strovel et al. 2016b; Burt et al. 2020). Although phase 0 investigations are conducted in people, they are not the same as the other stages of clinical trials. This phase's goal is to assist expedite and standardize the pharmaceutical drug approval process. Phase 0 trials may assist researchers in determining if the drugs perform as intended. This might save time and money that would otherwise be spent on later phase trials. In general, phase 0 trials have no therapeutic goal. These studies often involve a limited number of subjects and a very low non-pharmacologically active dose to understand the pharmacokinetics profile of a particular drug.

Clinical Phase I

Although in vitro experiments or animal models can provide helpful preclinical insights, data from people is required. People that take part in phase I trials are usually healthy volunteers, but they might also be patients who have tried and failed to recover on conventional therapy. These trials are frequently used to evaluate a dose level with tolerable side effects, as well as to examine the pharmacological features of a novel drug or its interaction with the human body. They can also provide early indications of effectiveness. They are concerned with issues such as drug and metabolite bioavailability and distribution in human body organs. Depending on the severity of the condition and the kind of intervention, phase I studies can be completed in a short period of time and with a small number of recruiting sites. Phase I studies can be completed in a short period of time and with a minimal number of recruitment sites, depending on the severity of the condition and the kind of intervention. Multiple phase I trials may be conducted, with the results used to design a phase II trial if the objectives are accomplished. A phase I study does not guarantee that a therapy will advance to a phase II trial.

Clinical Phase II

Once a drug has been demonstrated to be safe in people, it is tested in a phase II study to investigate its efficacy by evaluating important outcomes. They are used to assess safety, pharmacokinetics, and pharmacodynamics, but they may also be used to address issues about optimal dosages, dose frequencies, administration methods, and outcomes for phase III studies. Due to the ambiguity concerning dose-response, phase II studies may include several dosages, with as many as four or five intervention groups. These studies, which are also done on very small samples of 100–300 individuals, can take several years to complete. Most phase II studies are randomized, meaning that subjects are randomly assigned (rather than selective) to receive an experimental drug, standard of care, or a placebo (a harmless, inactive substance). People who receive standard treatment or a placebo are called controls.

Clinical Phase III

Phase III trials may be performed following the completion of phase II studies establishing drug safety and potential efficacy. These trials are also known as therapeutic confirmatory trials or randomized controlled trial (RCT). These studies are often undertaken with a higher sample size, involving thousands of patients across many sites. The sample size is influenced by the prevalence of illness in the study community. A large number of participants are needed to evaluate the effectiveness of drugs if the incidence of disease is low; however, when the incidence of a disease is higher, then fewer participants are required. This current clinical trial approach uses randomization to try to remove confounder imbalances and/or systematic disparities (or biases) across treatment groups.

Clinical Phase IV

A drug will be subjected to a phase IV clinical study once it has been licenced to be used in the broader population following phase I, II, and III studies to carefully assess its safety and effectiveness. The phase IV trial is also known as post-marketing surveillance. The primary goal of the phase IV trial is to test the drug's effectiveness in real-world circumstances, to investigate the drug's long-term pros and cons, and to identify any unusual adverse effects.

3.2 Field Trials

Field trials, sometimes known as preventative or prophylactic studies, are distinct from clinical trials in that they involve people who have not yet developed the condition and are thus not patients. A field trial also comprises contacting participants at their homes, offices, or schools. Salk immunization for poliomyelitis prevention is recognized as one of the largest field trials done in 1955, with roughly one million children participating. Levels of exposure in field studies should indeed be given as in clinical trials to increase companionability of groups while removing any flexibility from the intervener. Although random allocation may be the best option for this situation, its application is limited by the high sample size. Field trials are frequently logistically challenging with limitations in terms of ethics and practicality (Smith et al. 2015).

3.3 Community Trials

The community intervention trial is a follow-up to the field experiment that entails community-wide intervention.

4 Designing of Phase 0 Clinical Trials

Phase 0 clinical trial is the initial phase in the drug development process, and it establishes whether or not the novel chemical compound should be given serious attention. These studies are used to determine the proposed compound's pharmacodynamic and pharmacokinetic properties before moving on to the phase I study (Gupta et al. 2011; Smith et al. 2015; Burt et al. 2020). To be an appropriate candidate for phase 0 evaluation, essential considerations include pharmacodynamic activity, target and biomarker identification, a larger therapeutic window, nontoxicity, and the ability to regulate identified target and biomarker in a smaller sample size. ABT-888, a PARP inhibitor, passed aforementioned criteria and was suitable for the phase 0 clinical trials (Rose et al. 2020).

The design of a phase 0 trial varies based on the specific research objectives. The major goal of phase 0 studies is to collect meaningful information from a small sample size (1–10) and to aid in the advanced stages of drug development. The design of phase 0 trial depends on a gradual shift from preclinical to clinical which necessitates tight coordination between preclinical laboratory and clinical experts (Kunos et al. 2020).

In one type of experiment design used for a phase 0 clinical trial, it was demonstrated that the treatment impacted the target in human tumours in the same way as it did in preclinical models and followed the same mechanism of action. As a result, they cannot be 'micro-dose' investigations, because pharmacodynamic effects need biologically active concentrations. Although this type of phase 0 design requires less preclinical toxicity data than for phase I studies, the preclinical evaluation, including in vitro and in vivo assay development, is extensive. The identical methodology was used in the phase 0 clinical study of ABT-888, a PARP inhibitor. The study's primary goal was to measure pharmacodynamic parameters. Prior to enrolling the first patient, the timing of peripheral blood mononuclear cells (PBMCs) sample and tumour biopsies, as well as tissue acquisition, processing, and storage methods, were thoroughly tested in preclinical models (Kummar et al. 2009).

The second type of experiment design looked at two or more structurally similar analogues with the same molecular target. In conventional methodology, findings from preclinical study are often used to choose a lead candidate from similar analogues for clinical development. However, selecting from a pool of analogues with highly equivalent pharmacological qualities remains a difficult issue for drug developers owing to preclinical models' incapacity to predict drug behaviour in humans. Phase 0 clinical trials are a safe way to test numerous analogues with similar pharmacological qualities in a limited number of individuals, resulting in clinical pharmacology data that may be used to design future phases of clinical trials. Apart from that, these studies are extremely helpful in setting the dosing schedule for a molecularly targeted therapy or a biomodulator that will be used in combination with other medicines, such as well-known chemotherapeutics. The phase 0 trial has the benefit of allowing an early identification of a drug dose that might be used in phase I trials. The dose is chosen based on the best target modulation rather than the maximum tolerated dose (MTD), because target modulation may need a significantly lower dose than MTD. The PARP inhibitor ABT-888 was successfully evaluated using this method in a phase 0 study. Researchers were able to discover a dosage range and time course that resulted in substantial PARP inhibition in as few as 14 patients, which was crucial information for the design of numerous phase I studies with ABT-888 in combination with several recognized chemotherapeutic drugs (Murgo et al. 2008).

5 Designing of Phase I Clinical Trials

Phase I clinical trials are designed to assess a compound's safety, tolerability, and pharmacokinetics (PK). Phase I trials are exploratory in nature, with the goal of determining a safe dosage. They entail administering a low dosage to a small sample of participants, with the following group receiving a greater dose if tolerated (Ivy et al. 2010; Yan et al. 2018). This process is repeated until the provided dose causes an intolerable degree of adverse effects. This is distinct from the goal of phase II and III studies, which is to discover the optimal dose. Despite the fact that each dosage group must include a limited number of individuals, the trial should offer adequate information on safety and efficacy to evaluate if a novel medicine should be further studied. It can be tough to strike this equilibrium.

In phase I trials, the sample comprises of the healthy volunteer except in the case of cancer drug trials where traditional anticancer drugs are first investigated in cancer patients since the predicted harmful effects make testing in healthy volunteers unacceptable. Patients in phase I studies are cancer patients who have had previous cancer therapies but have not shown any improvement.

The key outcomes of phase I trials are frequently one or more toxicity measurements. A significant adverse event in healthy volunteers might be any reaction to the trial medicine that necessitates treatment and the discontinuation of the new therapy. A dose-limiting toxicity is what it's termed DLT should occur within a few hours after taking the medicine. Some adverse effects are predicted in phase I trials involving patients who are already sick and hence may not be considered as a DLT. Toxicity should be defined clearly in the study protocol.

The main goal is to determine the maximum tolerated dosage (MTD), which can be defined in a variety of ways (Le Tourneau et al. 2009). It is sometimes the dose at which a certain number of people have a serious adverse event, suggesting that the dose is too dangerous and that the next lowest dose should be studied further. At other instances, the MTD may be the dose with the least number of adverse effects, which is then employed in subsequent research. It's important to understand the definitions used in each study report.

There are a variety of designs to choose from, ranging from simple to complex. A number of dose-escalation algorithms are used to establish the dosage, with a toxicity rate of 33% or less being the goal. This goal is attained by raising the study drug's dose until the toxicity rate reaches 33%. A 3 + 3 design is a basic dose-escalation design. A 3 + 3 strategy makes sense when it comes to assessing toxicity-based dose escalation (Hansen et al. 2014). Three individuals are necessary for the initial sample size because one or two patients are inadequate to assess whether a 33% toxicity rate has been attained and dose escalation should be stopped.

More complicated dose-escalation strategies are thought to be more efficient. The continuous reassessment approach based on Bayesian methodologies is an example. They are based on statistical modelling and presume a sigmoid (flattened S-shaped) mathematical connection between dosage and the likelihood of experiencing a DLT at each dose.

Biological endpoints or pharmacological measurements, such as indices of therapeutic activity, may become more significant when new safer drugs are developed. Minimum biologically active dose (MBAD) is considered as the minimum dose having effects on biological parameters. Toxicity must still be monitored closely, but there may be other indicators that determine which dose is carried forward to a phase II study.

Pharmacodynamics and pharmacokinetics are the other parameters that are considered in the phase I trials. Pharmacodynamics is the study of physical and biological indicators of a drug's impact on the body. Pharmacokinetics are markers of how a medicine interacts with the body.

6 Designing of Phase II Clinical Trials

Phase II studies are used to see if a treatment has a strong enough signal of activity or other significant benefits to support continuing research in a final phase III study. There are several phase II designs, and most methods are intended to identify if a novel intervention is likely to be superior than the existing treatments in terms of disease status improvement or fewer adverse effects.

6.1 Nonrandomized Trials

Single-Arm Trials

A single-arm study is the most basic study design. In this study, an experimental therapy is given to a group of people with the specified medical condition, and then they are reviewed on a regular basis to see how they respond. This design is adopted when the aim of the trial is to collect initial evidence of the treatment's efficacy as well as further safety data, although it is not commonly used to confirm efficacy (Evans 2010; Ivy et al. 2010; Zheng et al. 2013; Yan et al. 2018). When the patient pool is restricted, it may be preferable to use this design rather than randomly assigning many people to the control arm. It is critical to explicitly describe the aim or hypothesis of interest while planning single-arm experiments. The goal might be to investigate any form of influence in a scenario of binary decision, such as response vs. non-response.

Single-Arm Two-Stage Design

Despite the fact that single-arm phase II trials typically contain 30–70 participants, it may be desirable to end the experiment early. The intervention is initially tried on a limited number of participants in a two-stage design, and the individuals are assessed at the conclusion of this step (Dutton and Holmes 2018). If a particular number of people respond, the experiment proceeds, and a subsequent set of people is enrolled; if not, the study ends here, and this is known as a stopping rule. Despite its simplicity, this trial design has a number of faults, and the study's results can be difficult to understand. The study is unable to distinguish whether the effect is due to therapy or placebo or if it is due to the subjects' natural history. It may be more challenging if the participants have shown no influence and the experiment has been conducted since there is no frame of reference for the comparison and the control arm in the single-arm study is missing. This strategy may be appropriate in cases when the natural history of the disease is well understood and the placebo effect is minimal.

In a phase II trial conducted by Evans et al., low-dose oral etoposide was tested for the treatment of AIDS-related Kaposi's sarcoma in patients who have relapsed or progressed following systemic chemotherapy. The trial's main purpose was to look at the tumour response rate. The investigation was done in a two-stage single-arm experiment, with 41 patients originally enrolled. If no response was obtained from the first 14 patients studied after the first stage, the trial would be terminated. The trial continued after the first 14 subjects exhibited response, and etoposide was found to be effective.

6.2 *Randomized Trials*

Randomized Trials with Control Arm

There are two groups of participants in randomized controlled trials: one for the intervention and the other for the control (placebo or traditional treatment). Many trials are designed as placebo controlled. The patients with targeted disease are randomly assigned in two or more treatments. Throughout the course of the experiment, a randomized participant receives just one treatment (or treatment plan). The responses of the participants are compared between groups after they are tracked over time.

To investigate the safety and effectiveness of modafinil for the treatment of tiredness in primary biliary cirrhosis, researchers conducted a randomized, double-blind, placebo-controlled trial. For 12 weeks, 40 patients were randomly assigned to either modafinil ($n = 20$) or placebo ($n = 20$). After 12 weeks of therapy, the primary result was a 50% improvement in fatigueness severity compared to baseline values.

Randomized Trials with Multi-Arms

Two or more novel therapies might be tested at the same time. Each arm is constructed as single-arm trial, with patients being randomly assigned to the various groups, providing the same benefits as before. Within a single study, a multi-arm trial compares several different treatment groups against a common control group. When compared to separate two-arm studies, this setup has the immediate benefit of requiring only one control group, resulting in a reduction in the number of patients on the control therapy.

7 Designing of Phase III Clinical Trials

A phase III study's primary goal is generally to determine efficacy or safety or both. Knowing the kind of intervention and the desired result, participant blinding (single or double), participants getting only one treatment (parallel or ungrouped data), and receiving all therapies (factorial or paired) are all critical aspects to consider when planning phase III studies.

Patients are divided into parallel groups in most experiments and only receive one treatment. This design is used when treating diseases that have long-term consequences, such as cancer, or when disease prevention or cure is required.

In a crossover study, it is conceivable to assign both the novel and traditional treatments to the same person in sequence for chronic illnesses when the target outcome is symptom alleviation rather than disease cure. Bioequivalence experiments also employ this strategy. In a crossover design, the patients are not assigned to treatment groups at random; instead, the treatment order is assigned at random, guaranteeing that an equal number of participants get one therapy first and next another treatment first. In a two-period, two-sequence (2 2) crossover study, for example, a volunteer is randomized into one of two sequences: A then B or B then A. The treatment sequence is randomized, which helps account for temporal patterns. There are certain drawbacks to crossover designs. One therapy should not modify the impact of another, as this would make comparing and identifying them more difficult. A sufficiently long washout will assist to alleviate this problem. There should be a pause between therapies, and no therapies should be provided during that period.

A factorial design is a type of multi-arm research in which multiple combinations of therapy and medication dosages are evaluated. When the interventions are expected to have independent effects or when the effects are thought to be complementary and there is interest in measuring their interplay, factorial designs are appealing. There are two scenarios: one in which the therapies should not interact with one another and one in which interaction is expected. If the combined impact of A and B varies from what would be predicted even in the case of the effects of A and B which were provided separately, there is an interaction.

8 Observational (Descriptive) Study Designs

Observational studies are crucial in clinical research for a number of reasons. They are frequently the initial expedition into a new disease or field of study. These studies delve into a variety of issues around the specific topic, with a focus on what, who, why, where, and when (Hennekens et al. 1987; Strom 2005; Vetter 2017). These studies are simply intended to explain the naturally existing variables and do not include any purposeful intervention (Vetter 2017). With the limitation of therapies with moderate or small impact, observational studies may be beneficial in assessing therapies with significant or major impacts, but the exact extent of the influence may still be uncertain. Case reports, case series, ecologic studies, cross-sectional studies, cohort studies, and case-control studies are examples of descriptive research designs (*Kelsey JL, Whittemore AS, Evans AS, Thompson WD. Methods in observational epidemiology, second edn. New York: Oxford University Press, Kelsey* et al. 1996).

8.1 Case Reports and Case Series

Occasionally, during clinical practice, observant clinicians notice uncommon illness or association, prompting additional examination and more stringent study design. They present these unusual results in the form of a case report, which includes a complete and thorough explanation of the situation. Case reports are the backbone of clinical research because they allow the researcher to discuss the results that set the case apart from others. The biggest drawback of this study design is that the results or findings cannot be extrapolated to the full population since the conclusion is unclear (Albrecht et al. 2000).

A case series is a collection of case reports that includes unique findings discovered in a group of patients. This raises the possibility of a novel disease entity, prompting the investigator to seek at mechanistic research opportunities to investigate further. However, in a case series, no direct comparison is made between the cases and the persons who do not have any symptoms, making it impossible to determine which aspects of the diagnosis are specific to the disease. For example, a case report based on the presence of metastatic Kaposi's sarcoma in a homosexual man and a case series of such individuals with *Pneumocystis carinii* pneumonia were the first to identify HIV/AID (Gottlieb et al. n.d.; Adler 2001).

8.2 Cross-Sectional Study

Cross-sectional studies are often known as prevalence studies or illness frequency surveys (Kesmodel and Kesmodel 2018). Cross-sectional studies gather information based on the presence of an exposure and the health status of a subset of individuals at the same time. These studies are particularly useful in determining the illness's prevalence and symptoms in a group or community, as well as providing critical information on the disease burden (Setia 2016b). These studies can be either descriptive or analytic in the terms of inferences. Analytical cross-sectional studies give information regarding the relationship between exposure and outcome, whereas descriptive cross-sectional studies just look at the distribution of one or more elements. HIV prevalence in sex workers was determined to be 33% in a cross-sectional research based on a questionnaire of sociodemographic and behavioural data and clinical and serological screening for sexually transmitted illnesses (Shinde et al. 2009). When compared to cohort studies, cross-sectional studies may typically be completed more quickly and at a lower cost (prospective). The presence of certain biases may serve as limitation for the study. The study is limited by the presence of certain biases (Ressing et al. 2010).

8.3 Ecologic Correlation Study

Ecologic studies look at the general incidence of disease in a group of people rather than in an individual and see whether there's a link to the exposure in population (Aggarwal and Ranganathan 2019). Information on disease and exposure is frequently extracted from public statistics, obviating the need for costly and time-consuming data collection. For example, in an ecologic study, no correlation is found between the use of statin and coronary mortality in Europe (Nilsson et al. 2011). In another example, high-altitude regions in the United States have a reduced risk of heart disease than lower-altitude regions. Both instances suggest that ecologic study designs have their own limitations in terms of being unable to link exposure to result in people and controlling for confounding and error. Individual features of ecologic study designs are unclear since they are assigned to the entire group or population (Sedgwick 2015).

8.4 Case-Control Study

A case-control study is a form of descriptive research that examines variables linked to diseases or outcomes. Case-control studies compare two groups of people, such as those who have disease termed as cases and people who don't have disease termed as controls, to see if there are any variations in risk factors (Lewallen and Courtright 1998). Case-control study is always retrospective by definition since it begins with a result and then goes back to assess exposures. In early research attempting to comprehend the case of HIV, the strength of a case-control study may be understood. Both risk categories (blood transfusion recipients, homosexuality) and risk factors (unnatural sex, no use of condoms) were identified in the case of AIDS, and the likelihood of disease transmission was lowered as a consequence (Peng et al. 2011).

When compared to other study designs, case-control studies provide a number of advantages. They are quick, economical, and simple in comparison. They're especially useful for analysing epidemics and investigating rare diseases. The possibility for recall bias is the most widely mentioned drawback in case-control study. In a case-control study, recall bias refers to the chance that individuals who have the outcome would recall and report exposures more often than those who do not. The researcher must select an adequate control group while planning a case-control study (Lewallen and Courtright 1998). The case group (those who received the outcome) and the control group (those who did not get the outcome) should have almost identical demographics, such as age, gender, overall health status, and other criteria.

8.5 Cohort Study

A cohort study is a type of observational or descriptive research in which participants are followed over time. Participants in cohort studies are chosen based on their amount of exposure rather than the desired outcome. Then they're tracked throughout time to see if the desired effect occurs. The term 'cohort' arises from the Latin word *cohors*, which means 'a troop of warriors' (Setia 2016a). Cohort is a group of people that were selected on the basis of an occurrence chosen by the researcher. People who smoke, for example, form a smoking cohort. Cohort studies are distinguished by the direction in which they followed people from exposure to outcome. Cohorts can be found retrospectively or prospectively, but the outcome status must be established at least twice in any instance (Hulley et al. n.d.; Elwood 2017). In prospective cohort studies, researchers plan and organize the study, recruit participants, and gather initial exposure data on all participants before any of them acquire any of the desired outcomes. The individuals are then tracked into the future to track the progression of any of the desired outcomes. A prospective cohort design has a number of drawbacks, including the fact that it is time intensive and expensive. In retrospective cohort studies, the researchers go back through time to find a suitable cohort that was previously disease-free and possibility to develop an outcome. They next assess each subject's exposure status at the start of the observation period using the data present at that time and then determine what occurred to the participants in the two or more groups. Another type of cohort research known as ambidirectional research includes investigations that begin after exposure but before the consequence has formed. Cohort studies are good for determining disease incidence and natural history, as well as different outcomes following a single exposure. Selection bias is the main limitation for the study.

8.6 Nested Case-Control Study

This study type is originated from the cohort study. The individuals from the cohort study who developed outcomes became the case in this study, which is similar to a case-control study. Participants who had not yet developed any result were chosen at random as controls. On the basis of important variables such as age and sex, controls and cases are matched to each other.

9 Methodologies Used in Clinical Research Design

9.1 Randomization

Randomization is the most well-researched strategy for structuring research plan to avoid selection biases in the study's conclusion. Randomization, in its most basic form, is a method in which each participant has an equal probability of being allocated to either the intervention or the control group. It should guarantee that individual characteristics are comparable across trial groups (i.e., confounding is minimized) and that bias is minimized. This is accomplished by guaranteeing that neither the trial personnel nor the individuals themselves are able to guess the treatment randomization (Bland: How to use randomize - Google Scholar n.d.; Frane 1998; Altman and Bland 1999). The treatment arms are intended to contain identical numbers of individuals otherwise stated somewhere. In 1931, the concept of randomization was first used in clinical research with the investigation of sanocrysin for tuberculosis patients. There are several approaches for randomly assigning participants in study groups. We will go through some of the most often used randomization strategies, as well as their benefits and drawbacks (Venables et al. n.d.; Domanski and McKinlay 2009; Muthén and Muthén 2017).

Simple Randomization

The most often used and easiest approach for allocating participants to various groups is simple randomization. Simple randomization is a technique in which there are no limits on the structure of the randomization process other than the number of people needed to achieve the suitable statistical power. Subjects have an equal and fair probability of being chosen in either group using this assignment approach. Randomization may be done in its most simple form by tossing a coin: if it comes up heads, provide treatment A, and if it comes up tails, provide treatment B. Using this strategy, almost half of the individuals were randomly assigned to treatment A, while the other half were randomly assigned to treatment B. Rather than employing the coin-flipping approach, a table of random numbers from 0 to 9 is employed for the lower sample size. In a larger sample size, computer software or a random number generator can be employed for random allocation. This simple randomization process has the benefit of being simple to apply. This strategy has a significant disadvantage in smaller trials, but the potential of a significant impact of severe imbalances is low in larger trials. In the case of smaller trials including 20 individuals to be divided into two groups, the possibility of a 12:8 split (12 in group A and 8 in group B) or worse is 50%; however for 100 participants, the same probability is just 5%. These irregularities in participant assignment to distinct groups do not invalidate the statistical tests, but they do lessen the likelihood of identifying the difference between the two groups. Furthermore, such imbalances appear strange and may lead to a loss of confidence in the investigation, particularly among individuals unfamiliar with statistics.

Block Randomization

Block randomization, sometimes known as permuted block randomization, was first introduced by Hill in 1951. This strategy overcomes one of the key drawbacks of simple randomization: the uneven distribution of participants. Blocked randomization ensures that the imbalance is never substantial throughout randomization and that at specific moments, the number of participants in each group is equal. The researcher choses the block size, and it should be the multiple of groups. For example, if there are two groups, the block size may be 4, 6, or 8. Let's take a look at the experimental setup: we create ten blocks of two participants, one treatment and the other one is control, to guarantee that both treatments are evenly represented throughout the experiment. These are the tiniest blocks we can produce, with each therapy represented appropriately. For each block, the order of the treatments (treatment first or control first) is decided at random. Thereafter participants are assigned randomly to the blocks. As a result, we combine the small representative units (blocks) together, and the treatment order within each block is determined at random, guaranteeing that participants are assigned to blocks at random. When the research groups are identified, one downside of block randomization is that the participant allocation may be anticipated, resulting in selection bias.

Stratified Randomization

In stratified randomization, participants in clinical research are separated into strata based on clinical variables that may impact the outcome and then assigned to one of many groups using various randomized methods. For example, a patient at the age of 70 should be first placed in the group under the age of 70, depending on the age factor. Despite the fact that stratification is widely employed in clinical trials, there is still the confusion about its significance. When a certain trait is suspected of being a confounder in a treatment comparison, stratifying the sample on that variable is an effective method. Gender, for example, becomes a confounding variable when an intervention has differing effects on men and women. Because simple random assignment will not be able to allocate participants based on these confounding variables, it is necessary to divide individuals into strata to assure equal allocation. This approach is especially effective when the sample size is limited and basic randomization cannot evenly distribute individuals into groups.

9.2 *Blinding*

Although randomization is relatively successful in avoiding biases in study design, it may not be completely successful owing to human judgement in assessment, reporting, and statistical analysis due to the known identities of groups. Because this subjective and judgemental bias is directly or indirectly connected to therapy, statistical inference on the treatment impact might be substantially distorted. In fact, assessing the bias and its effect on the evaluation of the treatment outcome is quite challenging. As a result, it is vital to reduce bias in clinical research by disguising the identification of drugs, a process known as blinding. The term 'blinding' refers to an experimental setting in which different groups of people engaged in the study are kept in the dark about the therapies given to patients and other pertinent information. Some research institutions, such as the National Institutes of Health, refer to blinding as masking. Blinding is divided into four types: (i) unblinded or open trial, (ii) single blind, (iii) double blind, and (iv) triple blind. In unblinded or open trial, there is no blinding present as both the participant and the researcher are aware of the therapy administered. A single-blind study is defined as one in which only the participants know the treatment they are getting. The benefits of this design are comparable to those of an unblinded study: it is typically easier to conduct than a double-blind trial, and knowing what the intervention is may allow the investigators to use their best judgement when caring for the participants. In a double-blind study, neither the participant nor the researcher knows the treatment they are getting. Typically, such approaches are limited to medication or biologic studies. The fundamental benefit of a genuinely double-blind study is that it reduces the likelihood of bias. A triple-blind study is an expansion of a double-blind research in which the committee evaluating response variables is not informed of the identities of the groups. The statistics for groups A and B are simply presented to the committee. Triple-blind research has the potential benefit of allowing the monitoring committee to more objectively examine the response variable data.

9.3 *Placebo*

There is no active ingredient in a placebo. Because many treatment studies entail taking pills, it is sometimes referred to as a sugar pill. A placebo, on the other hand, may be a saline injection, a fake surgical operation, a sham medical gadget, or any other treatment that is intended to mimic the trial treatment but has no proven impact on the condition of interest and no harmful effects. In recent years, the use of a placebo in clinical research has sparked disagreement in the medical community. Some believe that using placebos is generally unethical since other study designs might yield comparable results while posing less danger to individual research participants. Others say that using placebos is necessary to safeguard society from the dangers that may arise from the widespread use of inadequate medical therapies.

9.4 *Bias*

In clinical trials, bias is defined as a set of systemic flaws that favour one result over another. Bias has the ability to lead scientists to incorrect conclusions regarding the positive and detrimental impacts of treatments. Biases are sometimes mixed up with random error and cannot be distinguished. Random error is associated with sampling variability, which is an inherent occurrence when working with a sample of patients rather than the entire population. The sample size can be increased to decrease random error. Some types of bias, on the other hand, can be systemic and irrespective of sample size. Bias can occur at three stages of a study: at the initial enrolment of participants, during the study's execution, and during the analysis of the results. Descriptive studies don't involve control group as these types of studies enrolled the subjects based on the defined characteristics and there is the first source of the biases.

10 Conclusion

In recent years, attrition rate in clinical trials poses a serious risk to the society of pharmacy, and it must be addressed. High trial failure rates are caused by poor patient selection and recruitment strategies, as well as an inability to adequately monitor and train patients during clinical trials. To lower the attrition rate, it's critical that the proposed research's purpose and methods are crystal clear. The many designs and methodologies utilized in clinical research were covered in this chapter. It is very useful to know what kind of study has been employed for a clear comprehension of the research. As a result, it's critical to plan the study thoroughly so that the results are obvious to other researchers. Design, implementation, analytics, and evaluation are all closely linked, and each benefits from being evaluated in conjunction rather than alone. The many types of designs used in clinical trial phases were discussed. The goal of this chapter is to present the many types of clinical research designs and procedures used in clinical trials in order to assist researchers and practitioners in properly designing their trials.

References

Adler MW (2001) ABC of AIDS: development of the epidemic. BMJ Br Med J 322(7296):1226. https://doi.org/10.1136/BMJ.322.7296.1226

Aggarwal R, Ranganathan P (2019) Study designs: part 2 – descriptive studies. Perspect Clin Res 10(1):34. https://doi.org/10.4103/PICR.PICR_154_18

Akhondzadeh S (2016) The importance of clinical trials in drug development. Avicenna J Med Biotechnol 8(4):151. Available at: /pmc/articles/PMC5124250/. Accessed 23 Dec 2021

Albrecht J, Meves A, Bigby M (2000) Case reports and case series from Lancet had significant impact on medical literature. J Clin Epidemiol *58*(12):1227–1232. https://doi.org/10.1016/j.jclinepi.2005.04.003

Altman DG, Bland MJ (1999) Statistics notes: treatment allocation in controlled trials: why randomise? BMJ Br Med J 318(7192):1209. https://doi.org/10.1136/BMJ.318.7192.1209

Bland: How to use randomize - Google Scholar (n.d.). Available at: https://scholar.google.com/scholar_lookup?journal=BMJ&title=How+to+use+randomize&author=DG+Altaman&

author=JM+Bland&volume=319&publication_year=1999&pages=703-4&pmid=10480833&. Accessed 15 Feb 2022
Burt T et al (2020) Phase 0/microdosing approaches: time for mainstream application in drug development? Nature Rev Drug Discov 19(11):801–818. https://doi.org/10.1038/s41573-020-0080-x
Caparrotta TM et al (2019) Pharmacoepidemiology: using randomised control trials and observational studies in clinical decision-making. Br J Clin Pharmacol 85(9):1907. https://doi.org/10.1111/BCP.14024
Chiang I-CA, Jhangiani RS, Price PC (2015) Quasi-Experimental Research. BCcampus
Chidambaram AG, Josephson M (2019) Clinical research study designs: the essentials. Pediatr Investig 3(4):245–252. https://doi.org/10.1002/PED4.12166
Deaton A, Cartwright N (2018) Understanding and misunderstanding randomized controlled trials. Soc Sci Med Plastic Reconstruct 210:2. https://doi.org/10.1016/J.SOCSCIMED.2017.12.005
Domanski M, McKinlay S (2009) Successful randomized trials: a handbook for the 21st century. Available at: https://books.google.com/books?hl=en&lr=&id=J1V8qd-5nocC&oi=fnd&pg=PP11&ots=YN_rXhXy1c&sig=HXO3_naiQ45vaXfOULZJHxjZ7KM. Accessed 15 Feb 2022
Dutton P, Holmes J (2018) Single arm two-stage studies: improved designs for molecularly targeted agents. Pharmaceut Stat 17(6):761–769. https://doi.org/10.1002/PST.1896
Elwood M (2017) Critical appraisal of epidemiological studies and clinical trials. Available at: https://books.google.com/books?hl=en&lr=&id=ct0-DgAAQBAJ&oi=fnd&pg=PP1&ots=zK25lZiy7G&sig=gt_DDXocjS7VatB1NyVvD9RbNFc. Accessed 15 Feb 2022
Evans SR (2010) Clinical trial structures. J Exp Stroke Transl Med 3(1):8. https://doi.org/10.6030/1939-067X-3.1.8
Fogel DB (2018) Factors associated with clinical trials that fail and opportunities for improving the likelihood of success: a review. Contemp Clin Trials Commun 11:156. https://doi.org/10.1016/J.CONCTC.2018.08.001
Frane JW (1998) A method of biased coin randomization, its implementation, and its validation. Ther Innov Regul Sci 32(2):423–432. https://doi.org/10.1177/009286159803200213
Gal J et al (2018) Optimizing drug development in oncology by clinical trial simulation: why and how? Briefings Bioinformat 19(6):1203–1217. https://doi.org/10.1093/BIB/BBX055
Gottlieb G et al. (n.d.) A preliminary communication on extensively disseminated Kaposi's sarcoma in young homosexual men. journals.lww.com. Available at: https://journals.lww.com/amjdermatopathology/citation/1981/00320/a_preliminary_communication_on_extensively.2.aspx. Accessed 15 Feb 2022
Grimes DA, Schulz KF (2002) An overview of clinical research: the lay of the land. Lancet 359(9300):57–61. https://doi.org/10.1016/S0140-6736(02)07283-5
Gupta U et al (2011) Phase 0 clinical trials in oncology new drug development. Perspect Clin Res 2(1):13. https://doi.org/10.4103/2229-3485.76285
Hansen AR et al (2014) Phase 1 trial design: is 3 + 3 the best? Cancer Control 21(3):200–208. https://doi.org/10.1177/107327481402100304
Hariton E, Locascio JJ (2018) Randomised controlled trials—the gold standard for effectiveness research. BJOG 125(13):1716. https://doi.org/10.1111/1471-0528.15199
Harrer S et al (2019) Artificial intelligence for clinical trial design. Trends Pharmacol Sci 40(8):577–591. https://doi.org/10.1016/J.TIPS.2019.05.005
Hennekens CH, Buring J, E. (1987) Epidemiology in medicine, 1st edn. Little Brown, Boston Massachusetts
Hulley S et al. (n.d.) Designing clinical research: an epidemiologic approach. pesquisa.bvsalud.org. Available at: https://pesquisa.bvsalud.org/portal/resource/pt/lil-766545. Accessed 15 Feb 2022
Ivy SP et al (2010) Approaches to phase 1 clinical trial design focused on safety, efficiency and selected patient populations: a report from the clinical trial design task force of the National Cancer Institute investigational drug steering committee. Clin Cancer Res 16(6):1726. https://doi.org/10.1158/1078-0432.CCR-09-1961
Karlberg JP, Speers MA (2010) Reviewing Clinical Trials: A Guide for the Ethics Committee

Kelsey JL, Whittemore AS, Evans AS, Thompson WD. Methods in observational epidemiology, 2nd edn. New York: Oxford University Press, 1996. Google Search (no date). Available at: https://www.google.com/search?q=Kelsey+JL%2C+Whittemore+AS%2C+Evans+AS%2C+Thompson+WD.+Methods+in+observational+epidemiology%2C+2nd+edn.+New+York%3A+Oxford+University+Press%2C+1996.&rlz=1C1CHZL_enIN832IN832&oq=Kelsey+JL%2C+Whittemore+AS%2C+Evans+AS%2C+Thompson+WD.+Methods+in+observational+epidemiology%2C+2nd+edn.+New+York%3A+Oxford+University+Press%2C+1996.&aqs=chrome..69i57.1251j0j7&sourceid=chrome&ie=UTF-8. Accessed 14 Feb 2022

Kesmodel US, Kesmodel S (2018) Cross-sectional studies – what are they good for? Acta Obstetr Gynecol Scandinavica 97(4):388–393. https://doi.org/10.1111/AOGS.13331

Kummar S et al (2008) Phase 0 clinical trials: conceptions and misconceptions. Cancer J (Sudbury, Mass.) 14(3):133. https://doi.org/10.1097/PPO.0B013E318172D6F3

Kummar S et al (2009) Phase 0 clinical trial of the poly (ADP-ribose) polymerase inhibitor ABT-888 in patients with advanced malignancies. J Clin Oncol 27(16):2705. https://doi.org/10.1200/JCO.2008.19.7681

Kunos CA et al (2020) Phase 0 Radiopharmaceutical–Agent Clinical Development. Front Oncol 10:1310. https://doi.org/10.3389/FONC.2020.01310

Le Tourneau C, Lee JJ, Siu LL (2009) Dose escalation methods in phase I cancer clinical trials. JNCI J Natl Cancer Inst 101(10):708. https://doi.org/10.1093/JNCI/DJP079

Lewallen S, Courtright P (1998) Epidemiology in practice: case-control studies. Community Eye Health 11(28):57. Available at: /pmc/articles/PMC1706071/. Accessed: 15 Feb 2022

Marchetti S, Schellens JHM (2007) The impact of FDA and EMEA guidelines on drug development in relation to phase 0 trials. Br J cancer 97(5):577–581. https://doi.org/10.1038/sj.bjc.6603925

Murgo AJ et al (2008) Designing phase 0 cancer clinical trials. Clin Cancer Res 14(12):3675. https://doi.org/10.1158/1078-0432.CCR-07-4560

Muthén B, Muthén L (2017) Mplus. Available at: https://www.taylorfrancis.com/chapters/edit/10.1201/9781315117430-28/mplus-bengt-muthén-linda-muthén. Accessed 15 Feb 2022

Nair B (2019) Clinical Trial Designs. Indian Dermatol Online J 10(2):193. https://doi.org/10.4103/IDOJ.IDOJ_475_18

Nilsson S et al (2011) No connection between the level of exposition to statins in the population and the incidence/mortality of acute myocardial infarction: an ecological study based on Sweden's municipalities. J Negative Results Biomed 10(1):6. https://doi.org/10.1186/1477-5751-10-6

Peng EYC et al (2011) A case-control study of HIV infection among incarcerated female drug users: impact of sharing needles and having drug-using sexual partners. J Formosan Med Assoc 110(7):446–453. https://doi.org/10.1016/S0929-6646(11)60066-1

Preziosi P (2007) Drug development. Comprehens Med Chem II:173–202. https://doi.org/10.1016/B0-08-045044-X/00047-X

Ressing M, Blettner M, Klug SJ (2010) Data analysis of epidemiological studies: part 11 of a series on evaluation of scientific publications. Deutsches Arzteblatt Int 107(11):187. https://doi.org/10.3238/ARZTEBL.2010.0187

Rose M et al (2020) PARP inhibitors: clinical relevance, mechanisms of action and tumor resistance. Front Cell and Develop Biol 8:879. https://doi.org/10.3389/FCELL.2020.564601/BIBTEX

Schwartz D, Lellouch J (2015) Types of intervention and their development. J Chronic Dis 20(8): 637–648. https://doi.org/10.1016/0021-9681(67)90041-0

Sedgwick P (2015) Understanding the ecological fallacy. BMJ (clinical research ed.) 351. https://doi.org/10.1136/BMJ.H4773

Setia MS (2016a) Methodology series module 1: cohort studies. Indian J Dermatol 61(1):21. https://doi.org/10.4103/0019-5154.174011

Setia MS (2016b) Methodology series module 3: cross-sectional studies. Indian J Dermatol 61(3): 261. https://doi.org/10.4103/0019-5154.182410

Shinde S et al (2009) Male sex workers: are we ignoring a risk group in Mumbai, India? Indian J Dermatol Venereol Leprol 75(1):41–46. https://doi.org/10.4103/0378-6323.45219

Smith PG, Morrow RH, Ross DA (2015) Introduction to field trials of health interventions. OUP Oxford. Available at: https://www.ncbi.nlm.nih.gov/books/NBK305510/. Accessed 7 Feb 2022

Song JW, Chung KC (2010) Observational studies: cohort and case-control studies. Plastic Reconstruct Surg 126(6):2234. https://doi.org/10.1097/PRS.0B013E3181F44ABC

Strom BL (2005) Pharmacoepidemiology. J Wiley, p 889

Strovel J et al. (2016a) Early drug discovery and development guidelines: for academic researchers, collaborators, and start-up companies. Assay Guidance Manual. Available at: https://www.ncbi.nlm.nih.gov/books/NBK92015/. Accessed 20 Dec 2021

Strovel J et al. (2016b) Early drug discovery and development guidelines: for academic researchers, collaborators, and start-up companies. Assay Guidance Manual. Available at: https://www.ncbi.nlm.nih.gov/books/NBK92015/. Accessed 7 Feb 2022

Taylor D (2015) The pharmaceutical industry and the future of drug development. Issues Environ Sci Technol 41:1–33. https://doi.org/10.1039/9781782622345-00001

Torre K, Shahriari M (2017) Clinical trials in dermatology. Int J Women's Dermatol 3(3):180–183. https://doi.org/10.1016/J.IJWD.2016.12.001

Trials I of M. (US) C. on S. for S.-N.-P. C. R., Charles H, Evans J, Ildstad ST (2001) 'Design of Small Clinical Trials'. National Academies Press (US). Available at: https://www.ncbi.nlm.nih.gov/books/NBK223329/. Accessed 7 Feb 2022

Venables W et al. (n.d.) The R development core team. musicbrainz.org. Available at: https://scholar.google.com/ftp://ftp.musicbrainz.org/.1/cran/doc/contrib/manuals-jp/R-intro-170.jp.ps.gz. Accessed 15 Feb 2022

Vetter TR (2017) Magic mirror, on the wall-which is the right study design of them all?-part I. Anesth Analg 124(6):2068–2073. https://doi.org/10.1213/ANE.0000000000002117

Yan F et al (2018) Phase I–II clinical trial design: a state-of-the-art paradigm for dose finding. Ann Oncol 29(3):694–699. https://doi.org/10.1093/ANNONC/MDX795

Yildirim O et al (2016) Opportunities and challenges for drug development: public-private partnerships, adaptive designs and big data. Front Pharmacol 7(DEC):461. https://doi.org/10.3389/FPHAR.2016.00461/BIBTEX

Zheng L et al (2013) The design of single-arm clinical trials of combination antiretroviral regimens for treatment-naive HIV-infected patients. AIDS Research and Human Retroviruses 29(4):652. https://doi.org/10.1089/AID.2012.0180

Trailblazing Contemporary Frameworks for Drug Repurposing: A Saga on Drugs' Expedition to Disinter the Veiled Destiny

Kshreeraja S. Satish, Ganesan Rajalekshmi Saraswathy, G. N. S. Hemasree, Kamatchi Sundara Saravanan, V. Lakshmi Prasanna Marise, Mamatha Krishna Murthy, and Manikanta Murahari

Abstract Huge fiscal investments in conjunction with high attrition rates encountered during ***de novo*** drug discovery delay the timelines required for development and entry of novel drug molecules into the pharmaceutical market. This situation mandates the initiation of contemporary drug repurposing research strategies to reconnoiter the hidden off-label indications of de-risked compounds or existing approved drugs with established safety profile. This chapter leverages case studies to provide a comprehensive insight into state-of-the-art drug repurposing strategies. At the outset, the application of multifaceted omics data for identifying repurposable drug candidates in various disorders is highlighted. These include genomics-, transcriptomics-, proteomics-, epigenetics-, metabolomics-, microbomics-, pregnomics-, and multiomics- based approaches. Further, the utilization of massive dataretrieved from side effect databases and electronic health records to generate evidences in Fframing drug repurposing hypotheses is elaborated. Finally, text-

K. S. Satish
Department of Pharmacy Practice, Faculty of Pharmacy, M. S. Ramaiah University of Applied Sciences, Bangalore, Karnataka, India

G. R. Saraswathy · V. L. P. Marise (✉) · M. K. Murthy
Department of Pharmacy Practice, Faculty of Pharmacy, M. S. Ramaiah University of Applied Sciences, Bangalore, Karnataka, India

Pharamacological Modeling and Simulation Centre, Faculty of Pharmacy, M. S. Ramaiah University of Applied Sciences, Bangalore, Karnataka, India

G. N. S. Hemasree
Pharamacological Modeling and Simulation Centre, Faculty of Pharmacy, M. S. Ramaiah University of Applied Sciences, Bangalore, Karnataka, India

K. Sundara Saravanan
Department of Pharmacognosy, Faculty of Pharmacy, M. S. Ramaiah University of Applied Sciences, Bangalore, Karnataka, India

M. Murahari (✉)
Department of Pharmacy, Koneru Lakshmaiah Education Foundation, Vaddeswaram, Guntur District, Andhra Pradesh, India

V. S. Rajput, A. Runthala (eds.), *Drugs and a Methodological Compendium*,
https://doi.org/10.1007/978-981-19-7952-1_9

mining techniques, computational methods, and artificial intelligence subsets deployed in effectuating prolific drug repurposing are deliberated.

Abbreviations

AChE	Acetylcholinesterase
AD	Alzheimer's disease
ADHD	Attention-deficit/hyperactivity disorder
AGPS	AD-induced Gene Perturbation Signatures
AML	Acute myelogenous leukemia
ANZ	Anxiety disorders
APP	Amyloid precursor protein
APPS	AD-induced Protein Perturbation Signatures
ASD	Autistic spectrum disorders
ATC	Anatomical Therapeutic Chemical classification
AUC	Area under the curve
AUROC	Area under the receiver operating characteristics
BD	Bipolar disorder
BP	Biological processes
CCLE	Cancer Cell Line Encyclopedia
cGMP	Cyclic guanosine monophosphate
CGPS	Cancer-induced Gene Perturbation Score
CIC	Cell Identity Code
CID	Compound identifiers
CMap	Connectivity map
CPA	Core pharmacophore anchor
CpG	5′-Cytosine-phosphate-guanosine-3′ sites
CRLS1	Cardiolipin synthase 1
CTSA	Clinical and Translational Science Awards
CWR	Cures Within Reach
DEG	Differentially expressed genes
DEP	Differentially expressed proteins
DESSCD	Dragon Exploration System for Sickle Cell Disease
DGPS	Drug-induced Gene Perturbation Signatures
DGPSD	Drug-induced Gene Perturbation Signature Database
DRPC	Drug Repositioning Perturbation Classes
DRPS	Drug Repositioning Perturbation Score
DTN	Drug-target network
EASE	Expression Analysis Systematic Explorer
EGR-1	Early growth response-1
EHR	Electronic health record
EMR	Electronic medical record
EP-DTN	Epigenetic DTN

EP-PPIN	Epigenetic PPIN
ET-3	Endothelin-3
FAERS	FDA Adverse Event Reporting System
GBM	Glioblastoma multiforme
GC-MS	Gas chromatography-mass spectrometry
GDM	Gestational diabetes mellitus
GEM	Genome-scale metabolic model
GO	Gene ontology
GOEA	Gene Ontology Enrichment Analysis
GOF	Gain of function
GSEA	Gene set enrichment analysis
GTT	Glucose tolerance test
GWASs	Genome-wide association studies
HCC	Hepatocellular carcinoma
HFD	High-fat diet
HMDB	Human Metabolome Database
HN	Heterogeneous network
HPRD	Human Protein Reference Database
IBD	Inflammatory bowel disease
ICD-9	International Classification of Diseases 9
ID	Identity documents
JCO	*Journal of Oncology*
KBMF	Kernelized Bayesian Matrix Factorization
LC-MS	Liquid chromatography-MS
lnc-FFL	Long noncoding RNA-associated feedforward loops
lncRNA	Long noncoding RNA
LOF	Loss of function
LRDI	Literature-related discovery and innovation
MDA	Multimodal deep autoencoder
MDD	Major depressive disorder
MDS	Minimum driver set nodes
MEDI-HPS	MEDication-Indication High-Precision Subset database
MIC	Minimum inhibitory concentration
ML	Machine learning
MMN	Metabolite-Metabolite Network
NAFLD	Non-alcoholic fatty liver disease
NCATS	National Center for Advancing Translational Sciences
NCI	National Cancer Institute
ncRNA	Noncoding RNA
NDCG	Normalized Discounted Cumulative Gain
NeoDTI	NEural integration of neighbOr information for DTI prediction
NIH	National Institutes of Health
NIV	Neovascular inflammatory vitreoretinopathy
NLP	Natural language processing

NMR	Nuclear magnetic resonance
PA	Pharmacophore anchor
PARP	Poly(ADP-ribose) polymerase
PGC	Psychiatric Genomics Consortium
PGS1	Phosphatidylglycerophosphate synthase 1
PharmGKB	Pharmacogenomics Knowledge Base
PheWAS	Phenome-wide association studies
PPIN	Protein-Protein Interaction Network
PPMI	Positive Pointwise Mutual Information
PRKACA	Protein kinase cAMP-activated catalytic subunit alpha
ReDIReCT	Repurposing of Drugs: Innovative Revision of Cancer Treatment
RF	Random forest
RRN	Reaction-Reaction Network
RWR	Random walk with restart
SAM	S-adenosyl methionine
SCA	Sickle cell anemia
SCD	Sickle cell disease
SCN	Stem Cell Network
SCZ	Schizophrenia
SD	Synthetic Derivative
SE	Side effect
SEA	Similarity ensemble approach
SIDER	Side Effect Resource
SMPDB	Small Molecule Pathway Database
SVM	Support vector machine
T2DM	Type 2 diabetes mellitus
TCGA	The Cancer Genome Atlas
TM	Text mining
TTD	Therapeutic Target Database
UMLS	Unified Medical Language System
VAE	Variational autoencoder
VEGF	Vascular endothelial growth factor
VUMC	Vanderbilt University Medical Center

1 Introduction

Recent technological advancements invigorate translational drug discovery research to bring about successful transformation of a drug from research laboratory to clinical practice. However, this translational expedition is hampered by numerous factors i.e. massive fiscal and resource demands, prolonged timelines, lack of skilled research personnel, inadequate knowledge and experience in handling intricate techniques (Pushpakom et al. 2018). Despite the arduous efforts put forth in de novo drug discovery and developmental processes, novel chemical entities under clinical investigation fail to make a successful debut into the pharmaceutical field owing to their unreliable safety profiles and lack of therapeutic benefits (Breckenridge and Jacob 2018). Moreover, dwindling approval rates for new drugs and withdrawal of existing drugs from the market deters pharmaceutical industries from continuing clinical research. On the contrary, a substantial rise in disease burden attributed to emergence of new diseases, unmet medical conditions, and epidemic and pandemic outbreaks impose immense pressure on pharmaceutical industries to design and develop safe and effective novel therapeutic agents (Sharlow 2016). To tackle the incessantly evolving therapeutic requirements, a paradigm transition towards trailblazing drug repositioning stratagems to unveil unidentified indications for pre-approved or banned drugs is a need of the hour. This contemporary strategy with a serendipitous origin has garnered substantial attention in the recent decades. Currently, newfangled artificial intelligence techniques are also being incorporated to drug repurposing endeavors to achieve fascinating results.

Drug repurposing hypotheses are entrenched within the concepts of polypharmacology or network pharmacology, wherein multiple targets underlying disease pathways and drug actions are not entirely unearthed. Hence, an in-depth exploration into these gray areas uplifts mysterious mechanisms lurking behind a drug action and serves as a glimmer of hope to disinter novel indications for already existing drugs with established safety profiles (Nowak-Sliwinska et al. 2019). Furthermore, the advent of omics databases over the recent past has brought forth incredible acumen with regard to molecular and metabolic variations underlying disease pathology. However, analysis of this publicly available humungous data necessitates amalgamation of principles of computational biology and bioinformatics to ascertain target-specific drug repositioning. This mandates supercomputing facilities, sophisticated laboratories with state-of-the-art infrastructure, and a dedicated team with excellent skill sets.

Considering the perks and exceptional opportunities offered by drug repurposing strategies to unravel new therapeutic agents, various government, private, and non-profit organizations have initiated funding avenues to uphold research standards in academia and industry. In this regard, the National Institutes of Health successfully inaugurated the National Center for Advancing Translational Sciences (NCATS) with an objective to financially support researchers pursuing scientific projects that are intended to reveal novel indications for existing investigational drugs, FDA-approved drugs, and/or licensed biological agents. The Bench-to-Clinic Repurposing initiative of NCATS nurtures preclinical studies and proof-of-concept clinical trials by evaluating computational algorithm-generated drug repurposing hypotheses (NCATS 2017; Wu et al. 2013).

Further, NCATS, in collaboration with the world's premier biopharmaceutical companies such as Pfizer, Eli Lilly, and AstraZeneca, investigated numerous compounds pertinent to diverse therapeutic areas. As a pilot initiative, the National Institutes of Health has released around 12.7 million dollars in year 2013 to leverage drug repurposing research for Alzheimer's disease (AD), schizophrenia (SCZ), calcific aortic valve stenosis, peripheral artery disease, lymphangioleiomyomatosis, Duchenne muscular dystrophy, and nicotine and alcohol dependence. Subsequently, around three million dollars were granted by NCATS in 2015 to support drug repurposing research (NCATS 2014; Funding Opportunities n.d.). Most recently, NCATS announced fund release to repurpose prevailing phase I cleared drugs to explore their possible applications in the treatment of coronavirus disease 2019 and its associated sequelae (RFA-TR-20-003: Urgent Phase I/II Clinical Trials to Repurpose Existing Therapeutic Agents to Treat COVID-19 Sequelae (U01 Clinical Trial Required) n.d.).

Apart from NCATS, there is an ever-increasing expansion of funding agencies that motivate and promote research standards across the globe with respect to drug repurposing. For instance, Clinical and Translational Science Awards (CTSA) (Marusina et al. 2011), Cures Within Reach (CWR) (2015), Findacure (Findacure n.d.), Global Cures (n.d.), ReDIReCT (Drug repurposing | Anticancerfund n.d.), and Stem Cell Network (SCN) (Draper and Murray 2020) are few of the funding initiatives to enrich oncology-related drug repurposing. The dawn of several such streams of funding initiatives expedited drug repurposing research from scratch to success. The advent of numerous cutting-edge technologies acquired via research grants, alongside accelerated pharmaceutical industrial growth, has successfully driven drug repurposing research from virtual to real world.

Drugs are repurposed through hypotheses generated from various omics data resources such as genomics (Zhang et al. 2015), transcriptomics (Zhao et al. 2013), proteomics (Abbruzzese et al. 2017), epigenomics (Holder et al. 2017), metabolomics (Armitage and Southam 2016), microbomics (Moreira et al. 2018), and pregnomics (Fu et al. 2020). Of late, a combination of two or more omics data resources has also emerged as a drug repurposing strategy which is referred to as a multiomics approach. In conjunction with omics databases, readily available electronic medical records (EMR) (Zhou et al. 2021) and side effect data (Lounkine et al. 2012) are the recently discovered and least-explored healthcare data resources for tracking individual patient-specific information which can be later extrapolated to personalize treatment regimens via application of suitable drug repurposing strategies (Cha et al. 2018). Though availability and accessibility to big data is a great boon to scientific society, it presents certain limitations in terms of curing, processing, and analyzing enormous data. These constraints encountered in achieving precise and rapid drug repurposing decisions are deftly confronted by an umbrella of artificial intelligence technologies. The multifaceted repurposing strategies are depicted in Fig. 1, and the resources pertinent to these strategies are represented in Fig. 2.

This chapter encapsulates the avant-garde drug repurposing stratagems through case studies. In addition, it showcases the integrative amalgamation of a myriad of technologies to cope up with rising challenges in drug discovery.

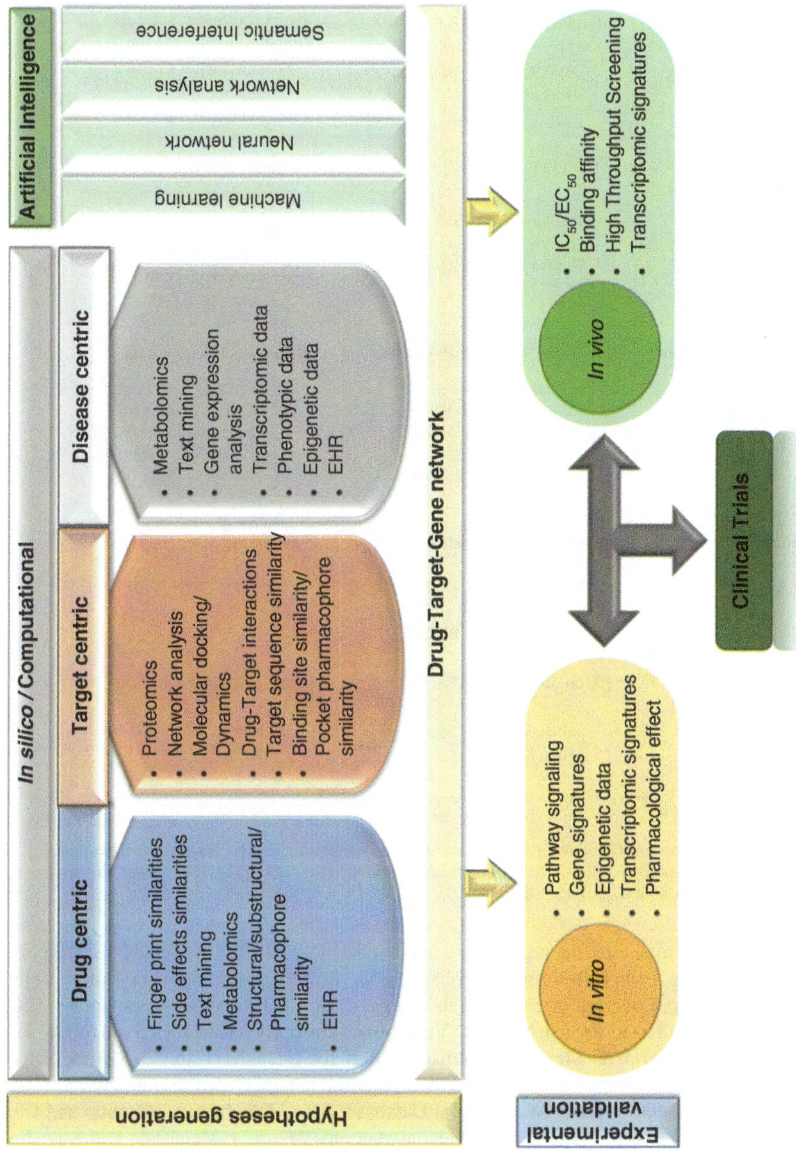

Fig. 1 Multifaceted drug repurposing approaches. *EHR* electronic health records, IC_{50} half inhibitory concentration, EC_{50} half maximal response

Fig. 2 Resources pertinent to drug repurposing approaches. *FAERS* FDA Adverse Event Reporting System, *SIDER* Side Effect Resource, *LINCS* Library of Integrated Network-Based Cellular Signatures, *ROCS* rapid overlay of chemical structures, *CDK* Chemistry Development Kit, *FLAP* Fingerprints for Ligands and Proteins, *VUMC* Vanderbilt University Medical Center, *JDMC* Japan Medical Data Center, *MIMIC* Medical Information Mart for Intensive Care, *PCORnet* National Patient-Centered Clinical Research Network, *NLP* natural language processing, *NLTK* Natural Language Toolkit, *HMDB* Human Metabolome Database, *BIGG* Biochemical Genetic and Genomic Knowledgebase of large-scale metabolic reconstructions, *MetaCyc* Metabolic Pathways From all Domains of Life, *HumanCyc* Encyclopedia of Human Genes and Metabolism, *Proteomics DB* Proteomics Database, *PDB* Protein Data Bank, *UniProtKB* Universal Protein Knowledgebase, *KEGG* Kyoto Encyclopedia of Genes and Genomes, *BioGRID* The Biological General Repository for Interaction Datasets, *NCBI BLAST* National Centre for Biotechnology Information-Basic Local Alignment Search Tool, *STRING* Search Tool for the Retrieval of Interacting Genes/Proteins, *HAPPI* Human Annotated and Predicted Protein Interactions, *PIR* Protein Information Resource, *PRODIGY-LIG* PROtein binDIng enerGY prediction-LIGand, *AMBER* Assisted Model Building

2 Genomics: Inferring Gene Expression Changes to Unveil Disease-Specific Targets

Genome refers to a complete complement of genetic information pertinent to all developmental stages of an organism. Genomics, an intriguing scientific discipline, encompasses structural genomics and functional genomics. The former constructs high-resolution genetic, phenotypic, and transcript maps, while the later amasses partially sequenced complementary DNA clones. This approach provides tools for comprehensive characterization of gene expression patterns (Emilien 2000).

The application of genomics to biomedical research focuses on correlations and influences of disease conditions and drugs on the genome of an organism. This research has resulted in aggregation of a vast amount of data with unlimited potential, which can be funneled into different research areas. One such area is drug repurposing, where the data interlinking disease-to-gene and drug-to-gene associations can be pooled in and utilized to spot novel indications for prevailing or even shelved drugs. Genomics, although a newly old concept, is an essential discipline with a consistently growing scope in the field of medical research. Genomics in conjunction with advanced bioinformatics approaches lay foundation for current drug repurposing research (Casares-Marfil et al. 2020).

The underpinnings of genomic research are primarily supported by collection of genotyping data. Such data are piled and made retrievable via databases/tools such as genome-wide association studies (GWASs) (Welter et al. 2014), phenome-wide association studies (PheWAS) (Denny et al. 2016), ExSNP (Yu et al. 2016), Blood eQTL browser (Westra et al. 2013), DAGS (Battle et al. 2014), GTEx (Lonsdale et al. 2013), Braineac (Ramasamy et al. 2014), NCBI GEO (Barrett et al. 2013), ENCODE (Sloan et al. 2016), Roadmap Epigenomics Project (Chadwick 2012), GWAS3D (Li et al. 2013), HumanBase (Greene et al. 2015), EnhancerAtlas (Gao et al. 2016), 3D Genome Browser (Wang et al. 2018), 4DGenome (Teng et al. 2015), 3DSNP (Lu et al. 2017), Chromatin-Chromatin Space Interaction (Xie et al. 2016), IM-PET (He et al. 2014), PreSTIGE (Corradin et al. 2014), OMIM (Hamosh et al. 2002), GWASdb (MacArthur et al. 2017), Open Targets (Koscielny et al. 2017), etc.

Fig. 2 (continued) with Energy Refinement, *GROMACS* GROningen MAchine for Chemical Simulation, *VMD* Visual Molecular Dynamics, *TTD* Therapeutic Targets Database, *BindingDB* Binding Database, *STITCH* search tool for interactions of chemicals, *MATADOR* Manually Annotated Targets and Drugs Online Resource, *PheWAS* phenome-wide association studies, *CPD* Cellular Phenotype Database, *CMPO* Cellular Microscopy Phenotype Ontology, *HPO* Human Phenotype Ontology, *JGA* Japanese Genotype-phenotype Archive, *NLP4J* Natural Language Processing for JVM languages, *NDB* Nucleic Acid Database, *ENCODE* Encyclopedia of DNA Elements, *TCGA* The Cancer Genome Atlas, *CMap* connectivity map, *dbSNP* Single Nucleotide Polymorphism Database, *SRA* Sequence Read Archive, *OMIM* Online Mendelian Inheritance in Man, *dbGaP* Database of Genotypes and Phenotypes, *DisGeNET* Database of gene-disease associations

GWASs unearth genetic variants underlying common diseases and provide valuable insights pertinent to systems biology, while PheWAS utilizes EMR and other relevant documented evidences to evaluate the associations among genetic variants with various clinical indices. GWAS studies are employed in multiple ways to identify repurposing candidates via detection of shared gene targets between diseases and drugs (Pushpakom et al. 2018; Casares-Marfil et al. 2020; Pritchard et al. 2017).

It is interesting to note that therapeutic agents correlated to disease characters via genetic studies carry higher probability to gain approval than those agents lacking established associations. Hence, excavating the gene expressions using genomic insights offers colossal support for discovering novel therapeutic disease targets for drug repurposing.

2.1 Drug Repurposing Based on Human Gene Expression Data to Confront Aging

Dönertaş et al. (2018) have adopted a drug repurposing framework based on a unique systems-level approach which directly employed human ageing gene expression data of the brain tissues to reveal novel potential geroprotective drugs. The investigators hypothesized that this method would find its applicability in predicting agents that influence lifespan and health span by utilizing human age-series data. Further, the researchers appreciated the robustness of this approach owing to negligible biases from literature and databases, as the study was independent of previous knowledge and assumptions. The methodology involved a complex process chunked into the following steps for ease of comprehension.

Step 1: Obtaining and Pre-Processing Microarray and RNA-Seq Datasets in the Brain Tissues

For microarray data, the investigators identified seven GEO datasets (Barrett et al. 2013). The datasets possessed expression details across different ages rather than comparing young and old groups. This aided in assessing the changes that occur throughout the process of aging. The seven datasets contained expression data from 22 different regions of the brain from individuals aged between 20 and 106 years. Separate analysis of the data from each brain region resulted in 26 datasets.

For RNA-seq data, transcriptome data encompassing all tissues with at least 20 samples based on the cause of death were obtained from GTEx (Lonsdale et al. 2013) project. This resulted in 35 datasets with 17 major tissue types inclusive of 13 brain tissues. Downstream analysis of these datasets exposed 2152 samples from 120 individuals, among which 623 samples were from brain regions of 99 individuals.

Step 2: Analysis of Age-Related Gene Expression Changes

Initially, the seven microarray datasets containing age-related RNA expression changes were analyzed to measure the connotation between age and gene expression using Spearman rank correlation coefficient. First, the association between age and degree of genetic expression of every gene within individual dataset was analyzed independently, and the number of significant changes within each dataset was estimated. To assess the gene expression pattern changes related to age, two measures for each gene were calculated, i.e., (1) correlation coefficient and (2) p-value. Magnitude of correlation coefficient assessed the strength of association, while sign represented the direction. A positive correlation was referred to as upregulation, and a negative correlation as downregulation of the gene throughout the lifespan. This analysis resulted in two datasets with significant changes, while others lacked substantial significant changes. The investigators cited low power due to small sample sizes as one of the reasons for this observation. Another reason was that Spearman correlation test computes only monotonic alterations which are significant. However, the changes throughout the aging processes may not necessarily be only monotonic. In view of this, another approach with correlation coefficient was applied across the datasets rather than within each dataset to check for consistent age-related gene expression changes. To identify these consistent patterns of upregulated and downregulated genes, the sign of correlation coefficient was used in conjunction with an appropriate testing scheme. Upon application of these tests and adopting appropriate statistical methods, the datasets were clustered based on data source instead of brain region. This indicates that aging-associated changes are heterogeneous and small, which makes it difficult to identify them. Further, when tested across datasets, significant correlations among diverse data sources were also obtained. These associations were then assessed and later used to develop an aging signature.

Step 3: Development of Aging Signature

Developing a robust aging signature primarily involved identification of genes whose age-related expression showed a consistently changing pattern in all data sources and then calculating the statistical significance of its consistency to establish its biological relevance. The set of significant genes that demonstrated such consistently changing expression patterns in all data sources was considered as aging signature and was grouped based on the direction of change. This resulted in an aging signature that encompassed 100 upregulated and 117 downregulated genes, which was corroborated by statistical permutation scheme.

Further, the reproducibility of the aging signature was tested. Herein, the pipeline that followed to identify the aging signature from microarray datasets (from analysis of expression changes to identifying and testing the significance of consistently changing genes) was repeated on an independent dataset generated by the GTEx consortium (Lonsdale et al. 2013) previously described (see Step 1). Since the datasets were generated with different platforms, i.e., microarray and RNA-seq, the robustness of their aging signature was tested across different platforms as well. Upon subjecting the RNA-seq dataset to the same pipeline, 1189 upregulated and 1352 downregulated genes were spotted. Furthermore, 50 upregulated genes and 48 downregulated genes were found to be common and significant to both GTEx (Lonsdale et al. 2013) and microarray signatures, demonstrating the reproducibility of the signature.

Step 4: Gene Ontology Enrichment Analysis (GOEA)

Using the microarray aging signature as background, a GOEA was performed for upregulated and downregulated genes separately, and the biological processes (BP) associated with this signature were explored. This revealed the enrichment of upregulated genes in differentiation- and proliferation-related processes while downregulated genes in biosynthetic and synaptic functions.

The GOEA was carried out by using GTEx (Lonsdale et al. 2013) aging signature. This resulted in a higher number of significant associations owing to more power as a result of a higher number of genes. A total of 194 and 256 gene ontology (GO) BP categories significantly interrelated with upregulated and downregulated genes, respectively, were identified. Enrichment of BP pertinent to autophagy, post-translational modifications, translation, and neuronal and synaptic functions was observed with downregulated genes, while enrichment of BP categories linked to response pathways, lipid metabolism, immune response, and macromolecule organization was observed with upregulated genes.

Step 5: Connectivity Map (CMap) Analysis

CMap (Lamb et al. 2006) is a resourceful database which provides information on drug-perturbed gene expression profiles. Querying this database with aging signature resulted in identification of 13 and 18 drugs that were significantly associated with microarray aging signature and GTEx aging signature, respectively. Seven drugs were found to be common among microarray and GTEx results. Therefore, 24 drugs were found to be significant. Finally, this method was precise enough to capture seven well-established prolongevity drugs as per DrugAge database, viz., resveratrol, levothyroxine, LY-294002, sirolimus, geldanamycin, wortmannin, and trichostatin A.

Step 6: Construction of Drug-Target Network (DTN)

Exploration of ChEMBL (Gaulton et al. 2017), PubChem (Kim et al. 2016), and DrugBank (Wishart et al. 2018a) databases alongside manual curation of literature was performed to investigate targets of the aforesaid 24 drugs. The identified targets were then tested for their previous implications in the aging process by means of GenAge human and model organism databases. All 24 drugs or their target genes other than thioridazine, rifabutin, trifluoperazine, and securinine were previously implicated in aging. Further, many of these drugs were found to share common targets. Comparable CMap similarity scores revealed their biological significance and potential mechanisms to confront aging.

Step 7: Characterization of Prolongevity Drugs

The aforementioned seven known prolongevity drugs identified in DrugAge were used to discover the BP associated with them. These drugs exert their effect either through the reversal of aging changes or by mimicking cellular defense mechanisms to prevent age-related damage. The downregulated genes were associated with BP categories such as metabolic processes and autophagy, while genes upregulated in aging were associated with reversal of changes witnessed in protein localization, protein complex or cellular complex assembly, and immunological functions. These results are in correlation with the mechanism of already known prolongevity drugs.

Step 8: Gene Set Enrichment Analysis (GSEA)

GSEA was performed to scrutinize drug-induced expression profiles, for all the genes regardless of their presence in the aging signature. This was done in order to analyze similarities among the drugs depending on changes in the degree of expression. The similarities existing between the drugs were measured, and the drugs were subsequently clustered based on Spearman rank correlation coefficient. It was found that drugs acting on the same proteins or pathways were often clustered together. Further, based on this approach, the drugs were categorized as (1) previously known prolongevity drugs, (2) drugs that were clustered together with a minimum of single prolongevity drug, (3) drugs that were clustered together yet without any already revealed prolongevity drugs, and (4) drugs that were not clustered with other drugs.

Step 9: Comparison of Aging Signature of the Brain with Other Tissues

As the aging signature compiled in this study consisted solely of the brain tissue, the investigators attempted to correlate the representativeness of the same for other tissues. Owing to the limited availability of datasets and inconsistent age-linked changes in other tissues, the investigators could not use the same pipeline. Therefore, a different approach was adopted for comparing only the direction of change in aging signature from this study to the direction of change in other tissues. By taking into account the significant similarity, through this step of analysis, the investigators concluded that the aging signature of the brain may also be a representative of other tissues.

3 Transcriptomics: Leveraging RNA Expression Data to Forecast Novel Repurposable Drug Targets

The overall content of information pertinent to an organism is detailed in DNA of its genome. Transcriptome, an intermediate between DNA and protein, is a snapshot of the total set of RNA transcripts expressed by a cell under specific circumstances. On the other hand, transcriptomics, a powerful approach, employs cutting-edge technologies to investigate the total RNA transcripts prevailing in an organism's transcriptome, besides comparing significant gene expression changes/perturbations elicited in an organism owing to functional and environmental aberrations (Lowe et al. 2017). These perturbations arising in the transcriptional program are considered as transcriptional signatures. The aforesaid signatures have demonstrated effectiveness in unveiling the association between drugs, genes, and diseases alongside a generation of proxies for clinicopathological phenotypes (Iorio et al. 2013). Most importantly, quantification of these gene expression patterns of a transcriptome in different organisms at diverse time frames is instrumental in predicting key players entangled in gene regulatory networks relevant to disease development, functions of previously unannotated genes, dynamics of BP, and functional drug discovery (Lowe et al. 2017; Chang et al. 2019).

Contemporary next-generation RNA-sequencing techniques deploy high-throughput sequencing of cDNA fragments to capture sequences, while microarray technologies via hybridization measure a set of predetermined transcripts. Technological advancements improve the specificity and allow large-scale testing of genes on a single array (Lowe et al. 2017; Liang 2013).

The transcriptome-based drug repurposing is categorized into (1) disease-based datasets (The Cancer Genome Atlas (TCGA) (Tomczak et al. 2015) and Cancer Cell Line Encyclopedia (CCLE) (Barretina et al. 2012) that include clinical or preclinical study gene expression data; (2) drug-based datasets (CMap (Lamb et al. 2006), LINCS (Keenan et al. 2018), and CTRP (Basu et al. 2013)) that store drug perturbation gene expression profiles, already recognized targets, and diverse drug-associated features; and (3) knowledge-based datasets (gene ontology, MSigDB (Liberzon et al. 2011), and KEGG (Kanehisa and Goto 2000)) that encompasses functional annotations of protein or genes. The ease of access to these datasets which contain immense data aids in formulating advanced artificial intelligence techniques to identify disease, gene, and drug connotations (Iorio et al. 2013; Kwon et al. 2019).

3.1 Retrieval of Imputed Transcriptomic Profiles Via GWAS Data Mining to Highlight Drug Repositioning Candidates for Psychiatric Disorders

So et al. *(*2017*)* demonstrated the usefulness of GWAS in imputing transcriptomic profiles for drug repurposing in psychiatric disorders. The methodology of the proposed study can be utilized to identify potentially repurposable drug candidates for various disorders.

The new strategy compares drug-induced expression changes derived from CMap2 (Lamb et al. 2006) database version 2 with imputed transcriptome profiles obtained from GWAS summary statistics using MetaXcan (Barbeira et al. 2016). This strategy is quite contrary to the routine use of RNA-sequencing technology-derived gene expression information for drug repurposing.

Initially, various statistical tools were employed to identify reversed expression patterns by comparing drug expression profiles with that of diseases. Herein, eight sets of GWAS summary statistics belonging to ten brain regions of seven psychiatric disorders were downloaded and compared with drug-induced transcriptomes from CMap (Lamb et al. 2006). These disorders include SCZ, bipolar disorder (BD), anxiety disorders (ANX), major depressive disorder (MDD), autistic spectrum disorders (ASD), AD, and attention-deficit/hyperactivity disorder (ADHD). Subsequently, based on the strength of association, the top 15 repositioning hits for each of the chosen diseases were shortlisted. Later, a thorough manual systematic literature search was carried out to curate and inspect drug disorder pairs in order to reveal a therapeutic potential of the hits. This approach was then validated through enrichment analyses of the drug sets that were (1) indicated for each disorder as per the MEDication-Indication High-Precision Subset (MEDI-HPS) database (Wei et al. 2013) and Anatomical Therapeutic Chemical (ATC) classification system and (2) in clinical trials. Further, meta-analysis was utilized to combine results obtained from these analyses across different brain regions.

Application of this framework to the aforementioned disorders resulted in identification of many drug candidates, which had both known and unknown indications pertinent to those disorders. On further inspection, those with known uses were strongly supported by existing preclinical and clinical information. For instance, some of the top repositioning candidates for SCZ included antipsychotics such as thiethylperazine, thioproperazine, pimozide, and spiperone which were already in use for treating SCZ. The strategy being blind to any information regarding existing drugs is a proof of its validity. Similarly, the study was precise in capturing idazoxan and paroxetine, which were reported to alleviate negative symptoms in clinical studies. Apart from the aforementioned psychotropic drugs, candidates exhibiting different mechanisms were also identified. For example, a diuretic, bumetanide, that acts via Na-K-Cl cotransporter-1 inhibition supported in clinical studies for treating SCZ was discovered.

In the case of MDD, application of described framework to two GWAS datasets, i.e., PGC (Psychiatric Genomics Consortium) (Sullivan 2010) and CONVERGE consortium (Cai et al. 2015), yielded different repositioning hits from each of the datasets. In results based on PGC, an antidepressant, namely, fluoxetine, was shortlisted among the top hits along with antipsychotics sulpiride, promazine, perphenazine, and loxapine. Moreover, candidates with different mechanisms such as phosphodiesterase inhibitors (papaverine), antimuscarinics (scopolamine), and antifungals (ketoconazole) were also selected as potential repurposable hits against MDD. On the other hand, results based on the CONVERGE dataset revealed zimelidine, nomifensine, and isocarboxazid as the top repositioning hits. Comparison of outcomes from both the datasets highlighted the occurrence of antidepressants and antipsychotics in common.

On similar grounds, this study exposed the repurposable potential of paroxetine, protriptyline, and ivermectin for ANX; naftidrofuryl, vinpocetine, meclofenamic acid, and ketorolac for AD; carbamazepine, lobeline, and tranylcypromine for ADHD; and lastly risperidone, pentoxifylline, amantadine, and ribavirin for ASD.

The superiority of this approach lies in the fact that it excludes confounding effects linked with the study such as samples with a previous history of medication or exposure to other environmental factors, as these elements do not influence imputed transcriptomes. As an added advantage, the application of this framework can be extended to any chemicals with easily available expression profiles, including drugs that are shelved as a result of unsuccessful trials. The versatility of this approach to pick novel mechanisms from known drugs further aids in drug repositioning.

3.2 AD Drug Repositioning Via Construction of Drug-Induced Gene Perturbation Signature Database: A Proteo-Transcriptomics Stratagem

Lee et al. (2020) ascertained the anti-AD potential of 31 drugs by constructing a Drug-induced Gene Perturbation Signature Database (DGPSD) after extracting Drug-induced Gene Perturbation Signatures (DGPS) from 26 cell lines which were exposed to 1520 compounds.

Firstly, 312, 159, and 1387 compounds were retrieved from LINCS (Keenan et al. 2018), TCGA (Tomczak et al. 2015), and CMap (Lamb et al. 2006) databases, respectively. Subsequent mapping of the compound names with their corresponding PubChem compound identifiers (CID) led to the collation of 1608 compounds. Secondly, exploration of CMap and L1000 databases had resulted in 74,171 gene expression profiles that were induced by 2021 compounds. Thereafter, the gene expression profiles associated with drugs that possessed CID were compiled to generate DGPS data. Thus obtained DGPS data was pre-processed and curated to construct DGPSD which encompassed 61,019 DGPS induced by 1520 compounds in 26 cell lines.

The next phase of the study involved a generation of Drug-induced Gene Perturbation Score for a particular disease. Herein, Cancer-induced Gene Perturbation Score (CGPS) was calculated for 152 drugs induced from 4948 gene expression profiles in nine cancer types.

Later, Drug Repositioning Perturbation Score (DRPS) was estimated for each compound that exhibited a reverse correlation between DGPS and CGPS. Drugs with high DRPS demonstrate proportional inverse signature expression pattern and also induce perturbations in a number of influential pharmacogenes. Post-DRPS calculation, the compounds were arranged in an ascending order and then were categorized into three Drug Repositioning Perturbation Classes (DRPC), i.e., high, intermediate, and low. This method was evaluated in all nine cancer types by computing area under the receiver operating characteristics (AUROC) curve of every individual DRPC, using DRPS-forecasted repositioning candidate drugs and TCGA drugs with CID as a gold standard.

High area under the curve (AUC) was observed in glioblastoma multiforme (GBM), a brain-related cancer that belonged to high-class DRPC. Therefore, this method was extrapolated to identify repositioning candidates for AD. As a part of this, 159 mRNA expression profiles from synapse database were exploited to calculate AD-induced Gene Perturbation Signatures (AGPS) of 9603 differentially expressed genes (DEGs). Comparison of AGPS of AD and CGPS of GBM displayed more number of commonly shared DEGs.

Thereafter, 175 differentially expressed proteins (DEPs) retrieved from human UniProt database were used to compute AD-induced Protein Perturbation Signatures (APPS). The aforementioned cancer drug repositioning method was then applied to AD drug repositioning, wherein DRPS was calculated using AGPS and DGPSD. Post-DRPS calculations, it was found that 1047 drugs were ranked as high class at least once. Further, the drugs that were present in the intersection between transcriptomic and proteomic data of high DRPCs and those drugs that do not fall under low DRPC were shortlisted. This stringent selection criterion yielded 31 drugs; of these, four drugs, bupivacaine, topiramate, selegiline, and iproniazid, were already reported to influence AD-related pathways. This proves the validity of this computational proteo-transcriptomic cutting-edge technology.

4 Proteomics: Highlighting the Active Domain of a Native Protein to Unravel Novel Indications for Existing Drugs

Proteome refers to the complete protein content of a cell with regard to its highly dynamic characteristics such as interactions, post-translational modifications, turnover, localization, functions, regulation, and degradation. Proteomics, on the other hand, denotes identification and quantification of overall protein signatures pertinent to a genome. This intricate methodology demands an in-depth characterization to cognize the roles of various proteoforms in living cells/tissues to comprehend molecular mechanisms underlying normal physiology, disease pathology, differential gene expression patterns, discovery of novel diagnostic markers, and functional pathways (Aslam et al. 2017).

The analysis of such complex native proteins and their structural folding within the whole proteome is successfully carried out by implementing mass spectrometry. Recent studies tend to employ advanced instrumentation to figure out peptides and post-translational modifications (Yates 2019).

Apart from this, chemical proteomics design probes that are capable of immobilizing small molecules are present in biological systems to scrutinize the proteome and link them specifically to their target proteins. The chemical proteomic strategies imperative in unravelling the drug off-targets to understand the mechanisms underlying side effects are activity-based protein profiling and compound-centric chemical proteomics (Song et al. 2020).

4.1 Personalized Proteomics of Liquid Biopsies to Disinter Potential Therapeutic Options for Neovascular Inflammatory Vitreoretinopathy (NIV)

Velez et al. (2017) developed a personalized proteome of liquid biopsies that can unveil aberrantly expressed cytokine-signaling proteins underlying different phases of NIV, an autosomal dominant genetic disorder. NIV is characterized by progressive uveitis, pigmentary retinal degeneration, cystoid macular edema, neovascularization of the iris and retina, vitreous hemorrhage, and tractional retinal detachment.

Initially, patients belonging to various stages of NIV and those who had experienced conventional therapeutic failure were recruited for the study. Thereafter, vitreous samples of four patient eyes with non-inflammatory disease, and eight NIV eyes were collected, centrifuged, and stored at −80 °C. Quantibody Human Cytokine Array 4000 proteomic platform was then used to screen and measure abnormally expressed cytokine signaling proteins. Comparison of the control and NIV eyes are via the following: (1) one-way ANOVA revealed 64 aberrantly expressed proteins (61 upregulated and 3 downregulated), (2) hierarchical heatmap clustering showed specific protein expression patterns of each group, and (3) pathway analyses identified mTOR and class I PI3K signaling molecular pathways. This cytokine expression analyses not only facilitated the identification of biologics and small molecules that can target upregulated proteins but also aided in creating a protein list that can act as targets for already existing injectable antibodies.

The analyses revealed no significant alteration in TNF-α levels in the eyes of all NIV patients and demonstrated overexpression of vascular endothelial growth factor (VEGF) in intermediate NIV stages. This explained the therapeutic failure of conventional NIV therapies with infusions of infliximab. The overexpression of VEGF led to the testing of an anti-angiogenic drug, bevacizumab, for the treatment of retinal neovascularization in NIV patients. A single anti-VEGF intravitreal injection of bevacizumab has ameliorated vitreous hemorrhages and surmounted the requirement for vitrectomy.

The identified signaling pathways of mTOR and class I PI3K in vitreous fluid of NIV patients play a crucial role in developmental fate of T cells. The post-surgical inflammatory cascades in NIV patients were previously antagonized with intraocular corticosteroid injections; however, this treatment often increased the risk of glaucoma. To alleviate the T cell-mediated post-surgical inflammation in NIV patients, intravitreal injection of methotrexate was successfully repurposed. Though methotrexate caused a significant decrease in NIV-specific T cell proliferation, disruption in blood-ocular barrier remained unresolved. To resolve this disruption, RETISERT that delivers fluocinolone acetonide continuously for 2 years was surgically implanted into NIV eyes at stages II and III. Proteomic analysis in post-surgical NIV patients highlighted the increase in IL-6 levels and decrease in FGF-4, FGF-7, PDGFRβ, VEGF, and VEGFR3 levels. These implants have contributed clinically to the reversal of retinal neovascularization but not retinal fibrosis. This evidence suggested the insensitivity of IL-6 to corticosteroid treatment. Therefore, intravenous infusions of tocilizumab, an anti-IL-6 agent, was repurposed with an intention to prevent retinal detachment and recurrent fibrosis in NIV patients.

In conclusion, characterization of biomarkers relevant to cell death apart from cytokines can enhance understanding of the neurodegenerative aspects related to NIV. Application of the same approach to liquid biopsies of other inflammatory diseases would yield promising research outcomes with respect to pathological changes and personalize therapeutic options.

4.2 Drug Repositioning through Substructure-Domain-Indication Correlation Analysis

Yang et al. (2020) framed a drug repositioning modus operandi to analyze associations between substructures of drugs and their indications. This method was developed as an attempt to predict all possible therapeutic effects of the drugs so as to aid in repurposing. The study explored interactions among substructure of a drug and functional domain of a protein, as opposed to studying the interactions between whole structures of drugs and proteins. The relationships between substructure-domain and indication-domain were analyzed using support vector machine (SVM) in order to construct a substructure-domain-indication network and derive substructure-indication correlations.

The methodology involved collection of data regarding drug, target, substructures, and indications from various databases. Initially, DrugBank (Wishart et al. 2018a) database was explored to retrieve 2021 approved drugs and 2670 targets with a UniProt link and 9797 drug-target pairs. Out of the 2021 approved drugs, only 1781 drugs with available substructure information were acquired from PubChem (Kim et al. 2016) database. Further, querying the advanced search interface of the DrugBank (Wishart et al. 2018a) database for extracting drugs with ATC-encoded indications had resulted in 4025 drug-indication relationships. Subsequent UniProt database (Apweiler et al. 2004) exploration of 2670 targets obtained from DrugBank database (Wishart et al. 2018a) had reported 2431 targets with domain information and 4430 target-domain relationships. Later, the known drug-target relationships were represented in the form of substructure-domain associations and indication-domain associations by binary vector representation. This yielded 1131 significant substructure-domain associations and 2788 indication-domain associations.

Next, the investigators employed logistic regression and SVM to unveil features specific to substructure and indications. Substructure-indication associations were identified based on the hypothesis that a minimum of one common domain is shared between the substructure feature and indication feature and were further quantified with a correlation score. This led to the identification of 1748 substructure-indication associations. Substructures that exhibited indication association scores above average were considered to offer significant contribution as dominant substructures for that indication, while substructures with single indication were defined as specific substructures for the above indication. Similarly, for dominant indications of related substructures and specific indications of a single substructure, the same definitions apply.

Finally, to predict indications, four steps were followed. First, specific substructures were extracted, and these were found to be rare and thus offer more information regarding specific activities. The strength of specificity greatly contributes to substructure-indication correlations. Next, investigators used predictive association scores to identify drugs' effective therapeutic effects by factoring in the correlation scores of substructure-indication associations and specific weight value scores for substructures. Thereafter, contribution ranks of drugs to a specific indication was calculated based on predictive association scores, number of substructures linked with an indication, substructure and indication correlation score, and weight value scores for specific substructures. On a similar note, contribution ranks of indication to a drug were calculated based on predictive association score and overall indications for a drug. Finally, significant drug-indication pairs were identified, and the strength of their relationships was assessed using product value of contribution rank of indication to drug and drug to indication. This revealed 1479 drugs and 178 indications contributing to 83,205 noteworthy drug-indication pairs. Factoring in all the aforementioned substantial interactions, a drug-substructure-indication network was framed in order to examine the intricate connections of drug-indication.

Validation of the outcome of this approach was done based on four aspects: (1) DrugBank database (Wishart et al. 2018a), (2) PubMed literature, (3) CTD database (Davis et al. 2019), and (4) comparison with other methods. Out of 2156 recognized drug-ATC relationships retrieved from DrugBank, 1464 associations were predicted by the novel methodology. In the validation approach based on PubMed biomedical literature, verification frequency for 83,205 relationships was 71.44%, and with the approach based on CTD database, the verification rate was 73.03%. In the last validation approach, three methods, viz., predictions of first-level, second-level, and third-level ATC code of drugs, were compared with the method of interest. It was found that this method developed by Yang et al. was superior to the first- and second-level ATC code prediction accuracy but inferior to the third-level ATC code prediction accuracy. However, it still remained superior in capturing the explicit structural features of therapeutic effects, in addition to identifying the key action spectrum with merely the substructure, and thus revealed the strength of drugs for one indication, in order to disinter the most appropriate drug for the indication.

This section highlights the application of the above methodology to predict indications for olaparib, a known anti-neoplastic and immunomodulatory agent. The method discovered 61 indications and 420 targets for olaparib. The aforementioned acknowledged actions of olaparib ranked 29 in their list of predicted indications. Pathway analysis by KEGG (Kanehisa and Goto 2000) has found the enrichment of 11 targets in apoptotic pathway (hsa04210). Their substructure association evaluation predicted that SUB494 (O = C–C: C) of olaparib interacts with the receptors of inositol 1,4,5-triphosphate present in the calcium channels of endoplasmic reticulum which might lead to caspase cascade activation by augmenting the release of calcium ions. In addition, it was found that SUB494 targets PARPs, which upon hydrolysis causes apoptosis by inhibiting DNA damage repair. The latter finding was confirmed with the existing concept that olaparib impairs DNA repair mechanisms and replication process by impeding PARP1, PARP2, and PARP3. Furthermore, code J05A of ATC, which describes antivirals for systemic use, was revealed as a dominant ATC analogous to olaparib. This was supported by the fact that 21 of the 420 envisaged targets of olaparib were enriched in two main pathways, both of which were proved to possess antiviral activity.

The investigators also tested this approach to predict indications for antiarrhythmic drugs such as amiodarone, quinidine, milrinone, and fosinopril on the basis of substructure-indication associations with correlation scores. Their results yielded in the discovery of SUB410 (O(:C)(:C)) for amiodarone, SUB182 (≥1 saturated/aromatic heteroatom-containing ring size 6) for quinidine, SUB673(O = C–C = C–[#1]) for milrinone, and SUB4(≥32 H) for fosinopril, as the dominant substructures for antiarrhythmic drugs. The significance of these results lies in the fact that the presence of these specific substructures in other drugs implies the possibility of those drugs having an antiarrhythmic activity, thus broadening the scope for drug discovery and repositioning. Furthermore, this approach can be extrapolated to optimize drug function by retaining substructures responsible for therapeutic activity and eliminating those that can cause undesirable activity.

5 Epigenomics: Post-Translational Modifications to Uncover Novel Therapeutic Possibilities

Epigenomics, an emerging omics science, studies the mitotically or meiotically inherent alterations in genetic expression without causing changes in DNA sequence (Chatterjee et al. 2018). These intricate, dynamic, reversible structural and chemical modifications occurring at the genome level within the nucleosome are implicated in identifying the influence of age, gender, nutritional status, stress, socioeconomic status, and environmental and chemical factors on gene expression (Chatterjee et al. 2018; Ganesan et al. 2019; Prachayasittikul et al. 2017).

DNA methylation mechanism uses DNA methyltransferases (DNMTs) to catalyze the addition of methyl groups to CpG sites. This process plays an important role in genomic imprinting, retrotransposon suppression, genome stability maintenance, inactivation of X-chromosome, and other gene controlling actions. On the other hand, DNA demethylation mechanism removes methyl group from CpG to reprogram the genes (Ganesan et al. 2019; Prachayasittikul et al. 2017). Histones are conserved and are the most abundant proteins that bind to DNA. The core histones wrap DNA to form nucleosome, while linker histone binds nucleosomes together to form chromatin. Cell differentiation can induce changes in the structural and chemical composition of chromatin. These post-translational modifications result in transcriptional repression, thereby modifying gene expression pattern (Tan et al. 2010). The post-translational modifications so far identified include acetylation, phosphorylation, methylation, mono-ubiquitination, SUMOylation, and ADP ribosylation (Hosseini and Minucci 2018). ncRNAs being functional RNA molecules do not code for proteins. Nevertheless, they can cause conformational changes in DNA by binding to them. Thereby, they control gene expression and mRNA stability at post-transcriptional levels.

These chemical alterations are catalyzed by three epigenetic enzymes referred as writers, readers, and erasers which act on DNA, histones, and noncoding RNAs (Prachayasittikul et al. 2017). The writers add methyl group or ubiquitin protein to DNA or histones. These modifications catalyzed by writers at the molecular level directly influence the affinity between DNA and histones alongside recruiting ncRNAs and chromatin remodelers. The readers control the binding interactions by recognizing specific features present within the modified nucleic acid and proteins. Ultimately, a series of erasers remove the written information (Ganesan et al. 2019).

The resilience of the epigenome is being explored for its ability to inversely modify epigenetic changes ensued in gene expression by working on epigenetic targets as treatment strategies to alleviate disease phenotypes. Of late, this approach reveals valuable clues to map phenotypic changes with genotypic changes for effective drug repurposing (Chatterjee et al. 2018; Ganesan et al. 2019; Raynal et al. 2017).

5.1 AD Drug Repositioning Via Construction of Human Interactome: An Epigenetic Approach

Paulami Chatterjee et al. (2018) identified 14 AD repurposable drugs based on epigenetic targets by constructing an extensive human interactome. Protein-Protein Interaction Network (PPIN).

Initially, a list of 54 AD-specific epigenetic genes was extracted from literature via PubMed text mining. Simultaneously, experimentally validated human protein-protein interaction data from Human Protein Reference Database (HPRD) (Peri et al.

2004), Bio GRID (Stark et al. 2006), and Mentha (Calderone et al. 2013) databases were combined to construct an extensive PPIN. Thus the obtained human interactome encompassed 190,771 protein-protein interactions. Collation of interacting partners of 54 epigenetic genes with PPIN generated epigenetic PPIN (EP-PPIN), which encompassed 8412 EP-PPIN. DrugBank was later used to retrieve drugs corresponding to EP-PPIN interactions which revealed 886 drugs for 419 interactions. Following this, an epigenetic-DTN (EP-DTN) consisting of 1920 drug-target interactions was constructed using the above-identified drugs and protein information. The constructed human interactome was further explored to unravel the interactions associated with epigenetic drugs and proteins. Consequently, these retrieved epigenetic drugs and protein interactions were analyzed in the already constructed EP-PPIN. The identified protein interactions were studied in EP-DTN to identify the initial phase of repositioning drugs. Around 249 drugs possessing 21 epigenetic targets were retrieved from EP-DTN. Of the total of 249 drugs, only 11 (acetyl salicylic acid, etoposide, tamoxifen, spironolactone, sorafenib, glyburide, diclofenac, caffeine, methotrexate, lamivudine, and ibuprofen) were shortlisted. Information with regard to target proteins, miRNAs, and long ncRNAs of these 11 drugs were retrieved from AlzGene (AlzGene | ALZFORUM n.d.), SM2miR (Liu et al. 2013), and DIANA-lncBase (Vlachos et al. 2015) databases, respectively. Around 313 miRNAs associated with target proteins of 11 drugs were identified, and the drugs specific to these miRNAs were retrieved from SM2miR database. A total of 53 drugs were retrieved, of which 43 drugs were present in EP-DTN. These 43 drugs encompassed three drugs (tamoxifen, sorafenib, and etoposide) which were already reported in the initial phase of drug repositioning. The final phase of drug repurposing involved functional enrichment analysis by mapping the drugs and targets with KEGG pathways relevant to AD. This resulted in the identification of 14 drugs (aspirin, sorafenib, spironolactone, glyburide, caffeine, nicotine, desipramine, bortezomib, ibuprofen, dexamethasone, atorvastatin, diclofenac, lamivudine, and metformin) that displayed significant involvement in KEGG pathways.

6 Metabolomics: Mapping Pathological Perturbations to Achieve Therapeutic Advances

Metabolites are the downstream output and upstream input of the genome and environment, respectively. These physiological, biochemical, and pathological markers display interactions occurring among different pathways within a cell (Pallares-Méndez et al. 2016). Metabolomics, a promising branch of omics science, focuses on characterization and mapping of low-molecular-weight endogenous molecules (<1500 Da) and metabolic processes occurring in biological systems. This approach captures and hoards information specific to perturbations of early biochemical changes in disease states, medical interventions, regulation of genetic activity, altered kinetics of enzymes, aberrant metabolic reactions, and interactions between genes and environment (Pallares-Méndez et al. 2016; Wishart 2016; Yeung 2018).

Metabolomics-based data-driven pipelines are exceptionally advantageous due to their unparalleled throughput and comprehensive interpretation of dynamic metabolome profiles of a wide range of solids, liquids, and gases by deploying advanced computational and analytical chemistry techniques (Wishart 2016; Dubuis et al. 2018).

In the wake of new horizon, the Canadian Institutes of Health Research sponsored Human Metabolome Project (Wishart et al. 2018b) in 2004 with a motto to expedite discovery of metabolites and biomarkers. The database designed for this project interlinks chemical, clinical, and molecular biology and biochemistry data. This data lays a background for metabolomics, clinical chemistry, and drug discovery research. The database stores around 115,434 metabolite (abundant (> 1 uM) and relatively rare (< 1 nM) entries that were linked to 5702 protein sequences. The data in data fields are hyperlinked to KEGG (Kanehisa and Goto 2000), GenBank (Benson et al. 2009), UniProt (Apweiler et al. 2004), PDB (Berman et al. 2000), MetaCyc (Karp et al. 2002), PubChem (Kim et al. 2016), and ChEBI (Hastings et al. 2016). The Human Metabolome Database (HMDB) (Wishart et al. 2018b) suite of databases are inclusive of T3DB (Wishart et al. 2015), which holds data of ~3670 common environmental pollutants and toxins; DrugBank (Wishart et al. 2018a), which contains ~2280 xenobiotics and their metabolite information; SMPDB (Frolkis et al. 2009), which encompasses ~25,000 human disease and metabolic pathways; and FooDB (n.d.), which consists of ~28,000 food additives and component data.

This extensive profiling of metabolites provides countless opportunities to discover and develop clinically relevant novel biomarkers which serve as quantitatively measurable indicators of physiological and pathological states, early diagnosis, prognosis, and prediction of therapeutic responses which can aid in invigoration of personalized or precision medicine (Pallares-Méndez et al. 2016; Wishart 2016; Yeung 2018).

6.1 Reconnoitering Anticancer Targets for Hepatocellular Carcinoma (HCC) Via Metabolic Networks

Bidkhori et al. (2018) designed an algorithm which integrates genome-scale metabolic models (GEMs), metabolite reaction association networks, transcriptomics, and protein expression data to prioritize anticancer genes/metabolites and drugs. The step-wise study methodology is mentioned below.

Step 1: Potential Target Fishing

Initially, an antimetabolite which displayed structural similarity with endogenous metabolites alongside the potential to halt cell growth was identified using previously designed algorithm tINIT (Agren et al. 2014). Gene silencing is performed in silico for genes which catalyze metabolic reactions. The forecasted antimetabolites along with silenced genes were then screened for their presence in HCC-specific GEMs.

Step 2: Construction of Metabolite-Metabolite and Reaction-Reaction Networks

In this step, Metabolite-Metabolite Network (MMN) and Reaction-Reaction Network (RRN) were constructed using personalized GEMs. In the MMN network, metabolites were considered as nodes, and they were connected by edges, if they were tangled in the same reaction, whereas RRN network considers reactions, if one product serves as a substrate for another reaction.

Step 3: Controllability Investigation

Minimum driver set nodes (MDS) in MMN and RRN regulate network dynamics upon perturbation. Popov-Belevitch-Hautus (PBH) rank condition (Yuan et al. 2013; Sontag 1998) was leveraged to identify MDS. This was entailed by identification of indispensable nodes, i.e., genes or metabolites, from MMN and RRN and computation of maximum geometric multiplicity for assessing node dispensability. Utilization of MDS in conjunction with indispensable nodes was referred to as controlling nodes. Later, nodes recognized concurrently in HCC and non-cancerous networks were excluded.

Step 4: Prioritization of Target

In this step, the controlling nodes with anticancer properties were shortlisted and were then ranked according to their node degree and betweenness centrality for MMN and RRN. MDS were ranked as per betweenness centrality, while indispensable nodes were ranked as per degree centrality. As a result, non-toxic anticancer-controlling node list was generated.

Step 5: Transcriptomic Data Retrieval and Modeling for HCC Patients

Transcriptomic data of 50HCC and 50 adjacent non-cancerous liver samples were extracted from NCI's Genome Data Commons (Jensen et al. 2017). Personalized GEMs were built using tINIT (Agren et al. 2014) and RAVEN (Agren et al. 2013) algorithms with human GEM HMR2 as reference. Biomass production and ATP utilization were considered as prime functions for HCC and non-cancerous GEMs. The model functionality for the aforesaid personalized GEMs was determined by considering metabolic tasks and objective functions. Fifty-seven and 56 metabolic tasks were retrieved for HCC and non-cancerous GEMs, respectively. Cell growth was also considered as a metabolic task for HCC.

Step 6: Identification of Antimetabolites, Anticancer Genes, and Networks

HCC-specific personalized GEMs consist of 2892 genes which control 7780 reactions and 2857 metabolites. For 2857 HCC-specific metabolites, 374 antimetabolites which were involved in acyl-CoA hydrolysis, fatty acid activation, and carnitine shuttle, as well as metabolism of proteins, amino acid, nucleotide, pyrimidine, purine, glycan, glycerolipid, and cholesterol, were identified. In silico gene silencing identified 2204 genes which were not involved in 56 metabolic reactions in non-cancerous GEMs. Also, 283 genes which inhibited cell growth at least in one HCC and eight genes which inhibited cell growth in all HCC were shortlisted.

Step 7: Determination of Non-toxic Anticancer-Controlling Metabolites Via MMN Controllability

Around 201 MDS and 578 indispensable nodes were identified. Finally, 142 cancer-controlling nodes were detected after excluding the common controlling nodes in HCC and non-cancerous models. Subsequently, 374 antimetabolites were screened considering the identified 142 cancer-controlling nodes. This resulted in 74 non-toxic anticancer metabolites with network controlling properties, viz., 54 were indispensable and 20 were MDS. The indispensable antimetabolites displayed a significantly higher degree in comparison with mean metabolite degree.

The identified non-toxic anticancer metabolites are entailed in metabolism of amino acids, DNA, and fatty acids. The antimetabolites predicted by the algorithm also exhibited a good agreement with FDA-approved cancer drugs obtained from DrugBank. Amid all amino acids, leucine, isoleucine, and glutamine were finalized as non-toxic anticancer metabolite targets.

Step 8: HCC-Specific Gene Target Identification Via RRN and Experimental Validation

Silencing of RRN controlling genes showed specific features with respect to HCC and non-cancerous networks. Out of the eight genes identified, only three genes, namely, protein kinase cAMP-activated catalytic subunit alpha (PRKACA), phosphatidylglycerophosphate synthase 1 (PGS1), and cardiolipin synthase 1 (CRLS1), which possessed non-toxic anticancer-controlling properties were selected for experimental validation.

The Hep3B and HepG2 human HCC cell lines were infected by predesigned targeted silencer siRNAs for each gene. Later, the expression profiles of PRKACA, CRLS1, and PGS1 were measured and analyzed in Hep3B and HepG2 via quantitative real-time polymerase chain reaction. The silencing of PGS1 and PRKACA decreased 22–24% growth in both the cell lines, whereas silencing of CRLS1 decreased 30% growth in HepG2.

7 Microbomics: Visualization of Invisible to Elucidate Therapeutic Responses

The human microbiome approach that involves "visualization of invisible" is a newfangled gateway for drug repurposing by redefining pharmacokinetics of drugs. Microbes existing in the human body are crucial in multiple tasks like digestion and metabolism of endogenous and exogenous substances. During the said processes, these microbes also suffer certain levels of insult as and when they get exposed to toxic molecules and/or metabolites. Subsequently, a decline in the population of mutualistic bacteria involved in drug metabolism is experienced that in turn alters drugs' pharmacokinetics, thus influencing efficacy and safety. At the same time, the inhabitant bacteria can cause a waning in an anticipated drug effect owing to the variation in xenobiotic concentration and reduction in bioavailability. This abate in expected drug effects can be attributed to various mechanisms including hydrolysis, deacylation, decarboxylation, dehydroxylation, demethylation, and dehalogenation. These mechanisms can contribute to reduction in effective target binding as well. Additionally, microbiome also competes with drug molecules for active sites, thereby inhibiting numerous hepatic enzymes associated with xenobiotic metabolism. Inhibition of these enzymes impairs elimination and hence augments the possibility of cytotoxicity. Reports revealed that one in four non-antibiotic-approved drugs out of 1000 substantially affected the gut microbiota. This fact elucidates the pathways involved in untoward effects of already existing drugs (Pulley et al. 2020; Taroncher-Oldenburg et al. 2018; Iqbal et al. 2016; Lee et al. 2019).

Recently, research strategy encompassing microbiome manipulation is well appreciated, apart from its effect on ADMET (pharmacokinetic) properties and efficacy of medications. For instance, successful outcomes of fecal transplantation in patients treated with *Clostridium difficile* colitis brought the ability of microbomics approach to limelight (Ooijevaar et al. 2019). This novel approach may be extrapolated to a wide range of pathological conditions such as obesity, metabolic syndrome, and graft-versus-host disease, besides amelioration of multidrug-resistant organisms.

7.1 Liraglutide Manipulates Microbiota to Alleviate Non-alcoholic Fatty Liver Disease (NAFLD) in Obese Mice

Moreira et al. (Moreira et al. 2018) predicted that gut microbiota can influence metabolic alterations in the initial stages of diseases like NAFLD and obesity. Herein, the influence of GLP-1 analogue liraglutide on gut microbiota of 8-week-old genetically modified obese mice and high-fat diet (HFD) obese mice groups was studied. The rationale behind this study is evinced from the fact that GLP-1 plays a key role in governing the functions of the intestinal epithelium that may be correlated with alterations in gut microbiota.

The methodology of this in vivo study involved subcutaneous administration of liraglutide twice a day for 2 weeks in genetically modified ob/ob mice and HFD obese mice at a dose of 200 μg/kg body weight. Liraglutide administration was observed to lower-fasting blood sugar levels and enhanced insulin sensitivity, alongside a 7% reduction in body fat in both the groups. In addition, significant reduction in adipocyte size, retroperitoneal fat, subcutaneous fat, perigonial fat, and intensity of inflammatory cell infiltration with no epithelial changes were noticed in HFD mice along with reversal of fatty liver, whereas ob/ob mice, besides demonstrating similar changes like HFD mice, displayed a significant rise in the number of epithelial cells. The fecal microbial analysis revealed a wide diversity between the groups. In the HFD group, there was a decrease in population of *Proteobacteria* and *Helicobacter* phyla, while an increase in *Verrucomicrobia* and *Akkermansia* phyla was noted. In ob/ob mice, there was an increase in *Thermotogota* phylum and decrease in *Proteobacteria* phylum. A strong link between the influence of liraglutide on gut microbiota with its anti-obesity property and reversal of fatty liver suggested the repurposable ability of an anti-diabetic drug for NAFLD.

8 Pregnomics: Exceptional Data from Special Population to Disclose Drug Repurposing Hints

Pregnant women are considered vulnerable to a plethora of pathological conditions, and the use of drugs in any trimester to treat the morbidities poses a significant risk to both mother and the developing fetus. In general, therapeutic interventions are expected to act on altered perinatal and fetal physiology, alongside biological features of the placenta (Feghali et al. 2015). Inadequate knowledge pertinent to physiological variations, dearth in literature elucidating the pathological mechanisms of common disease conditions, and stringent restrictions to conduct drug trials in expectant mothers narrow down the therapeutic options to prevent and confront obstetrics and gynecology-related morbidity (Fogel 2018). In the wake of these challenges, drug development and clinical trials remain a mirage during gestation. Nevertheless, drug repurposing becomes an attractive proposal owing to its use of drugs with established safety profiles, yet not been tested in pregnancy (Goldstein et al. 2018a). Therefore, acquisition of an in-depth knowledge on fetal toxicity and pregnant physiology is essential to formulate novel therapeutic strategies.

8.1 Design of lncRNA-Mediated Feedforward Loop Networks to Reveal Drug Repurposing Candidates for Gestational Diabetes Mellitus

Long noncoding RNA-associated feedforward loops (lnc-FFLs) are regulatory elements which are composed of genes, microRNAs (miRNAslp), and long noncoding RNAs (Jiang et al. 2019). Recently, the role of lnc-FFLs in BP inclusive of cell development and differentiation, alongside their involvement in disease pathology, is highlighted.

On this grounds, Xuelian Fu et al. (2020) derived valuable clues for unravelling repurposable candidates for gestational diabetes mellitus (GDM) by extracting information on dysregulated lnc-FFL involved in glycometabolism and thyroid hormone signaling pathways.

Initially, a global lnc-FFL network was constructed to identify interactions among gene-miRNA and gene-lnc-FFL by thorough utilization of experimentally validated data retrieved from miRTarBase7.0 (Huang et al. 2020) and RAID 2.0 (Yi et al. 2017) databases, respectively. As a result, a large-scale free network containing 1347 lnc-FFLs, 114 coding genes, 164 lncRNAs, and 154 miRNAs was generated. The second step involved extraction of genes, lnc-FFLs, and miRNAs specific to GDM by systematically analyzing GSE92772 dataset (Stirm et al. 2018). In the third step, detailed analysis of 1552 hormone-related genes and 1845 glycometabolism genes from AmiGO database (Carbon et al. 2009) revealed 435 genes at the intersection.

After this methodical procedure, lnc-FFL networks specific to hormone and glycometabolism were built by taking into account the aforesaid global lnc-FFL. This network showcased 20 coding genes, 229 lnc-FFLs, 58 miRNAs, and 44 lncRNAs for glycometabolism and and 42 coding genes, 530 lnc-FFLs, 111 miRNAs, and 69 lncRNAs linked to hormones.

Later, an algorithm to predict the impaired glycometabolism and hormone-related lnc-FFLs in GDM was developed. This algorithm captured 11 glycometabolism-linked and 29 hormone-related lnc-FFLs, respectively, with 10 lnc-FFLs in common. These common lnc-FFLs displayed significant interactions with other components of the network and thus were considered key modules. Subsequently, extraction of one key module resulted in unveiling of highly interactive SP1 gene that connects seven lnc-FFLs and seven miRNAs. These miRNAs were found to be enriched in pathways belonging to fatty acid biosynthesis, thyroid hormone signaling, and lysine degradation.

Following network construction and key module identification, drug molecules that interact with specific lnc-FFLs and drug molecules related to the identified miRNAs were retrieved by exploring DrugBank (Wishart et al. 2018a) and SM2miR (Liu et al. 2013) databases, respectively.

Thereafter, a drug network related to dysregulated lnc-FFLs and miRNAs specific to GDM was constructed which encompassed 62 drugs, 4 genes, and 15 miRNAs. From this network, troglitazone (an anti-diabetic drug, withdrawn owing to its hepatotoxicity) that exhibits interaction with ESRRA gene (steroid hormone receptor gene) and miR-125b-5pmiRNA of dysregulated lnc-FFL was predicted to be a potential candidate for GDM.

9 Multiomics: A Multifaceted Approach to Invigorate Drug Repurposing Predictions

Current trends in drug repurposing or discovery research are oriented toward amalgamation of multiple omics branches in order to acquire multifaceted concepts that enhance therapeutic success rates. For instance, genomic data is crucial for identification of genes associated with diseases, which could be used as potential targets for which drugs can be repurposed. On a similar note, metabolomic data analysis aids in preclinical research and biomarker detection, where biomarkers are commonly used as impending targets for repurposing studies. Transcriptomics, on the other hand, is used to gain precise lead compounds in drug discovery process, as it takes into account the genomic mutations and its analogous epigenetic changes that modify gene expression and functions. Consequently, genes that are differentially co-expressed in the disease are identified, and these genes are used as targets for drug repurposing. Proteomics characterizes total protein content of an organism and investigates protein-protein and protein-nucleic acid interactions along with post-translational modifications. It provides detailed information with regard to disease states and untoward effects of drugs and assists in reconnoitering significant biomarkers. Multiomics approach toward drug repurposing is elaborated in this section.

9.1 Integrative Systematic Multiomics Data Mining and Computational Approach to Highlight Repurposable Drugs for AD

Zhang et al. (2016) adopted a multiomics approach to repurpose drugs for AD by combining information retrieved from genomics, epigenomics, proteomics, and metabolomics databases. Details from these databases were collated in order to generate a metabolite-protein network, after which the proteins were linked to their existing drugs. Later, potential anti-AD targets based on protein function and drug mode of action were shortlisted. A ranking algorithm was developed subsequently to prioritize repurposable AD-specific drugs.

Multiomics data mining was carried out in various databases to obtain the details pertinent to respective omics data. Thirty-one GWASs which encompassed AD-associated genetic information were retrieved from NHGRI-EBI GWAS Catalog (Welter et al. 2014); 18 metabolomic studies were extracted from HMDB (Wishart et al. 2018a); and 4 epigenetic and 7 proteomic AD-related studies were obtained from PubMed database. The analysis of these studies revealed 244 genetic variations implicating 220 genes, 86 metabolites, 14 epigenetic events, and 98 proteins that were associated with AD.

Proteins that affect AD-associated metabolites were extracted from HMDB (Wishart et al. 2018a). Following which, a metabolite-protein network was constructed and visualized via Cytoscape v3.3.0 (Shannon et al. 2003). This network led to the identification of 1179 metabolite-protein pairs, where 200 proteins were associated with ≥ two metabolites linked with AD. Next, proteins pertinent to AD that were extracted from GWAS, epigenetic, and proteomic studies and those obtained from the metabolite-protein network were linked to the existing drugs taken from Therapeutic Target Database (TTD) (Zhu et al. 2012) and DrugBank databases (Wishart et al. 2018a). Upon collation of the protein- and drug-related data (drug name, target, indication, clinical trial status, and modes of action), potential drug-target pairs were shortlisted. Further, the drug-target pairs were narrowed down to include drugs which were either approved or under clinical trials, so as to identify the most promising agents to be repurposed for AD.

Additional information regarding the gain of function (GOF) or loss of function (LOF) of the potential targets in disease pathogenesis were obtained from OMIM database and PubMed literature search. This information along with the previously extracted drugs' modes of action was leveraged to capture enticing anti-AD drugs.

At this stage of the study, 524 unique AD-related proteins were identified, out of which, eight proteins were found to be common in any two of the omics platforms searched. Among these eight, four had reports on AD-related functional studies (ABCA7, APOE, BIN1, and PICALM), while findings related to the other four, i.e., CELF1, INPP5D, SPON1, and SOD3, suggested a need for further analysis to establish their role in AD pathogenesis.

Out of the total 524 proteins identified, it was found that only 19 proteins were related to 92 approved or drugs shelved in clinical trials. Two of these 19 proteins were previously established anti-AD targets, i.e., acetylcholinesterase (AChE) and amyloid precursor protein (APP). Further, the drugs obtained for these targets, i.e., galantamine, rivastigmine, and donepezil, are already in use for AD management. These findings validated the methodology used in this study and its potential to identify relevant novel agents that can be repurposed for AD. The study revealed 75 existing drugs for 18 potential targets that can be repurposed for AD. Additionally, the study revealed seven novel agents for AChE (pyridostigmine, demecarium, physostigmine, gallamine triethiodide, edrophonium, ambenonium, isoflurophate) that were not previously used for AD therapy and thus carry a potential for repurposing.

An algorithm to prioritize anti-AD targets was generated to rank the drugs by considering a target score which was computed using a weighted sum model. Three main criteria that were given equal weights were used to rank the drugs. The criteria include (1) the level of fold changes of AD-related proteins, (2) the number of Google Scholar citations of the article that revealed the target's role in AD pathogenesis, and (3) the number of PubMed publications that linked the target to AD. Thereafter, internal controls were used to adjust target scores to the known AD-related proteins/genes. This entire process was then evaluated using three well-recognized anti-AD drug targets, i.e., AChE, TREM2, and APOE, that demonstrated a medium/high target score of 0.384, 0.459, and 0.887, respectively.

Out of the 17 novel anti-AD targets scored, CD33 and MIF showed scores higher than that of AChE. CD33 and MIF which were previously tested for acute myelogenous leukemia (AML) were also found to be linked with microglial activation and neuroinflammation. Thus they may also serve as good candidates for treating AD-related neuroinflammation. Additionally in this study, a bioinformatic analysis was performed via construction of a PPIN via Search Tool for the Retrieval of Interacting Genes/Proteins (STRING) tool (von Mering et al. 2003). As a result, two core hub proteins, namely, APP and CAD, were identified.

In the final part of the study, the identified anti-AD targets were validated using ToppGene tool (Chen et al. 2009). This tool works by comparing candidate genes with training genes on the basis of functional similarity. Herein, five training genes, i.e., PSEN1, PSEN2, APP, APOE, and TREM2, were chosen based on the strongest AD risk effect. Eight of the top 10 anti-AD drug targets demonstrated medium/high ToppGene scores and ToppNet scores. This indicates that their ranking algorithm corresponds well to other ranking methods.

Further, to evaluate small molecule candidate drugs that could be potentially repurposed for AD, two online resources, i.e., CMap (Lamb et al. 2006) and C2Maps (Huang et al. 2012), were used. CMap (Lamb et al. 2006) was used to compare the similarity of altered gene expression patterns between repurposed drugs and known anti-AD drugs (memantine and galantamine), while C2Maps (Huang et al. 2012) appraised the anti-AD drug and gene association on the basis of network mining, literature mining, and drug effect annotation. CMap revealed a positive correlation of edrophonium with memantine, while C2Maps revealed only one drug with a high protein ranking score, i.e., physostigmine. The latter resource also validated three known anti-AD drugs, such as galantamine, rivastigmine, and donepezil.

To summarize, CD33 and MIF particularly emerged as strong candidates for which seven existing drugs could be repurposed. Also, this study adds value to the growing body of evidence by stating the manipulation of the immune system and neuroinflammation as promising research areas for anti-AD drug development.

9.2 Leveraging Multiomics Data to Unravel Repurposable Drug Candidates for Melanoma

Khosravi et al. (2019) applied multiomics data for drug repurposing to unravel melanoma-related genes, proteins, and metabolites. This multiomics approach retrieved genomic, phenomic, metabolomic, and transcriptomic data from their respective databases alongside data from other biomedical resources.

To obtain genomics data associated with melanomas, the investigators explored GWAS (Welter et al. 2014) and PheWAS (Denny et al. 2016) catalogs. Data was extracted from GWAS using "melanoma" and "cutaneous malignant melanoma" as the disease/trait attributes selection. Similarly, to reveal the entire set of melanoma-associated genes corresponding to variant alleles from diverse ethnicities, "melanoma" and "skin cancer" were set as phenotypic queries in the PheWAS catalog. Data on melanoma-related metabolites was mined from HMDB (Wishart et al. 2018a) through the following search terms, "melanoma" and "malignant melanoma." The proteins associated with the extracted metabolites were manually curated. For transcriptomic data, Khosravi et al. (2019) selected a study with a dataset consisting of differentially co-expressed genes, which were considered as lead targets against which drugs could be repurposed. In addition to these omics resources, gene associations were also extracted from EDGAR database (Babbi et al. 2017). This database derives data from various biomedical databases such as OMIM (Hamosh et al. 2002), UniProt (Apweiler et al. 2004), and CLINVAR (Landrum et al. 2016), by querying phenotypic traits for disease-based gene associations. In this case, phenotype OMIM ID 155600 analogous to cutaneous malignant melanoma was used to retrieve skin cancer-associated genes with annotated relationships.

This data mining revealed 55 unique genes mapped with different SNP-risk alleles through GWAS data, 765 alleles which correspond to 260 unique associated genes via PheWAS data, 23 metabolites linked with 616 unique proteins from metabolomics data, 27 validated differentially expressed genes from transcriptome data, and lastly 11 unique genes extracted from various biomedical resources. The total set of unique genes and proteins identified from this multiomics and biomedical data mining included 1178 entities, with some of them being identified by more than one approach.

The next step involved construction and analysis of a metabolite-protein network using Cytoscape tool (Shannon et al. 2003). Interconnections between shortlisted enzymes, transporters, or other protein types with metabolites were identified and mapped. The constructed network with proteins/metabolites as nodes and their associations as edges revealed 23 unique metabolites connecting to 617 associated enzymes or transporters.

The study after excluding experimental, illicit, withdrawn, and investigational molecules proceeded to map drugs obtained from DrugBank 5.0 (Wishart et al. 2018a) and to 1178 targets identified through the aforementioned methodology. These targets were then grouped according to four types, i.e., protein, enzyme, transporter, and carrier. Mapping to find druggable targets revealed the association of (1) 215 drugs with 127 protein targets, (2) 604 drugs with 56 enzyme targets, (3) 362 drugs with 33 transporter targets, and (4) six drugs with four carrier targets.

Later, all the 1178 targets were analyzed for druggability, but only 193 could be mapped. Ultimately, 731 potential drug molecules were found in this step. Additionally, the researchers searched for drugs with known indications of melanoma, i.e., melanoma-related drugs, and found that 75 drugs were common in both the proposed drug list and the melanoma-related drug list. Removing the common drugs resulted in 658 drug molecules associated with 184 targets.

The researchers conducted a thorough analysis for pathogenesis with the help of OMIM database (Hamosh et al. 2002) and other scientific reports in order to identify functions of pathogenic genes in humans such as LOF or GOF. In turn, the data was correlated with the antagonistic and agonistic actions of the proposed drugs to assure anti-melanoma efficacy. After correlation, genes whose activities do not match their pathogenetic information were removed. This finally resulted in 277 potential drugs interacting with 74 targets.

CMap (Lamb et al. 2006) used expression profiles of genes to analyze all the potential drug candidates to primarily confirm drug-target associations linked to melanoma. Upon obtaining the drug expression profiles through a signature query in CMap, and using the correlation between known anti-melanoma drugs and the proposed drugs, a final list containing 35 potentially repurposable drugs was prepared by considering mean and enrichment scores. Khosravi et al. (2019) stated that this approach could be used for identifying repurposing candidates for other diseases and disorders as well.

10 Side Effects: Redirecting Adverse Event Signals to Explore off-Target Therapeutic Indications

Small molecules, proteins, and FDA-approved drugs exhibit molecular interactions with on- or off-targets than anticipated in clinical trials, owing to the intricacies in cellular environments. Such unintended interactions are attributed to perturbations in the downstream signaling pathways that modulate off-target activity, thereby leading to side effects (Chartier et al. 2017). These phenotypic expressions of drugs at the body system level promote better understanding of complex and delicate pathways underlying drug-induced toxicities. These toxicities remain as major concerns across the globe for 30% failure of drug discovery and developmental programs. Therefore, polypharmacology concept that proposes the modulation of multiple targets by a single drug can be meticulously implemented to seemingly unrelated drugs that are indicated for treating a disease with similar side effect profile. The mysterious mechanism of action interlinking the side effects and disease is essential for designing promising strategies to reposition existing drugs for new therapeutic indications (Xu and Wang 2015; Yang and Agarwal 2011).

Target-based and ligand-based methods are adopted to identify off-targets of drugs. Similarity ensemble approach (SEA), a ligand-based prediction, is currently deployed for 3665 approved or investigational drugs against 246 targets. Side effect data attained from Side Effect Resource (SIDER) (Kuhn et al. 2016) encompasses 996 drugs and 4192 side effects (Ye et al. 2014). FDA Adverse Event Reporting System (FAERS) (Fang et al. 2014) yields data relevant to frequently reported adverse drug reactions as signals. Upon subsequent screening through Pharmacogenomics Knowledge Base (PharmGKB) (Thorn et al. 2013), a genomics database can abet in the identification of targets and their associated diseases.

10.1 Application of Drug Network Designed Based on Side Effects for Drug Repositioning

Hao Ye et al. *(*Ye et al. 2014*)* hypothesized that drugs which share similar side effects may also possess similar therapeutic properties. Based on this assumption, a novel network-based drug repurposing approach was designed by exploiting similar side effects shared by drugs.

Initially, side effect data was retrieved from Meyler's Side Effects of Drugs 15th edition (Aronson 2015) and Side Effects of Drugs Annuals (2007–2012) (Ray 2020). The data was then mapped in accordance with MedDRA terms (MedDRA n.d.). On similar grounds, 98 drugs from SIDER (Kuhn et al. 2016) which were not incorporated in the above databases were collated and mapped with MedDRA terms to frame external test sample sets. FDA-approved indications were acquired from Pipeline, Thomson Reuters Partnering, and GeneGO databases and were modified according to the fourth level of MeSH (Bodenreider 2004) headings. This revealed 2183 drugs associated with 6495 side effects.

For the construction of drug-drug network, a binary vector termed as "side effect fingerprint" was generated for each drug. According to this, a drug will be labeled as "1" if it displays a particular side effect from above-mentioned 6495 side effects or else "0." This was generated for all 2183 drugs with respect to 6495 side effects. Next, a similarity index was created for each pair of drugs using Jaccard index (J). Drug-drug pairs with $J \geq 0.275$ were selected for network construction. To rule out similarity bias, random side effect fingerprint for a drug was created by randomly choosing the same number of side effects from side effect pool 10,000 times. This was repeated for every drug, and the J was calculated. Later, to identify the similarities between a pair of drugs and random distribution set, Z score was calculated with a threshold of $Z \geq 2.576$. Two criteria were considered while setting thresholds: (1) chosen drug-drug pairs must possess as many as drugs possible in the network, and (2) sharing of indications between chosen drug-drug pairs must be as high as possible. Expression Analysis Systematic Explorer (EASE) and Normalized Discounted Cumulative Gain (NDCG) parameters were used to evaluate the drug's indication enrichment and ranking of drugs, respectively. Drug-drug network was constructed between 1647 drugs with 17,400 drug-drug pairs which were mapped to 81 ATC therapeutic categories and 584 FDA-approved indications. Ninety-eight drugs retrieved from SIDER were tested for similarity with each drug in the aforementioned network. This revealed that 84.69% of drugs from SIDER exhibited similarity with network drugs ($J \leq 0.275$). Upon application of this method, parecoxib was found to be enriched in pain relief (EASE score $= 1.766 * 10^{-14}$) and rheumatoid arthritis (RA) (EASE score $= 1.03 * 10^{-21}$). This drug was also sharing similar side effects with other 33 RA drugs. Similarly, tolcapone was found to share side effects with other four drugs used to treat Parkinson's disease (EASE $= 0.39$). It was also found to be enriched with antidepressants (EASE $= 5.09 * 10^{-9}$). Tramadol was found to share side effects with 22 pain management drugs (EASE $= 0.0063$) and 13 antidepressants (EASE $= 9.06 * 10^{-5}$). This indicates a significant antidepressant potential of tramadol.

10.2 Reclamation of Anticancer Drug-Side Effect Pairs by Merging Automatic Table Classification and Relationship Extraction

Xu and Wang (2015) developed an automatic approach with an intricate yet straightforward methodology in an attempt to create a comprehensive knowledge base for computational target discovery and toxicity prediction.

This approach fundamentally combined automatic table classification and relationship extraction to draw out drug-side effect (SE) pairs. The process began by downloading full-text articles from the *Journal of Oncology* (*JCO*) and extracting tables embedded in them. *JCO*, a leading journal in oncology, consists of various cancer-related research publications, including but not limited to case reports, meta-analyses, and clinical trials. A total of 31,255 tables were extracted from 13,855 downloaded articles. The required content associated with each table was drawn out. Publicly available information retrieval system named library Lucene (Mohd 2011) was used to create a search engine for article titles, table contents, and table legends. After data retrieval, a comprehensive drug lexicon consisting of 791,848 terms with respect to cancer and non-cancer drugs was compiled from the Unified Medical Language System (UMLS). The lexicon comprises FDA-approved, investigational, and failed drugs. The latter two were keys in identifying drug-SE pairs from downloaded articles. From this lexicon, a cancer-specific drug lexicon was created based on automatic filtering and manual curation. Automatic filtering was done by searching each term in their local search engine and then removing drugs that were not present in any of the tables. This step was followed by manually curating the terms to remove ambiguous and mis-classified terms. Their effort finally resulted in a clean anticancer drug lexicon of 4256 drugs. For SE lexicon, investigators used a previously developed clean SE lexicon of 49,625 terms.

Next, the drug-SE tables were classified as related and unrelated. To remove unrelated tables, i.e., those without drug or SE terms, the investigators manually classified 700 randomly selected tables. This data was then used to train SVM classifiers using the bag-of-words feature. The performance of the SVM classifier was evaluated on 100 randomly selected full-text articles. Upon evaluation, their SVM classifier had attained a recall of 0.941, precision of 0.711, and F1 value of 0.810.

Subsequently, to draw out drug-SE pairs from the extracted tables, article identity documents (ID) for the positively classified tables were noted. Then individual drug from anticancer drug lexicon was used as a query term in their local search engine for *JCO* tables. If the table title/content/legend showed the term, the article ID was documented. On a similar note, each SE term from the clean SE lexicon was used as a search term in the same search engine, and if the term appeared in any of the table content, the article ID and SE were documented. Finally, to draw out drug-SE pairs, the article IDs were matched. Post-manual curation resulted in identification of 588 drug-SE pairs which included 32 drugs and 205 SEs from 34 positively classified tables. These pairs were then used as a yardstick to compare the pair extraction performance of the SVM-classified tables to that of manually curated and unclassified tables. The result of the comparison demonstrated a similar performance between automatically classified tables and manually classified tables. However, the recall and precision of both methods were moderate, which implies that the coverage of the SE lexicon or underlying drug is imperfect.

The next phase compared drug-SE relationships extracted from *JCO* tables to those extracted from SIDER database (Kuhn et al. 2016). A total of 26,918 and 58,454 drug-SE pairs were extracted from *JCO* articles and SIDER, respectively, by using the previously created lexicon. The investigators compared the two resources for their coverages of SE associated with targeted novel anticancer drugs which have high toxicity rates. Another aim of this comparison was to evaluate the ability of data extracted from *JCO* tables to capture additional information regarding these agents, which the FDA labeling could not. A list containing 45 targeted anticancer drugs was compiled from NCI. Using drugs from this list, the drug-SE pairs from both *JCO* tables and SIDER database were filtered. The data overlapped between them was assessed with respect to targeted drugs as well as anticancer drugs in general. Ultimately, out of the 45 drugs from NCI, 31 drugs were seen in *JCO* tables and only 15 in SIDER.

The last phase of the study evaluated the capability of anticancer drug-SE pairs to build computational approaches that can unveil anticancer drug targets, forecast unidentified side effects, and repurpose drugs. The hypothesis that drugs with same SE relationships also share similar target genes, metabolism genes, or disease indications. This was further explored by calculating an average number of shared gene targets, average number of shared metabolism genes, and average number of shared disease indications for drug-drug pairs that share SEs at different cutoffs.

The study demonstrated a positive correlation between drug-SE and drug targets. The average number of shared gene targets for all drug-drug combinations was 0.608, which substantially increased when the drug combinations shared at least 30 SEs and increased even more when the shared number of SEs was 50. On assessing this parameter for targeted anticancer drugs, a stronger positive correlation was found between side effect and gene target. Conversely, non-specific cytotoxic anticancer agents, although having distinct targets at which they act, do not show much variation in the type of side effects expressed. Thus, they are implausible in forecasting drug targets on the basis of shared side effects.

On a similar note, drug-SE and drug metabolism gene associations also showed a positive correlation. However, with targeted anticancer agents, an opposite trend was observed where a weaker correlation was observed for targeted anticancer agents. The former positive correlation demonstrates the need to incorporate drug metabolism genes in addition to known drug on-target genes in computational models to predict drug toxicities.

Drug-SE and their disease indication associations showed an outcome identical to that of metabolism gene associations, where a strong correlation for all anticancer drugs was revealed, besides a weaker association for targeted anticancer drugs. This positive correlation achieved depending on clinically observed side effects associated with drug demonstrates the repurposing potential of anticancer drugs for non-cancer treatments and vice versa.

11 Electronic Health Records: Amassed Patients' Clinical Data Paves Way for Drug Repurposing

Electronic health record (EHR) is an electronic application through which individuals can access, manage, and share their health information (Tang et al. 2006). EHRs being an integral part of the healthcare system collect and store enormous amounts of quantitative, qualitative, and transactional data digitally during clinical practice and are the least-explored medical data sources for a successful drug repurposing endeavor. Expansion of this stored data has resulted in advancements of data management and analysis which converts this massive data to knowledge and information. This evolutionary technology termed big data analyzes unstructured data confined within EHRs and permits automated data acquisition by utilizing natural language processing (NLP). Moreover, big data not only creates an observational evidence base for clinical questions but also analyzes existing EHRs to generate a dash board that assists physicians in clinical decision-making. The clinical decisions driven by data support tools decrease the economic burden and will improve the standards of care. Lastly, this big data transforms health care in a patient-directed fashion by integrating conventional models with social determinants of health.

Current clinical studies collecting data from a single EHR system lack agreeable power for analyzing drugs and their corresponding indications. This challenge can be vanquished by integrating EHR data belonging to multiple institutions (Xu et al. 2020; Pagliari et al. 2007; Than Win and Cooper 2004).

11.1 Harnessing EHR and Genomic Data to Unveil the Potential of Calcium Channel Blockers against GDM: An Informatics-Based Approach

Goldstein et al. *(*2018b*)* used an informatics-based approach to identify drugs that could be potentially repurposed for GDM by harnessing de-identified EMR data from Vanderbilt University Medical Center (VUMC) (Roden et al. 2008) with drug and target data from various online sources. For ease of comprehension, this study is split into multiple phases.

Phase I: Identification of Safe Drugs in Pregnancy

Initially, 332 safe drugs in pregnancy that were rated A and B1 as per the Australian Prescribing Medicines in Pregnancy database and two other guides for US practitioners were extracted. Consequently, the study included only active oral and subcutaneous drugs and excluded illicit, withdrawn, contraindicated drugs and those devoid of a human protein target, like antibiotics. The final list after systematic exclusion comprised 129 drugs from 65 mechanistic classes.

Further, these drugs were subjected to target mining in DrugBank.ca and Psychoactive Drug Screening Program to identify targets based on pharmacological action and receptor profiling, respectively. These targets were subsequently mapped to their corresponding genes. To unveil associated genes of those drugs whose target data was unavailable in the previous approach and to identify relevant targets, a literature search on PubMed and Google Scholar was carried out. This search revealed 196 genes, which included drugs with multitarget activity and multiple drugs acting on a single target. Later, a gene set for each drug was developed in order to establish correlation with SNPs derived from systematic genetic analysis of VUMC's EMR for GDM.

Phase II: Leveraging VUMC's EMR Data to Unravel Genetic Associations for GDM

Synthetic Derivative (SD), a de-identified image of VUMC's EMR, identified patients and extracted their International Classification of Diseases 9 (ICD-9) diagnostic information. BioVU, a biorepository of DNA samples of 37,380 subjects tested using Illumina Infinium Human Exome Bead Chip, is linked with SD with an intention to unravel comprehensive genome phenome associations. Of the 196 genes of interest, 130 were denoted on the chip by 306 SNPs.

To predict disease SNP associations, 1281 women aged between 15 and 50 years with SNP genotyping who visited a clinic for supervision of a normal or high-alert pregnancy (ICD-9 V22.x or V23.x) were identified. Thereafter, the patients with at least one diagnosis of diabetes in pregnancy (ICD-9648.0x) or glucose intolerance in pregnancy (ICD-9648.8x) were recruited. While women with a diagnosis of type I diabetes mellitus (phecode 250.1) were excluded, those with type 2 diabetes mellitus (T2DM) were retained, as GDM shares pathophysiological features with the latter. All other women in the cohort were considered as controls. Ultimately, the study contained 130 cases and 950 controls. Later, the sample was narrowed down by retaining only European Ancestry. This yielded 85 GDM cases and 622 controls for further analysis. Using age as a covariate, logistic regression was performed to identify the existence of GDM in PLINK version 1.9.

Parallelly, to reveal SNPs associated with T2DM, the investigators re-analyzed the phenome screening data reported in their previous study (Denny et al. 2010) and shortlisted 4372 cases and 17,646 controls. Cases were defined as patients who were diagnosed with T2DM. The genetics of cases was then compared to controls. Logistic regression was performed by considering age and current sex as covariates for the presence of T2DM in PLINK version 1.9. Post-logistic regression analyses revealed 305 SNP associations for T2DM versus 215 for GDM.

Phase III: Gene Set Analysis to Identify Repurposable Drugs in GDM

To test the reliability of this analysis, the investigators compared their results with previously published nine SNPs associated with GDM. Herein, a SNP was considered as well replicated if it demonstrated an unadjusted p-value of <0.05. Out of the total, two SNPs were replicated for GDM, while six SNPs were replicated for T2DM. However, as a combination, seven of the nine SNPs were replicated. Further, the SNPs and p-values for GDM and T2DM were collated into a single dataset. The strongest single SNP per gene in the gene sets developed for each drug in phase I of the study was identified using Fisher's method. This gene set analysis revealed calcium channel blockers to exhibit the strongest association.

Phase IV: Variant Analysis for Validation of the Study

Next, the robustness of the results was tested using three approaches. Two approaches used different gene sets, i.e., pharmacologically active set and a complete set derived from Drugbank.ca instead of the custom gene set generated in this study. The third approach involved changing the statistical analysis method used while retaining the same custom gene set. Calcium channel blockers earned its position in the top 5 hits for all three approaches, whereas 5HT-3 antagonists were ranked lower.

Phase V: Evaluation of the Influence of the Hits on Glucose Tolerance Test (GTT)

The last section evaluated the effect of the identified hits on glucose tolerance during pregnancy. Initially, a cohort of 9960 pregnant women with reports of 50-gram GTT during their prenatal visit were included from the SD. Women with a diagnosis of pre-pregnancy diabetes or type 1 diabetes were excluded. Of the 9960 patients selected, 6390 pregnancy records had complete data. Using this data, patients were then categorized as "exposed," if there was a mention of hit drug in their record or as "not exposed" if it was not present.

The hit drugs were further grouped into 65 mechanistic drug classes. Classes that were not exposed to more than 50 patients in the cohort, those lacking FDA approval or an established FDA contraindication, classes with questionable ascertainment regarding usage (over-the-counter medications), and those with significant addiction potential were excluded. Therefore, only 20 out of the 65 mechanistic classes were analyzed. Based on the result of the multivariate analysis, calcium channel blocker exposure was linked with a 3.18 mg/dL reduction in blood glucose during the GTT. Unsurprisingly, 5HT-3 antagonists, which showed up as a hit with lower association in variant analysis, were connected with a 3.54 mg/dL increase in blood glucose.

In conclusion, calcium channel blockers demonstrated an improvement in glucose tolerance, unlike 5HT-3 antagonists. The combination of EHR and genomics augments the reliability of study results and thus can be employed for generating hypotheses which might be extrapolated to other disease conditions as well. Limitations of this study include the following: (1) selection of cases based on ICD-9 criteria is unreliable, 2) restricted number of cases and low sample numbers reduce the power of the study, and (3) cases with missing or incomplete information cannot be included.

12 Text Mining: A Methodical Literature-Based Approach to Curate Drug Repurposing Candidates

Exponential growth in scientific literatures across the research community has resulted in massive data generation. Text-mining (TM) approaches have become an imperative tool to facilitate research, use this huge unstructured or semi-structured literature sources to extract data, and comprehend them with an attempt to discover innovative information. When a set of data is provided as input, TM approaches make an effort to derive possible novel patterns, relationships, and trends within the enclosed literatures. This approach goes hand in hand with NLP algorithm that uses relatively simple text processing tasks to Named Entity Recognition algorithm that extracts relatively complex information. TM approaches find their application in the field of biomedicine toward drug development, DR, and adverse drug event prediction. Innumerable breakthrough researches for identifying new treatment options for diseases such as cataracts, multiple sclerosis, Parkinson's disease, cancer, etc. are some of the successful outcomes of TM concept (Gonzalez et al. 2016; Henry and McInnes 2017).

12.1 Dragon Exploration System for Sickle Cell Disease (DESSCD): TM Tool for Sickle Cell Disease (SCD) Drug Repurposing

Essack et al. (2013) developed a knowledge-based TM tool DESSCD, by integrating TM and data-mining approaches. This project was carried out in three steps:

Step 1: Construction of DESSCD

At the outset, specialized keyword search in the query interface of PubMed revealed 419,612 MEDLINE abstracts pertinent to SCD. Thus extracted abstracts were analyzed by employing DES using the concepts from dictionaries such as "Chemicals with Pharmacological Effects," "Human Genes and Proteins," "Human Anatomy Related Concepts," "Pathways," "Metabolites and Enzymes," and "Disease Related Concepts" which were cross-referenced to the following databases inclusive of UniProt (Apweiler et al. 2004), KEGG Pathway (Kanehisa and Goto 2000), Entrez Gene, and REACTOME (Croft et al. 2011). This analysis led to the generation of DESSCD data files, which encompassed biomedical concepts recovered from the previously compiled dictionaries, alongside other relevant information obtained from analyzed text.

Users can retrieve SCD-specific information from DESSCD data files by using specific concept, keyword, and phrase searches in the query interface in order to generate probable association networks and hypotheses.

Step 2: Validation of DESSCD

DESSCD was validated with a pre-established concept as hypothesis, i.e., hydroxyurea's association with SCD targets.

Herein, sickle cell anemia (SCA) was used as input query under "Disease Related Concepts," hydroxyurea under "Chemicals with Pharmacological Effects," and potential target under "Human Genes and Proteins." These connecting sets of concepts involving disease-drug-target were chosen for hypothesis generation. This evaluation process revealed early growth response-1 (EGR-1) and endothelin-3 (ET-3) as the target concepts. These concepts which were being claimed as therapeutic targets for hydroxyurea in confronting SCD provide evidence for reproducibility and literature validation of DESSCD.

Step 3: Application of DESSCD for Hypothesis Generation to Unravel Repurposable SCD Drugs

Subsequent to the validation procedure, researchers explored DESSCD to identify repurposable drugs for SCD. Initially, specific connecting concepts, for instance, SCA from "Disease Related Concepts" and blood from "Human Anatomy Related Concepts," were manually selected, while "Chemicals with Pharmacological Effects" was chosen as the target dictionary to generate hypothesis. This process unveiled an inherent relationship between SCA and hydroxyfasudil, which is considered to be a novel finding.

Hydroxyurea, the drug of choice in SCD, acts by upregulating nitric oxide-cyclic guanosine monophosphate (cGMP) signaling pathway for increasing HbF production. The same mode of action is observed in hydroxyfasudil, a Rho kinase inhibitor, that attenuates pulmonary hypertension by inducing eNOS, NO, and cGMP. However, this mechanism is not proved experimentally with hydroxyfasudil. This common feature shared between hydroxyfasudil and hydroxyurea renders valuable clues for investigating the repurposable potential of hydroxyfasudil against SCD.

In addition, the researchers witnessed the propensity of hydroxyfasudil in decreasing the levels of pro-inflammatory cytokine IL-6, surmounting the inflammatory sequelae encountered in SCD after hypoxia or reoxygenation injury. On the contrary, hydroxyurea is reported to stimulate the genetic expression of IL-6, IL-8, IL-10, IL-1β, IL-1α, CCL2, CCL5, CCL8, and CCL20 in endothelial cells, thereby resulting in exacerbation of chronic inflammatory condition in SCD. In addition, hydroxyurea also decreases the levels of GM-CSF and TNF-α. On the other hand, hydroxyfasudil also decreases the levels of ET1, ET3, sICAM-1, sVICAM, sSELE, and sSELP which are known to be involved in ischemia-reperfusion injury. Yet, its role in modulating GM-CSF, IL-8, and IL-3 needs further investigation. Besides its anti-inflammatory role, hydroxyfasudil downregulates FN1 gene that encodes for vaso-occlusive glycoprotein fibronectin, which is not evinced with hydroxyurea.

Ultimately, the ability of hydroxyfasudil in reducing pro-inflammatory mediators, increasing anti-inflammatory markers in conjunction with its protective role in combating vaso-occlusive crisis, makes it a more suitable candidate for current SCD therapy.

12.2 Literature-Related Discovery and Innovation (LRDI): A Novel Drug Repurposing Approach for Inflammatory Bowel Disease (IBD)

An innovative methodology to identify novel potential repurposable therapeutic agents for IBD was developed via LRDI-based text-mining approach (Kostoff et al. 2020).

In this project, initially, the crucial biomarkers and their direction of change were acquired from prevailing IBD literatures. This process was carried out by scrutinizing the currently available IBD treatments from clinical trials and clinical practice. Subsequently, care was taken to simultaneously eliminate the modifiable confounding factors such as diet, exercise, smoking, alcohol consumption, iatrogenic and biotoxin exposures, environmental/occupational exposures, psychosocial stressors, etc. that majorly contribute to IBD pathogenesis. Later, a query was framed coalescing the above-identified key biomarkers with their preferred treatment-derived directions of change. This query was surfed in MEDLINE database for extracting literatures associated with non-IBD cases. This was performed with an intention to identify potential treatments that demonstrate similar biomarker changes documented in core IBD literature.

Around 9500 records were retrieved, among which 350 records which were considered to be the most recent were further analyzed to prove the robustness of the adopted approach to provide treatment-specific information for IBD. Thus, obtained potential treatments were validated independently to ensure that the acquired treatment was not being tested in any IBD models earlier.

All the aforesaid steps resulted in identification of 64 novel repurposable therapeutic agents for IBD, viz., fluprostenol, Liu Shen Wan, erdosteine, etc., alongside ten other investigational agents such as taraxasterol, *Rhodiola rosea* salidroside, VAS2870, etc. Retrieval of these recently explored ten treatment options for IBD corroborates the predictive potential of this LRDI approach. This text-mining approach can be further extrapolated and implemented to other fields.

Intriguingly, it is observed that many of the novel treatments attained by this approach were found to be already in use for the treatment of diseases such as diabetes, cancer, depression, and dementia with most of them being herbal in origin. The novel treatments queried from this study were based only on 20 biomarkers. It would be an added advantage if other biomarkers would have been included as query inputs to yield an array of other treatment options.

13 Computational Techniques to Visualize Intricate Molecular Interactions

Contemporary computational techniques have brought in remarkable advantages in designing small molecules. These techniques utilize drug-protein interaction data to reconnoiter crucial binding sites of targets, in order to unveil new indications. Molecular docking and pharmacophore modeling techniques are being extensively employed for target identification, hit-to-lead identification, lead optimization, and drug repurposing approaches. Docking studies predict the behavior of a molecule within the binding pocket of target protein and visualize the interactions, while pharmacophore modeling analyzes the chemical features and predicts those features which bring about effective interaction with target protein. Further, molecular dynamics simulation studies validate the stability of protein-ligand complex. On the other hand, homology modeling aids in figuring out the structures of proteins that are not readily available in repositories. The modeled protein may serve as target for molecular docking studies. These computational techniques are less tedious, cost-effective, and superior to traditional methods of drug discovery. This section of the chapter summarizes the computational drug repurposing strategies through case studies (Lagarde et al. 2018; Kilambi and Gray 2017; Kalyaanamoorthy and Chen 2011; Muchtaridi et al. 2017; Yang 2010).

13.1 Target-Specific Drug Repurposing against FtsZ Protein of Salmonella Typhi *Via Computational Techniques*

In silico computational tools have evidenced their competence in drug discovery processes to reduce the timelines, financial burden, and tackle crisis. An effort has been made by Naz et al. (2020) to reconnoiter potential drug molecules for FtsZ target of *S. Typhi* by adopting computational techniques such as pharmacophore modeling, molecular docking, and molecular dynamics.

This process was initiated by extracting chemical structures of FDA-approved drugs from DrugBank. 3D homology model of target protein was designed to execute the process using SWISS-MODEL tool (Bairoch 2000). Pharmacophore model was developed using Discovery Studio program by locating the amino acid residues in binding pocket. This process identified three hydrogen bond acceptors, three hydrogen bond donors, and two hydrophobic groups which are essential for significant inhibition. Further, the pharmacophore model was validated by reported inhibitors and decoy sets. Screening of 1551 drug molecules against validated pharmacophore model has identified 256 molecules with matching features. Later, molecular docking studies were performed to investigate the binding affinity and significant interactions at the active pocket. Following this, molecular dynamics simulations were carried out to analyze the stability of complex. This analyzes identified six drugs, which were ultimately subjected to experimental investigation to evaluate GTPase activity, polymerization inhibition, and minimum inhibitory concentration (MIC) determination. Out of the six drugs, benzethonium chloride has demonstrated promising results in cell-based in vitro assays. Benzethonium chloride has exhibited potent antibacterial activity against ciprofloxacin-resistant *S. Typhi* with an MIC of 8 μg/ml. Overall, the study suggested a structural modification of benzethonium chloride can be considered to design a potent and safe molecule that targets drug-resistant bacteria.

13.2 Leveraging Pharmacophore Anchor Models Pertinent to Flaviviral NS3 Proteases for Drug Repurposing against DENV Infection

Viruses from the *Flaviviridae* are known to cause major infections like dengue (DENV), Japanese encephalitis, hepatitis C (HCV), West Nile, etc., across the globe. Flaviviral NS3 proteases which play a crucial role in viral replication and survival are considered as potential targets for rational drug design and development. In this study (Pathak et al. 2017), pharmacophore anchor (PA)- and core pharmacophore anchor (CPA)-based models were developed for the discovery of potential drug molecules. A library of 187,740 compounds were docked against the NS3 proteases of all four viruses using GEMDOCK tool (Yang and Chen 2004). Compounds with good interaction energies for respective target protein were shortlisted and further utilized for construction of interaction profiles. The profiles of residue-compound interactions were analyzed using SiMMap analysis tool (Chen et al. 2010) for assigning anchors. The anchors with protein active site were signified as PA models for each of the four selected NS3 proteases. Later, these PA models were aligned to identify the conserved "core anchors." Core anchors in conjunction with aligned protease active sites were taken into account for creating a CPA model. PA/CPA models were validated by scrutinizing conservation and mutation activity for anchor residues alongside exploration of binding mechanistics of 89 well-known inhibitors of NS3 protease. Thereafter, an integrated anchor-based virtual screening was framed for DENV NS3 protease in order to virtually screen FDA drugs. Post screening, the potential candidates were subjected to in vitro anti-dengue activity testing, which was tailed by structure-anchor-activity relationship studies. Experimental investigation of anti-DENV inhibition identified asunaprevir and telaprevir with EC_{50} values of 10.4 and 24.5 μM, respectively. Surprisingly, both the molecules were pre-reported for their HCV NS3 protease inhibition with an IC50 of ~1 and 10 nM, respectively. Occupancy of chemical features in different sites of active pocket is expected to enhance the inhibitory potential of the aforesaid compounds. Therefore, structural modification of both drugs will pave way for designing potent and safe protease inhibitors for DENV. PA/CPA models can be extended to screen large chemical libraries to augment lead optimization in order to identify potential hits against various flaviviral infections.

14 Trailblazing Artificial Intelligence to Expedite Drug Repurposing Endeavor

The advent of fast computing in silico prediction algorithms expedites drug discovery processes. However, identification of therapeutic agents to manage diseases with complex multifactorial pathological mechanisms still remains challenging.

Emergence of big data is a boon to scientific community across the globe; nevertheless, handling massive heterogeneous data demands construction of supercomputing models with precision. In due course of this model construction, a number of critical issues are encountered in terms of retrieval of relevant datasets for training and subsequent testing, identification and application of suitable computational techniques, and finally selection of appropriate validation tools. These intricacies in handling multidimensional datasets via classic computational methods are being surmounted with advanced artificial intelligence and machine learning (ML) techniques (Toh et al. 2019; Bishop 2006; Topol 2019).

14.1 Deep Learning Pipeline Based on Semantic Relationship between Drug and Disease Features

Moridi et al. (2019) exemplified a novel non-linear method for drug repurposing by developing a semantic relationship between drug and disease using deep learning methods. They employed deep neural network for construction of drug-disease associations.

Construction of Drug-Drug Similarity Matrix

Drug-drug similarity matrix was created by gathering information related to chemical structures, gene expression profiles, and protein and enzyme sequences of drug targets corresponding to the drugs from DrugBank, PubChem, and CMap databases. Variational autoencoder (VAE) was used to clean the SMILES format of all the extracted drugs. A deep neural network, "ProtVec," was used to impute protein and enzyme sequences into the vectors. CMap datasets were used to retrieve gene expression data. A stacked autoencoder containing five layers was designed to generate a vector corresponding to gene co-expression data, pathways, and BP. A vector called Cell Identity Code (CIC) was created to resist the noise and missing data. Bayesian approach was employed for hyper-parameter optimization, and the data was distributed into training (60%), test (25%), and validation (15%) and sets. The final hyper-parameters were selected after performing 100 iterations. Finally, the network was constructed after performing 300 iterations of selected hyper-parameters.

Construction of Disease-Disease Similarity Matrix

For construction of disease-disease matrix, two features, namely, phenotypes and genotypes, were selected. Herein, monarch platform possessing integrative data that connects phenotypes to genotypes across species was used to extract 8662 phenotypes corresponding to 10,881 diseases and 10,764 genotypes corresponding to 7217 diseases. Among the above-retrieved data, 5955 diseases were found to possess both phenotypes and genotypes. Disease-disease similarity matrix was constructed by imputing phenotypic and genotypic features generated through one-hot encoders.

Construction of Drug-Disease Association Matrix

Finally, drug-disease association matrix was constructed by assembling known drug-disease associations. The known drug-disease associations were hidden, and other known associations were utilized to score the hidden drug-disease associations. They extracted 585 diseases pertaining to 146 drugs. AUC of each disease was calculated for each subset of drugs' features. AUC of each drug was computed for individual subset of feature, and finally average AUC was considered.

Comparison with Other Models and New Drug-Disease Predictions

AUC of this model was found to be 0.935 when compared with other recent prediction tools such as PREDICT (0.902) (Gottlieb et al. 2011), SCMFDD (0.920) (Zhang et al. 2018), and DisDrugPred(0.922) (Xuan et al. 2019). This pipeline is efficient in predicting the following drug-disease pairs: mycophenolate mofetil for lupus nephritis, sirolimus for paroxysmal nocturnal hemoglobinuria, ramipril for peripheral arterial disease, and cladribine for multiple sclerosis.

14.2 deepDR: A Deep Learning Network-Based Approach for Drug Repurposing

Xiangxiang Zeng et al. created "deepDR" (Zeng et al. 2019), an innovative drug repositioning model based on network semantics. Initially, they collated the drugs from DrugBank and RepoDB and mapped them with corresponding experimental data in order to construct heterogeneous networks such as drug-target network, clinically validated drug-drug interactions, drug-side-effect network, chemically similar drug network, therapeutic similarities based on ATC, drugs' target sequence similarities, GO BP, cellular component, and molecular function networks. Later, a deep learning predictive model was built by congregating clinically validated 6677 drug-disease pairs linking 1519 drugs with 1229 diseases. Later, external validation set was created by collating the updated drug-disease association information available in ClinicalTrials.gov (https://clinicaltrials.gov/) (NIH n.d.), by eliminating the data from databases that were exploited for construction of the aforementioned networks, i.e., DrugBank (Wishart et al. 2018a) and RepoDB (Brown and Patel 2017).

Following network construction, a transition model was designed to predict the structural details of the network and explore the topological properties of individual drug within the networks. As a pre-processing step, this transition model was used to calculate Positive Pointwise Mutual Information (PPMI) matrix through random walk with restart (RWR). Later, fusion of multiple networks was enabled via multimodal deep autoencoder (MDA) in order to exploit premium-quality drug features by integrating the hidden layers into single bottleneck layer. The extracted features were then fed into VAE to envisage drug-disease connotations. Finally, deepDR was validated through fivefold cross-validation by randomly selecting 20% of clinically reported drug-disease associations with a matching number of randomly sampled unknown pairs as test set, and the rest were reserved as training set. The overall credibility of this model was evaluated based on the parameters such as AUROC and the area under the precision-recall curve (AUPR). This process revealed the highest level of accuracy demonstrated by deepDR (AUROC = 0.908 and AUPR = 0.923) and was considered to exhibit superior performance over other existing models like DTINet, Kernelized Bayesian Matrix Factorization (KBMF), random forest (RF), RWR, and Katz.

After analyzing the merits of deepDR, it was employed to predict potential drug candidates for AD and Parkinson's disease. Isoprenaline (β-adrenergic receptor agonist), risperidone, and aripiprazole (antipsychotics) were predicted as potential candidates for AD by deepDR which is supported by literature evidence. Similarly, orphenadrine (anticholinergic), pergolide (dopamine receptor agonist), and methylphenidate (a stimulant used in narcolepsy) were predicted for Parkinson's disease.

14.3 Reconnoitering New Drug-Target Interactions from Heterogeneous Networks Via NeoDTI: A Neural Network-Based Prediction Tool

Fangping Wan et al. (2019) proposed a novel drug repurposing framework, namely, "NEural integration of neighbOr information for DTI prediction (NeoDTI)," which predicts new indications for existing drugs by integrating and analyzing heterogeneous data derived from diverse sources.

Initially, a heterogeneous network (HN) was constructed with nodes and edges, wherein the nodes represent either drug, target, side effect, or disease and edges epitomize the relationships between nodes. This network elucidates drug-target, drug-drug, and disease-protein interactions. Subsequently, NeoDTI was designed to map the nodes to their respective feature representations besides preserving the drugs' original topological characteristics to ease the prediction of DTIs.

Later, a projection matrix was constructed by extracting topological features from HN. Known drug-target pairs were considered as positive pairs and unknown were considered as negative pairs. Hyper-parameters pertinent to this ML tool were determined by running a tenfold cross-validation on all positive sets and randomly chosen negative sets. AUPR and AUROC curve were used for assessing the performance of NeoDTI. To avoid the redundancy of DTI, evaluation patterns were performed by removing DTIs sharing similar drugs, proteins, drug-drug interactions, drug-SE interactions, and drug-disease similarities. This framework exhibited the best performance in terms of AUPR (86.2%) and AUROC (95.1%). Ultimately, NeoDTI predicted sorafenib's (FDA-approved drug for liver cancer) interaction with colony-stimulating factor 1 receptor which is crucial for mammary gland carcinogenesis. Likewise, NeoDTI predicted the interaction of acetazolamide (a diuretic) with carbonic anhydrase 6, which is supported by literature evidence, thereby adding value to the validation of this prediction tool.

15 Conclusion

The quantum of data generated in the field of biomedical research is ever rising. This data can be transformed into operational knowledge by integrating and applying suitable multiomics and AI techniques that collect and analyze data efficiently to develop novel therapeutic agents to accomplish successful translational research.

References

Abbruzzese C, Matteoni S, Signore M, Cardone L, Nath K, Glickson JD et al (2017) Drug repurposing for the treatment of glioblastoma multiforme. J Exp Clin Cancer Res 36:36. https://doi.org/10.1186/s13046-017-0642-x

Agren R, Liu L, Shoaie S, Vongsangnak W, Nookaew I, Nielsen J (2013) The RAVEN toolbox and its use for generating a genome-scale metabolic model for Penicillium chrysogenum. PLoS Comput Biol 9:e1002980. https://doi.org/10.1371/journal.pcbi.1002980

Agren R, Mardinoglu A, Asplund A, Kampf C, Uhlen M, Nielsen J (2014) Identification of anticancer drugs for hepatocellular carcinoma through personalized genome-scale metabolic modeling. Mol Syst Biol 10:721. https://doi.org/10.1002/msb.145122

AlzGene | ALZFORUM (n.d.). https://www.alzforum.org/alzgene. Accessed 3 Apr 2021.

Apweiler R, Bairoch A, Wu CH, Barker WC, Boeckmann B, Ferro S et al (2004) UniProt: the universal protein knowledgebase. Nucleic Acids Res 32:D115–D119. https://doi.org/10.1093/nar/gkh131

Armitage EG, Southam AD (2016) Monitoring cancer prognosis, diagnosis and treatment efficacy using metabolomics and lipidomics. Metabolomics 12:12. https://doi.org/10.1007/s11306-016-1093-7

Aronson JK (2015) Meyler's side effects of drugs, 16th edn. Elsevier

Aslam B, Basit M, Nisar MA, Khurshid M, Rasool MH (2017) Proteomics: technologies and their applications. J Chromatogr Sci 55:182–196. https://doi.org/10.1093/chromsci/bmw167

Babbi G, Martelli PL, Profiti G, Bovo S, Savojardo C, Casadio R (2017) eDGAR: a database of disease-gene associations with annotated relationships among genes. BMC Genomics 18:554. https://doi.org/10.1186/s12864-017-3911-3

Bairoch A (2000) The SWISS-PROT protein sequence database and its supplement TrEMBL in 2000. Nucleic Acids Res 28:45–48. https://doi.org/10.1093/nar/28.1.45

Barbeira A, Shah KP, Torres JM, Wheeler HE, Torstenson ES, Edwards T et al (2016) MetaXcan: summary statistics based gene-level association method infers accurate PrediXcan results. BioRxiv

Barretina J, Caponigro G, Stransky N, Venkatesan K, Margolin AA, Kim S et al (2012) The cancer cell line encyclopedia enables predictive modelling of anticancer drug sensitivity. Nature 483: 603–607. https://doi.org/10.1038/nature11003

Barrett T, Wilhite SE, Ledoux P, Evangelista C, Kim IF, Tomashevsky M et al (2013) NCBI GEO: archive for functional genomics data sets--update. Nucleic Acids Res 41:D991–D995. https://doi.org/10.1093/nar/gks1193

Basu A, Bodycombe NE, Cheah JH, Price EV, Liu K, Schaefer GI et al (2013) An interactive resource to identify cancer genetic and lineage dependencies targeted by small molecules. Cell 154:1151–1161. https://doi.org/10.1016/j.cell.2013.08.003

Battle A, Mostafavi S, Zhu X, Potash JB, Weissman MM, McCormick C et al (2014) Characterizing the genetic basis of transcriptome diversity through RNA-sequencing of 922 individuals. Genome Res 24:14–24. https://doi.org/10.1101/gr.155192.113

Benson DA, Karsch-Mizrachi I, Lipman DJ, Ostell J, Sayers EW (2009) GenBank. Nucleic Acids Res 38:D46. https://doi.org/10.1093/nar/gkp1024

Berman HM, Westbrook J, Feng Z, Gilliland G, Bhat TN, Weissig H et al (2000) The protein data bank. Nucl Acids Res 28:235–242

Bidkhori G, Benfeitas R, Elmas E, Kararoudi MN, Arif M, Uhlen M et al (2018) Metabolic network-based identification and prioritization of anticancer targets based on expression data in hepatocellular carcinoma. Front Physiol 9:1–11. https://doi.org/10.3389/fphys.2018.00916

Bishop C (2006) Pattern recognition and machine learning, 1st edn. Springer-Verlag, New York

Bodenreider O (2004) The unified medical language system (UMLS): integrating biomedical terminology. Nucleic Acids Res 32:D267–D270. https://doi.org/10.1093/nar/gkh061

Breckenridge A, Jacob R (2018) Overcoming the legal and regulatory barriers to drug repurposing. Nat Rev Drug Discov 18:1–2. https://doi.org/10.1038/nrd.2018.92

Brown AS, Patel CJ (2017) A standard database for drug repositioning. Sci Data 4:1–7. https://doi.org/10.1038/sdata.2017.29

Cai N, Bigdeli TB, Kretzschmar W, Lei Y, Liang J, Song L et al (2015) Sparse whole-genome sequencing identifies two loci for major depressive disorder. Nature 523:588–591. https://doi.org/10.1038/nature14659

Calderone A, Castagnoli L, Cesareni G (2013) Mentha: a resource for browsing integrated protein-interaction networks. Nat Methods 10:690–691. https://doi.org/10.1038/nmeth.2561

Carbon S, Ireland A, Mungall CJ, Shu S, Marshall B, Lewis S et al (2009) AmiGO: online access to ontology and annotation data. Bioinformatics 25:288–289. https://doi.org/10.1093/bioinformatics/btn615

Casares-Marfil D, Martín J, Acosta-Herrera M (2020) Genomic opportunities for drug repositioning in systemic seropositive rheumatic diseases. Expert Rev Clin Immunol 16:343–346. https://doi.org/10.1080/1744666X.2020.1738926

Cha Y, Erez T, Reynolds IJ, Kumar D, Ross J, Koytiger G et al (2018) Drug repurposing from the perspective of pharmaceutical companies. Br J Pharmacol 175:168–180. https://doi.org/10.1111/bph.13798

Chadwick LH (2012) The NIH Roadmap Epigenomics Program data resource. Epigenomics 4:317. https://doi.org/10.2217/epi.12.18

Chang YM, Lin HH, Liu WY, Yu CP, Chen HJ, Wartini PP et al (2019) Comparative transcriptomics method to infer gene coexpression networks and its applications to maize and rice leaf transcriptomes. Proc Natl Acad Sci U S A 116:3091–3099. https://doi.org/10.1073/pnas.1817621116

Chartier M, Morency LP, Zylber MI, Najmanovich RJ (2017) Large-scale detection of drug off-targets: hypotheses for drug repurposing and understanding side-effects. BMC. Pharmacol Toxicol 18:18. https://doi.org/10.1186/s40360-017-0128-7

Chatterjee P, Roy D, Rathi N (2018) Epigenetic drug repositioning for Alzheimer's disease based on epigenetic targets in human Interactome. J Alzheimers Dis 61:53–65. https://doi.org/10.3233/JAD-161104

Chen J, Bardes EE, Aronow BJ, Jegga AG (2009) ToppGene suite for gene list enrichment analysis and candidate gene prioritization. Nucleic Acids Res 37:W305. https://doi.org/10.1093/nar/gkp427

Chen YF, Hsu KC, Lin SR, Wang WC, Huang YC, Yang JM (2010) SiMMap: a web server for inferring site-moiety map to recognize interaction preferences between protein pockets and compound moieties. Nucleic Acids Res 38:38. https://doi.org/10.1093/nar/gkq480

Corradin O, Saiakhova A, Akhtar-Zaidi B, Myeroff L, Willis J, Cowper-Sallari R et al (2014) Combinatorial effects of multiple enhancer variants in linkage disequilibrium dictate levels of gene expression to confer susceptibility to common traits. Genome Res 24:1–13. https://doi.org/10.1101/gr.164079.113

Croft D, O'Kelly G, Wu G, Haw R, Gillespie M, Matthews L et al (2011) Reactome: a database of reactions, pathways and biological processes. Nucleic Acids Res 39:D691–D697. https://doi.org/10.1093/nar/gkq1018

Cures Within Reach (2015) Cures Within Reach - Repurposing a Vaccine for Type I Diabetes 2015. https://www.cureswithinreach.org/research/search-complete-research/research-projects/498-repurposing-a-vaccine-for-type-i-diabetes. Accessed 25 July 2018.

Davis AP, Grondin CJ, Johnson RJ, Sciaky D, McMorran R, Wiegers J et al (2019) The comparative Toxicogenomics database: update 2019. Nucleic Acids Res 47:D948–D954. https://doi.org/10.1093/nar/gky868

Denny JC, Bastarache L, Roden DM (2016) Phenome-wide association studies as a tool to advance precision medicine. Annu Rev Genomics Hum Genet 17:353–373. https://doi.org/10.1146/annurev-genom-090314-024956

Denny JC, Ritchie MD, Basford MA, Pulley JM, Bastarache L, Brown-Gentry K et al (2010) PheWAS: demonstrating the feasibility of a phenome-wide scan to discover gene-disease associations. Bioinformatics 26:1205–1210. https://doi.org/10.1093/bioinformatics/btq126

Dönertaş HM, Fuentealba Valenzuela M, Partridge L, Thornton JM (2018) Gene expression-based drug repurposing to target aging. Aging Cell 17:1–14. https://doi.org/10.1111/acel.12819

Draper J, Murray C (2020) Stem Cell Network. Stem Cell Res 47. https://doi.org/10.1016/j.scr.2020.101890

Drug repurposing | Anticancerfund (n.d.). https://www.anticancerfund.org/en/drug-repurposing. Accessed 12 Dec 2019.

Dubuis S, Ortmayr K, Zampieri M (2018) A framework for large-scale metabolome drug profiling links coenzyme a metabolism to the toxicity of anti-cancer drug dichloroacetate. Commun Biol 1:101. https://doi.org/10.1038/s42003-018-0111-x

Emilien G (2000) Impact of genomics on drug discovery and clinical medicine. QJM 93:391–423. https://doi.org/10.1093/qjmed/93.7.391

Essack M, Radovanovic A, Bajic VB (2013) Information exploration system for sickle cell disease and repurposing of Hydroxyfasudil. PLoS One 8:8. https://doi.org/10.1371/journal.pone.0065190

Fang H, Su Z, Wang Y, Miller A, Liu Z, Howard PC et al (2014) Exploring the FDA adverse event reporting system to generate hypotheses for monitoring of disease characteristics. Clin Pharmacol Ther 95:496–498. https://doi.org/10.1038/clpt.2014.17

Feghali M, Venkataramanan R, Caritis S (2015) Pharmacokinetics of drugs in pregnancy. Semin Perinatol 39:512–519. https://doi.org/10.1053/j.semperi.2015.08.003

Findacure | 7,000 rare diseases, 1 common goal (n.d.). https://www.findacure.org.uk/. Accessed 12 Dec 2019.

Fogel DB (2018) Factors associated with clinical trials that fail and opportunities for improving the likelihood of success: a review. Contemp Clin Trials Commun 11:156–164. https://doi.org/10.1016/j.conctc.2018.08.001

FooDB (n.d.). https://www.foodb.ca/. Accessed 3 Apr 2021.

Frolkis A, Knox C, Lim E, Jewison T, Law V, Hau DD et al (2009) SMPDB: the small molecule pathway database. Nucleic Acids Res 38:38. https://doi.org/10.1093/nar/gkp1002

Fu X, Cong H, Zhao S, Li Y, Liu T, Sun Y et al (2020) Construction of Glycometabolism- and hormone-related lncRNA-mediated feedforward loop networks reveals global patterns of lncRNAs and drug repurposing in gestational diabetes. Front Endocrinol (Lausanne) 11:1–12. https://doi.org/10.3389/fendo.2020.00093

Funding Opportunities (n.d.). https://www.nia.nih.gov/research/grants-funding/announcements. Accessed 23 Aug 2019.

Ganesan A, Arimondo PB, Rots MG, Jeronimo C, Berdasco M (2019) The timeline of epigenetic drug discovery: from reality to dreams. Clin Epigenetics 11:1–17. https://doi.org/10.1186/s13148-019-0776-0

Gao T, He B, Liu S, Zhu H, Tan K, Qian J (2016) EnhancerAtlas: a resource for enhancer annotation and analysis in 105 human cell/tissue types. Bioinformatics 32:3543–3551. https://doi.org/10.1093/bioinformatics/btw495

Gaulton A, Hersey A, Nowotka ML, Patricia Bento A, Chambers J, Mendez D et al (2017) The ChEMBL database in 2017. Nucleic Acids Res 45:D945–D954. https://doi.org/10.1093/nar/gkw1074

Global Cures (n.d.). https://www.global-cures.org/. Accessed 12 Dec 2019.

Goldstein JA, Bastarache LA, Denny JC, Pulley JM, Aronoff DM (2018a) PregOMICS—Leveraging systems biology and bioinformatics for drug repurposing in maternal-child health. Am J Reprod Immunol 80:e12971. https://doi.org/10.1111/aji.12971. Blackwell Publishing Ltd

Goldstein JA, Bastarache LA, Denny JC, Roden DM, Pulley JM, Aronoff DM (2018b) Calcium channel blockers as drug repurposing candidates for gestational diabetes: mining large scale genomic and electronic health records data to repurpose medications. Pharmacol Res 130:44–51. https://doi.org/10.1016/j.phrs.2018.02.013

Gonzalez GH, Tahsin T, Goodale BC, Greene AC, Greene CS (2016) Recent advances and emerging applications in text and data mining for biomedical discovery. Brief Bioinform 17:33–42. https://doi.org/10.1093/bib/bbv087

Gottlieb A, Stein GY, Ruppin E, Sharan R (2011) PREDICT: a method for inferring novel drug indications with application to personalized medicine. Mol Syst Biol 7:496. https://doi.org/10.1038/msb.2011.26

Greene CS, Krishnan A, Wong AK, Ricciotti E, Zelaya RA, Himmelstein DS et al (2015) Understanding multicellular function and disease with human tissue-specific networks. Nat Genet 47:569–576. https://doi.org/10.1038/ng.3259

Hamosh A, Scott AF, Amberger J, Bocchini C, Valle D, McKusick VA (2002) Online Mendelian inheritance in man (OMIM), a knowledgebase of human genes and genetic disorders. Nucleic Acids Res 30:52–55

Hastings J, Owen G, Dekker A, Ennis M, Kale N, Muthukrishnan V et al (2016) ChEBI in 2016: improved services and an expanding collection of metabolites. Nucleic Acids Res 44:D1214–D1219. https://doi.org/10.1093/nar/gkv1031

He B, Chen C, Teng L, Tan K (2014) Global view of enhancer-promoter interactome in human cells. Proc Natl Acad Sci U S A 111:111. https://doi.org/10.1073/pnas.1320308111

Henry S, McInnes BT (2017) Literature based discovery: models, methods, and trends. J Biomed Inform 74:20–32. https://doi.org/10.1016/j.jbi.2017.08.011

Holder LB, Haque MM, Skinner MK (2017) Machine learning for epigenetics and future medical applications. Epigenetics 12:505–514. https://doi.org/10.1080/15592294.2017.1329068

Hosseini A, Minucci S (2018) Alterations of histone modifications in cancer. In: Epigenetics in human disease. Elsevier, pp 141–217. https://doi.org/10.1016/b978-0-12-812215-0.00006-6

Huang H, Wu X, Pandey R, Li J, Zhao G, Ibrahim S et al (2012) C2Maps: a network pharmacology database with comprehensive disease-gene-drug connectivity relationships. BMC Genomics 13 (Suppl 6):S17. https://doi.org/10.1186/1471-2164-13-s6-s17

Huang HY, Lin YCD, Li J, Huang KY, Shrestha S, Hong HC et al (2020) MiRTarBase 2020: updates to the experimentally validated microRNA-target interaction database. Nucleic Acids Res 48:D148–D154. https://doi.org/10.1093/nar/gkz896

Iorio F, Rittman T, Ge H, Menden M, Saez-Rodriguez J (2013) Transcriptional data: a new gateway to drug repositioning? Drug Discov Today 18:350–357. https://doi.org/10.1016/j.drudis.2012.07.014

Iqbal J, Yuen T, Sun L, Zaidi M (2016) From the gut to the strut: where inflammation reigns, bone abstains. J Clin Invest 126:2045–2048. https://doi.org/10.1172/JCI87430

Jensen MA, Ferretti V, Grossman RL, Staudt LM (2017) The NCI genomic data commons as an engine for precision medicine. Blood 130:453–459. https://doi.org/10.1182/blood-2017-03-735654

Jiang L, Yu X, Ma X, Liu H, Zhou S, Zhou X et al (2019) Identification of transcription factor-miRNA-lncRNA feed-forward loops in breast cancer subtypes. Comput Biol Chem 78:1–7. https://doi.org/10.1016/j.compbiolchem.2018.11.008

Kalyaanamoorthy S, Chen Y-PP (2011) Structure-based drug design to augment hit discovery. Drug Discov Today 16:831–839. https://doi.org/10.1016/j.drudis.2011.07.006

Kanehisa M, Goto S (2000) KEGG: Kyoto Encyclopedia of Genes and Genomes. Nucl Acids Res 28:27

Karp PD, Riley M, Paley SM, Pellegrini-Toole A (2002) The MetaCyc database. Nucleic Acids Res 30:59–61. https://doi.org/10.1093/nar/30.1.59

Keenan AB, Jenkins SL, Jagodnik KM, Koplev S, He E, Torre D et al (2018) The library of integrated network-based cellular signatures NIH program: system-level cataloging of human cells response to perturbations. Cell Syst 6:13–24. https://doi.org/10.1016/j.cels.2017.11.001

Khosravi A, Jayaram B, Goliaei B, Masoudi-Nejad A (2019) Active repurposing of drug candidates for melanoma based on GWAS, PheWAS and a wide range of omics data. Mol Med 25:30. https://doi.org/10.1186/s10020-019-0098-x

Kilambi KP, Gray JJ (2017) Structure-based cross-docking analysis of antibody-antigen interactions. Sci Rep 7:1–15. https://doi.org/10.1038/s41598-017-08414-y

Kim S, Thiessen PA, Bolton EE, Chen J, Fu G, Gindulyte A et al (2016) PubChem substance and compound databases. Nucleic Acids Res 44:D1202–D1213. https://doi.org/10.1093/nar/gkv951

Koscielny G, An P, Carvalho-Silva D, Cham JA, Fumis L, Gasparyan R et al (2017) Open targets: a platform for therapeutic target identification and validation. Nucleic Acids Res 45:D985–D994. https://doi.org/10.1093/nar/gkw1055

Kostoff RN, Briggs MB, Shores DR (2020) Treatment repurposing for inflammatory bowel disease using literature-related discovery and innovation. World J Gastroenterol 26:4889–4899. https://doi.org/10.3748/wjg.v26.i33.4889

Kuhn M, Letunic I, Jensen LJ, Bork P (2016) The SIDER database of drugs and side effects. Nucleic Acids Res 44:D1075–D1079. https://doi.org/10.1093/nar/gkv1075

Kwon OS, Kim W, Cha HJ, Lee H (2019) In silico drug repositioning: from large-scale transcriptome data to therapeutics. Arch Pharm Res 42:879–889. https://doi.org/10.1007/s12272-019-01176-3

Lagarde N, Carbone A, Sacquin-Mora S (2018) Hidden partners: using cross-docking calculations to predict binding sites for proteins with multiple interactions. Proteins Struct Funct Bioinforma 86:723–737. https://doi.org/10.1002/prot.25506

Lamb J, Crawford ED, Peck D, Modell JW, Blat IC, Wrobel MJ et al (2006) The connectivity map: Using gene-expression signatures to connect small molecules, genes, and disease. Science (80-) 313:1929–1935. https://doi.org/10.1126/science.1132939

Landrum MJ, Lee JM, Benson M, Brown G, Chao C, Chitipiralla S et al (2016) ClinVar: public archive of interpretations of clinically relevant variants. Nucleic Acids Res 44:D862–D868. https://doi.org/10.1093/nar/gkv1222

Lee P, Yacyshyn BR, Yacyshyn MB (2019) Gut microbiota and obesity: An opportunity to alter obesity through faecal microbiota transplant (FMT). Diabetes Obes Metab 21:479–490. https://doi.org/10.1111/dom.13561

Lee SY, Song MY, Kim D, Park C, Park DK, Kim DG et al (2020) A proteotranscriptomic-based computational drug-repositioning method for Alzheimer's disease. Front Pharmacol 10:1–11. https://doi.org/10.3389/fphar.2019.01653

Li MJ, Wang LY, Xia Z, Sham PC, Wang J (2013) GWAS3D: detecting human regulatory variants by integrative analysis of genome-wide associations, chromosome interactions and histone modifications. Nucleic Acids Res 41:41. https://doi.org/10.1093/nar/gkt456

Liang K-H (2013) Transcriptomics. In: Bioinformatics for biomedical science and clinical applications. Elsevier, pp 49–82. https://doi.org/10.1533/9781908818232.49

Liberzon A, Subramanian A, Pinchback R, Thorvaldsdóttir H, Tamayo P, Mesirov JP (2011) Molecular signatures database (MSigDB) 3.0. Bioinformatics 27:1739–1740. https://doi.org/10.1093/bioinformatics/btr260

Liu X, Wang S, Meng F, Wang J, Zhang Y, Dai E et al (2013) SM2miR: a database of the experimentally validated small molecules' effects on microRNA expression. Bioinformatics 29: 409–411. https://doi.org/10.1093/bioinformatics/bts698

Lonsdale J, Thomas J, Salvatore M, Phillips R, Lo E, Shad S et al (2013) The genotype-tissue expression (GTEx) project. Nat Genet 45:580–585. https://doi.org/10.1038/ng.2653

Lounkine E, Keiser MJ, Whitebread S, Mikhailov D, Hamon J, Jenkins JL et al (2012) Large-scale prediction and testing of drug activity on side-effect targets. Nature 486:361–367. https://doi.org/10.1038/nature11159

Lowe R, Shirley N, Bleackley M, Dolan S, Shafee T (2017) Transcriptomics technologies. PLoS Comput Biol 13:e1005457. https://doi.org/10.1371/journal.pcbi.1005457

Lu Y, Quan C, Chen H, Bo X, Zhang C (2017) 3DSNP: a database for linking human noncoding SNPs to their three-dimensional interacting genes. Nucleic Acids Res 45:D643–D649. https://doi.org/10.1093/nar/gkw1022

MacArthur J, Bowler E, Cerezo M, Gil L, Hall P, Hastings E et al (2017) The new NHGRI-EBI catalog of published genome-wide association studies (GWAS catalog). Nucleic Acids Res 45: D896–D901. https://doi.org/10.1093/nar/gkw1133

Marusina K, Welsch DJ, Rose L, Brock D, Bahr N (2011) The CTSA pharmaceutical assets portal - a public-private partnership model for drug repositioning. Drug Discov Today Ther Strateg 8: 77–83. https://doi.org/10.1016/j.ddstr.2011.06.006

MedDRA (n.d.). https://www.meddra.org/. Accessed 12 Dec 2019

von Mering C, Huynen M, Jaeggi D, Schmidt S, Bork P, Snel B (2003) STRING: a database of predicted functional associations between proteins. Nucleic Acids Res 31:258–261

Mohd M (2011) Development of search engines using Lucene: an experience. Procedia Soc Behav Sci 18:282–286. https://doi.org/10.1016/j.sbspro.2011.05.040

Moreira GV, Azevedo FF, Ribeiro LM, Santos A, Guadagnini D, Gama P et al (2018) Liraglutide modulates gut microbiota and reduces NAFLD in obese mice. J Nutr Biochem 62:143–154. https://doi.org/10.1016/j.jnutbio.2018.07.009

Moridi M, Ghadirinia M, Sharifi-Zarchi A, Zare-Mirakabad F (2019) The assessment of efficient representation of drug features using deep learning for drug repositioning. BMC Bioinform 20: 1–11. https://doi.org/10.1186/s12859-019-3165-y

Muchtaridi M, Syahidah HN, Subarnas A, Yusuf M, Bryant SD, Langer T (2017) Molecular docking and 3D-pharmacophore modeling to study the interactions of chalcone derivatives with estrogen receptor alpha. Pharmaceuticals 10:1–12. https://doi.org/10.3390/ph10040081

Naz F, Mashkoor M, Sharma P, Haque MA, Kapil A, Kumar M et al (2020) Drug repurposing approach to target FtsZ cell division protein from salmonella Typhi. Int J Biol Macromol 159: 1073–1083. https://doi.org/10.1016/j.ijbiomac.2020.05.063

NCATS. NCATS Announces Funding Opportunities to Repurpose Drug Candidates from Industry | National Center for Advancing Translational Sciences 2014. https://ncats.nih.gov/news/releases/2014/ntu-funding-2014. Accessed 24 July 2018.

NCATS 2017 Bench-to-Clinic Projects | National Center for Advancing Translational Sciences 2017. https://ncats.nih.gov/ntu/projects/2017. Accessed 24 July 2018.

NIH (n.d.) Home - ClinicalTrials.gov. https://clinicaltrials.gov/ct2/home. Accessed 24 July 2018.

Nowak-Sliwinska P, Scapozza L, Altaba AR, i. (2019) Drug repurposing in oncology: compounds, pathways, phenotypes and computational approaches for colorectal cancer. Biochim Biophys Acta – Rev Cancer 1871:434–454. https://doi.org/10.1016/j.bbcan.2019.04.005

Ooijevaar RE, Terveer EM, Verspaget HW, Kuijper EJ, Keller JJ (2019) Clinical application and potential of fecal microbiota transplantation. Annu Rev Med 70:335–351. https://doi.org/10.1146/annurev-med-111717-122956

Pagliari C, Detmer D, Singleton P (2007) Potential of electronic personal health records. Br Med J 335:330–333. https://doi.org/10.1136/bmj.39279.482963.ad

Pallares-Méndez R, Aguilar-Salinas CA, Cruz-Bautista I, del Bosque-Plata L (2016) Metabolomics in diabetes, a review. Ann Med 48:89–102. https://doi.org/10.3109/07853890.2015.1137630

Pathak N, Lai ML, Chen WY, Hsieh BW, Yu GY, Yang JM (2017) Pharmacophore anchor models of flaviviral NS3 proteases lead to drug repurposing for DENV infection. BMC Bioinform 18:548. https://doi.org/10.1186/s12859-017-1957-5

Peri S, Navarro JD, Kristiansen TZ, Amanchy R, Surendranath V, Muthusamy B et al (2004) Human protein reference database as a discovery resource for proteomics. Nucleic Acids Res 32:D497–D501. https://doi.org/10.1093/nar/gkh070

Prachayasittikul V, Prathipati P, Pratiwi R, Phanus-umporn C, Malik AA, Schaduangrat N et al (2017) Exploring the epigenetic drug discovery landscape. Expert Opin Drug Discov 12:345–362. https://doi.org/10.1080/17460441.2017.1295954

Pritchard J-LE, O'Mara TA, Glubb DM (2017) Enhancing the promise of drug repositioning through genetics. Front Pharmacol 8:896. https://doi.org/10.3389/fphar.2017.00896

Pulley JM, Rhoads JP, Jerome RN, Challa AP, Erreger KB, Joly MM et al (2020) Using what we already have: uncovering new drug repurposing strategies in existing omics data. Annu Rev Pharmacol Toxicol 60:333–352. https://doi.org/10.1146/annurev-pharmtox-010919-023537

Pushpakom S, Iorio F, Eyers PA, Escott KJ, Hopper S, Wells A et al (2018) Drug repurposing: Progress, challenges and recommendations. Nat Rev Drug Discov 18:41–58. https://doi.org/10.1038/nrd.2018.168

Ramasamy A, Trabzuni D, Guelfi S, Varghese V, Smith C, Walker R et al (2014) Genetic variability in the regulation of gene expression in ten regions of the human brain. Nat Neurosci 17:1418–1428. https://doi.org/10.1038/nn.3801

Ray SD (ed) (2020) Side effects of drugs annual | a worldwide yearly survey of new data in adverse drug reactions, vol 42. Elsevier

Raynal NJM, Da Costa EM, Lee JT, Gharibyan V, Ahmed S, Zhang H et al (2017) Repositioning FDA-approved drugs in combination with epigenetic drugs to reprogram colon cancer epigenome. Mol Cancer Ther 16:397–407. https://doi.org/10.1158/1535-7163.MCT-16-0588

RFA-TR-20-003: Urgent Phase I/II Clinical Trials to Repurpose Existing Therapeutic Agents to Treat COVID-19 Sequelae (U01 Clinical Trial Required) n.d.. https://grants.nih.gov/grants/guide/rfa-files/rfa-tr-20-003.html. Accessed 1 Apr 2021.

Roden DM, Pulley JM, Basford MA, Bernard GR, Clayton EW, Balser JR et al (2008) Development of a large-scale de-identified DNA biobank to enable personalized medicine. Clin Pharmacol Ther 84:362–369. https://doi.org/10.1038/clpt.2008.89

Shannon P, Markiel A, Ozier O, Baliga NS, Wang JT, Ramage D et al (2003) Cytoscape: a software environment for integrated models of biomolecular interaction networks. Genome Res 13:2498–2504. https://doi.org/10.1101/gr.1239303.metabolite

Sharlow ER (2016) Revisiting repurposing. Assay Drug Dev Technol 14:554–556. https://doi.org/10.1089/adt.2016.766

Sloan CA, Chan ET, Davidson JM, Malladi VS, Strattan JS, Hitz BC et al (2016) ENCODE data at the ENCODE portal. Nucleic Acids Res 44:D726–D732. https://doi.org/10.1093/nar/gkv1160

So HC, Chau CKL, Chiu WT, Ho KS, Lo CP, Yim SHY et al (2017) Analysis of genome-wide association data highlights candidates for drug repositioning in psychiatry. Nat Neurosci 20: 1342–1349. https://doi.org/10.1038/nn.4618

Song Y, Luo L, Wang K (2020) Off-target identification by chemical proteomics for the understanding of drug side effects. Expert Rev Proteomics 17:695–697. https://doi.org/10.1080/14789450.2020.1873134

Sontag ED (1998) Mathematical Control Theory - Deterministic Finite Dimensional Systems, vol 6, 2nd edn. Springer, New York

Stark C, Breitkreutz B-J, Reguly T, Boucher L, Breitkreutz A, Tyers M (2006) BioGRID: a general repository for interaction datasets. Nucleic Acids Res 34:D535–D539. https://doi.org/10.1093/nar/gkj109

Stirm L, Huypens P, Sass S, Batra R, Fritsche L, Brucker S et al (2018) Maternal whole blood cell miRNA-340 is elevated in gestational diabetes and inversely regulated by glucose and insulin. Sci Rep 8:8. https://doi.org/10.1038/s41598-018-19200-9

Sullivan PF (2010) The psychiatric GWAS consortium: big science comes to psychiatry. Neuron 68:182–186. https://doi.org/10.1016/j.neuron.2010.10.003

Tan J, Cang S, Ma Y, Petrillo RL, Liu D (2010) Novel histone deacetylase inhibitors in clinical trials as anti-cancer agents. J Hematol Oncol 3:3. https://doi.org/10.1186/1756-8722-3-5

Tang PC, Ash JS, Bates DW, Overhage JM, Sands DZ (2006) Personal health records: definitions, benefits, and strategies for overcoming barriers to adoption. J Am Med Informatics Assoc 13: 121–126. https://doi.org/10.1197/jamia.M2025

Taroncher-Oldenburg G, Jones S, Blaser M, Bonneau R, Christey P, Clemente JC et al (2018) Translating microbiome futures. Nat Biotechnol 36:1037–1042. https://doi.org/10.1038/nbt.4287

Teng L, He B, Wang J, Tan K (2015) 4DGenome: a comprehensive database of chromatin interactions. Bioinformatics 31:2560–2564. https://doi.org/10.1093/bioinformatics/btv158

Than Win K, Cooper J (2004) Information age, electronic health record and australia healthcare. Int J Comput Internet Manag 12(14):121

Thorn CF, Klein TE, Altman RB (2013) PharmGKB: the pharmacogenomics Knowledge Base. Methods Mol Biol 1015:311–320. https://doi.org/10.1007/978-1-62703-435-7_20

Toh TS, Dondelinger F, Wang D (2019) Looking beyond the hype: applied AI and machine learning in translational medicine. EBioMedicine 47:607–615. https://doi.org/10.1016/j.ebiom.2019.08.027

Tomczak K, Czerwińska P, Wiznerowicz M (2015) The cancer genome atlas (TCGA): An immeasurable source of knowledge. Wspolczesna Onkol 1A:A68–A77. https://doi.org/10.5114/wo.2014.47136

Topol EJ (2019) High-performance medicine: the convergence of human and artificial intelligence. Nat Med 25:44–56. https://doi.org/10.1038/s41591-018-0300-7

Velez G, Bassuk AG, Colgan D, Tsang SH, Mahajan VB (2017) Therapeutic drug repositioning using personalized proteomics of liquid biopsies. JCI Insight 2:2. https://doi.org/10.1172/jci.insight.97818

Vlachos IS, Paraskevopoulou MD, Karagkouni D, Georgakilas G, Vergoulis T, Kanellos I et al (2015) DIANA-TarBase v7.0: indexing more than half a million experimentally supported miRNA:mRNA interactions. Nucleic Acids Res 43:D153–D159. https://doi.org/10.1093/nar/gku1215

Wan F, Hong L, Xiao A, Jiang T, Zeng J (2019) NeoDTI: neural integration of neighbor information from a heterogeneous network for discovering new drug-target interactions. Bioinformatics 35:104–111. https://doi.org/10.1093/bioinformatics/bty543

Wang Y, Song F, Zhang B, Zhang L, Xu J, Kuang D et al (2018) The 3D genome browser: a web-based browser for visualizing 3D genome organization and long-range chromatin interactions. Genome Biol 19:151. https://doi.org/10.1186/s13059-018-1519-9

Wei WQ, Cronin RM, Xu H, Lasko TA, Bastarache L, Denny JC (2013) Development and evaluation of an ensemble resource linking medications to their indications. J Am Med Inform Assoc 20:954–961. https://doi.org/10.1136/amiajnl-2012-001431

Welter D, MacArthur J, Morales J, Burdett T, Hall P, Junkins H et al (2014) The NHGRI GWAS catalog, a curated resource of SNP-trait associations. Nucleic Acids Res 42:D1001–D1006. https://doi.org/10.1093/nar/gkt1229

Westra HJ, Peters MJ, Esko T, Yaghootkar H, Schurmann C, Kettunen J et al (2013) Systematic identification of trans eQTLs as putative drivers of known disease associations. Nat Genet 45: 1238–1243. https://doi.org/10.1038/ng.2756

Wishart D, Arndt D, Pon A, Sajed T, Guo AC, Djoumbou Y et al (2015) T3DB: the toxic exposome database. Nucleic Acids Res 43:D928–D934. https://doi.org/10.1093/nar/gku1004

Wishart DS (2016) Emerging applications of metabolomics in drug discovery and precision medicine. Nat Rev Drug Discov 15:473–484. https://doi.org/10.1038/nrd.2016.32

Wishart DS, Feunang YD, Guo AC, Lo EJ, Marcu A, Grant R et al (2018) DrugBank 5. 0: a major update to the DrugBank database for 2018. Nucl Acids Res 46:1074–1082. https://doi.org/10.1093/nar/gkx1037

Wishart DS, Feunang YD, Marcu A, Guo AC, Liang K, Vázquez-Fresno R et al (2018) HMDB 4.0: the human metabolome database for 2018. Nucl Acids Res 46:D608–D617. https://doi.org/10.1093/nar/gkx1089

Wu C, Gudivada RC, Aronow BJ, Jegga AG (2013) Computational drug repositioning through heterogeneous network clustering. BMC Syst Biol 7:S6. https://doi.org/10.1186/1752-0509-7-S5-S6

Xie X, Ma W, Songyang Z, Luo Z, Huang J, Dai Z et al (2016) CCSI: a database providing chromatin-chromatin spatial interaction information. Database 2016. https://doi.org/10.1093/database/bav124

Xu H, Li J, Jiang X, Chen Q (2020) Electronic health Records for Drug Repurposing: current status, challenges, and future directions. Clin Pharmacol Ther 107:712–714. https://doi.org/10.1002/cpt.1769

Xu R, Wang QQ (2015) Combining automatic table classification and relationship extraction in extracting anticancer drug-side effect pairs from full-text articles. J Biomed Inform 53:128–135. https://doi.org/10.1016/j.jbi.2014.10.002

Xuan P, Cao Y, Zhang T, Wang X, Pan S, Shen T (2019) Drug repositioning through integration of prior knowledge and projections of drugs and diseases. Bioinformatics 35:4108–4119. https://doi.org/10.1093/bioinformatics/btz182

Yang J, Zhang D, Liu L, Li G, Cai Y, Zhang Y et al (2020) Computational drug repositioning based on the relationships between substructure–indication. Brief Bioinform 22. https://doi.org/10.1093/bib/bbaa348

Yang JM, Chen CC (2004) GEMDOCK: a generic evolutionary method for molecular docking. Proteins Struct Funct Genet 55:288–304. https://doi.org/10.1002/prot.20035

Yang L, Agarwal P (2011) Systematic drug repositioning based on clinical side-effects. PLoS One 6:6. https://doi.org/10.1371/journal.pone.0028025

Yang SY (2010) Pharmacophore modeling and applications in drug discovery: challenges and recent advances. Drug Discov Today 15:444–450. https://doi.org/10.1016/j.drudis.2010.03.013

Yates JR (2019) Recent technical advances in proteomics. F1000Research 8:10.12688/f1000research.16987.1

Ye H, Liu Q, Wei J (2014) Construction of drug network based on side effects and its application for drug repositioning. PLoS One 9:e87864. https://doi.org/10.1371/journal.pone.0087864

Yeung PK (2018) Metabolomics and biomarkers for drug discovery. Meta 8:8. https://doi.org/10.3390/metabo8010011

Yi Y, Zhao Y, Li C, Zhang L, Huang H, Li Y et al (2017) RAID v2.0: An updated resource of RNA-associated interactions across organisms. Nucleic Acids Res 45:D115–D118. https://doi.org/10.1093/nar/gkw1052

Yu CH, Pal LR, Moult J (2016) Consensus genome-wide expression quantitative trait loci and their relationship with human complex trait disease. Omi A J Integr Biol 20:400–414. https://doi.org/10.1089/omi.2016.0063

Yuan Z, Zhao C, Di Z, Wang WX, Lai YC (2013) Exact controllability of complex networks. Nat Commun 4:1–9. https://doi.org/10.1038/ncomms3447

Zeng X, Zhu S, Liu X, Zhou Y, Nussinov R, Cheng F (2019) DeepDR: a network-based deep learning approach to in silico drug repositioning. Bioinformatics 35:5191–5198. https://doi.org/10.1093/bioinformatics/btz418

Zhang J, Jiang K, Lv L, Wang H, Shen Z, Gao Z et al (2015) Use of genome-wide association studies for cancer research and drug repositioning. PLoS One 10:e0116477. https://doi.org/10.1371/journal.pone.0116477

Zhang M, Schmitt-Ulms G, Sato C, Xi Z, Zhang Y, Zhou Y et al (2016) Drug repositioning for Alzheimer's disease based on systematic "omics" data mining. PLoS One 11:1–15. https://doi.org/10.1371/journal.pone.0168812

Zhang W, Yue X, Lin W, Wu W, Liu R, Huang F et al (2018) Predicting drug-disease associations by using similarity constrained matrix factorization. BMC Bioinformatics 19. https://doi.org/10.1186/s12859-018-2220-4

Zhao H, Jin G, Cui K, Ren D, Liu T, Chen P et al (2013) Novel modeling of cancer cell signaling pathways enables systematic drug repositioning for distinct breast cancer metastases. Cancer Res 73:6149–6163. https://doi.org/10.1158/0008-5472.CAN-12-4617

Zhou M, Wang QQ, Zheng C, John Rush A, Volkow ND, Xu R (2021) Drug repurposing for opioid use disorders: integration of computational prediction, clinical corroboration, and mechanism of action analyses. Mol Psychiatry 26:5286. https://doi.org/10.1038/s41380-020-01011-y

Zhu F, Shi Z, Qin C, Tao L, Liu X, Xu F et al (2012) Therapeutic target database update 2012: a resource for facilitating target-oriented drug discovery. Nucleic Acids Res 40:D1128–D1136. https://doi.org/10.1093/nar/gkr797

Small Molecules as Promising Tool for Targeted Cancer Therapies: An Overview of the Twenty-First Century

Saima Shakil Malik and Nosheen Masood

Abstract It has been an established fact now from more than 50 years that genetic differences among individuals are responsible for inter-individual variations as a response to most of the generally used drugs. Pharmacogenetics is the scientific discipline that deals with the genetic affects and their response to certain drugs (Cascorbi 2018). It is an emerging field comprehending the systematic recognition of the genes, their functioning, products and intra-individual differences in function and expression over time that may lead to the advent of novel therapeutics (Nishant et al. 2012). Toxicity of therapeutic agents and lack of effectiveness curb the survival period and quality of life among cancer patients (Pirisinu et al. 2020). These are the crucial challenges in cancer treatment as majority of the anticancer drugs have limited therapeutic index effective against few individuals causing serious toxicities leading to death in some cases (Kummar et al. 2006). However, targeted therapeutics are with far greater specificity compared to cytotoxic drugs, still linked through substantial hostile event profiles from both on- and off-target effects (Masood and Malik 2020). Therefore, it is the need of the hour to develop and practise novel therapeutic options with better treatment outcomes, prolonged survival rates, and lessened toxicities. In these circumstances, the basic goal of personalized medicine is to tailor the treatment decisions on an individual basis so that therapeutic responses can be improved with minimal adverse effects (Hassan et al. 2020). Pharmacogenomics might help in accomplishing these targets by identifying molecular

---

Saima Shakil Malik and Nosheen Masood contributed equally with all other contributors.

S. S. Malik (✉)
Centre of Biotechnology and Genomic Medicine, Medical College of Georgia, Augusta University, Augusta, GA, USA

N. Masood (✉)
Department of Biotechnology, Fatima Jinnah Women University, The Mall Rawalpindi, Rawalpindi, Pakistan
e-mail: dr.nosheen@fjwu.edu.pk

V. S. Rajput, A. Runthala (eds.), *Drugs and a Methodological Compendium*,
https://doi.org/10.1007/978-981-19-7952-1_10

markers that would be capable of predicting drug response and serving as the basis for patient stratification (Kalinin et al. 2018; Udagawa and Zembutsu 2020).

1 Introduction

It has been an established fact now from more than 50 years that genetic differences among individuals are responsible for inter-individual variations as a response to most of the generally used drugs. Pharmacogenetics is the scientific discipline that deals with the genetic affects and their response to certain drugs (Cascorbi 2018). It is an emerging field comprehending the systematic recognition of the genes, their functioning, products and intra-individual differences in function and expression over time that may lead to the advent of novel therapeutics (Nishant et al. 2012). Toxicity of therapeutic agents and lack of effectiveness curb the survival period and quality of life among cancer patients (Pirisinu et al. 2020). These are the crucial challenges in cancer treatment as majority of the anticancer drugs have limited therapeutic index effective against few individuals causing serious toxicities leading to death in some cases (Kummar et al. 2006). However, targeted therapeutics are with far greater specificity compared to cytotoxic drugs, still linked through substantial hostile event profiles from both on- and off-target effects (Masood and Malik 2020). Therefore, it is the need of the hour to develop and practise novel therapeutic options with better treatment outcomes, prolonged survival rates, and lessened toxicities. In these circumstances, the basic goal of personalized medicine is to tailor the treatment decisions on an individual basis so that therapeutic responses can be improved with minimal adverse effects (Hassan et al. 2020). Pharmacogenomics might help in accomplishing these targets by identifying molecular markers that would be capable of predicting drug response and serving as the basis for patient stratification (Kalinin et al. 2018; Udagawa and Zembutsu 2020).

It is imperative to mention about the latest developments in genome-wide tools that have noticeably improved the screening practices to recognize predictive biomarkers. Therefore, high-throughput sequencing techniques can present a broad genome-wide viewpoint of various molecular alterations among cancer cases and cancer cells in a cost-effective manner (Masood and Malik 2020). Such important information leads to a plethora of potential novel diagnostic and prognostic biomarkers, along with prospective druggable targets which would have potent future clinico-pathological applications (Fig. 1). Predictive biomarkers may derive from tumour cells and the host, referred as somatic and germline variations, respectively (Lauschke et al. 2018).

Cancer is spreading enormously since the last few decades, and with the advancement in cancer aetiology, many drugs have been designed to treat this disease with respect to stage and tumour type (Malik et al. 2019a). However, drug resistance as well as recurrence of disease had been frequently reported, and researchers brought up the concept of personalized medicine to overcome these issues. Due to

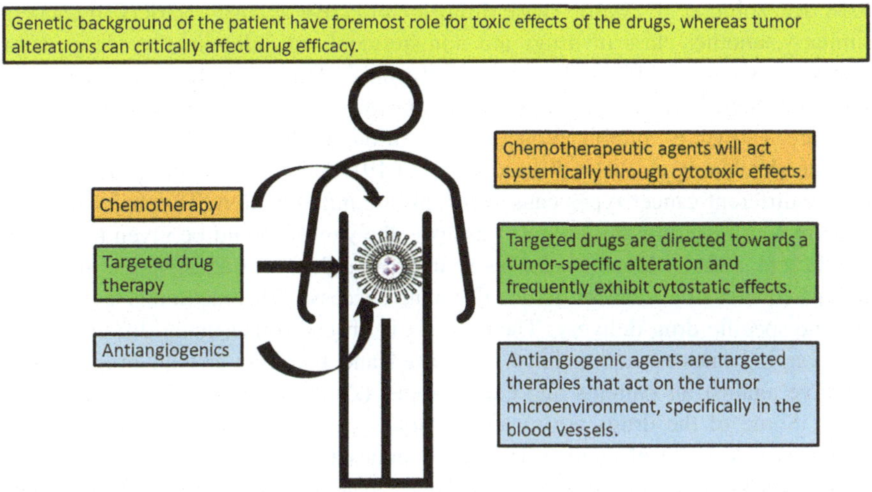

Fig. 1 Potential anticancer treatment options with their mode of action

characterization of metabolome, proteome, and epigenome, studies with human samples along with completion of human genome project have opened up a new field of pharmacogenomics. Now science has been moving a step ahead towards the availability of tailored medicine for a particular patient. The term has been the outcome of a single patient's molecular profiling causing administration of respective medicines. Researches have taken it to another level and start cancer screening together with early-stage diagnosis with the help of nanobiotechnology, being more effective and which may lead towards progression-free survival (Malik et al. 2019b; Mallick 2019).

In this chapter, we are going to discuss the role of small molecules as potential tools for targeted cancer therapies. There are a hundred of drugs used to treat cancer. However, each group of drugs has a different potential and molecular basis towards cancer treatment. Generally, cancer therapeutic drugs not only destroy cancer cells but also damage the normal ones too; however tailored medicine will be focused only on the cancer cell molecular profiling and immune response (Lee et al. 2018). Some drugs such as doxorubicin, daunorubicin, epirubicin, capecitabine, carmofur, etc. are antineoplastic and are given to breast, colorectal, leukaemia, and ovarian cancer patients capable of halting the cancer cell progression (Wang et al. 2017b); some are antimitotic drugs as cancer cells have abnormally higher rate of cell division (Olziersky and Labidi-Galy 2017). Two major categories of antimitotic drugs are vinca alkaloids (e.g. eribulin, vintafolide) and taxanes (e.g. paclitaxel) which both act at microtubule stabilization and are effective against pancreatic and prostatic cancer. Histone deacetylase inhibitors are another option for cancer treatment as they lead towards normal gene expression that may be changed due to mutations in specific genes and in turn inhibit multiplication of cancer cells by repressing transcription causing higher chromatin condensation (Tsilimigras et al.

2018; Ruzzolini et al. 2020). These drugs are effective against a variety of cancers. Similarly, another class of drugs are non-steroidal anti-inflammatory drugs commonly known as NSAIDs which are majorly anti-inflammatory but also involved in treating cancer. However, there is much controversy regarding its side effects, but they are known to inhibit prostaglandin through blockage of cyclooxygenase enzyme (Vallée et al. 2019; Tomić et al. 2019). This class of drugs is effective against different cancer types caused by chronic inflammation. The major problem faced in administering any group of drugs is the way it should be given to patients (Bindu et al. 2020). Liposomes are also currently studied for transferring of medicine to targeted sites in the cancer tissue. The main success of this method is less toxicity and site-specific drug delivery. Therefore, it is observed that many different drugs with varying sites of action have been used and therefore not all drugs can be effective against all patients and cancer types (Zaimy et al. 2017; Roovers et al. 2019). Some of the drugs are meant to target specific genetic mutations or their protein expression levels, and every individual has different mutational spectrum, so its not possible to use one single drug for every patient based on genetic differences. Therefore, personalized medicine is the hour of need and gathering researcher's attention around the globe (Sharma et al. 2019). There is a need of time to move these small drug molecules from the literature to the market with information about genome of individuals.

2 Cytotoxic Drugs

Advancement in the cytotoxic agents has revolutionized the field of cancer therapeutics in the last century. They have the potential to cure various carcinomas such as Hodgkin's disease, gestational trophoblastic disease, childhood acute leukaemia, non-Hodgkin's lymphoma, and germ cell tumours (Wijaya et al. 2020; Nikolaou et al. 2018). Cytotoxic drugs along with adjuvant therapy for a variety of cancer types presented better survival outcomes in addition to the favourable effects achieved through surgical treatment alone (Abdallah et al. 2021). They have paved a new pathway to deliver explicit tumour management with better quality of life by relieving major disease symptoms among patients with metastasis or recurrence (Diana et al. 2020). However, cytotoxic therapy developments have faced many challenges due to its narrow therapeutic index and the fact that it does not only damage cancer cells but also poses harm to normal body cells (Suh et al. 2020; Falzone et al. 2018). Cytotoxic drugs come up with certain limitations particularly for advanced stage tumours where they are unable to deliver expected benefits because of their potential adverse effects (Gillessen et al. 2018). Resistance to many cytotoxic drugs is also another limitation towards their efficacy. Therefore, it is the need of the hour to develop safe, efficient, and most importantly convenient drugs as most of the drugs are administered intravenously and sometime necessitate continuous intravenous infusion which entails hospitalization and higher costs (Ismael et al. 2008).

Nab-paclitaxel (nanoparticle albumin-bound paclitaxel) is a novel drug formed by the homogenization of paclitaxel with a nanoparticle and human serum albumin resulting in a colloidal suspension (Adrianzen Herrera et al. 2019). Nab-paclitaxel is confirmed to have reduced toxicity and greater efficiency than paclitaxel in various clinical trials (Macarulla et al. 2019; Lee et al. 2020).

Another cytotoxic drug, docosahexaenoic acid-paclitaxel (DHA-paclitaxel/ Taxoprexin), is prepared by conjugating a natural fatty acid DHA to paclitaxel through a covalent linkage which is meant to work (Sun et al. 2017; Fattahi et al. 2020) as a prodrug by specifically accumulating in the tumour tissue. This conjugate is also the most effective than paclitaxel and extensively studied in almost all tumour types via phase I and II clinical trials (Bernabeu et al. 2017; Liu et al. 2019; Eltweri et al. 2019).

Paclitaxel poliglumex is an additional novel therapeutic agent formed by modulating the common paclitaxel (Wang et al. 2017a) with greater retention and tumour permeability, enhanced solubility of hydrophobic drugs (Liu et al. 2019), and minimal exposure of the healthy tissue to drug along with evasion of multidrug resistance efflux pumps through pinocytotic tumoural acceptance (Masood and Malik 2020; Pillai 2019).

Epothilones signify another novel class of cytotoxic drug function by inhibiting the cell cycle and significantly disrupting the process of cell division (Mukhtar et al. 2014; Sacco and Gridelli 2017). Epothilones are capable of possibly evading the underlying mechanisms of multidrug resistance phenotype development, generally coupled with recurrent cancers (Gasch et al. 2017).

Kinesin spindle proteins being motor proteins (Huszar et al. 2009) performed crucial function in mitotic spindle formation. Ispinesib, used as kinesin spindle protein's inhibitor, interferes with bipolar spindle formation leading to its consumption as potential anticancer agent (Park et al. 2017; Jungwirth et al. 2021).

Camptothecins comprises of pentacyclic ring structures which have shown interaction with topoisomerase I enzyme, with specificity for the S-phase of the cell cycle (Hu et al. 2018). Camptothecin along with its analogues such as irinotecan and topotecan possesses anticancer potential against a variety of solid tumours (Ruan et al. 2021) and numerous haematological malignancies as well (Wahid and Bano 2014). Gimatecan is an orally administered, lipophilic-adapted camptothecin analogue (Zou et al. 2018) having a beneficial therapeutic index with verified antitumoural activity in lung, endometrial, gastric, and breast carcinomas (Chen et al. 2017; Yuan et al. 2018; Qi et al. 2017; Heitz et al. 2014).

Pemetrexed is an antifolate agent reported to be active in numerous malignancies such as lung cancer and mesothelioma (Lievense et al. 2017). It works by inhibiting thymidylate synthase enzyme leading to reduced thymidine levels which is the basic requirement for pyrimidine synthesis disrupting the process of DNA synthesis (García-Fernández et al. 2020). Pemetrexed is also involved in the inhibition of glycinamide ribonucleotide formyl transferase and dihydrofolate reductase, whereas the former one is a folate-dependent enzyme responsible for purine synthesis (Wallace-Povirk et al. 2021).

Despite explosive and emerging increase in the knowledge of molecular events triggering cancer diagnosis and progression, and the subsequent growth of molecular targeted anticancer agents, treatment of various malignancies is still dependent primarily on cytotoxic agents. Therefore, there is a huge need for more effective and safer anticancer therapeutics that exclusively target specific cancer cells which can overcome the resistance pathways. Development of new taxanes is much efficacious in decreasing the normal tissue exposure to drug, circumventing hypersensitivity reactions due to conventional paclitaxel consumption and evading the multidrug resistance efflux pumps. Innovative cytotoxic agents have considerably improved the therapeutic portfolio available for cancer patients, leading to greater efficacy and reduced adverse outcomes. Various novel cytotoxic agents are in the process of phase III clinical trials and are believed to bring advancement in cancer therapeutics.

3 Antineoplastic Drugs

Antineoplastic drugs come up with two different classes from different generations of explicitly conventional molecules and drugs from targeted therapy. The beginning of the twentieth century has laid the foundation of conventional chemotherapy with the expansion of chemical weapons. One of the earliest members of antineoplastic drug family, yet administered now, is established with the help of certain molecules resembling in structure to mustard gas applied on the battlefields of World War I. One of the major side effects observed due to this gas was myelosuppressive action leading to the foundation of the very first antineoplastic drugs against leukaemia (Guichard et al. 2017; Aristizabal-Pachon and Castillo 2020). This success has initiated the launch of various national and international programmes aiming the development of novel molecules as cancer therapeutics results in an uncontrolled race for the discovery of novel anticancer drugs such as pyrimidine/purine analogues, antifolate molecules, and antibiotics. Most of the present-day used conventional antineoplastic drugs were established in that era (Simon et al. 2020). These drugs are popular due to their anticancer activity because of their strong chemical nature but pose detrimental health effects too. Therefore, a continuous struggle is in the process of exploring less toxic, efficient, and safer options. Researchers have divided antineoplastic drugs into three broader categories depending upon their action (Sauter and Gillingham 2020) on deoxyribonucleic acid (DNA):

1. Antimetabolites: molecules targeting DNA synthesis (MAROSI and FÜGGER 2020).
2. Topoisomerase inhibitors, intercalating and alkylating agents: molecules directly targeting DNA (Dehshahri et al. 2020).
3. Antitubilin agents: molecules targeting cell cycle precisely mitosis (Naaz et al. 2019).

4 Tyrosine Kinase Inhibitors

Tyrosine kinase inhibitors are specifically responsible for the inhibition of tyrosine kinase enzymes. They are low-molecular-weight compounds catalysing phosphate transfer to proteins from adenosine triphosphate (Qin et al. 2019). It plays a key role in cell cycle regulation such as differentiation, proliferation, migration and survival. Tyrosine kinase inhibitors induces the diffusion of molecules through the cell membrane by reacting with enzymes of cytoplasmic or intracellular membrane (Pottier et al. 2020). Clinical oncology has explored the potential of tyrosine kinase inhibitors when they are triggered by mutation and lead to tumour progression. In 1990's first tyrosine kinase inhibitor, imatinib was discovered as an anticancer agent, derived from drug development based on BCR-ABL protein downregulation, and finds its potential in chronic myeloid leukaemia treatment (Liao et al. 2019). Due to resistance development with the passage of time, two more analogues nilotinib and dasatinib have taken the attention of pharmaceutical market (Ongoren et al. 2018; Abu Rmilah et al. 2020). Later, erlotinib (Ding et al. 2017) and lapatinib (Xuhong et al. 2019), targeting epidermal growth factor receptors, were introduced to treat lung (Zhong et al. 2019) and breast cancers (Widatalla et al. 2019), respectively. Sunitinib is another popular member of this family (Paludetto et al. 2018) playing an important role in renal, lung, and gastrointestinal cancers, by acting as vascular endothelial growth factor receptors (Jiao et al. 2018; Murtuza et al. 2019).

5 Histone Inhibitors

Addition of acetyl moiety to lysine amino group from acetyl coenzyme A is known as lysine acetylation, mostly lysine deacetylase and lysine acetylase, and these are commonly known as histone acetylases and histone deacetylases. They reverse acetylation of many cytoplasmic and nuclear proteins. Histones are positively charged proteins attached to negatively charged DNA in a nucleosome and are responsible for condensation of chromatin in a supercoiled conformation. Lysine and arginine residues of histones are mostly methylated, but acetylation is more dynamic in its modulations (Wysocka et al. 2006). Remodelling of nucleosome and transcription activation is associated with acetylation, and histone deacetylation is known to control repression of transcription by condensation of chromatin. Acetylation causes neutralization of positively charged histone and allows the transcription apparatus to attach at the promotor site and change the conformation of chromatin to the relaxed state. Therefore, histone acetylation is directly related to gene expression; the more the gene is active, the more the acetylation of histones. Histone acetyltransferases (HATs) and histone deacetylases (HDACs) are two enzymes involved in this process of acetylation. Broadly HDAC inhibitors are divided into five classes, majorly having reversible or irreversible action (Makkar et al. 2020):

1. Small-molecular-weight carbohydrates: This group includes both aliphatic and aromatic agents, and it encompasses the set of HDAC inhibitors, majorly studied till date through in vitro and in vivo experiments.
2. Hydroxamic acids: Examples of such agents are TSA (trichostatin A) and SAHA (suberoylanilide hydroxamic acid), having hydroxamic acid, reacting with zinc activity, and removing water molecules that are nucleophilic at the active site of the enzyme.
3. Epoxyketone-based HDAC inhibitors.
4. Benzamides: They have a good reactive ability even at micromolar concentrations.
5. Cyclic peptides.

When a cell is transformed into a malignant form, many morphological and cytoskeletal frameworks are altered. Loss of actin stress fibres due to this transformation can be reversed HDAC inhibitors. Oncogenes such as v-ras, v-src and v-sis become activated in many types of cancers such as leukaemia, bladder cancer, colorectal cancer, breast cancer, and neuroblastoma. Gelsolin is a key player in organizing actin cytoskeleton, and its levels increase due to HDAC inhibitors and result in reassembling and reorganizing transformed phenotype (Wurzer et al. 2019). It was found that if antigelsolin antibodies are injected, and it will suppress HDAC inhibitors' morphological effects.

Mainly HDAC inhibitors arrest the cell cycle and ultimately control proliferation. The reason for such a response might be its action upstream of that Rb/E2F and Myc/Mad pathways in the G1 phase of cell cycle. Chromatin of p21 is hyperacetylated due to HDAC inhibitors (José-Enériz et al. 2019). Although the precise mechanism of activation is unclear, it is certainly p53-independent and requires Sp-1 binding sites within the p21 gene promoter. HDAC inhibition effects few genes not all with unknown selectivity. Many non-histone proteins also play role in acetylation such as p53, a tumour suppressor gene. The expressional increase of p53 reduces T cell factor production whereas co-activator ACTR gets involved in nuclear receptor signalling. HDAC inhibition causes apoptosis by relaxing chromatin and allowing endonuclease enzymes to access DNA for apoptosis (Nikolova et al. 2017). In vivo it was found that millimolar concentrations of weak HDAC inhibitors such as sodium phenylacetate, sodium phenylbutyrate and sodium butyrate are efficient in leukaemias and solid tumours at a relatively lower rate, whereas strong HDAC inhibitors such as depsipeptide reduce APL growth in immunodeficient mice. It was reported that All-trans retinoic acid (ATRA) and depsipeptide synergistically control antitumor activity in mice models (Ercolano et al. 2019).

6 Antimetabolites

It's impossible to treat cancer with any specific single drug; therefore a combination of drugs has been used, each having a different activity. In case of ALL (acute lymphoblastic leukaemia), the survival rate is 80%, if the combination of chemotherapeutic drugs is used (Jabbour et al. 2018). But in many cases, recurrence rate is very high that leads to development of new drugs, acting on a particular pathway, involved in cancer incidence. Most of the combination treatment drugs used in cancer have an antimetabolite along with anticancer agents. Some modulators such as purine and pyrimidine intermediates are also used. Selection of different drugs and antimetabolites is very important for effective treatment. Antimetabolites were introduced in the 1950s, and since then, many new drugs in this class have been discovered depending upon two main factors such as efficiency and no toxicity (Gilad et al. 2021), while three widely used metabolites are:

1. Methotrexate.
2. 5-Fluorouracil (5FU)
3. Thiopurines.

Most commonly 5-fluorouracil (5FU) is used in combination with leucovorin, as it showed more toxicity in vitro in cancer cells, and it is now commonly used to treat colon cancer. The reports on such combinations are contradictory, and in vitro results are promising, but survival rates seem to be purported. In addition to leucovorin, interferons were also considered beneficial in the 1990s, but now they are not prescribed anymore due to their complexities. It is shown that PALA (N-phosphonacetyl-L-aspartate) increases the 5FU activity by reducing UTP levels (Peters 2018).

Uridine interacts with 5FU, inhibiting its attachment in RNA synthesis, and is given after 5FU treatment for its efficacy. DPD (dihydropyrimidine dehydrogenase) is involved in the catabolism of 5FU. Usually, inhibition of DPD leads to effective 5FU working by preventing its degradation in the GI tract. Topoisomerase I inhibitors and platinum analogues are also known to show activity against tumour in cancer of the colon and rectum. Platinum adducts are formed between A and G as well as between two G nucleotides resulting in abnormal DNA functioning (Kong et al. 2020). But it may not be effective in all types of cancer and used depending upon several other factors. Currently oxaliplatin is a new platinum-based drug which is a third generation with no reported toxicity and few effects haematologically. It was found to enhance the effect as antitumour agent when coadministered with 5FU. Its best response was observed when given along with LV in preclinical trials. Topoisomerase are responsible for normal topology of DNA when the cell is replicating (Lee et al. 2019). Inhibitors of topoisomerase increase DNA degradation and can be used as an antitumour agent.

Another analogue of deoxynucleoside is gemcitabine found in DNA linkages with inter-nucleotide. It may interfere with RNA and lead to apoptosis and is used in combination with other drugs efficiently. All of these and many other gemcitabine combinations are used to treat different cancer types with progression-free survival (Ueno et al. 2019). In conclusion different antimetabolites in combination with other chemical agents are used in chemotherapy considering their mode of action. The type of interaction between drugs should also be considered either they are synergistic, additive or antagonist. New combinations should be sufficiently screened in animal models before human application and serum profile should be evaluated to identify the best dose and combination of drugs.

7 NSAIDs

NSAIDs are known as non-steroidal anti-inflammatory drugs widely taken throughout the world for fever and inflammation and as pain reliever. They basically inhibit production of prostaglandin due to blockage of cyclooxygenase enzyme. Since the major part played by NSAIDs is for inflammation, it is a known fact that cancer is associated with inflammation. Cancers of the head and neck, ovary, colorectal, prostate and breast are found to be reduced by the use of NSAIDs in epidemiological studies, but reports do not have consistent findings (Cai et al. 2020). But it seems logical that since cancer is linked with inflammation therefore preventing inflammation may reduce cancer risk. NSAIDs are known to boost immune response of cells, increase apoptosis and reduce angiogenesis. But long-term usage of NSAIDs is itself linked with many disorders. The first connection between cancer and chronic inflammation was proposed by Virchow in the 1860s (Phillips et al. 2019). It was proposed that cancer originated from the inflamed area occurred due to cellular injury caused by cell proliferation. Cancer is marked by angiogenesis, as well as apoptosis; abnormalities leading to disturbed signalling pathways and similar mechanism are found in inflammation. In this regard cytokines are the key regulators that change the microenvironment into the tumourous tissue if inflammation is chronic, and the most prominent cytokines contributing to this transformation are TNF-α (tumour-necrosis factor-alpha), IL-6 (interleukin-6), TGF-ß (transforming growth factor-beta) and IL-10 (interleukin-10) (Wang et al. 2019). In cancer cell angiogenesis, cell migration, cell proliferation and metastasis are due to expression of different chemokines and cytokines, and all these events are also a hallmark of chronic inflammation. Therefore, in order to treat inflammation, drugs such as diclofenac, aspirin, celecoxib, mefenamic acid and ibuprofen are all administered because they acquire properties of blocking COX (cyclooxygenase enzyme) and PGHS (prostaglandin endoperoxide H synthase). They may also be given for cancer treatment. In addition to this, another important player of inflammation are platelets that are also involved in cancer aetiology (Wong 2019). TXA2 is a PGH2-derived substance produced by activated platelets, which exerts a potent vasoconstrictor effect and a stimulatory effect on platelet aggregation. Platelets basically produce products of lipid that increase tumour vascularization and ultimately discharge them into the blood vessel system. They also cause tumour extravasation by cell cycle arrest caused by adhesion of tumour cells to the endothelial wall. Platelets are also known to enhance cell survival of tumours by establishing aggregates of platelet cell around tumour cells (Naderi-Meshkin and Ahmadiankia 2018). Much of the experimental data using cell lines and mouse models along with epidemiological one supports the association between the use of NSAIDs and cancer leading to antitumour activity with success. Studies involving the use of aspirin and non-aspirin NSAIDs have shown that in case of gastrointestinal cancer they exert anticancer effects (Mohammed et al. 2018). Although most of the studies have shown contrasting results, in conclusion the use of NSAIDs depends on the type of cancer. Like in case of breast cancer, the effect of NSAIDs differs from person to person due to differences in the molecular pattern of each individual. Therefore, there is a need to examine in detail their effects in experimental studies, as up till now most of the studies are epidemiological rather than experimental. Another important point to be addressed is that although the role of NSAIDs is established in cancer treatment, their use has their own side effects such as renal failure, gastrointestinal bleeding and myocardial infarction.

8 Natural Products

The use of allopathic drugs for cancer treatment has never been a choice for most people, but they prefer to use natural products such as fruits and vegetables and are beneficial in chronic diseases that are linked with oxidative stress such as cancer. Fruits and vegetables are rich in terpenoids, phenolic compounds, carotenoids and flavonoids that have proven anticancer activity (Turkiewicz et al. 2019). Curcumin, a polyphenol in turmeric rhizome, has anti-inflammatory and antioxidant properties. It suppresses cytochrome P450, increases phase II detoxifying enzyme production and exerts anticancer activity. Multiple signalling pathways such as protein kinase, mitochondrial, death receptor, tumour suppressor, caspase activation, cell survival and cell proliferation are regulated by these compounds and ultimately encounter cancer tissue (Tuli et al. 2021).

The skin of red wines and red grapes is rich in phytoalexin known as trans-3,5,4-tryhydroxystilbene (resveratrol) that is known to have an antitumour activity due to antioxidant and anti-proliferative activity. It has been shown to inhibit tumour necrosis factor-α-mediated matrix metalloproteinase-9 expression in HepG2 cells by downregulation of the nuclear factor-kB signalling pathway (Singh et al. 2019). Another plant flavone known as apigenin is present in fruits and vegetables. All these chemicals are known to have anti-progressive, anti-growth, anti-inflammatory, anti-carcinogenic, anti-mutagenic and antioxidant properties. Dimethyl benzanthracene-induced cancers of the skin can be cured with topical application of apigenin.

Quercetin is found in leaves, barks, nuts and even seeds of apples, tomatoes, tea, shallots, onions, grapes, vegetables of brassica and berries and is a flavonoid. Quercetin is a polyphenol and effective against cancer containing different enzymes such as nicotinamide adenine dinucleotide phosphate oxidase, LOX and xanthine oxidase that control cell cycle and inhibit tyrosine kinase and communication with estrogen type II binding sites (Patel et al. 2018).

Turnip, radish, horseradish, cabbage, broccoli, Brussel sprout and watercress are all cruciferous vegetables rich in isothiocyanates and are electrophilic compounds (Abbaoui et al. 2018). These vegetables are known to have anticancer effect since ages, and they reduce carcinogen activation by increasing detoxification of such agents.

Soya bean, soy sauce, soy milk and tofu have genistein that is a promising agent used in chemotherapy. Similarly, ursolic acids abundant in apple peels and medicinal herbs have anti-inflammatory, antitumour and antioxidant functions (Corrêa et al. 2021). Fruits and vegetables have anticancer activities and may be used as chemotherapeutic drugs by inhibiting tumour cell proliferation. They are accessible, acceptable, applicable, inexpensive and easy to use. However, their frequent use in normal individuals is more effective as reduced risk of cancer.

9 Drug Delivery Methods

One of the biggest challenges faced by oncologists and researchers while treating cancer tissues is the appropriate method for targeting cancer cells. As cancer cells have enormous similarity with normal cells, it poses a serious issue on how to target only cancer cells while sparing the normal ones. Once drugs are in contact with normal cells, it leads to their toxicity with other side effects. Although a combination of drugs is used to counter the effect of drug toxicity, due to the dissimilar physicochemical nature of tumour cells, uptake concentration of a drug varies in cancer cells (Mukherjee et al. 2020). In this regard, nanocarriers are used to deliver drug in different ways (Fig. 2). One possible way to deliver a combination of drugs is that one drug is given freely along with other drugs in nanocarrier or both drugs in nanocarriers or co-encapsulation of both drugs and then inside a nanocarrier. Once the problem of drug delivery to the exact site is resolved, the cancer treatment options are many. It will lead to EPR (enhanced permeability and retention) effect. Drug delivery can be actively or passively targeted or both ways (Zhou et al. 2021). In the active transport system, the drug is directly delivered while in a passive mode, and cancer cell properties such as temperature and pH are used to attract a drug carrier to the exact site.

Nanoparticles are available in different structures, materials, shapes and sizes that have differences in their stability, targeting, release and loading capacity. The

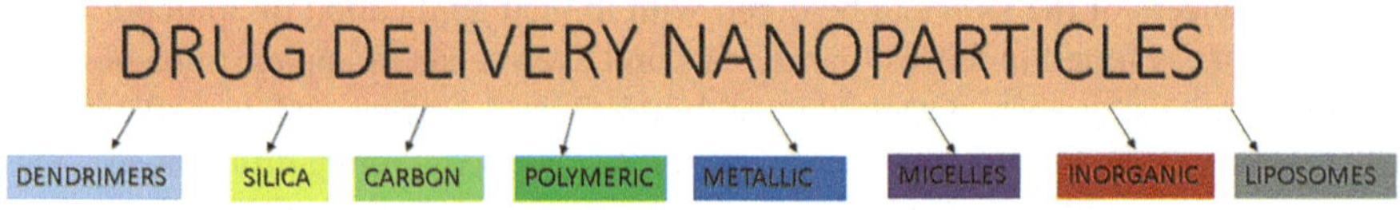

Fig. 2 An overview of different drug delivery methods used as nanocarriers

targeted therapy using nanocarriers is easier as compared to combination drug delivery. Different types of nanocarriers used for this method are:

1. Inorganic nanocarriers: They are best due to their accumulation capability without being recognizable by Pgp, targeted drug delivery, controllable release of drug at a particular site and biocompatibility. They may be gold-, silica-, clay- or graphene-based nanoparticles, but surface toxicity of these particles is the only side effect that can be overcome with agents that can reduce toxicity (Shi et al. 2020).
2. Carbon nanotubes: The sp^2-hybridized carbon atoms, arranged in hexagonal tubes or ball, are used to load drugs inside it. They are effective in delivering small molecules, DNA, RNA and proteins by forming complexities by entering the cell membrane and directly delivering the drug inside the cell or even in some cases nucleus (Berber et al. 2020).
3. Metallic nanoparticles: The particles with strong magnetic properties are used as nanoparticles, and the patient is placed in a room with magnetic field to direct its location to the target site; otherwise, they may block blood vessels and further complicate the condition. But nowadays, the non-toxic and easily degraded material iron oxide nanoparticles are used. Therefore, rapid removal of these nanoparticles frm the blood stream is the matter of concern and need to be explored otherwise it can cause other health complications.
4. Silica nanoparticles: The hollow mesoporous silica nanoparticles, mesoporous materials, nanotubes, hollow silica and silica particles are used for nanoparticle formation. Mesoporous particles are preferred either hollow or normal due to their functionality adjustment, biocompatibility, chemical stability, adjustable pore size, high-pore volume and large surface area, aiding the loading of hydrophilic and hydrophobic drugs (Cheng et al. 2020).
5. Dendrimer nanocarriers: These have branch structures due to polymerization forming layers and effective in the loading of hydrophilic and hydrophobic drugs (Hanurry et al. 2020).
6. Liposomes: They are amphiphilic lipids, divided into monolayer and multilayer types. Lipid bilayers provide protection against the environment. They can be easily removed from the blood after delivery. Different liposomes have been prepared by different researchers, each having their own significance and methods, but overall they are effective in delivering drugs to their target sites (Kommineni et al. 2019).
7. Micelles: They are made up of self-assembling random/block amphiphilic co-polymers in an aqueous medium. They are more stable compared to liposomes, and drugs may be covalently attached or physically loaded for delivery.
8. Polymeric nanoparticles: Polymeric nanoparticles are submicron colloidal systems, and their mechanisms of drug loading include dissolution, entrapment, adsorption, binding or encapsulation of drugs (Li et al. 2017).

Therefore, the nanoparticles, for their key carrier role for a chemotherapeutic drug, upbring a new research field, with promising results. Hopefully in the future, the personalized drugs will be delivered to the target cells using nanocarriers.

10 Conclusion

Despite the contest from macromolecule drugs, small-molecule targeted drugs will maintain their importance in cancer therapeutics due to their distinctive benefits. With the evolution of novel drug-based research and comprehensive perception of tumour pathology, it has been believed that newer small-molecule anticancer drugs able to target novel genes or their mechanism of action will be established soon. Additionally, new research areas such as combination of small-molecule targeted drugs with proteolysis-targeting chimera (PROTAC), tumour antibody-drug conjugate drugs and immunotherapy are expected to gain a substantial progress in the near future.

References

Abbaoui B, Lucas CR, Riedl KM, Clinton SK, Mortazavi A (2018) Cruciferous vegetables, isothiocyanates, and bladder cancer prevention. Mol Nutr Food Res 62(18):1800079

Abdallah HM, Martinez-Meehan D, Lutfi W, Dhupar R, Grenda T, Schuchert MJ, Christie NA, Luketich JD, Okusanya OT (2021) Adjuvant chemotherapy for pulmonary sarcomatoid carcinoma: a retrospective analysis of the National Cancer Database. J Thoracic Cardiovasc Surg 163(5):1669–1681

Abu Rmilah AA, Lin G, Begna KH, Friedman PA, Herrmann J (2020) Risk of QTc prolongation among cancer patients treated with tyrosine kinase inhibitors. Int J Cancer 147(11):3160–3167

Adrianzen Herrera D, Ashai N, Perez-Soler R, Cheng H (2019) Nanoparticle albumin bound-paclitaxel for treatment of advanced non-small cell lung cancer: an evaluation of the clinical evidence. Expert Opin Pharmacother 20(1):95–102

Aristizabal-Pachon AF, Castillo WO (2020) Genotoxic evaluation of occupational exposure to antineoplastic drugs. Toxicol Res 36(1):29–36

Berber MR, Elkhenany H, Hafez IH, El-Badawy A, Essawy M, El-Badri N (2020) Efficient tailoring of platinum nanoparticles supported on multiwalled carbon nanotubes for cancer therapy. Nanomedicine 15(08):793–808

Bernabeu E, Cagel M, Lagomarsino E, Moretton M, Chiappetta DA (2017) Paclitaxel: what has been done and the challenges remain ahead. Int J Pharm 526(1–2):474–495

Bindu S, Mazumder S, Bandyopadhyay U (2020) Non-steroidal anti-inflammatory drugs (NSAIDs) and organ damage: a current perspective. Biochem Pharmacol 180:114147

Cai Y, Yousef A, Grandis JR, Johnson DE (2020) NSAID therapy for PIK3CA-altered colorectal, breast, and head and neck cancer. Adv Biol Regul 75:100653

Cascorbi I (2018) Significance of pharmacogenomics in precision medicine. Wiley Online Library

Chen Z, Liu Z, Huang W, Li Z, Zou J, Wang J, Lin X, Li B, Chen D, Hu Y (2017) Gimatecan exerts potent antitumor activity against gastric cancer in vitro and in vivo via AKT and MAPK signaling pathways. J Transl Med 15(1):1–11

Cheng Y-J, Hu J-J, Qin S-Y, Zhang A-Q, Zhang X-Z (2020) Recent advances in functional mesoporous silica-based nanoplatforms for combinational photo-chemotherapy of cancer. Biomaterials 232:119738

Corrêa JGDS, Bianchin M, Lopes AP, Silva E, Ames FQ, Pomini AM, Carpes ST, de Carvalho Rinaldi J, Melo RC, Kioshima ES (2021) Chemical profile, antioxidant and anti-inflammatory properties of Miconia albicans (Sw.) Triana (Melastomataceae) fruits extract. J Ethnopharmacol:273, 113979

Dehshahri A, Ashrafizadeh M, Afshar EG, Pardakhty A, Mandegary A, Mohammadinejad R, Sethi G (2020) Topoisomerase inhibitors: pharmacology and emerging nanoscale delivery systems. Pharmacol Res 151:104551

Diana A, Carlino F, Franzese E, Oikonomidou O, Criscitiello C, De Vita F, Ciardiello F, Orditura M (2020) Early triple negative breast cancer: conventional treatment and emerging therapeutic landscapes. Cancers 12(4):819

Ding PN, Lord SJ, Gebski V, Links M, Bray V, Gralla RJ, Yang JC-H, Lee CK (2017) Risk of treatment-related toxicities from EGFR tyrosine kinase inhibitors: a meta-analysis of clinical trials of gefitinib, erlotinib, and afatinib in advanced EGFR-mutated non–small cell lung cancer. J Thorac Oncol 12(4):633–643

Eltweri AM, Thomas AL, Chung WY, Morgan B, Thompson J, Dennison AR, Bowrey DJ (2019) The effect of supplementary Omegaven® on the clinical outcome of patients with advanced esophagogastric adenocarcinoma receiving palliative epirubicin, oxaliplatin, and capecitabine chemotherapy: a phase II clinical trial. Anticancer Res 39(2):853–861

Ercolano G, De Cicco P, Ianaro A (2019) New drugs from the sea: pro-apoptotic activity of sponges and algae derived compounds. Mar Drugs 17(1):31

Falzone L, Salomone S, Libra M (2018) Evolution of cancer pharmacological treatments at the turn of the third millennium. Front Pharmacol 9:1300

Fattahi N, Shahbazi M-A, Maleki A, Hamidi M, Ramazani A, Santos HA (2020) Emerging insights on drug delivery by fatty acid mediated synthesis of lipophilic prodrugs as novel nanomedicines. J Control Release 326:556–598

García-Fernández C, Fornaguera C, Borrós S (2020) Nanomedicine in non-small cell lung cancer: from conventional treatments to immunotherapy. Cancers 12(6):1609

Gasch C, Ffrench B, O'Leary JJ, Gallagher MF (2017) Catching moving targets: cancer stem cell hierarchies, therapy-resistance & considerations for clinical intervention. Mol Cancer 16(1): 1–15

Gilad Y, Gellerman G, Lonard DM, O'Malley BW (2021) Drug combination in cancer treatment—from cocktails to conjugated combinations. Cancers 13(4):669

Gillessen S, Attard G, Beer TM, Beltran H, Bossi A, Bristow R, Carver B, Castellano D, Chung BH, Clarke N (2018) Management of patients with advanced prostate cancer: the report of the advanced prostate cancer consensus conference APCCC 2017. Eur Urol 73(2):178–211

Guichard N, Guillarme D, Bonnabry P, Fleury-Souverain S (2017) Antineoplastic drugs and their analysis: a state of the art review. Analyst 142(13):2273–2321

Hanurry EY, Mekonnen TW, Andrgie AT, Darge HF, Birhan YS, Hsu W-H, Chou H-Y, Cheng C-C, Lai J-Y, Tsai H-C (2020) Biotin-decorated PAMAM G4. 5 dendrimer nanoparticles to enhance the delivery, anti-proliferative, and apoptotic effects of chemotherapeutic drug in cancer cells. Pharmaceutics 12(5):443

Hassan R, Allali I, Agamah FE, Elsheikh SS, Thomford NE, Dandara C, Chimusa ER (2020) Drug response in association with pharmacogenomics and pharmacomicrobiomics: towards a better personalized medicine. Brief Bioinform 22(4):bbaa292

Heitz F, Amant F, Fotopoulou C, Battista MJ, Wimberger P, Traut A, Fisseler-Eckhoff A, Harter P, Vandenput I, Sehouli J (2014) Synchronous ovarian and endometrial cancer—an international multicenter case-control study. Int J Gynecol Cancer 24(1):54–60

Hu W, Huang X-S, Wu J-F, Yang L, Zheng Y-T, Shen Y-M, Li Z-Y, Li X (2018) Discovery of novel topoisomerase II inhibitors by medicinal chemistry approaches. J Med Chem 61(20): 8947–8980

Huszar D, Theoclitou M-E, Skolnik J, Herbst R (2009) Kinesin motor proteins as targets for cancer therapy. Cancer Metastasis Rev 28(1):197–208

Ismael GF, Rosa DD, Mano MS, Awada A (2008) Novel cytotoxic drugs: old challenges, new solutions. Cancer Treat Rev 34(1):81–91

Jabbour E, Sasaki K, Ravandi F, Huang X, Short NJ, Khouri M, Kebriaei P, Burger J, Khoury J, Jorgensen J (2018) Chemoimmunotherapy with inotuzumab ozogamicin combined with mini-hyper-CVD, with or without blinatumomab, is highly effective in patients with Philadelphia

chromosome–negative acute lymphoblastic leukemia in first salvage. Cancer 124(20): 4044–4055
Jiao Q, Bi L, Ren Y, Song S, Wang Q, Wang Y-s (2018) Advances in studies of tyrosine kinase inhibitors and their acquired resistance. Mol Cancer 17(1):1–12
José-Enériz S, Gimenez-Camino N, Agirre X, Prosper F (2019) HDAC inhibitors in acute myeloid leukemia. Cancers 11(11):1794
Jungwirth G, Yu T, Cao J, Eddine MA, Moustafa M, Warta R, Debus J, Unterberg A, Abdollahi A, Herold-Mende C (2021) KIF11 inhibitors filanesib and ispinesib inhibit meningioma growth in vitro and in vivo. Cancer Lett *506*:1–10
Kalinin AA, Higgins GA, Reamaroon N, Soroushmehr S, Allyn-Feuer A, Dinov ID, Najarian K, Athey BD (2018) Deep learning in pharmacogenomics: from gene regulation to patient stratification. Pharmacogenomics 19(7):629–650
Kommineni N, Mahira S, Domb AJ, Khan W (2019) Cabazitaxel-loaded nanocarriers for cancer therapy with reduced side effects. Pharmaceutics 11(3):141
Kong YW, Dreaden EC, Morandell S, Zhou W, Dhara SS, Sriram G, Lam FC, Patterson JC, Quadir M, Dinh A (2020) Enhancing chemotherapy response through augmented synthetic lethality by co-targeting nucleotide excision repair and cell-cycle checkpoints. Nat Commun 11(1):1–12
Kummar S, Gutierrez M, Doroshow JH, Murgo AJ (2006) Drug development in oncology: classical cytotoxics and molecularly targeted agents. Br J Clin Pharmacol 62(1):15–26
Lauschke VM, Milani L, Ingelman-Sundberg M (2018) Pharmacogenomic biomarkers for improved drug therapy—recent progress and future developments. AAPS J 20(1):1–16
Lee CM, Wang G, Pertsinidis A, Marians KJ (2019) Topoisomerase III acts at the replication fork to remove precatenanes. J Bacteriol 201(7):e00563–e00518
Lee H, Park S, Kang JE, Lee HM, Kim SA, Rhie SJ (2020) Efficacy and safety of nanoparticle-albumin-bound paclitaxel compared with solvent-based taxanes for metastatic breast cancer: a meta-analysis. Sci Rep 10(1):1–9
Lee YT, Tan YJ, Oon CE (2018) Molecular targeted therapy: treating cancer with specificity. Eur J Pharmacol 834:188–196
Li B, Li Q, Mo J, Dai H (2017) Drug-loaded polymeric nanoparticles for cancer stem cell targeting. Front Pharmacol 8:51
Liao H-C, Chou Y-J, Lin C-C, Liu S-H, Oswita A, Huang Y-L, Wang Y-L, Syu J-L, Sun C-M, Leu C-M (2019) Andrographolide and its potent derivative exhibit anticancer effects against imatinib-resistant chronic myeloid leukemia cells by downregulating the Bcr-Abl oncoprotein. Biochem Pharmacol 163:308–320
Lievense LA, Sterman DH, Cornelissen R, Aerts JG (2017) Checkpoint blockade in lung cancer and mesothelioma. Am J Respir Crit Care Med 196(3):274–282
Liu Y, Khan AR, Du X, Zhai Y, Tan H, Zhai G (2019) Progress in the polymer-paclitaxel conjugate. J Drug Deliv Sci Technol 54:101237
Macarulla T, Pazo-Cid R, Guillén-Ponce C, López R, Vera R, Reboredo M, Muñoz Martin A, Rivera F, Díaz Beveridge R, La Casta A (2019) Phase I/II trial to evaluate the efficacy and safety of nanoparticle albumin-bound paclitaxel in combination with gemcitabine in patients with pancreatic cancer and an ECOG performance status of 2. J Clin Oncol 37(3):230–238
Makkar R, Behl T, Arora S (2020) Role of HDAC inhibitors in diabetes mellitus. Curr Res Transl Med 68(2):45–50
Malik SS, Masood N, Asif M, Ahmed P, Shah ZU, Khan JS (2019a) Expressional analysis of MLH1 and MSH2 in breast cancer. Curr Probl Cancer 43(2):97–105
Malik SS, Masood N, Sherrard A, Bishop PN (2019b) Small non-coding RNAs as a tool for personalized therapy in familial cancers. In: AGO-driven non-coding RNAs. Elsevier, pp 179–208
Mallick B (2019) AGO-driven non-coding RNAs: codes to decode the therapeutics of diseases. Academic Press

Marosi C, Függer R (2020) Drug therapy against cancer. Effects Cancer Treatment Nervous System 1:62

Masood N, Malik SS (2020) Essentials of cancer genomic, computational approaches and precision medicine. Springer

Mohammed A, Yarla NS, Madka V, Rao CV (2018) Clinically relevant anti-inflammatory agents for chemoprevention of colorectal cancer: new perspectives. Int J Mol Sci 19(8):2332

Mukherjee S, Mehta D, Dhangar K, Kumar M (2020) Environmental fate, distribution and state-of-the-art removal of antineoplastic drugs: a comprehensive insight. Chem Eng J 407:127184

Mukhtar E, Adhami VM, Mukhtar H (2014) Targeting microtubules by natural agents for cancer therapy. Mol Cancer Ther 13(2):275–284

Murtuza A, Bulbul A, Shen JP, Keshavarzian P, Woodward BD, Lopez-Diaz FJ, Lippman SM, Husain H (2019) Novel third-generation EGFR tyrosine kinase inhibitors and strategies to overcome therapeutic resistance in lung cancer. Cancer Res 79(4):689–698

Naaz F, Haider MR, Shafi S, Yar MS (2019) Anti-tubulin agents of natural origin: targeting taxol, vinca, and colchicine binding domains. Eur J Med Chem 171:310–331

Naderi-Meshkin H, Ahmadiankia N (2018) Cancer metastasis versus stem cell homing: role of platelets. J Cell Physiol 233(12):9167–9178

Nikolaou M, Pavlopoulou A, Georgakilas AG, Kyrodimos E (2018) The challenge of drug resistance in cancer treatment: a current overview. Clin Exp Metastasis 35(4):309–318

Nikolova T, Kiweler N, Krämer OH (2017) Interstrand crosslink repair as a target for HDAC inhibition. Trends Pharmacol Sci 38(9):822–836

Nishant T, Bindu H, Kumar S, Kumar A (2012) Pharmacogenomics-personalized treatment of cancer, diabetes and cardiovascular diseases. J Pharmacogenom Pharmacoproteomics 3(107): 2153. 0645.1000107

Olziersky A-M, Labidi-Galy SI (2017) Clinical development of anti-mitotic drugs in cancer. Cell Div Mach Dis 1002:125–152

Ongoren S, Eskazan AE, Suzan V, Savci S, Erdogan Ozunal I, Berk S, Yalniz FF, Elverdi T, Salihoglu A, Erbilgin Y (2018) Third-line treatment with second-generation tyrosine kinase inhibitors (dasatinib or nilotinib) in patients with chronic myeloid leukemia after two prior TKIs: real-life data on a single center experience along with the review of the literature. Hematology 23(4):212–220

Paludetto M-N, Bijani C, Puisset F, Bernardes-Génisson V, Cc A, Robert A (2018) Metalloporphyrin-catalyzed oxidation of sunitinib and pazopanib, two anticancer tyrosine kinase inhibitors: evidence for new potentially toxic metabolites. J Med Chem 61(17): 7849–7860

Park H-W, Ma Z, Zhu H, Jiang S, Robinson RC, Endow SA (2017) Structural basis of small molecule ATPase inhibition of a human mitotic kinesin motor protein. Sci Rep 7(1):1–12

Patel RV, Mistry BM, Shinde SK, Syed R, Singh V, Shin H-S (2018) Therapeutic potential of quercetin as a cardiovascular agent. Eur J Med Chem 155:889–904

Peters GJ (2018) Antipyrimidine effects of five different pyrimidine de novo synthesis inhibitors in three head and neck cancer cell lines. Nucleosides Nucleotides Nucleic Acids 37(6):329–339

Phillips CM, Chen L-W, Heude B, Bernard JY, Harvey NC, Duijts L, Mensink-Bout SM, Polanska K, Mancano G, Suderman M (2019) Dietary inflammatory index and non-communicable disease risk: a narrative review. Nutrients 11(8):1873

Pillai G (2019) Nanotechnology toward treating cancer: a comprehensive review. Appl Target Nano Drugs Deliv Syst:221–256;doi.org/10.1016/B978-0-12-814029-1.00009-0

Pirisinu M, Pham TC, Zhang DX, Hong TN, Nguyen LT, Le MT (2020) Extracellular vesicles as natural therapeutic agents and innate drug delivery systems for cancer treatment: recent advances, current obstacles, and challenges for clinical translation. In: Seminars in cancer biology. Elsevier

Pottier C, Fresnais M, Gilon M, Jérusalem G, Longuespée R, Sounni NE (2020) Tyrosine kinase inhibitors in cancer: breakthrough and challenges of targeted therapy. Cancers 12(3):731

Qi X, Wagenaar E, Xu W, Huang K, Schinkel AH (2017) Ochratoxin a transport by the human breast cancer resistance protein (BCRP), multidrug resistance protein 2 (MRP2), and organic anion-transporting polypeptides 1A2, 1B1 and 2B1. Toxicol Appl Pharmacol 329:18–25

Qin S, Li A, Yi M, Yu S, Zhang M, Wu K (2019) Recent advances on anti-angiogenesis receptor tyrosine kinase inhibitors in cancer therapy. J Hematol Oncol 12(1):1–11

Roovers S, Deprez J, Priwitaningrum D, Lajoinie G, Rivron N, Declercq H, De Wever O, Stride E, Le Gac S, Versluis M (2019) Sonoprinting liposomes on tumor spheroids by microbubbles and ultrasound. J Control Release 316:79–92

Ruan Q, Patel G, Wang J, Luo E, Zhou W, Sieniawska E, Hao X, Kai G (2021) Current advances of endophytes as a platform for production of anti-cancer drug camptothecin. Food Chem Toxicol 151:112113

Ruzzolini J, Laurenzana A, Andreucci E, Peppicelli S, Bianchini F, Carta F, Supuran CT, Romanelli MN, Nediani C, Calorini L (2020) A potentiated cooperation of carbonic anhydrase IX and histone deacetylase inhibitors against cancer. J Enzyme Inhib Med Chem 35(1):391–397

Sacco PC, Gridelli C (2017) An update on the developing mitotic inhibitors for the treatment of non-small cell carcinoma. Expert Opin Emerg Drugs 22(3):213–222

Sauter B, Gillingham D (2020) DNA damaging agents in chemical biology and cancer. CHIMIA Int J Chem 74(9):693–698

Sharma P, Mehta M, Dhanjal DS, Kaur S, Gupta G, Singh H, Thangavelu L, Rajeshkumar S, Tambuwala M, Bakshi HA (2019) Emerging trends in the novel drug delivery approaches for the treatment of lung cancer. Chem Biol Interact 309:108720

Shi Z, Zhou Y, Fan T, Lin Y, Zhang H, Mei L (2020) Inorganic nano-carriers based smart drug delivery systems for tumor therapy. Smart Mater Med 1:32–47

Simon N, Guichard N, Odou P, Decaudin B, Bonnabry P, Fleury-Souverain S (2020) Efficiency of four solutions in removing 23 conventional antineoplastic drugs from contaminated surfaces. PLoS One 15(6):e0235131

Singh V, Kumar D, Chowdhary S, Maniar K, Narwal M, Bhattacharyya R, Banerjee D (2019) Ligand-based designing of natural products. In: Bioactive natural products for the management of cancer: from bench to bedside. Springer, pp 167–175

Suh JH, Kotecha R, Chao ST, Ahluwalia MS, Sahgal A, Chang EL (2020) Current approaches to the management of brain metastases. Nat Rev Clin Oncol 17(5):279–299

Sun B, Luo C, Cui W, Sun J, He Z (2017) Chemotherapy agent-unsaturated fatty acid prodrugs and prodrug-nanoplatforms for cancer chemotherapy. J Control Release 264:145–159

Tomić T, Domínguez-López S, Barrios-Rodríguez R (2019) Non-aspirin non-steroidal anti-inflammatory drugs in prevention of colorectal cancer in people aged 40 or older: a systematic review and meta-analysis. Cancer Epidemiol 58:52–62

Tsilimigras DI, Ntanasis-Stathopoulos I, Moris D, Spartalis E, Pawlik TM (2018) Histone deacetylase inhibitors in hepatocellular carcinoma: a therapeutic perspective. Surg Oncol 27(4):611–618

Tuli HS, Mittal S, Loka M, Aggarwal V, Aggarwal D, Masurkar A, Kaur G, Varol M, Sak K, Kumar M (2021) Deguelin targets multiple oncogenic signaling pathways to combat human malignancies. Pharmacol Res 166:105487

Turkiewicz IP, Wojdyło A, Tkacz K, Nowicka P, Hernández F (2019) Antidiabetic, anticholinesterase and antioxidant activity vs. terpenoids and phenolic compounds in selected new cultivars and hybrids of artichoke Cynara scolymus L. Molecules 24(7):1222

Udagawa C, Zembutsu H (2020) Pharmacogenetics for severe adverse drug reactions induced by molecular-targeted therapy. Cancer Sci 111(10):3445

Ueno M, Ikeda M, Morizane C, Kobayashi S, Ohno I, Kondo S, Okano N, Kimura K, Asada S, Namba Y (2019) Nivolumab alone or in combination with cisplatin plus gemcitabine in Japanese patients with unresectable or recurrent biliary tract cancer: a non-randomised, multicentre, open-label, phase 1 study. Lancet Gastroenterol Hepatol 4(8):611–621

Vallée A, Lecarpentier Y, Vallée J-N (2019) Targeting the canonical WNT/β-catenin pathway in cancer treatment using non-steroidal anti-inflammatory drugs. Cell 8(7):726

Wahid M, Bano Q (2014) Camptothecin and its analogs antitumor activity by poisoning topoisomerase I, their structure activity relationship and clinical development perspective of analogs. J App Pharm 6:286–295

Wallace-Povirk A, Tong N, Wong-Roushar J, O'Connor C, Zhou X, Hou Z, Bao X, Garcia GE, Li J, Kim S (2021) Discovery of 6-substituted thieno [2, 3-d] pyrimidine analogs as dual inhibitors of glycinamide ribonucleotide formyltransferase and 5-aminoimidazole-4-carboxamide ribonucleotide formyltransferase in de novo purine nucleotide biosynthesis in folate receptor expressing human tumors. Bioorgan Med Chem 37:116093

Wang F, Porter M, Konstantopoulos A, Zhang P, Cui H (2017a) Preclinical development of drug delivery systems for paclitaxel-based cancer chemotherapy. J Control Release 267:100–118

Wang J, Li D, Cang H, Guo B (2019) Crosstalk between cancer and immune cells: role of tumor-associated macrophages in the tumor microenvironment. Cancer Med 8(10):4709–4721

Wang L, Liang C, Li F, Guan D, Wu X, Fu X, Lu A, Zhang G (2017b) PARP1 in carcinomas and PARP1 inhibitors as antineoplastic drugs. Int J Mol Sci 18(10):2111

Widatalla SE, Korolkova OY, Whalen DS, Goodwin JS, Williams KP, Ochieng J, Sakwe AM (2019) Lapatinib-induced annexin A6 upregulation as an adaptive response of triple-negative breast cancer cells to EGFR tyrosine kinase inhibitors. Carcinogenesis 40(8):998–1009

Wijaya J, Gose T, Schuetz JD (2020) Using pharmacology to squeeze the life out of childhood Leukemia, and potential strategies to achieve breakthroughs in medulloblastoma treatment. Pharmacol Rev 72(3):668–691

Wong RS (2019) Role of nonsteroidal anti-inflammatory drugs (NSAIDs) in cancer prevention and cancer promotion. Adv Pharmacol Sci 2019:1–10

Wurzer H, Hoffmann C, Al Absi A, Thomas C (2019) Actin cytoskeleton straddling the immunological synapse between cytotoxic lymphocytes and cancer cells. Cell 8(5):463

Wysocka J, Allis CD, Coonrod S (2006) Histone arginine methylation and its dynamic regulation. Front Biosci 11(2006):344–355

Xuhong J-C, Qi X-W, Zhang Y, Jiang J (2019) Mechanism, safety and efficacy of three tyrosine kinase inhibitors lapatinib, neratinib and pyrotinib in HER2-positive breast cancer. Am J Cancer Res 9(10):2103

Yuan S, Qiao T, Li X, Zhuang X, Chen W, Chen X, Zhang Q (2018) Toll-like receptor 9 activation by CpG oligodeoxynucleotide 7909 enhances the radiosensitivity of A549 lung cancer cells via the p53 signaling pathway. Oncol Lett 15(4):5271–5279

Zaimy M, Saffarzadeh N, Mohammadi A, Pourghadamyari H, Izadi P, Sarli A, Moghaddam L, Paschepari S, Azizi H, Torkamandi S (2017) New methods in the diagnosis of cancer and gene therapy of cancer based on nanoparticles. Cancer Gene Ther 24(6):233–243

Zhong W-Z, Chen K-N, Chen C, Gu C-D, Wang J, Yang X-N, Mao W-M, Wang Q, Qiao G-B, Cheng Y (2019) Erlotinib versus gemcitabine plus cisplatin as neoadjuvant treatment of stage IIIA-N2 EGFR-mutant non–small-cell lung cancer (EMERGING-CTONG 1103): a randomized phase II study. J Clin Oncol 37(25):2235–2245

Zhou X, Huang X, Wang B, Tan L, Zhang Y, Jiao Y (2021) Light/gas cascade-propelled Janus micromotors that actively overcome sequential and multi-staged biological barriers for precise drug delivery. Chem Eng J 408:127897

Zou J, Li S, Chen Z, Lu Z, Gao J, Zou J, Lin X, Li Y, Zhang C, Shen L (2018) A novel oral camptothecin analog, gimatecan, exhibits superior antitumor efficacy than irinotecan toward esophageal squamous cell carcinoma in vitro and in vivo. Cell Death Dis 9(6):1–10

Nanoparticles for Targeted Drug Delivery Systems with Cancer Therapy in Perspective

Shweta Paroha, Vikas Jain, Laxmi Rani, S. L. Neha, Arzoo Pannu, Bhumika Kumar, Phool Singh Yaduwanshi, Rajni Kant Panik, and Pravat K. Sahoo

Abstract Cancer is a lethal disease and is one of the leading causes of morbidity worldwide. Chemotherapy is successful to combat cancer but medical applicability of many chemotherapeutic agents is limited by the non-specific targeting of malignant cells, leading to unwanted side effects. Nanoparticles (NPs) are well-established nano-carriers that can protect drugs from enzymatic degradation and unspecific release in the circulatory system, and deliver it into tumor cells which can modulate pharmacokinetic parameters and therapeutic efficacy of a drug. NPs represent a successful strategy toward targeted drug delivery to treat cancer. The specific characteristics of NPs include biodegradability, non-toxicity, biocompatibility, prolonged circulation time, and a wide range of drug encapsulation capacities. Owing to its small size and spherical shape, it can encapsulate a higher amount of drug and deliver it specifically to the target size, either by passive or by active targeting, and reduce the drug-associated toxicity. Conjugation of specific ligands on the surface of NPs can enhance therapeutic efficacy and reduce unwanted side effects of the drug.

Keywords Cancer · Nanoparticles · Chemotherapy · Pharmacokinetic · Efficacy · Toxicity

---

S. Paroha · L. Rani · S. L. Neha · B. Kumar · P. K. Sahoo (✉)
Department of Pharmaceutics, Delhi Institute of Pharmaceutical Sciences and Research (DIPSAR), Delhi Pharmaceutical Sciences and Research University, New Delhi, India

V. Jain
Injectable Pharma Research, Lupin Research Park, Lupin Limited, Pune, India

A. Pannu
Department of Pharmacology, Delhi Pharmaceutical Sciences and Research University, New Delhi, India

P. S. Yaduwanshi
IES Institute of Pharmacy, IES University Kalkheda, Bhopal, India

R. K. Panik
Columbia Institute of Pharmacy, Raipur, Chhattisgarh, India

V. S. Rajput, A. Runthala (eds.), *Drugs and a Methodological Compendium*,
https://doi.org/10.1007/978-981-19-7952-1_11

1 Introduction

The term "nano" comes from the Greek word "nanos," meaning dwarf. So, a nanoparticle (NPs) is a very small particle. NPs are nano-sized particles with a size range of 1–1000 nm. For drug delivery applications, 5–200 nm is a preferred size to overcome different biological barriers. Due to their unique physiochemical properties, NPs demonstrate extensive biomedical utilities as carriers for drugs, genes, and imaging agents (Petros and DeSimone 2010). Due to their specific physicochemical properties, NPs have a significant impact on half-life and biodistribution of therapeutic moiety (Jadidi-Niaragh et al. 2017). The particle size of NPs is one of the critical quality attributes which significantly affect the pharmacokinetic property of the drug (Dubey et al. 2016a; Paroha et al. 2021). For instance, the NPs having a smaller size (less than 7 nm) can undergo renal filtration and urinary excretion (Najafi-Hajivar et al. 2016; Soo Choi et al. 2007). At the same time, larger-size NPs (more than 200 nm) are more quickly engulfed by the phagocytic cells and get cleared from the circulatory system (Davis et al. 2008; Decuzzi et al. 2008). Depending on the type of materials utilized, the NPs can be divided into four major classes: (a). NPs fabricated with a polymer; (b). NPs fabricated with lipids; (c). NPs fabricated with inorganic materials; and (d). NPs are inspired by biological materials.

In general, the structures of NPs simply consist of a core, shell, and surface. The core of NPs determines the size of particles having an inactive matrix and active drug while the shell of NPs chemically or physically bound to the core. The principal mechanism for the internalization of NPs into target cells is endocytosis which can be further divided into phagocytosis and pinocytosis. Phagocytosis is a process by which target cells engulf solid NPs while pinocytosis, otherwise, known as fluid endocytosis (Yameen et al. 2014). In the last four decades, several research articles published which have shown the potential of nano-sized materials in cancer targeting, diagnosis, or imaging of tumors and therapy (Dubey et al. 2017; Farokhzad and Langer 2009). NPs offer several advantages, including (a). Increased bioavailability; (b). Enhanced residence time in the circulatory system; and (c). Selective targeting of the desired site (Mudshinge et al. 2011). NPs have a large surface area because of their small size leading to increased solubility of hydrophobic drugs, enhanced bioavailability, can cross blood–brain barrier (BBB), cross pulmonary system, and cross endothelial cells (Rizvi and Saleh 2018).

In a traditional medication, which utilized a non-targeted approach of oral or intravascular drug administration, a very small part of the drug reaches the tumor site, resulting in unwanted side effects. In contrast to this, NPs-based drug delivery approach could lead to selective deposition of drugs at cancer sites and the healthy cells remain untouched. Novel drug delivery approaches enhance treatment efficacy and reduce unwanted effects of the drug (Bertrand and Leroux 2012). Several designs of NPs have been proposed for cancer therapy and diagnosis, based on route of administration, particle size, and biodistribution. For efficient delivery of a drug to the targeted tumor, the NPs must have the ability to sustain in the circulatory system for a longer period of time without changing particle integrity. In addition to

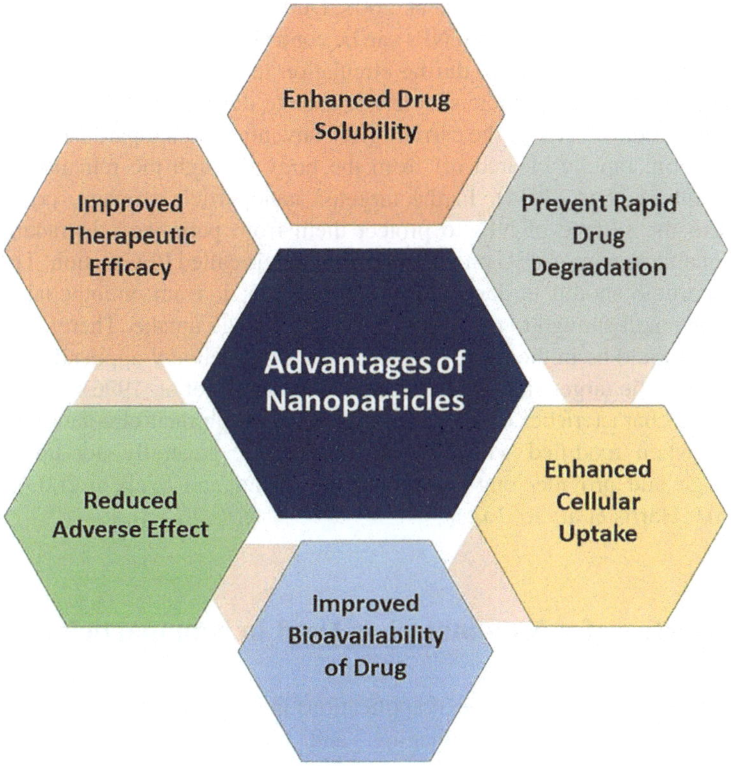

Fig. 1 Advantages of nanoparticles in drug delivery system

their size, the surface property plays an important role in their life span and circulation time (Cho et al. 2008). To avoid macrophage uptake in circulatory system, the hydrophilic surface of NPs is a prerequisite, which can be archived in two ways: surface coating of NPs with poly (ethylene glycol) (PEG) or fabrication of NPs with block copolymers having hydrophilic and hydrophobic polymer (Harris et al. 2001b; Moghimi and Szebeni 2003). NPs offer several advantages in targeted delivery system which is summarized in Fig. 1. In this chapter, we have described the implications of different NPs for targeted drug delivery systems with cancer therapy in perspective.

2 Size and Surface Characteristics of Nanoparticles

NPs have to sustain the plasma circulation for a longer period of time to get successfully absorbed into the target tissue. The traditional particles and unmodified NPs are often trapped in the liver and spleen through the reticuloendothelial system, due to their size and surface (Cho et al. 2008; Dubey et al. 2016a). By adjusting their size and surface, the administered NPs can be controlled. For effective drug delivery, the NPs must be in intact form during circulation in the bloodstream. The size and morphology of the nanoparticle should not change during circulation and the drug must be encapsulated within the carrier. The conventional nanoparticles, due to lack of PEGylation, rapidly cleared off from the body through the reticuloendothelial system (Abbasi et al. 2014). In the targeted nanoparticle strategy, poly-PEG is attached to the surface of NPs to protect them from peripheral degradation. The process of attachment of PEG on the surface of NPs is called PEGylation. The size of the nanoparticle should be large enough to prevent it from endothelial capillary leakage but small enough to prevent it from macrophage uptake. Therefore, the size of the NPs should be in the range of 100 nm for drug delivery applications to allow them to reach the target site (Dellian et al. 2000; Wisse et al. 1996). In addition to size, surface characteristics of the NPs also play a role in their circulation behavior. The NPs when modified with hydrophilic ligands, generally not up taken by macrophage and provide long circulation (Moghimi and Szebeni 2003); Adams et al. 2003; Harris et al. 2001a).

3 Properties of a Nanomaterials Used in Nanoparticles

Nanomaterials have a wide range of applications in pharmaceutical fields, especially in drug delivery, diagnostic, imaging, and biosensor. The conversion of a nanomaterial to nanoscale particle, can change its physicochemical properties, such as increase in surface area, pharmacokinetics, and drug loading capacity. There are some nanomaterials like PEG, which are used to alter the release characteristics of their cargo based on biological or physicochemical stimulus. Various natural polymers used for NPs preparation are alginate, chitosan, or albumin (Alam et al. 2015; Dubey et al. 2015). NPs are prepared from several materials such as natural polymers, proteins, polysaccharides, and synthetic polymers. Commonly used synthetic polymers include PEG, polylactide (PLA), polystyrene (PS), poly (lactide-co-glycolide) (PLGA), poly(cyanoacrylate) (PCA), polycaprolactone (PCL), poly(vinylpyrrolidone), and their copolymers (Dubey et al. 2016b; Gupta et al. 2014; Palombo et al. 2014). They are widely used because of their biocompatible and biodegradable nature as well as their ability to encapsulate hydrophobic drugs (Duan et al. 2006; Zhang and Feng 2006). The methods employed for the encapsulation of drugs in these polymers are spontaneous emulsification, solvent diffusion, polymerization, solvent evaporation, etc. The selection of a material is based on several factors (Palombo et al. 2014; Paroha et al. 2020), such as:

(a) Size of nanoparticles required.
(b) Surface characteristics (e.g., charge on the surface).
(c) Membrane permeability.
(d) The inherent properties of the drug (e.g., aqueous solubility and stability).
(e) The degree of biocompatibility, biodegradability, and toxicity.

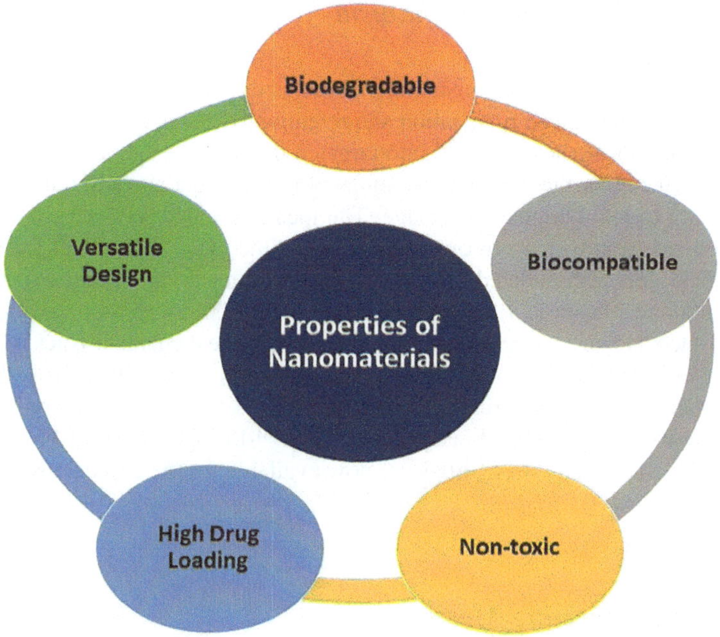

Fig. 2 Properties of nanomaterials used for preparation of nanoparticles

(f) The need for desired drug release profile.
(g) Targeting ability to desired site.
(h) The antigenicity of the final drug formulation.

A nanomaterial should not obstruct blood vessels and be able to target the specific diseased tissue in a safe concentration (Webster et al. 2013). The nanomaterials used for the preparation of a NP should be biocompatible, biodegradable, non-toxic, versatile design, and capable of high drug loading. The properties of nanomaterials used for preparation of NPs are summarized in Fig. 2.

4 Nanoparticles Targeting Mechanism

The administration of a drug-loaded NP directly to the specific site is generally difficult. This is because there is a systemic spreading of a disease all over the cells and tissues, and the nature of each disease is different. Paul Ehrlich's concept of a magic bullet for targeted drug delivery proved as a boon for homing a drug to desired site and reducing non-specific drug distribution, leading to reduced side effects (Imae 2012). Based on the physiological characteristics of the tumor microenvironment, there are two major strategies for drug delivery, e.g., passive targeting and active targeting.

4.1 Passive Targeting Mechanism

In the passive targeting mechanism, nanoparticles targeted to cancer sites are based on particle size. Enhanced permeation and retention (EPR) is the main phenomenon by which the NPs are accumulated into targeted tissues, like the tumor. Due to high vascular permeation, the NPs get accumulated in the target site and remain trapped there, owing to low lymphatic drainage. The locally trapped NPs release drugs in a controlled manner over a prolonged period of time. The EPR effect by which a nanoparticle gets accumulated into the inflamed target tissue is first explained by Matsumura and Maeda in 1986 (Haley and Frenkel 2008; Torchilin 2011). Hydrophilic surface and size range of less than 200 nm are ideal characteristics of NPs for drug delivery application (Fang et al. 2011). Some of the clinically approved passive targeted nanocarrier-based drug products are Doxil™, Onivyde™, DaunoXome™--US, Marqibo™, Mepact™, Abraxane™, Myocet™-Europe, Genexol-PM™-Korea; SMANCS™-Japan, and these drugs are now available for the clinical use (Shi et al. 2017).

4.2 Active Targeting Mechanism

In active targeting, the NPs' surface, attached with small ligands, whose selective binding ability to the specific receptor gets assimilated by the target cells. The ligand attached to the surface of NPs has the ability to bind specific receptors which are overexpressed at the target cells (Nacev et al. 2010). The targeting ligands, which can be attached to the surface of NPs are small molecules, peptides (arginyl glycylaspartic acid), antibodies, and its fragments, lipoproteins, lectins, glycoproteins (transferrin), hormones, folic acid, polysaccharides, growth factors, and nucleic acids (Danhier et al. 2010). After interacting with the receptor, the NPs are taken up by cells through endocytosis for cancer treatment. This is due to the overexpression of specific receptors on cancer cells (Patil et al. 12009), for example, transferrin receptor, overexpressed in breast cancer, epidermal growth factor receptor (EGFR), and folate in ovarian or lung cancer (Pirollo and Chang 2008).

5 Nanoparticles in Clinical Use

A nanomedicine may take up to two decades to enter the market after its initial discovery (Bawa et al. 2008). The pharmaceutical company should file a patent to protect intellectual property during the drug discovery and development process. The development and commercialization of a new nano pharmaceutical drug product requires time and money from ideation and preclinical research, industrial development, human clinical trials, regulatory approval, and industrial production and marketing (Conde and Artzi 2015). As per FDA, each pharmaceutical product should be evaluated for its prescribed medical application. For instance, an oral dosage form like tablet and capsules should be evaluated in the simulated digestive tract and an injectable drug should be tested in the bloodstream. Several nano-based pharmaceutical medicines have been successfully developed and are now available in the market. Nanomedicine is utilized by many patients in day-to-day use. Some of the polymer-based nanoparticles are summarized in Table 1 (Farjadian et al. 2019).

6 Types of Nanoparticles Used as Drug Delivery System to Cancer

6.1 Polymeric Nanoparticles

In the case of chemotherapy, polymeric nanomaterial has been found to have great potential to improve delivery systems. The polymeric NPs offer a favorable drug delivery method that enhances therapeutic efficacy and decreases toxicity of therapeutic agent by limiting the interaction of the drug with healthy cells via forming a protective layer around it. Therefore, they effectively decrease the side effects, more dosage can be loaded on NPs to increase the bioavailability (Paroha et al. 2018), providing continuous and controlled release, increased patient compliance, and deliver more than one drugs to a single carrier (Abbruzzese et al. 1991; Gabizon 1992; Northfelt et al. 1996; ten Hagen et al. 2002). All these indicate that pharmacokinetic properties of the final drugs are usually based on the pharmacokinetic properties of the particles until the drug remains entrapped and gets released (Allen and Cullis 2004; Brewer et al. 2011). The delivery devices used are made of erodible polymers, which are more better option than the non-erodible ones, because they usually get degraded and slowly disappear after delivery. In recent years most commonly used polymers for NPs preparation are polylactic acid (PLA), hyaluronic acid (HA), poly (D, L-lactic-co-glycolic acid) (PLGA), chitosan (CS), and PEG (Saneja et al. 2018a; Saneja et al. 2018b; Saneja et al. 2019). Different studies have been done to prove the effectiveness of an anticancer drug carrier by reducing renal secretion and preventing uptake by reticuloendothelial system (Saneja et al. 2017a; Saneja et al. 2017b). Therefore, in the current scenario, the research is focused on the advancement of these polymers with "smart" technologies that respond to environmental conditions.

This can be divided into two categories: (1). Site-targeting, where the particles seek out and attach themselves to specific and diseased cells using molecules such as ligands, aptamers, and antibodies (Brewer et al. 2011); (2). Site triggering, in this category physical or chemical changes in the environment result in rapid release of the drug. In site targeting, polymeric NPs are decorated on the outer surface to attain suitable functions to suit the environment of the tumor or drug properties. Typical modifications on the surface include charge modification, siRNA amphipathic compound graft, and bioactive peptide insertion. These modified polymeric

Table 1 Examples of polymer-based marketed nanoparticles for clinical use

Drug/generic name	Brand name	Company name	Technology	Indication	FDA approval
Certolizumabpegol	Cimzia®	UCB, Brussels, Belgium	Pegylated blocker of tumor necrosis factor-alpha (TNF-α)	Rheumatoid arthritis	2008
Pegademase bovine	Adagen®	Enzon, Inc., NJ, USA	Pegylated adenosine deaminase	Immunodeficiency disorder	1990
Pegfilgrastim	Neulasta®	Amgen, Inc., CA, USA	Pegylated form of filgrastim	Febrile neutropenia	2002
L-asparaginase	Oncaspar®	Enzon Pharmaceuticals Inc., NJ, USA	Pegylated-l-asparaginase	Acute lymphoblastic leukemia, and chronic myelogenous leukemia	1994
Recombinant human alfa-2a interferon	Pegasys®®	Genentech USA, INC., CA, USA	Conjugated branched peg (40 kDa)	Hepatitis c and HBeAg positive chronic hepatitis-B	2002
Pegvisomant (b2036-peg)	Somavert®	Pfizer Pharmaceuticals, CT, USA	Pegylated Analog of human growth hormone (gh)	Acromegaly	2003
Pegatinib sodium	Macugen®	Eyetech Pharmaceuticals marketed by Pfizer Inc	Aptamer-based nano-vectors	Anti-angiogenesis agent	2004
Epoetin β	Mircera®	European Commission and the FDA	Epoetin β (epo) conjugated to methoxy-peg	Anemia	2007
Alpha interferon (INF)	Peg-intron®	Merck Sharp & Dohme B. V.	INF molecule Conjugated to a mono peg	Chronic hepatitis-C	2001
Pegloticase	Krystexxa®	Puricase, NJ, USA	Pegylated amino acid	Refractory chronic gout	2010
Peg-IFN-β-1a	Plegridy®	Biogen Inc.	PEG conjugated recombinant IFN-β	Relapsing remitting Multiple sclerosis (RRMS)	2014
Anti-hemophilic factor	Adynovate®	Takeda Pharma	Recombinant pegylated anti-hemophilic factor	Hemophilia A	2016
Paclitaxel	Abraxane®	Celgene Pharmaceutical Co. Ltd	Albumin–NPs bound to paclitaxel	Metastatic breast cancer	2005

nanoparticles accumulate the loaded drugs within the cancer tissue precisely and thereby protect the damage to nearby healthy cells or tissues (Abbruzzese et al. 1991; Brewer et al. 2011; Cummings and McArdle 1986; Lu and Low 2002; Yoo and Park 2004). In site-triggered nanomaterials, pH and thermo-sensitive particles have been used. In a pH-sensitive method, the release of drugs at the tumor site is done by the advantage of the low pH value of the tumor environment.

6.2 Solid Lipid Nanoparticles

Solid lipid nanoparticles (SLNs) have emerged as a promising and effective alternative drug delivery system for cancer treatment. It is composed of colloidal particles of the size of a submicron, having a diameter between 50 and 1000 nm, and contains physiological lipids that remain stable at room temperature and body temperature (Garcia-Pinel et al. 2019). In solid lipid NPs, the used solid lipid forms a matrix to encapsulate the drug and contains mixtures of mono-, di-, or triglycerides, complex glycerides, and fatty acids. This matrix is strengthened or stabilized by a combination of polymers or surfactants. SLNs have certain important benefits, such as long-term physical stability, site-specific identification, and possible controlled release of lipophilic as well hydrophilic drugs, low cost, labile drug protection, non-toxicity, and ease of preparation (Mydin and Moshawih 2019). Furthermore, in terms of toxicity, these NPs possess lower toxic effects in contrast to human granulocytes. All these significant benefits make it an important component of drug delivery systems. In contrast to advantages, solid lipid NPs exert some disadvantages also, such as medium drug load and drug withdrawal because of the crystallization process during storage conditions. All the unique features of solid lipid NPs makes it a standout method for drug delivery of therapeutic anticancer drugs via encapsulation to enhance oral bioavailability of anticancer drug, giving protection to drug and alleviating adverse effect by reducing dose by active targeting at the action site (Bayon-Cordero et al. 2019; Garcia-Pinel et al. 2019; Mydin and Moshawih 2019). In 2018, Pindiprolu et al. proved that improved niclosamide-induced SLNs, improved uptake by cells, and anticancer efficacy in the case of breast cancer cells. Further in 2018, Eskiler et al. formulated SLNs of talazoparib that improve the therapeutic index of talazoparib against breast cancer cells. This development was a result of reducing toxicity and homologous resistance. In many studies, researchers have found that SLNs of resveratrol have shown great potential in the treatment of breast cancer as compared to nascent resveratrol (GuneyEskiler et al. 2018). Also, some other studies have shown that SLNs could also be utilized as a vehicle for increasing the effectiveness of floxuridine in the treatment of cancer. Another example showing the benefits of SLNs in cancer treatment is illustrated by indirubin, anticancer agent (hydrophobic) part of traditional Chinese medicine, which has shown remarkable improvement in the efficacy of anticancer effect of a hydrophobic drug after being loaded into SLNs (GuneyEskiler et al. 2018; Oliveira et al. 2018).

6.3 Magnetic Nanoparticles

Magnetic NPs have received a great deal of attention in anticancer therapy because of their distinctive physicochemical properties, facile synthesis, in contrast to magnetic resonance imaging, low toxicity, trouble-free surface decoration, as well as significant biodegradability that helps them work as excellent imaging agents, as well as cancer delivery vehicles for theranostics (Mukherjee et al. 2020). Magnetic NPs act as competent agents due to their increased magnetic field when using external magnetic fields and excellent resting skills. Therefore, magnetic NPs are widely used in a variety of cancer therapy including biosensors, MRI imaging, theranostics, magnetic hyperthermia, delivery, photothermal ablation, and photodynamic therapy. Surface charge, size, and shape of magnetic NPs can be used for many cancer treatment programs, such as "size-dependent hyperthermia treatment and theranostics." The final surface charge of composite NPs could be used to bind nucleic acids electrostatically or to increase the timing of systematic dispersion. Although magnetic NPs have been widely used in cancer treatment, some studies are done on the use of various shapes of NPs like nanocube morphologies and hollow rod morphologies for drug delivery for chemo-photothermal treatment (Gleich and Weizenecker 2005; Sun et al. 2008).

Magnetic NPs are developed using nickel, gadolinium, cobalt, and Prussian blue, but magnetic iron oxide NPs remain relatively extensively studied for magnetic NPs-based cancer therapy due to low toxicity and magnetic resonance imaging with strong comparative properties. Magnetic NPs usually consist of a polymer coating over a magnetic core shell. Therefore, magnetic NPs increase colloidal stability, permit for electrostatic or covalent binding of medical cargo, target-specific moieties and influence specific properties, i.e., pharmacokinetics, clearance rate, systemic toxicity, non-specific protein adsorption/cell interactions, and continuous drug release (Cole et al. 2011; Koenig and Kellar 1995; Laurent et al. 2008; Yang et al. 2017).

In addition, magnetic particle imaging was a major eye-catching tool using magnetic NPs. Many other studies have been assessing the performance of magnetic NPs, condition of the field used, and particle structure for magnetic resonance imaging. Importantly, the USA FDA has approved several magnetic NPs drugs such as Endorem®, Feraheme®, Gastromark®, Ferumoxytol®, Lumiren®, Combidex®, Feridex, and Radiogardase® for different uses in iron deficiency, imagining of lymph node metastases, iron replacement therapy, comparative magnetic resonance imaging agents and oral antidotes to heavy metal contamination in people. In addition, the "European Medicines Agency" recently approved NanoTherm® treatment for glioblastoma multiform (Harisinghani et al. 2003; Singh et al. 2008).

6.4 Mesoporous Silica Nanoparticles

Mesoporous silica NPs have transformed the vision of controlled drug delivery systems. Their internal mesoporous are usually 2–6 nm with large pore volume, i.e., 0.6–1 cm^3/g and surface area of 700–1000 m^2/g, with adjustable size of 50–200 nm. The shape, durability, and flexibility of the surface modification turned out to be the ideal advantages that make them efficient nanosystems. Vallet-Regi et al. first introduced the mesoporous silica NPs as a potent drug delivery system and strong efforts have been devoted to the development of multidisciplinary mesoporous silica NPs in the treatment of various diseases, with special stress on cancer treatment. High drug loading capacity and textural properties of mesoporous silica NPs have played an important role in the execution of these nanosystems as an ideal drug delivery system (Dilnawaz 2019). The pore width acts as a loading size selector for biologically active molecules within it, which controls the rate of release, thus acting as a boundary that controls the processes of molecular distribution in the body environment. The surface area ascertains the load capacity of the nanoplatform because the available high-contact area significantly increases the number of guest molecules included. The pore volume can also influence the number of drugs loaded where it is aimed at complete mesoporous filling. In addition, their hemocompatibility has also been proven (Pasqua et al. 2016).

Mesoporous silica NPs display unique features that make them suitable for handling, protecting, and transporting drugs in the target area or cancer cell site. It is possible to insert targeting agents into the external surface area of mesoporous silica NPs to allow them to target the unhealthy cancer tissues, and this increases specificity and reduces the undesirable side effects. Another important challenge is to inhibit premature disbursement of drugs before reaching their intended destination. For this problem, the pore penetration of mesoporous silica NPs can be protected using stimuli-responsive gate guards. Therefore, exposure to external or internal stimuli can stimulate pore openings to allow the flow of drugs. Stimuli-responsive behavior could be achieved by anchoring pore-blocking caps at all links that could be cleaved when a given stimulus is available. These triggers are classified as internal, such as pH, enzymes, redox energy, etc., and external, such as light or ultrasound and magnetic fields, which could be used remotely by a clinician (Dilnawaz 2019; Pasqua et al. 2016).

6.5 Gold Nanoparticles

Gold NPs have sizes ranging from 5 to 400 nm, and are used for their tumor diagnosis and antitumor roles. According to literature, the gold NPs have a tendency to deposit in tumor sites (Grigore et al. 2017), and are thus used for in vitro and in vivo imaging, in vitro assay of cancer cells, tumor therapy, and drug delivery. Different subtypes of gold NPs, used for cancer therapy, include gold nanorods, nanocage, nanospheres, and nanoshells (Grigore et al. 2017). These subtypes are prepared by different methods such as the Turkevich method, chemical methods, Brust-Schiffrin method, electrochemical method, biological method, and ionic liquid methods (Herizchi et al. 2016).

Tumor cells are killed by gold NPs by mechanical method, by photothermal therapy and they are also used as a delivery system for anticancer agents (Grigore et al. 2017). In photothermal therapy, the gold NPs absorb light and the surrounding temperature increases to 70–80 °C which results in the killing of cancer cells (Cai et al. 2008). A number of anticancer drugs are loaded covalently (prodrug form) and non-covalently (in active form) in gold NPs (Lim et al. 2011). For example, the tumor necrosis factor-alpha (TNF-α) has anticancer properties, but it is toxic systemically. However, when it is loaded with PEG-coated gold NPs, its systemic toxicity decreases, and its tumoricidal activity increases (Cai et al. 2008). The pictorial representation of different nanoparticles used in drug delivery has been summarized in Fig. 3.

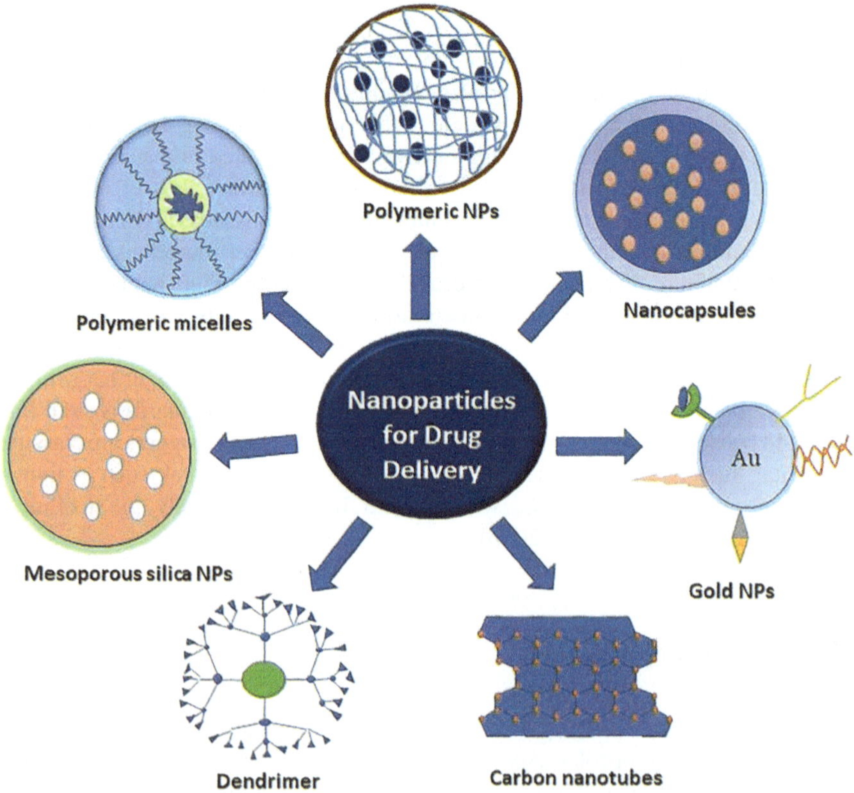

Fig. 3 The pictorial representation of various nanoparticles used in drug delivery

6.6 Dendrimers

Dendrimers are nanoscale tree-like structures. They have a polymeric core macromolecule surrounded by symmetric branching units. Dendrimers are synthesized by two methods: divergent and convergent methods. In a divergent method, the core of the dendrimer is synthesized first and after that branches are attached. In a convergent method, synthesis starts from the outermost branched arms. These days, novel dendrimers are used for cancer diagnosis and its treatment (Abbasi et al. 2014). Drugs can be entrapped non-covalently inside the interior region of dendrimers or attached covalently to the surface (Hu et al. 2011). Positively charged dendrimers show toxicity, and thus the dendrimers having neutral molecules are preferred (Kesharwani et al. 2017). In a distinct study, peptide-dendrimer-doxorubicin (DOX) delivery systems have been developed using pH low insertion peptides for pH-responsive drug delivery. They found that pH low insertion peptide DOX system showed pH dependent, temporally controlled release of DOX to cytosol and also enhances the DOX action, in contrast to free drug (Burns and Delehanty 2018). Ozturk et al. (2017) have synthesized the gemcitabine-loaded and PEG-cored PAMAM dendrimer-based delivery system to target Flt-1 receptor for enhancing the efficacy of gemcitabine in pancreatic cancer. They performed in vivo study in mice model and concluded that these complexes are engulfed by Flt-1 expressing pancreatic cancer cells, which amplify the activity of gemcitabine (Ozturk and Esendagri 2017). Dendrimer-based delivery system is a developing strategy for cancer therapy. The studies related to dendrimer's biocompatibility, metabolic mechanism, and enhanced efficacy of treatment still need to be explored (Xiong et al. 2018).

6.7 Polymeric Micelles

Self-assembled nanoscopic polymeric micelles consist of a hydrophobic core and hydrophilic shell. Polymeric micelles (PMs) are formed by the spontaneous arrangement of amphiphilic block copolymers above the critical micelle concentration (CMC) in an aqueous solution. The inner core which is the hydrophobic part of PMs encapsulates the poorly water-soluble drugs, whereas the outer hydrophilic shell stabilizes PMs against remembrance in vivo by the reticuloendothelial system (RES) (Avramovic et al. 2020; Wang et al. 2018). Polymeric micelles are nanosized (10–200 nm) and have a loading efficiency of about 80% (w/w). Block copolymers and graft copolymers are used for the formation of PMs. The first PMs are unstable and are used to solubilize the hydrophobic drugs to overcome this next generation PMs have been developed to achieve a high encapsulation and prolonged circulation. For the treatment of cancer, next-generation intelligent PMs have been developed by 3C approach (core-crosslinking, covalent drug entrapment, and combination therapies) (Varela-Moreira et al. 2017; Wan et al. 2020).

The core-crosslinking and covalent drug entrapment has improved the stability of polymeric micelles in systemic circulation by passive targeting through EPR effect. The passively targeted micelles are neutral in charge and large in size, resulting in less cellular uptake. However, the activity of passively targeted medicines can be enhanced by combination therapies or active targeting (Mochida et al. 2017; Yu et al. 2019). The length of a linker used for the conjugation of a drug molecule to polymer, also plays an important role in the drug release pattern (Bregadze et al. 2020; Dubey et al. 2021). Different types of ligand-targeting moieties such as small organic molecules (e.g., folate), sugar moieties, peptides, and monoclonal antibodies have been used so far. Ligand-modified polymeric micelles have improved the anticancer effect of drugs because of receptor-mediated transport mechanisms (Talelli et al. 2012).

6.8 Carbon Nanotubes

Carbon nanotubes (CNTs) are used as a novel carrier system for chemotherapeutic agents. Because of large surface area of CNTs, they are used as a carrier for both small and large therapeutic agents. CNTs can be engineered with some functional groups to change the surface to alter the physical and biological properties (Elhissi et al. 2012). CNTs can be used for the treatment of cancer as topoisomerase inhibitors, platinum-based drug carriers, anti-microtubule carriers, and inducers of photothermal and photodynamic effects. Due to photodynamic property, CNTs convert oxygen to singlet oxygen, and this singlet oxygen destroys cancer cells without damaging normal cells (Son et al. 2016). CNTs have been used for lymph node cancer because of their long-term retention ability in lymph nodes (Elhissi et al. 2012).

Yong et al. have been using gemcitabine-loaded magnetic multiwalled carbon nanotubes (MCNTs) in mice models and have found that after subcutaneous injection of both MCNTs formulation and conventional gemcitabine formulation, MCNTs have higher activity against lymph node metastasis (Yang et al. 2009). Dhar and co-workers have reported a complex of cisplatin and folic acid named as longboat delivery system, engulfed by the cancer cells via endocytosis mechanism. After engulfing the delivery system by cancer cells, the drug is released and it interacts with nuclear DNA which damages cancer cells (Dhar et al. 2008). Wu et al. have developed 10-hydroxycamptothecin (HCPT)-loaded MCNTs which show better antitumor activity and longer circulation time than conventional HCPT formulation, when tested in vivo by single photon emission computer tomography and ex vivo by gamma-scintillation counting analyses (Wu et al. 2009). Liu et al. have also reported that PEG-conjugated paclitaxel single wall CNTs, when tested in vivo have a better tumor-targeted accumulation with lower toxicity than Taxol (Liu et al. 2008). Thus, CNTs will become the strongest tool in cancer therapy.

6.9 Nano-Capsules

Nano-capsule (NC) is a vesicular system having a core containing an active molecule, which is surrounded by a polymer coating. NC delivers active molecules to the targeted tumor site by a change in the surrounding environment. Different types of NC such as lipid NC, layer-by-layer NC, core shell NC, liposome-like NC, protein NC, polymeric, and shell cross-linked NC have been used in chemotherapy (Yurgel et al. 2013).

Rebecca and co-workers have investigated that curcumin-loaded oil cored PEGylated NC has better pharmacokinetics and bioavailability at colon solid tumor sites than conventional curcumin after i.v. injection (Klippstein et al. 2015). Kheira and Delia et al. have reported that 5 FU-loaded hybrid magnetic NCs are pointed by an external magnetic field toward the tumor cells. This movement of magnetic NCs can induce the hyperthermic effect. This effect has been united with a controlled release of 5 FU, destroying the tumor cells (Dellali et al. 2020). Son et al. designed doxorubicin hydrochloride loaded hollow photosensitizer NC to damage cancer tissues. This study showed that combination of anticancer drugs and the photosensitizer in NC formulation is more fruitful than chemotherapy or Photodynamic therapy alone (Son et al. 2011). As shown by Hureaux, NC formulation shows a better chemotherapeutic effect when paclitaxel-loaded lipid nano-capsules are nebulized in malignant lung disease in place of conventional paclitaxel formulation (Hureaux et al. 2009).

6.10 Quantum Dots

Quantum dots (QDs) are semiconductor crystals with nearly spherical shape and have metalloid crystalline semiconductor that regulates fluorescence emission (Smith and Nie 2010). The main advantages of QDs are small size, good intracellular uptake, controlled drug release, and easy surface modifications. Apart from drug delivery application, the QDs are applied for biological optical detection (Michalet et al. 2005), and cellular and intracellular targeting (Hild et al. 2008). QDs have exhibited a unique physical, chemical, and optical property which facilitates an in-depth study of NPs interactions with biological proteins inside the body by real-time monitoring of NP, intracellular uptake, drug release, and biodistribution. QDs labelled NPs offer a powerful platform to track the behavior of novel drug delivery vehicles which is useful in the design and optimization of physicochemical properties for specific drug delivery applications (Probst et al. 2013). QDs can help in the targeting and improvement of the bioavailability of drugs when used for drug delivery applications. They can be used for localized treatments of specific disease sites because of their specific size and shape. In QDs, hydrophobic drugs can be incorporated between the inorganic cores of the amphiphilic polymer-coated layer, whereas hydrophilic therapeutic agents and biomolecules can be immobilized on top of the hydrophilic side of the amphiphiles (Cui et al. 2010). Recently, multifunctional QDs have been prepared for image-diagnosed chemotherapy. The polymer captured and bioconjugated QDs probes have been developed for tumor targeting and in vivo imaging (Bharali et al. 2005).

7 Conclusion

NPs have received tremendous attention from the scientific community in the field of drug delivery, diagnosis, and gene delivery. NPs for drug delivery applications are useful in improving the pharmacokinetic and therapeutic efficacy of conventional drugs. The critical quality attributes which need to be considered during development of efficient NPs are their shape, particle size, surface characteristics, solid-state properties or crystal structure, and physical stability. In the case of cancer therapy, the encapsulation of a drug into NPs not only protects the drug from enzymatic degradation in the circulatory system but also provides targeted release of drug at the desired site. The major beneficial outcome of NPs in drug delivery includes improved drug solubility, enhanced drug stability, favorable biodistribution, and reduced dose-dependent toxicity. Almost all types of NPs including, polymeric, solid lipid, magnetic, mesoporous silica, gold NPs, micelles, carbon nanotube, quantum dots, nano-capsules, and dendrimers have been evaluated for their applicability for multifunctional utilities that can be applied for selective treatment of cancers. After the advancement of ligand-targeted NPs one can detect cancer cells, detect their location in the body, killing specific to the cancer cells with no side effects in surrounding tissues. Despite great success, there is still scope for improvement in the development of effective NPs for clinical application. These improvements include control of particle size and protection of drug in the circulatory system, scale-up of formulation batches, cost-effectiveness, long-term safety, and in vivo biodegradability. Scientific community is actively involved to tackle these issues and bring this therapy to individual patients in clinics. NPs witness great clinical involvement in future decades for drug delivery for diagnosis and cure of several diseases while the main challenges are the barriers associated with the expansion of novel nanomaterials for targeted drug delivery.

References

Abbasi E, Aval SF, Akbarzadeh A, Milani M, Nasrabadi HT, Joo SW, Hanifehpour Y, Nejati-Koshki K, Pashaei-Asl R (2014) Dendrimers: synthesis, applications, and properties. Nanoscale Res Lett 9(1):247. https://doi.org/10.1186/1556-276X-9-247

Abbruzzese JL, Grunewald R, Weeks EA, Gravel D, Adams T, Nowak B, Mineishi S, Tarassoff P, Satterlee W, Raber MN (1991) A phase I clinical, plasma, and cellular pharmacology study of gemcitabine. J Clin Oncol 9(3):491–498. https://doi.org/10.1200/JCO.1991.9.3.491

Adams ML, Lavasanifar A, Kwon GS (2003) Amphiphilic block copolymers for drug delivery. J Pharm Sci 92(7):1343–1355. https://doi.org/10.1002/jps.10397

Alam N, Dubey RD, Kumar A, Koul M, Sharma N, Sharma PR, Chandan BK, Singh SK, Singh G, Gupta PN (2015) Reduced toxicological manifestations of cisplatin following encapsulation in folate grafted albumin nanoparticles. Life Sci 142:76–85. https://doi.org/10.1016/j.lfs.2015.10.019

Allen TM, Cullis PR (2004) Drug delivery systems: entering the mainstream. Science 303(5665): 1818–1822. https://doi.org/10.1126/science.1095833

Avramovic N, Mandic B, Savic-Radojevic A, Simic T (2020) Polymeric nanocarriers of drug delivery systems in cancer therapy. Pharmaceutics 12(4). https://doi.org/10.3390/pharmaceutics12040298

Bawa R, Melethil S, Simmons WJ, Harris D (2008) Nanopharmaceuticals: patenting issues and FDA regulatory challenges. SciTech Lawyer 5(2):10–15

Bayon-Cordero L, Alkorta I, Arana L (2019) Application of solid lipid nanoparticles to improve the efficiency of anticancer drugs. Nano 9(3). https://doi.org/10.3390/nano9030474

Bertrand N, Leroux J-C (2012) The journey of a drug-carrier in the body: an anatomo-physiological perspective. J Control Release 161(2):152–163. https://doi.org/10.1016/j.jconrel.2011.09.098

Bharali DJ, Lucey DW, Jayakumar H, Pudavar HE, Prasad PN (2005) Folate-receptor-mediated delivery of InP quantum dots for bioimaging using confocal and two-photon microscopy. J Am Chem Soc 127(32):11364–11371. https://doi.org/10.1021/ja051455x

Bregadze VI, Sivaev IB, Dubey RD, Semioshkin A, Shmal'ko AV, Kosenko ID, Lebedeva KV, Mandal S, Sreejyothi P, Sarkar A, Shen Z, Wu A, Hosmane NS (2020) Boron-containing lipids and liposomes: new conjugates of cholesterol with polyhedral boron hydrides. Chem Eur J 26(61):13832–13841. https://doi.org/10.1002/chem.201905083

Brewer E, Coleman J, Lowman A (2011) Emerging technologies of polymeric nanoparticles in cancer drug delivery. J Nanomater 2011:408675. https://doi.org/10.1155/2011/408675

Burns KE, Delehanty JB (2018) Cellular delivery of doxorubicin mediated by disulfide reduction of a peptide-dendrimer bioconjugate. Int J Pharm 545(1):64–73. https://doi.org/10.1016/j.ijpharm.2018.04.027

Cai W, Gao T, Hong H, Sun J (2008) Applications of gold nanoparticles in cancer nanotechnology. Nanotechnol Sci Appl 1:17

Cho K, Wang X, Nie S, Chen Z, Shin DM (2008) Therapeutic nanoparticles for drug delivery in cancer. Clin Cancer Res 14(5):1310–1316. https://doi.org/10.1158/1078-0432.CCR-07-1441

Cole AJ, Yang VC, David AE (2011) Cancer theranostics: the rise of targeted magnetic nanoparticles. Trends Biotechnol 29(7):323–332. https://doi.org/10.1016/j.tibtech.2011.03.001

Conde J, Artzi N (2015) Are RNAi and miRNA therapeutics truly dead? Trends Biotechnol 33(3): 141–144. https://doi.org/10.1016/j.tibtech.2014.12.005

Cui H, Webber MJ, Stupp SI (2010) Self-assembly of peptide amphiphiles: from molecules to nanostructures to biomaterials. Pept Sci 94(1):1–18. https://doi.org/10.1002/bip.21328

Cummings J, McArdle CS (1986) Studies on the in vivo disposition of adriamycin in human tumours which exhibit different responses to the drug. Br J Cancer 53(6):835–838. https://doi.org/10.1038/bjc.1986.141

Danhier F, Feron O, Preat V (2010) To exploit the tumor microenvironment: passive and active tumor targeting of nanocarriers for anti-cancer drug delivery. J Control Release 148(2): 135–146. https://doi.org/10.1016/j.jconrel.2010.08.027

Davis ME, Chen Z, Shin DM (2008) Nanoparticle therapeutics: an emerging treatment modality for cancer. Nat Rev Drug Discov 7(9):771–782. https://doi.org/10.1038/nrd2614

Decuzzi P, Pasqualini R, Arap W, Ferrari M (2008) Intravascular delivery of particulate systems: does geometry really matter? Pharm Res 26(1):235. https://doi.org/10.1007/s11095-008-9697-x

Dellali KZ, Rata DM, Popa M, Djennad M, Ouagued A, Gherghel D (2020) Antitumoral drug: loaded hybrid nanocapsules based on chitosan with potential effects in breast cancer therapy. Int J Mol Sci 21(16):5659. https://doi.org/10.3390/ijms21165659

Dellian M, Yuan F, Trubetskoy VS, Torchilin VP, Jain RK (2000) Vascular permeability in a human tumour xenograft: molecular charge dependence. Br J Cancer 82(9):1513–1518. https://doi.org/10.1054/bjoc.1999.1171

Dhar S, Liu Z, Thomale J, Dai H, Lippard SJ (2008) Targeted single-wall carbon nanotube-mediated Pt(IV) prodrug delivery using folate as a homing device. J Am Chem Soc 130(34): 11467–11476. https://doi.org/10.1021/ja803036e

Dilnawaz F (2019) Multifunctional mesoporous silica nanoparticles for cancer therapy and imaging. Curr Med Chem 26(31):5745–5763. https://doi.org/10.2174/0929867325666180501101044

Duan Y, Sun X, Gong T, Wang Q, Zhang Z (2006) Preparation of DHAQ-loaded mPEG-PLGA-mPEG nanoparticles and evaluation of drug release behaviors in vitro/in vivo. J Mater Sci Mater Med 17(6):509–516. https://doi.org/10.1007/s10856-006-8933-3

Dubey RD, Alam N, Saneja A, Khare V, Kumar A, Vaidh S, Mahajan G, Sharma PR, Singh SK, Mondhe DM, Gupta PN (2015) Development and evaluation of folate functionalized albumin nanoparticles for targeted delivery of gemcitabine. Int J Pharm 492(1):80–91. https://doi.org/10.1016/j.ijpharm.2015.07.012

Dubey RD, Klippstein R, Wang JT-W, Hodgins N, Mei K-C, Sosabowski J, Hider RC, Abbate V, Gupta PN, Al-Jamal KT (2017) Novel hyaluronic acid conjugates for dual nuclear imaging and therapy in CD44-expressing tumors in mice in vivo. Nano 1(1):59–79. https://doi.org/10.7150/ntno.17896

Dubey RD, Saneja A, Gupta PK, Gupta PN (2016a) Recent advances in drug delivery strategies for improved therapeutic efficacy of gemcitabine. Eur J Pharm Sci 93:147–162. https://doi.org/10.1016/j.ejps.2016.08.021

Dubey RD, Saneja A, Qayum A, Singh A, Mahajan G, Chashoo G, Kumar A, Andotra SS, Singh SK, Singh G, Koul S, Mondhe DM, Gupta PN (2016b) PLGA nanoparticles augmented the anticancer potential of pentacyclictriterpenediol in vivo in mice. RSC Adv 6(78):74586–74597. https://doi.org/10.1039/C6RA14929D

Dubey RD, Sarkar A, Shen Z, Bregadze VI, Sivaev IB, Druzina AA, Zhidkova OB, Shmal'ko AV, Kosenko ID, Mandal S, Hosmane NS (2021) Effects of linkers on the development of liposomal formulation of cholesterol conjugated cobalt Bis(dicarbollides). J Pharm Sci 110(3):1365–1373. https://doi.org/10.1016/j.xphs.2020.12.017

Elhissi AM, Ahmed W, Hassan IU, Dhanak VR, D'Emanuele A (2012) Carbon nanotubes in cancer therapy and drug delivery. J Drug Deliv 2012:837327. https://doi.org/10.1155/2012/837327

Fang J, Nakamura H, Maeda H (2011) The EPR effect: unique features of tumor blood vessels for drug delivery, factors involved, and limitations and augmentation of the effect. Adv Drug Deliv Rev 63(3):136–151. https://doi.org/10.1016/j.addr.2010.04.009

Farjadian F, Ghasemi A, Gohari O, Roointan A, Karimi M, Hamblin MR (2019) Nanopharmaceuticals and nanomedicines currently on the market: challenges and opportunities. Nanomedicine 14(1):93–126

Farokhzad OC, Langer R (2009) Impact of nanotechnology on drug delivery. ACS Nano 3(1): 16–20. https://doi.org/10.1021/nn900002m

Gabizon AA (1992) Selective tumor localization and improved therapeutic index of anthracyclines encapsulated in long-circulating liposomes. Cancer Res 52(4):891–896

Garcia-Pinel B, Porras-Alcala C, Ortega-Rodriguez A, Sarabia F, Prados J, Melguizo C, Lopez-Romero JM (2019) Lipid-based nanoparticles: application and recent advances in cancer treatment. Nano 9(4). https://doi.org/10.3390/nano9040638

Gleich B, Weizenecker J (2005) Tomographic imaging using the nonlinear response of magnetic particles. Nature 435(7046):1214–1217. https://doi.org/10.1038/nature03808

Grigore ME, Holban AM, Grumezescu AM (2017) Chapter 9 - Nanotherapeutics in the management of infections and cancer. In: Razavi M, Thakor A (eds) Nanobiomaterials science, Development and Evaluation. Woodhead Publishing, Duxford, pp 163–189

GuneyEskiler G, Cecener G, Egeli U, Tunca B (2018) Synthetically lethal BMN 673 (Talazoparib) loaded solid lipid nanoparticles for BRCA1 mutant triple negative breast cancer. Pharm Res 35(11):218. https://doi.org/10.1007/s11095-018-2502-6

Gupta PN, Jain S, Nehate C, Alam N, Khare V, Dubey RD, Saneja A, Kour S, Singh SK (2014) Development and evaluation of paclitaxel loaded PLGA:poloxamer blend nanoparticles for cancer chemotherapy. Int J Biol Macromol 69:393–399. https://doi.org/10.1016/j.ijbiomac.2014.05.067

ten Hagen TL, Seynhaeve AL, van Tiel ST, Ruiter DJ, Eggermont AM (2002) Pegylated liposomal tumor necrosis factor-alpha results in reduced toxicity and synergistic antitumor activity after systemic administration in combination with liposomal doxorubicin (Doxil) in soft tissue sarcoma-bearing rats. Int J Cancer 97(1):115–120. https://doi.org/10.1002/ijc.1578

Haley B, Frenkel E (2008) Nanoparticles for drug delivery in cancer treatment. Urol Oncol 26(1): 57–64. https://doi.org/10.1016/j.urolonc.2007.03.015

Harisinghani MG, Barentsz J, Hahn PF, Deserno WM, Tabatabaei S, van de Kaa CH, de la Rosette J, Weissleder R (2003) Noninvasive detection of clinically occult lymph-node metastases in prostate cancer. N Engl J Med 348(25):2491–2499. https://doi.org/10.1056/NEJMoa022749

Harris JM, Martin NE, Modi M (2001a) Pegylation. Clin Pharmacokinet 40(7):539–551. https://doi.org/10.2165/00003088-200140070-00005

Harris JM, Martin NE, Modi M (2001b) Pegylation: a novel process for modifying pharmacokinetics. Clin Pharmacokinet 40(7):539–551. https://doi.org/10.2165/00003088-200140070-00005

Herizchi R, Abbasi E, Milani M, Akbarzadeh A (2016) Current methods for synthesis of gold nanoparticles. Artif Cells Nanomed Biotechnol 44(2):596–602. https://doi.org/10.3109/21691401.2014.971807

Hild WA, Breunig M, Goepferich A (2008) Quantum dots - nano-sized probes for the exploration of cellular and intracellular targeting. Eur J Pharm Biopharm 68(2):153–168. https://doi.org/10.1016/j.ejpb.2007.06.009

Hu J, Su Y, Zhang H, Xu T, Cheng Y (2011) Design of interior-functionalized fully acetylated dendrimers for anticancer drug delivery. Biomaterials 32(36):9950–9959. https://doi.org/10.1016/j.biomaterials.2011.09.016

Hureaux J, Lagarce F, Gagnadoux F, Vecellio L, Clavreul A, Roger E, Kempf M, Racineux J-L, Diot P, Benoit J-P, Urban T (2009) Lipid nanocapsules: ready-to-use nanovectors for the aerosol delivery of paclitaxel. Eur J Pharm Biopharm 73(2):239–246. https://doi.org/10.1016/j.ejpb.2009.06.013

Imae T (2012) Physicochemical properties of dendrimers and dendrimer complexes. In: Dendrimer-based drug delivery systems: from theory to practice. Wiley, Hoboken, pp 55–92. https://doi.org/10.1002/9781118275238.ch2

Jadidi-Niaragh F, Atyabi F, Rastegari A, Kheshtchin N, Arab S, Hassannia H, Ajami M, Mirsanei Z, Habibi S, Masoumi F, Noorbakhsh F, Shokri F, Hadjati J (2017) CD73 specific siRNA loaded chitosan lactate nanoparticles potentiate the antitumor effect of a dendritic cell vaccine in 4T1 breast cancer bearing mice. J Control Release 246:46–59. https://doi.org/10.1016/j.jconrel.2016.12.012

Kesharwani P, Amin MCIM, Giri N, Jain A, Gajbhiye V (2017) Chapter 11 - Dendrimers in targeting and delivery of drugs. In: Mishra V, Kesharwani P, Mohd Amin MCI, Iyer A (eds) Nanotechnology-based approaches for targeting and delivery of drugs and genes. Academic Press, Amsterdam, pp 363–388

Klippstein R, Wang JT, El-Gogary RI, Bai J, Mustafa F, Rubio N, Bansal S, Al-Jamal WT, Al-Jamal KT (2015) Passively targeted curcumin-loaded PEGylated PLGA nanocapsules for colon cancer therapy in vivo. Small 11(36):4704–4722. https://doi.org/10.1002/smll.201403799

Koenig SH, Kellar KE (1995) Theory of 1/T1 and 1/T2 NMRD profiles of solutions of magnetic nanoparticles. Magn Reson Med 34(2):227–233. https://doi.org/10.1002/mrm.1910340214

Laurent S, Forge D, Port M, Roch A, Robic C, Vander Elst L, Muller RN (2008) Magnetic iron oxide nanoparticles: synthesis, stabilization, vectorization, physicochemical characterizations, and biological applications. Chem Rev 108(6):2064–2110. https://doi.org/10.1021/cr068445e

Lim Z-ZJ, Li J-EJ, Ng C-T, Yung L-YL, Bay B-H (2011) Gold nanoparticles in cancer therapy. Acta Pharmacol Sin 32(8):983–990. https://doi.org/10.1038/aps.2011.82

Liu Z, Chen K, Davis C, Sherlock S, Cao Q, Chen X, Dai H (2008) Drug delivery with carbon nanotubes for in vivo cancer treatment. Cancer Res 68(16):6652–6660. https://doi.org/10.1158/0008-5472.CAN-08-1468

Lu Y, Low PS (2002) Folate targeting of haptens to cancer cell surfaces mediates immunotherapy of syngeneic murine tumors. Cancer Immunol Immunother 51(3):153–162. https://doi.org/10.1007/s00262-002-0266-6

Michalet X, Pinaud FF, Bentolila LA, Tsay JM, Doose S, Li JJ, Sundaresan G, Wu AM, Gambhir SS, Weiss S (2005) Quantum dots for live cells, in vivo imaging, and diagnostics. Science (New York, NY) 307(5709):538–544. https://doi.org/10.1126/science.1104274

Mochida Y, Cabral H, Kataoka K (2017) Polymeric micelles for targeted tumor therapy of platinum anticancer drugs. Expert Opin Drug Deliv 14(12):1423–1438. https://doi.org/10.1080/17425247.2017.1307338

Moghimi SM, Szebeni J (2003) Stealth liposomes and long circulating nanoparticles: critical issues in pharmacokinetics, opsonization and protein-binding properties. Prog Lipid Res 42(6): 463–478. https://doi.org/10.1016/S0163-7827(03)00033-X

Mudshinge SR, Deore AB, Patil S, Bhalgat CM (2011) Nanoparticles: emerging carriers for drug delivery. Saudi Pharm J 19(3):129–141. https://doi.org/10.1016/j.jsps.2011.04.001

Mukherjee S, Liang L, Veiseh O (2020) Recent advancements of magnetic nanomaterials in cancer therapy. Pharmaceutics 12(2). https://doi.org/10.3390/pharmaceutics12020147

Mydin RBS, Moshawih S (2019) Nanoparticles in nanomedicine application: lipid-based nanoparticles and their safety concerns nanotechnology: applications in energy, drug and food. Springer, pp 227–232

Nacev A, Beni C, Bruno O, Shapiro B (2010) Magnetic nanoparticle transport within flowing blood and into surrounding tissue. Nanomedicine 5(9):1459–1466. https://doi.org/10.2217/nnm.10.104

Najafi-Hajivar S, Zakeri-Milani P, Mohammadi H, Niazi M, Soleymani-Goloujeh M, Baradaran B, Valizadeh H (2016) Overview on experimental models of interactions between nanoparticles and the immune system. Biomed Pharmacother 83:1365–1378. https://doi.org/10.1016/j.biopha.2016.08.060

Northfelt DW, Martin FJ, Working P, Volberding PA, Russell J, Newman M, Amantea MA, Kaplan LD (1996) Doxorubicin encapsulated in liposomes containing surface-bound polyethylene glycol: pharmacokinetics, tumor localization, and safety in patients with AIDS-related Kaposi's sarcoma. J Clin Pharmacol 36(1):55–63. https://doi.org/10.1002/j.1552-4604.1996.tb04152.x

Oliveira RR, Carriao MS, Pacheco MT, Branquinho LC, de Souza ALR, Bakuzis AF, Lima EM (2018) Triggered release of paclitaxel from magnetic solid lipid nanoparticles by magnetic hyperthermia. Mater Sci Eng C Mater Biol Appl 92:547–553. https://doi.org/10.1016/j.msec.2018.07.011

Ozturk K, Esendagri G (2017) Effective targeting of gemcitabine to pancreatic cancer through PEG-cored Flt-1 antibody-conjugated dendrimers. Int J Pharm 517(1):157–167. https://doi.org/10.1016/j.ijpharm.2016.12.009

Ozturk K, Esendagli G, Gurbuz MU, Tulu M, Calis S (2017) Effective targeting of gemcitabine to pancreatic cancer through PEG-cored Flt-1 antibody-conjugated dendrimers. Int J Pharm 517(1-2):157–167. https://doi.org/10.1016/j.ijpharm.2016.12.009

Palombo M, Deshmukh M, Myers D, Gao J, Szekely Z, Sinko PJ (2014) Pharmaceutical and toxicological properties of engineered nanomaterials for drug delivery. Annu Rev Pharmacol Toxicol 54(1):581–598. https://doi.org/10.1146/annurev-pharmtox-010611-134615

Paroha S, Chandel AKS, Dubey RD (2018) Nanosystems for drug delivery of coenzyme Q10. Environ Chem Lett 16(1):71–77. https://doi.org/10.1007/s10311-017-0664-9

Paroha S, Dewangan RP, Dubey RD, Sahoo PK (2020) Conventional and nanomaterial-based techniques to increase the bioavailability of therapeutic natural products: a review. Environ Chem Lett 18(6):1767–1778. https://doi.org/10.1007/s10311-020-01038-1

Paroha S, Verma J, Dubey RD, Dewangan RP, Molugulu N, Bapat RA, Sahoo PK, Kesharwani P (2021) Recent advances and prospects in gemcitabine drug delivery systems. Int J Pharm 592: 120043. https://doi.org/10.1016/j.ijpharm.2020.120043

Pasqua L, Leggio A, Sisci D, Ando S, Morelli C (2016) Mesoporous silica nanoparticles in cancer therapy: relevance of the targeting function. Mini Rev Med Chem 16(9):743–753. https://doi.org/10.2174/1389557516666160321113620

Patil Y, Sadhukha T, Ma L, Panyam J (2009) Nanoparticle-mediated simultaneous and targeted delivery of paclitaxel and tariquidar overcomes tumor drug resistance. J Control Release 136(1): 21–29. https://doi.org/10.1016/j.jconrel.2009.01.021

Petros RA, DeSimone JM (2010) Strategies in the design of nanoparticles for therapeutic applications. Nat Rev Drug Discov 9(8):615–627. https://doi.org/10.1038/nrd2591

Pirollo KF, Chang EH (2008) Does a targeting ligand influence nanoparticle tumor localization or uptake? Trends Biotechnol 26(10):552–558

Probst CE, Zrazhevskiy P, Bagalkot V, Gao X (2013) Quantum dots as a platform for nanoparticle drug delivery vehicle design. Adv Drug Deliv Rev 65(5):703–718. https://doi.org/10.1016/j.addr.2012.09.036

Rizvi SAA, Saleh AM (2018) Applications of nanoparticle systems in drug delivery technology. Saudi Pharm J 26(1):64–70. https://doi.org/10.1016/j.jsps.2017.10.012

Saneja A, Arora D, Kumar R, Dubey RD, Panda AK, Gupta PN (2018a) CD44 targeted PLGA nanomedicines for cancer chemotherapy. Eur J Pharm Sci 121:47–58. https://doi.org/10.1016/j.ejps.2018.05.012

Saneja A, Arora D, Kumar R, Dubey RD, Panda AK, Gupta PN (2018b) Therapeutic applications of betulinic acid nanoformulations. Ann N Y Acad Sci 1421(1):5–18. https://doi.org/10.1111/nyas.13570

Saneja A, Kumar R, Mintoo MJ, Dubey RD, Sangwan PL, Mondhe DM, Panda AK, Gupta PN (2019) Gemcitabine and betulinic acid co-encapsulated PLGA–PEG polymer nanoparticles for improved efficacy of cancer chemotherapy. Mater Sci Eng C 98:764–771. https://doi.org/10.1016/j.msec.2019.01.026

Saneja A, Kumar R, Singh A, Dhar Dubey R, Mintoo MJ, Singh G, Mondhe DM, Panda AK, Gupta PN (2017a) Development and evaluation of long-circulating nanoparticles loaded with betulinic acid for improved anti-tumor efficacy. Int J Pharm 531(1):153–166. https://doi.org/10.1016/j.ijpharm.2017.08.076

Saneja A, Sharma L, Dubey RD, Mintoo MJ, Singh A, Kumar A, Sangwan PL, Tasaduq SA, Singh G, Mondhe DM, Gupta PN (2017b) Synthesis, characterization and augmented anticancer potential of PEG-betulinic acid conjugate. Mater Sci Eng C 73:616–626. https://doi.org/10.1016/j.msec.2016.12.109

Shi J, Kantoff PW, Wooster R, Farokhzad OC (2017) Cancer nanomedicine: progress, challenges and opportunities. Nat Rev Cancer 17(1):20–37. https://doi.org/10.1038/nrc.2016.108

Singh A, Patel T, Hertel J, Bernardo M, Kausz A, Brenner L (2008) Safety of ferumoxytol in patients with anemia and CKD. Am J Kidney Dis 52(5):907–915. https://doi.org/10.1053/j.ajkd.2008.08.001

Smith AM, Nie S (2010) Semiconductor nanocrystals: structure, properties, and band gap engineering. Acc Chem Res 43(2):190–200. https://doi.org/10.1021/ar9001069

Son KH, Hong JH, Lee JW (2016) Carbon nanotubes as cancer therapeutic carriers and mediators. Int J Nanomedicine 11:5163–5185. https://doi.org/10.2147/IJN.S112660

Son KJ, Yoon HJ, Kim JH, Jang WD, Lee Y, Koh WG (2011) Photosensitizing hollow nanocapsules for combination cancer therapy. Angew Chem 50(50):11968–11971. https://doi.org/10.1002/anie.201102658

Soo Choi H, Liu W, Misra P, Tanaka E, Zimmer JP, Itty Ipe B, Bawendi MG, Frangioni JV (2007) Renal clearance of quantum dots. Nat Biotechnol 25(10):1165–1170. https://doi.org/10.1038/nbt1340

Sun C, Lee JS, Zhang M (2008) Magnetic nanoparticles in MR imaging and drug delivery. Adv Drug Deliv Rev 60(11):1252–1265. https://doi.org/10.1016/j.addr.2008.03.018

Talelli M, Rijcken CJF, Hennink WE, Lammers T (2012) Polymeric micelles for cancer therapy: 3 C's to enhance efficacy. Curr Opinion Solid State Mater Sci 16(6):302–309. https://doi.org/10.1016/j.cossms.2012.10.003

Torchilin V (2011) Tumor delivery of macromolecular drugs based on the EPR effect. Adv Drug Deliv Rev 63(3):131–135. https://doi.org/10.1016/j.addr.2010.03.011

Varela-Moreira A, Shi Y, Fens MHAM, Lammers T, Hennink WE, Schiffelers RM (2017) Clinical application of polymeric micelles for the treatment of cancer. Mater Chem Front 1(8):1485–1501. https://doi.org/10.1039/C6QM00289G

Wan D, Li C, Pan J (2020) Polymeric micelles with reduction-responsive function for targeted cancer chemotherapy. ACS Appl Bio Mater 3(2):1139–1146. https://doi.org/10.1021/acsabm.9b01070

Wang J, Li S, Han Y, Guan J, Chung S, Wang C, Li D (2018) Poly(ethylene glycol)–polylactide micelles for cancer therapy. Front Pharmacol 9(202). https://doi.org/10.3389/fphar.2018.00202

Webster DM, Sundaram P, Byrne ME (2013) Injectable nanomaterials for drug delivery: carriers, targeting moieties, and therapeutics. Eur J Pharm Biopharm 84(1):1–20. https://doi.org/10.1016/j.ejpb.2012.12.009

Wisse E, Braet F, Luo D, De Zanger R, Jans D, Crabbe E, Vermoesen A (1996) Structure and function of sinusoidal lining cells in the liver. Toxicol Pathol 24(1):100–111

Wu W, Li R, Bian X, Zhu Z, Ding D, Li X, Jia Z, Jiang X, Hu Y (2009) Covalently combining carbon nanotubes with anticancer agent: preparation and antitumor activity. ACS Nano 3(9):2740–2750. https://doi.org/10.1021/nn9005686

Xiong Z, Shen M, Shi X (2018) Dendrimer-based strategies for cancer therapy: recent advances and future perspectives. Sci China Mater 61(11):1387–1403. https://doi.org/10.1007/s40843-018-9271-4

Yameen B, Choi WI, Vilos C, Swami A, Shi J, Farokhzad OC (2014) Insight into nanoparticle cellular uptake and intracellular targeting. J Control Release 190:485–499. https://doi.org/10.1016/j.jconrel.2014.06.038

Yang F, Hu J, Yang D, Long J, Luo G, Jin C, Yu X, Xu J, Wang C, Ni Q, Fu D (2009) Pilot study of targeting magnetic carbon nanotubes to lymph nodes. Nanomedicine 4(3):317–330. https://doi.org/10.2217/nnm.09.5

Yang Y, Jing L, Li X, Lin L, Yue X, Dai Z (2017) Hyaluronic acid conjugated magnetic Prussian blue@quantum dot nanoparticles for cancer theranostics. Theranostics 7(2):466–481. https://doi.org/10.7150/thno.17411

Yoo HS, Park TG (2004) Folate receptor targeted biodegradable polymeric doxorubicin micelles. J Control Release 96(2):273–283. https://doi.org/10.1016/j.jconrel.2004.02.003

Yu G, Ning Q, Mo Z, Tang S (2019) Intelligent polymeric micelles for multidrug co-delivery and cancer therapy. Artif Cells Nanomed Biotechnol 47(1):1476–1487. https://doi.org/10.1080/21691401.2019.1601104

Yurgel V, Collares T, Seixas F (2013) Developments in the use of nanocapsules in oncology. Braz J Med Biol Res 46(6):486–501. https://doi.org/10.1590/1414-431X20132643

Zhang Z, Feng S-S (2006) Nanoparticles of poly(lactide)/vitamin E TPGS copolymer for cancer chemotherapy: synthesis, formulation, characterization and in vitro drug release. Biomaterials 27(2):262–270. https://doi.org/10.1016/j.biomaterials.2005.05.104

Biomimetic Approach for the Controlled Drug Delivery through 3D Bioactive Scaffolds: A Novel Strategy for Tissue Engineering Applications

Aggarapu Chandana, Sarada Prasanna Mallick, Bhisham Narayan Singh, Aditya Anand, Dheerendra Kumar Suman, Venkata Rajesh Yella, Rupita Ghosh, and S. R. Krishna Motukuri

Abstract A suitable drug delivery system may accelerate the growth and repair of a new tissue. Small-molecule chemicals, proteins, peptides, cytokines, etc. typically known as drugs have a positive effect on cellular function and tissue regeneration. Controlled drug delivery can be achieved by the physical or chemical adsorption of the drug to the scaffold matrix, thereby entrapping the drug inside the scaffold, and then release takes place by the process of diffusion or with the degradation of the scaffold. The minimum threshold is essential for drugs to be effective, but as a result of short half-life in vivo, the challenge lies in supplying the appropriate dose at the injured site for an increased period. To overcome the challenge, various biomimetic materials are being employed for delivering drugs in a controlled manner, which includes porous materials, hydrogels, nanofibrous scaffolds, etc. made from natural and synthetic polymers.

Keywords Biomimetic materials · Drug delivery · Scaffold · Degradation · Hydrogels

---

A. Chandana · S. P. Mallick (✉) · V. R. Yella · R. Ghosh
Department of Biotechnology, KoneruLakshmaiah Education Foundation, Guntur, India
e-mail: yourssarada@kluniversity.in

B. N. Singh
Department of Ageing Research, Manipal School of Life Sciences, Manipal Academy of Higher Education, Manipal, India

A. Anand
School of Biochemical Engineering, Indian Institute of Technology (Banaras Hindu University), Varanasi, India

D. K. Suman
Department of Biotechnology, National Institute of Technology, Tadepalligudem, India

S. R. K. Motukuri
Department of Agricultural Biotechnology, College of Agriculture, Koneru Lakshmaiah Education Foundation, Guntur, India

V. S. Rajput, A. Runthala (eds.), *Drugs and a Methodological Compendium*,
https://doi.org/10.1007/978-981-19-7952-1_12

1 Introduction

Tissue engineering exploits the innate potential of living cells to regenerate biologically when provided with the suitable substrate for cell attachment, proliferation, differentiation, and cell migration (Dhandayuthapani et al. 2011; Guimarães et al. 2020; Mabrouk et al. 2020). Tissue regeneration is a multifaceted process regulated by intricate crosstalk between cells, adjacent cells, and the ECM (Sodhi and Panitch 2021). In a living system, cells reside in a three-dimensional cushioned network protected against external mechanical stress and governed by various biochemical and biophysical signals like cell fate determination and contact guidance. This microenvironment that provides the cell with an appropriate niche for cellular functions makes mimicking the natural ECM architecture the key idea behind fabricating scaffolds and successful tissue engineering. Therefore, an ideal scaffold has to be bioactive to aid cellular functions and should mimic the native ECM structure, guiding cellular behavior by promoting mitogenic activity and neovascularization (Christy et al. 2020; Khalil et al. 2020).

It has been over three decades since porous biomaterials from different origins including metals, polymers, ceramics, and glasses have been intensively used as scaffolds for tissue regeneration (Koons et al. 2020). With advancements in the research of biomaterials and the development of specific biodegradable materials, implantable bioactive scaffolds were fabricated that not only act as a substrate but actively participate in cellular functions, thereby promoting cell attachment, proliferation, and migration (Doostmohammadi et al. 2020; He et al. 2020). These scaffolds are tuned to gradually biodegrade in the course of healing and tissue regeneration, thereby obviating the need for surgical removal of the implant. Moreover, these scaffolds also act as novel carriers and are nowadays used to deliver the desired cells and biological agents at the target site. Bioactive and biomimetic scaffolds serve as delivery vehicles that deliver bioagents at the site of injury with precision in a controlled manner and are tailored according to the need of the tissue that needs repair. Most of the designed scaffolds use a passive delivery mechanism like molecular diffusion, scaffold degradation, and cell migration to deliver biological agents at the target site. However, passive delivery mechanisms have several drawbacks and limit dynamic external regulations, making biomaterials that respond to external stimuli like temperature, pressure, pH, and enzymes particularly desirable (Christy et al. 2020).

2 Scaffolds as Cell Delivery Vehicles

Cells from varying differentiation stages harnessed from both adult and embryonic sources are being explored for therapeutic usage in humans and animals. These transplanted cells are usually short-lived due to their inability to cope up with the extreme environment of the target site and therefore die shortly after implantation.

This short life span of the transplanted cell can be attributed to (a) ischemia (lack or absence of vascularization), (b) anoikis (absence of anchorage sites), (c) host immune response, (d) low cell density, and (e) faulty differentiation (differentiation to an undesired lineage than the desired one).

Fabrication of bioactive scaffolds with interconnected porous that mimics the micro-architecture of the ECM can overcome ischemia and deliver cells and growth factors that promote cellular functions and vascularization (Chen and Mooney 2003; Ren et al. 2020). Moreover, these biocompatible scaffolds provide excellent anchorage sites and guided regulation to regulate cellular behavior, thereby maintaining appropriate cell density and desired differentiation. Non-immunogenicity, cytocompatibility, regulated degradation, optimum activity at physiological parameters, and its customizability to add desirable signal molecules/bioagents (e.g., growth factors, genes, ECM proteins) make the scaffold highly desirable as a cell delivery vehicle (Schneible et al. 2021).

The conventional cell-seeding methods into scaffolds were inefficient as they involved haphazard placement of cells and formed 2D cell culture, thereby limiting the spatial regulation needed for cells to attain physiologically relevant cell morphology (Cui et al. 2020; Rajab et al. 2020). In contrast, biomimetic scaffolds are 3D scaffolds that allow for efficient cell attachment and differentiation cues and provide the spatial regulatory environment to the cells to attain physiologically relevant morphology. Several methods are used to fabricate 3D scaffolds with living cells. One method that uses electrospinning allows periodic spraying of living cells throughout the process. Co-axial electrospinning involves direct electrospinning of live cells that forms nanofibrous scaffolds with fibers encapsulating live cells (Niemczyk-Soczynska et al. 2020).

Bioprinting, a rapid prototyping technique, is comparatively efficient in the precise positioning of cells and forming cell-laden scaffolds with the desired geometries. This method has a strong potential to produce tailorable and defect-induced geometrical scaffolds for tissue regeneration as it permits simultaneous printing of biocompatible polymers, bioactive molecules, and living cells (Agarwala 2016; Gu et al. 2015). Hydrogels are scaffolds of choice for bioprinting. Several hydrogels are currently used as cellular carriers in both clinical and pre-clinical research. Tisseel® is a fibrin-based hydrogel made from human plasma capable of delivering calcium at the target site (Ivica et al. 2020). It is used as a sealant to close incision wounds that cannot be closed by other existing means. HyStem® hydrogel is a cross-linked hyaluronate having a bacterial origin and is modified to provide cellular attachment sites and slow delivery of heparin at the target site (Yu et al. 2020).

In short, cells with bioagents are either encapsulated or seeded into the scaffold and then implanted or injected into the human/animal body (Singh et al. 2019). Tissue remodeling is regulated by regulating the various factors involved in cellular crosstalk either by inducing or inhibiting cellular processes. Several growth factors are incorporated into the scaffolds by encapsulation, surface adsorption, and the addition of microspheres for local and sustained delivery at the site of implantation (Narayanan and Calve 2021; San Antonio et al. 2021; Tong et al. 2021).

3 Scaffolds as Drug Delivery Vehicles

Scaffolds with drug delivery potential are well-turned-out alternatives to conventional formulations that permit the spatiotemporally regulated release of bioactive compounds (Oliveira et al. 2021). As discussed, various methods of fabricating scaffolds have emerged in the field of tissue engineering and regenerative medicine. A drug delivery vehicle is a carrier platform of on-site delivery which loads bioactive compounds and delivers them to a specific target site increasing the safety and efficacy of the drug (Alyassin et al. 2020; Jacob et al. 2018). These polymers can be from a natural source or synthetic source or composite scaffolds comprising of polymers from both sources (Puppi et al. 2010). Drug delivery vehicles are emerging platforms in the field of precision medicine that aims to customize therapy as per patient need (Duffy 2016). Depending upon the type of tissue intended to be repaired (hard/soft), the biomaterial is selected from the wide spectrum of biomaterials available (Wishart 2016). The scaffolds used as drug delivery vehicles can either be implantable or in an injectable platform (Talebian et al. 2018). Implantable scaffolds are pre-fabricated in 2D or 3D forms made up of single polymers or composite scaffolds, while injectable scaffolds are basically hydrogels that can be administered with minimal invasive surgery and small incision sizes, with the ability to form in situ 3D network. This level of efficiency is very difficult to attain while dealing with implantable scaffolds even with the 3D printing technique (Guvendiren et al. 2016).

The principal aim of a drug delivery platform is to maintain optimal supply at the site of injury and keep the environment conducive to regeneration. A platform like this prevents multiple drug administration and keeps the therapeutic level maintained. Moreover, it also provides the opportunity to deliver multiple drugs with different kinetics at the desired release rate. Biocompatible scaffolds used as drug delivery vehicles can be used not only to deliver growth factors and drugs but also cells, proteins, and genes (Batrakova et al. 2011; MaHam et al. 2009). The cell-scaffold interaction enhances cellular function and accelerates tissue remodeling. Several osteogenic and differentiation factors such as transforming growth factor-beta (TGF-β), vascular endothelial growth factor (VEGF), and various signaling molecules are associated with angiogenesis (El-FiqiA et al. 2021; Soker et al. 2000). Collagen implant loaded with growth factors, and bone morphogenetic protein 2 (BMP-2) named Medtronic's Infuse®, is approved by Food and drug administration (FDA) (Rico-Llanos et al. 2021). Silk fibroin scaffolds were recently used to deliver growth factors such as stromal-derived factor-1α (SDF-1α) and TGF-β1 to make conducive microenvironments for cartilage injury repair (Chen et al. 2019b). Several anti-cancer chemicals, including doxorubicin, camptothecin, rapamycin, and cisplatin, have been spatially released using a hyaluronic acid-based injectable hydrogel (Antimisiaris et al. 2021; Wang et al. 2021). Polycaprolactone (PCL) and polylactic acid (PLA) are two of the few FDA-approved polymers used for scaffold fabrication to deliver drugs on targeted sites. PCL scaffolds are used to deliver on-site drugs against osteosarcoma. PLA scaffolds are used as a delivery vehicle for prednisolone, dexamethasone, and various growth factors (Patel et al. 2019).

Summarizing scaffolds as drug delivery vehicles, these scaffolds show an immense potential and can be further used to develop customized smarter therapies not only for tissue regeneration but also to deal with certain illnesses such as cancer and cardiovascular diseases (Haraguchi et al. 2012; Horch et al. 2013).

4 Biomimetic Bioactive Scaffolds Types and Fabrication Methodologies

Tissue engineering applications involve the generation of transplantable cell-scaffold constructs mimicking physical, chemical, and biological properties of natural tissue extracellular matrix (Vacanti and Langer 1999). Moreover, designing aspects of scaffold plays an important role in enhancing its ability to mimic closely the tissue ECM architecture and thereby provides an optimum microenvironment for tissue reconstruction. Various scaffolding technologies were developed for the generation of functional scaffolds using suitable biomaterials for tissue regeneration, drug or bioactive factor delivery, and repair of damaged/diseased tissue. Scaffolds are artificial bio-substitutes designed to host cells and deliver growth factors or drugs for repair, maintenance, and regeneration of damaged or diseased tissues. Numerous scaffolds of various forms as shown in Fig. 1 such as particles, spheres, porous

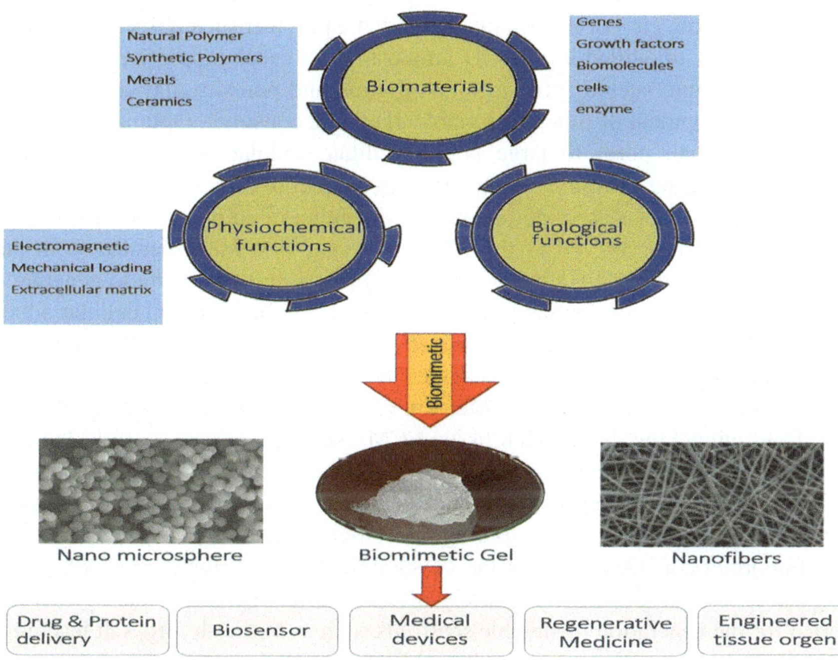

Fig. 1 Schematic representation of biomimetic materials, structures, functions, and their applications

materials, fibrous scaffolds, hydrogel, 3D-printed template, etc. were fabricated using suitable biomaterial and scaffolding techniques. Biomimetic scaffolds are those that closely resemble the ECM of the tissue, and the scaffold manufacturing tries to mimic the tissue ECM architecture and its many features, including cellular behavior, to maintain and heal the tissue. As a result, the rational design of biomimetic scaffolds is inspired by natural ECM scaffolds to give optimal physico-chemical and biological cues for cellular behavior such as growth, differentiation, and functioning (Owen and Shoichet 2010; Shoichet 2010). Moreover, biomimetic scaffolds capable of mimicking natural tissue ECM properties and able to facilitate tissue regeneration and integration are referred to as bioactive biomimetic scaffolds. However, for any scaffold application for tissue engineering regardless of tissue types, it is important that scaffold should be biocompatible and biodegradable and possess ideal morphological, mechanical, and biological properties. Biocompatible scaffolds usually do not elicit any severe inflammatory reaction leading to post-implantation rejection of scaffold. Also, such scaffold needs to be biodegradable and exhibit controlled degradation with gradual deposition of natural ECM by cells growing over the scaffold matrix (Babensee et al. 1998). The cellular behavior over scaffold surfaces is regulated not only by surface properties but also by mechanical properties of the scaffold system or biomaterial. Usually, scaffolds are designed to mimic tissue mechanical properties, which might be elastic like the muscle tissue to load-bearing like the bone tissue. Thus, the mechanical properties of scaffold play an important role in cell fate determination as well as ECM deposition (Kjær et al. 2006; Mizutani et al. 2009). Apart from the inherent (natural proteins such as silk, gelatin, etc.) or incorporated (synthetic polymer such polycaprolactone, polyurethane, etc. modified with RGD amino motifs) cells' supportive properties, scaffold architecture such as pore size distribution and porosity play an important role in the development of functional viable 3D tissue construct. Optimal pore sizes ranging from small pores to large pores facilitate cellular adhesion, migration, growth factor holding, vascularization, etc. (Ko et al. 2007). Moreover, the porosity of the scaffold also plays an important role in the influx of nutrients and distribution throughout the matrix as well as the efflux of waste or waste generated from the degradation of the scaffold (O'Brien et al. 2005). Thus, it is important to generate a biomimetic scaffold with optimal physico-chemical, mechanical, and cellular adhesion and drug/growth factor delivery properties to provide optimal cells supportive microenvironment for damaged or diseased tissue repair or regeneration (Fig. 1) (Das and Noh 2018; Han et al. 2019; Mercante et al. 2021). Design and fabrication of the scaffold mimicking the targeted tissue ECM using appropriate scaffold fabrication techniques, surface modification or functionalization to mimic tissue-specific microenvironment, and incorporation of drug or biological factors to modulate tissue regeneration or repair in response to drug or biological factor delivery are just a few of the important considerations for the development of biomimetic scaffolds replicating the natural tissue ECM. A variety of scaffold fabrication technologies were reported for the generation of suitable scaffolds such as freeze-drying, salt leaching, electrospinning, phase separation, melt molding, 3D printing, etc. Scaffolding techniques are specific to fabricate scaffold of particular architecture such as porous solid scaffold using freeze-drying, porous hydrogel using freeze gelation, fibrous scaffold using electrospinning, scaffold with predefined 3D geometry using 3D printer, and many more depending on the requirement for mimicking the natural tissue ECM architecture. Thus, biomimetic scaffold of particular type can be fabricated using specific scaffolding techniques or there combinations.

5 Fabrication of Porous Scaffold by Conventional Scaffolding Technique

Scaffold with porous interconnected architecture mimics various tissue porous ECM to a certain extent and plays a significant role in facilitating efficient cells seeding across the 3D structure. Most of the natural tissues including the bone, cartilage, skin, liver, etc. possess porous ECM to facilitate optimal mass transport of nutrients, gases, and metabolic waste for cellular growth or viability (Hollister 2005). Such interconnected porous architecture also facilitates coordination among cells and their functionality and thus maintains the overall function of the tissue or organ. Various biodegradable porous scaffolds were fabricated using conventional scaffolding techniques. Here we will discuss some of the important fabrication techniques such as particulate leaching, freeze-drying, phase separation, gas foaming, and expansion in the supercritical fluid.

5.1 Particulate or Porogen Leaching

The particulate or porogen leaching technique of scaffolding (Fig. 2a) is one of the most popular techniques of ease to fabricate interconnected porous scaffolds (Lebourg et al. 2008; Liao et al. 2002). Few of the most common porogens such as salt, sugar, and wax of the micrometric level are usually incorporated into the polymeric solution. The porogen-embedded polymeric solutions were kept for solidification preferably at low temperature, and then porogen was subjected to leach out using non-solvent for the polymer. The pore size, morphology, and interconnectivity are widely controlled by porogen size or shape. Various porous polymeric scaffolds made of PLGA, polyurethane, and other materials have previously been constructed and tested for tissue engineering purposes (Nam et al. 2000; Tran et al. 2011). Also, it has been observed that removal of porogen from the interior of thick scaffold poses certain difficulties, and thus it is difficult to make thick scaffold through particulate leaching scaffolding technique. However, attempts were made for further advancement of the particulate leaching technique such as the development of paraffin-based spherical porogen and porogen based on nucleation and crystallization science.

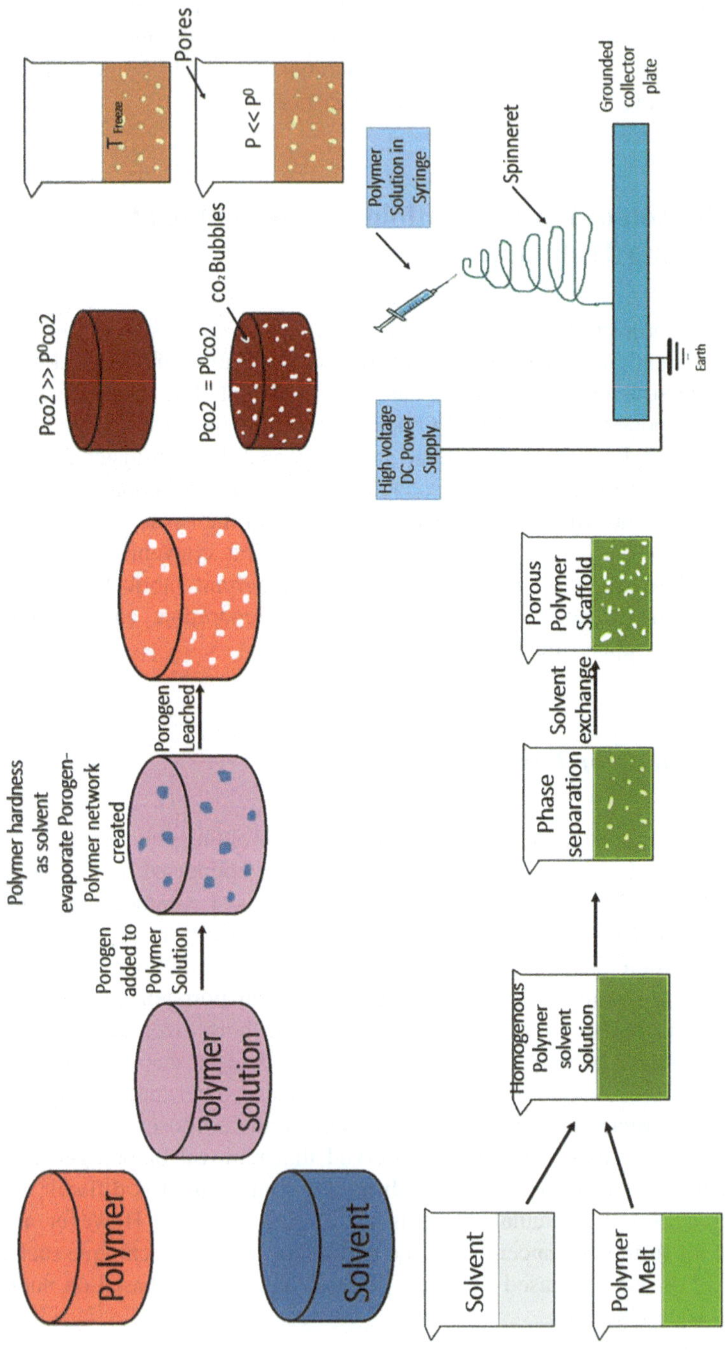

Fig. 2 Overview of scaffold fabrication technologies: **(a)** particulate leaching, **(b)** gas foaming, **(c)** phase separation, and **(d)** electrospinning

5.2 Freeze-Drying and Gas Foaming

The freeze-drying technique is widely used to fabricate scaffolds with a higher degree of porosity and controlled pore size distribution. Usually the desired polymeric biomaterial is dissolved in water and freezes at low temperature in the range of −20 °C to −80 °C. Because of the freezing process, the polymeric emulsions led to the crystallisation of the solvent, and the solute molecules or polymer were segregated around the ice crystals. This is followed by the removal of ice crystals acting as porogen through lyophilization at low temperature and applying lower pressure than the equilibrium vapor pressure of the frozen solvent (Deville et al. 2006; Thomson et al. 2000). Also, to improve the structural stability in wet conditions, obtained polymeric porous scaffolds were cross-linked using suitable cross-linker such as glutaraldehyde, genipin, ethanol, etc. Moreover, the pore size distribution and uniform porosity across the scaffold can be controlled through optimizing freezing and lyophilization temperature while scaffolding the process. Various porous scaffolds were previously developed through freeze-drying and tested for cartilage, skin, bone, and many more tissue engineering applications (Kumari et al. 2019; Singh and Pramanik 2018; Singh et al. 2020). Also, the low-temperature processing for freeze-dried scaffold fabrication shows aided advantage to incorporate heat-labile bioactive factors such as growth factors, drugs, etc., for controlled drug delivery (Chen et al. 2002). The gas foaming technique (Fig. 2b) is also used to fabricate porous scaffolds by injecting high-pressure gas in the polymeric emulsion. Injected gas in the polymeric emulsion induces phase separation with nucleation as a gas bubble followed by curing of a polymer leading to the solidification of foam. The gas bubble can be created by chemical reaction, by bubbling of inert gas, or by changing the physical condition leading to the lower solubility of the gas in the liquid to facilitate nucleation of the gas bubble. For gas foaming various effervescent salt porogens such as sodium bicarbonate or ammonium bicarbonate are used with polymeric emulsion to induce gas bubble formation under acidic conditions. Also, supercritical carbon dioxide is used along with various hydrophobic biopolymers such as PLGA, PLLA, and PGA for porous scaffold fabrication through gas foaming technique upon pressure reduction (Dehghani and Annabi 2011). The porous scaffold developed through freeze-drying or gas foaming technique shows porous architecture with pore size in the range of 100 to 500 μm and mimics the porous architecture of tissue ECM to facilitate cellular growth and tissue regeneration.

5.3 Sintering or Melt Molding and Phase Separation

In the melt molding method, appropriate mold is filled with a mixture of polymer and porogen followed by sintering at glass transition temperature of the polymer or porogen evaporation or melting temperature (Oh et al. 2003). The polymer will adopt the shape of mold, and porogen will be removed by washing the scaffold in an appropriate solvent. Synthetic polymers are more suitable as compared to heat-sensitive natural polymers for scaffold fabrication through melt molding. Also, the major limitation with this method is the inability to incorporate heat-labile bioactive factors in the scaffold as high temperature leads to denaturation of such bioactive factors. The phase separation technique (Fig. 2c) requires quenching of the polymeric solution leading to the two-phase formation one with polymer-rich phase and the other as polymer-poor phase (Nam and Park 1999). The architecture of porous scaffolds developed through phase separation depends on various factors such as polymeric concentration, solvent type, temperature, quenching rate, and distribution of solute molecules. Solid- and liquid-phase separation that occurs due to lowering of temperature induces solvent crystallization of a polymeric solution. Removal of crystallized solvent through solvent exchange or sublimation resulted in the generation of porous scaffold mimicking porous tissue ECM architecture. Moreover, the phase separation process is carried out at a low temperature, which facilitates the incorporation of heat-labile bioactive factors in the scaffold system (Akbarzadeh and Yousefi 2014).

5.4 *Fibrous Biomimetic Matrices by Electrospinning*

Electrospinning is a versatile scaffold fabrication technique used to fabricate fibrous scaffolds, which replicate structural and biological functions of natural tissue nanofibrous extracellular matrix. Moreover, nanofibrous scaffold provides a high surface-to-volume ratio, which facilitates enhanced bioactivity and more cellular responsive potential of the matrices. Varieties of polymers including natural and synthetic biopolymers can be processed into fibrous scaffolds using electrospinning. Electrospinning involves the application of Coulomb force to draw a charged thread from a polymer solution or melt in the range of micron to the nanoscale. In nozzle-based electrospinning, polymeric solution in the syringe at controlled flow rate is subjected to eject solution under applied electric potential between the needle tip and the collector (Fig. 2d). As the applied electric potential overcomes the surface tension of polymeric solution, it resulted in the formation of Taylor cone and finally generation of fiber over the collector fixed at a certain distance from the needle tip. Distance between the needle tip and collector, polymeric solution concentration, and applied voltage play an important role in fiber generation and defining fiber diameter (Teo and Ramakrishna 2006). The main benefit of electrospinning is that it produces a fibrous scaffold with a higher degree of porosity in the nanoscale range, whereas the main disadvantage is that it has a small pore size, which limits cellular penetration and makes it impossible to deposit thick 3D scaffold greater than 100-micron thickness. Furthermore, electrospun matrices enable tissue engineering applications to control medication distribution and cellular behavior (Chutipakdeevong et al. 2015; Tan and Zhou 2020).

5.5 3D Printing or Rapid Prototyping

The rapid prototyping technique involves automated layer-by-layer fabrication of scaffolds based on programmed 3D images. Rapid prototyping facilitates the generation of 3D scaffolds using varieties of polymer, ceramic, and composites. Biomimetic scaffold generated through rapid prototyping shows a better control over pore size, shape, connectivity, and overall dimension specifically in terms of organ-specific 3D designing and fabrication. Various types of 3D printing system have been developed such as selective laser sintering of powder at elevated temperature and removal of unbounded powder to fabricate the 3D structure, fused deposition modeling involving layer-by-layer deposition of the polymer through melt extrusion of thermal plastic filament, and more recently bioplotting of cells loading hydrogel into 3D construct in the layer-by-layer fashion (Billiet et al. 2012; Melchels et al. 2012; Silva et al. 2007). Bioplotting gains the attraction of various researchers to generate organ-specific tissue construct over the past decade. This method uses extrusion-based printing under the control of a robotic system with micron-scale resolution. Moreover, bioplotting does not involve high temperature and is thus favorable for using natural polymer and bioactive factor incorporation for tissue construct generation.

6 Bioactive Scaffolds Delivering Growth Factors for Various Tissue Engineering Applications

Tissue engineering utilizes the principles of engineering and life sciences in the development of damaged tissues. The triad of tissue engineering involves cells, scaffolds, and growth factors which upon combining help in the regeneration of the concerned tissue (Riha et al. 2005). The success of tissue engineering includes two methods being employed currently. The first strategy is the implantation of the cell-seeded scaffold at the defect site, the cells being pre-cultured and seeded on a three-dimensional (3D) scaffold (Saha et al. 2013). The basic feature of such scaffolds is that they are biodegradable which degrade while the tissue regenerates. The second strategy is to implant a scaffold without cells next to the injured site.

The principle behind this method is to use such scaffolds as a medium of delivering certain appropriate biomolecules at the defect site that too in a controlled manner helps in recruiting progenitor cells at the defect site and even assuring their proliferation and differentiation. Recently, the trend has shifted toward utilization of a combination of these two strategies which is advantageous as the scaffold provides controlled release of growth factors or biomolecules which makes the proliferation and differentiation of the cells seeded on it possible. This approach effectively boosts tissue regeneration, and bioactive scaffolds are those that serve as both physical support for cells and a source of biomolecules that influence tissue renewal in the surrounding environment (Mallick et al. 2015). The different biomolecules that frequently get incorporated in bioactive scaffolds are proteins which may include cytokines and growth factors, also growth factor coding genes. Growth factors can be seen as proteins that are endogenous and are able to bind the cell surface receptors and enable the cells to take part in regeneration of the new tissue. Growth factors exogenously added in terms of localized delivery are considered to be therapeutically effective in producing the cellular components essential for healing and tissue development. Nevertheless, the success of this localized delivery of growth factors depends widely on production of recombinant growth factors, and that is quite costly.

6.1 Growth Factors

Responsible for either stimulating or inhibiting cell proliferation, differentiation, migration, adhesion, and gene expression, growth factors can be defined as polypeptides that play an important role in transmitting signals to modulate cellular activities. The same growth factor can be produced by many cell types; also, many cell types can be acted upon by the same growth factor, being known as pleiotropism. Adding to this, the same biological effect can be shared by different growth factors, known as redundancy. Apart from this, growth factors can also influence the secretion and action of other growth factors, a process known as antagonizing or enhancing effect (Chen et al. 2010; Fiorillo et al. 2021).

The initiation of their action is upon binding to the specific receptors present on the cell surface. Classification of growth factors based on proximity of their synthesis site to their target site is as follows: endocrine, target cell is located at a distant site; paracrine, target cell is located nearby; autocrine, the cell secreting the growth factor is the target site itself, and proteins from the inducing cell interact with receptor proteins from nearby responding cells in juxtacrine interactions; juxtacrine, apposition of the target cell to growth factor-receptor complex (Hull and Harvey 2014). Usually existing as inactive or partially active precursors, growth factors require activation through proteolysis. They even need to attach to matrix molecules to get activated or stabilized.

7 Bioactive Scaffolds for Delivery of Therapeutic Growth Factors

Incorporation of growth factors on a polymer carrier or scaffold facilitates the sustained release of such biomolecules, thereby increasing the in vivo efficacy of growth factors; this sustained release is over an extended period and makes sure to provide the optimum concentration of biomolecules required for cellular activities (Ma and Elisseeff 2005). Such bioactive scaffolds when get implanted at the defect site facilitate localized growth factor delivery. The scaffolds come in a variety of shapes and combinations, and they're made from a variety of natural and synthetic polymers, depending on whether they're biodegradable or not. All these polymeric devices have the common characteristic of delivering controlled release of proteins or growth factors for an extended period differing only in the mechanism involved in such delivery. The factors regulating the controlled delivery of growth factors or drugs from bioactive scaffolds include growth factor or drug loading, polymer type used, and the conditions used in processing. By processing conditions, it is meant to avoid conditions causing protein aggregation or denaturation, maintain the stability of protein when being exposed to moisture, etc. Such alterations are substantially different from the ones to improve the productivity of an enzyme molecule through various strategies of directed evolution (Amrein et al. 2019; Runthala and Chowdhury 2019; Kamjula et al. 2020; Phulara et al. 2020; Runthala et al. 2020; Runthala 2021; Shukla et al. 2022).

Growth factors can be incorporated immediately into the scaffold or after it has been fabricated. While the tissue regenerates in a biodegradable scaffold, the growth factor is released. The diffusion-controlled mechanism, which is dependent on the pore size of the scaffold, regulates the release of growth factors in scaffolds where they are directly integrated. Mechanism of erosion or its combination with diffusion is also one of the factors by which growth factors or proteins are released from bioactive scaffolds. Another mechanism of release can be with the incorporation of growth factor delivery devices such as nanoparticles, microparticles, fibers, injectable complexes, etc. The rate by which the growth factor will release depends on the rate by which the scaffold degrades; in addition to this, it also depends on the diffusion rate of the growth factor through the pores of the scaffold. The design of these scaffolds for different tissues requires optimization of various parameters such as types of growth, dosage pattern, release pattern, release kinetics, and delivery duration (Soto-Gutierrez et al. 2010; Zhao et al. 2011). For each growth factor to be delivered, these parameters need to be optimized.

8 Principles Involved in Growth Factor Delivery

Upon physical and chemical processing, growth factors lose their activity quite easily. As a result, maintaining the protein activity is a need for the successful delivery of the growth factor. The stability of the protein or growth factor which has to be incorporated in the scaffold must be preserved at three stages, which are as follows: at the time of fabrication of scaffold, at the time of storage of scaffold, and at the time of degradation of the scaffold.

Another important issue that needs to be considered while delivering the growth factor from the designed 3D scaffold is its release profile. Most of the growth factors have very short half-lives in serum. Hence, it is essential on the part of the scaffold to maintain the desired concentration of growth factors that could direct the tissue growth.

8.1 Concerning Bone Tissue Engineering

The regeneration and repair of bone are difficult and time-consuming processes. When bone regeneration is required, several hormones, cytokines, and growth factors enter the picture. It is a complex cascade of different events at the molecular level governing the process of bone regeneration. Of all the various growth factors participating in bone regeneration, the one that has a vital role in repair and performs regeneration considerably includes BMP, i.e., the bone morphogenetic protein that belongs to the family of TGF-β, and examples of which are BMP-2, BMP-6, BMP-7, etc. BMPs increase the level of alkaline phosphate (ALP) expression that is an indication of early differentiation of cells toward osteoblast phenotype. Apart from this BMPs are also capable of differentiation of osteoblast in vitro and formation of bone in vivo. Bone formation and regeneration require the growth factor BMP widely required; still, BMP's complex structure, their shorter half-life, and rapid exiting property from the material are seen as few obstacles in their usage as growth factors. The scaffold carrying BMPs for its delivery should show an increase in the amount of release along with a sustained release for a prolonged time. In addition to this, the scaffold should also be having the capacity to protect the growth factor from denaturation. Given what has been said, scaffolds based on chitosan can be suitable for immobilizing and delivering BMP-2 as BMP-2 is known for maintaining its structure near the isoelectric point of chitosan, i.e., between 5 and 6. In pre-osteoblast cells, Nath et al. reported that a polymeric scaffold containing chitosan and hyaluronic acid can release growth factors in a regulated manner (Nath et al. 2015). The analysis of real-time polymerase chain reaction revealed a surge in the expression of those genes responsible for inducing osteoblast cells attached to BMP-2-loaded chitosan-based scaffold for bone regeneration.

Suitable mechanical property is one of the most significant properties of an ideal scaffold for bone tissue engineering applications as it provides a microenvironment well suited for the growth of osteogenic cells. In this regard, Niu et al. worked on a polymeric 3D scaffold comprising of collagen, nanohydroxyapatite, and PLA, i.e., polylactic acid (Gu et al. 2019; Niu et al. 2009). The research team loaded the polymeric scaffold with chitosan microparticles with growth factors encapsulated in it. The work proved to be promising in making an efficient scaffold for bone tissue regeneration applications. The growth factors which were loaded on the scaffold were BSA, i.e., bovine serum albumin, and BMP-2, and the loading onto the chitosan microparticles was performed by emulsion method having thymidine pyrophosphate (TPP) as the agent responsible for cross-linking. On analyzing the characteristics of the bioactive material, a significant increase in the mechanical property and degradation rate of the scaffold was observed owing to the presence of chitosan microparticles. The bioactive scaffold showed considerable bioactivity along with mimicking the physiological bone microenvironment (Vasile et al. 2020).

8.2 Concerning Nerve Tissue Engineering

Because of its intricate architecture and limited capacity to regenerate on its own, the regeneration of the animal brain becomes one of the most complicated processes following any accident (Chen et al. 2019a; Modo 2019). Consequently, transporting neural stem cells to the site of injury is important for brain cell regeneration. In this regard, as per the work reported by Shia et al., BDNF (brain-derived neurotrophic factor) is essential for neuron differentiation because, upon its action, neuronal stem cells could participate in the process of differentiation into neurons in both the conditions, be it in vitro or in vivo (Shi et al. 2012). Both in vitro and in vivo studies have been done with the nerve conduits made up of chitosan and loaded with the growth factor BDNF along with hUCMSCs, human umbilical cord mesenchymal stromal cells in the treating cases of traumatic brain injury. Chitosan scaffolds were created by the process of freeze-drying and the loading of BDNF onto the scaffold followed by mixing was performed by using genipin, one of the natural cross-linking agents. It was observed that the growth factor BDNF release from the scaffolds was in a controlled manner and well suited for improving the differentiation of hUCMSCs. Another growth factor that has been observed in neural tissue engineering to be very efficient in ensuring the growth and survival of neural stem cells is FGF-2 loaded on scaffolds which are chitosan-based. In another study reported by Skop et al., he prepared a nerve conduit/scaffold comprising of chitosan and heparin in which the role of cross-linking agent was prepared by genipin (Skop et al. 2013). The microspheres of chitosan and heparin then were loaded with the growth factor FGF-2, and the resultant scaffold was observed to be very effective in providing the optimum environment for the growth of neural stem cells in comparison to that of the control. Regarding the growth factor FGF-2, there was no issue related to its loss of activity as the loading of FGF-2 in the presence of heparin helped retain its biological activity.

Peripheral nerve regeneration, unlike other body tissues, necessitates the use of the tissue from the same individual's body, i.e., autografts. The reconstruction of peripheral nerve damage is sluggish and insufficient due to the scarcity of autografts. As an alternative to such autografts, scientists have reported as well as proposed certain nerve conduits which are chitosan-based.

The fragility of chitosan-based nerve conduits and the inability to control their rate of degradation are some of the disadvantages. However, as seen in the nerve conduits created by mixing chitosan and gelatin, these limits can be overcome. This scaffold's biochemical research revealed that it has considerable mechanical capabilities and a regulated rate of deterioration. When Schwann cells and the transforming growth factor (TGF) were added to the nerve conduit made from chitosan and gelatin, the rate of recovery in the in vivo animal model was observed to be satisfactory. Furthermore, the overall effect of this scaffold, like that of the autograft, was shown to be extremely good.

8.3 Concerning Skin Tissue Engineering

The inherent ability of chitosan to promote wound healing makes it an important component of skin tissue engineering (Deng et al. 2021; Zhang et al. 2021). One of the significant roles includes the activation of platelets whenever it encounters blood. The polymeric 3D scaffolds made up of chitosan and loaded with various growth factors have become common in recent years for the healing and regeneration of wounds. These growth factors have long been recognized as the most powerful and expedient biomaterial for wound healing. In a study reported by Mizuno et al., they developed scaffolds of hydroxypropyl combined with chitosan (Mizuno et al. 2003). The developed hydroxypropyl chitosan scaffolds were then loaded with a growth factor bFGF. The study was conducted in diabetic mice. The scientists on analyzing the developed scaffolds found them to have a substantial impact on wound healing in comparison to the scaffolds devoid of the growth factor bFGF. In another study reported by Kweon et al., scaffolds consisting of chitosan and heparin were synthesized which on further biochemical analysis were found to be very efficient in terms of healing the wound (Kweon et al. 2003). The presence of heparin ensures unrestricted binding of the growth factor to the 3D chitosan heparin complex. To test the healing capacity of wound of the rat model, the developed scaffold complex was added to the defect site, and a histological examination was performed after 15 days. When comparing the wounds treated with the scaffold made out of combining chitosan and heparin, loaded with the growth factor to those treated with the control, the findings revealed that the regeneration of the wound took less time in case of the defect treated with the developed complex scaffold. In recent times pressure ulcers were treated in aged mice using microparticles of gelatin encapsulated with the growth factor bFGF, thereafter loading the gelatin microparticles onto the scaffolds comprising chitosan.

Pressure ulcer healing was increased, and angiogenesis was enhanced with the chitosan scaffolds loaded with bFGF. Cao et al. created hybrid scaffolds comprising collagen, chitosan, and chondroitin sulfate (Cao et al. 2015). The developed scaffold complex was encased with microspheres of PLGA loaded with the growth factor bFGF. The team was successful in developing a scaffold for skin tissue engineering with regulated drug release capacity. The rate of diffusion of the synthesized scaffold along with the release of the encased growth factor bFGF was observed to be high because of their higher rate of swelling and degradation. According to cell proliferation reports, the polymeric scaffold complex consisting of chitosan, collagen, and chondroitin sulfate had significant biocompatibility. In addition to this, it also had a significant ability to promote the proliferation of fibroblast cells, thereby regenerating the skin tissue.

8.4 Concerning Cartilage Tissue Engineering

Because the cartilage lacks blood vessels and has a lower cell capacity, it has a limited potential to regenerate (Mallick et al. 2019). Recently, scientists have been working on several polymeric 3D scaffolds for cartilage repair and regeneration. Glycosaminoglycans (GAGs) are well-known for their ability to regenerate cartilage tissue because they play an important role in the process of chondrogenesis or cartilage creation. Chitosan has been discovered to be structurally comparable to GAG, making it a viable choice for cartilage regeneration. Tigli et al. reported that scaffold was produced by covalently immobilizing the RGD sequence (arginine, glycine, and aspartate) coupled with the epithelial growth factor (EGF) on it in one of their chondrogenic research (Tiğli and Gümüşderelioğlu 2008). The biochemical assays, particularly MTT assay, revealed a significant increase in the proliferation of cells attached to the polymeric EGF-loaded chitosan scaffold. An increase in the GAG concentration and DNA amount was also observed with time which pointed out the ability of the developed scaffold in regenerating the cartilage.

Other growth factors include TGF-β, IGF, FGF-2, etc., which are also known to play a significant role in cartilage tissue engineering by healing the wound with their controlled release from the scaffold on which they are loaded, thus ensuring efficient proliferation and differentiation of chondrocytes (Green et al. 2015; Zhang et al. 2009).

9 Conclusion

Scaffolds along with growth factors are the need of the hour as far as tissue engineering is concerned. The role of polymeric 3D scaffolds comes into the picture in almost all types of tissue engineering, namely, bone, cartilage, nerve, skin, etc., as they provide proper support to the cells for their proliferation, migration, and differentiation. Growth factors on the other hand function as an inducing signal for the cells to undergo growth and differentiation, thereby ultimately regenerating the damaged or defected tissue.

References

Agarwala S (2016) A perspective on 3D bioprinting technology: present and future. Am J Eng Applied Sci 9:985–990

Akbarzadeh R, Yousefi AM (2014) Effects of processing parameters in thermally induced phase separation technique on porous architecture of scaffolds for bone tissue engineering. J Biomed Mater Res Part B Appl Biomater 102:1304–1315

Alyassin Y, Sayed EG, Mehta P et al (2020) Application of mesoporous silica nanoparticles as drug delivery carriers for chemotherapeutic agents. Drug Discov Today 25:1513–1520

Amrein BA, Runthala A, Kamerlin SCL (2019) In silico-directed evolution using CADEE. In: Sikosek T (ed) Computational methods in protein evolution, methods in molecular biology, vol 1851. Humana Press, New York, pp 381–415
Antimisiaris S, Marazioti A, Kannavou M et al (2021) Overcoming barriers by local drug delivery with liposomes. Adv Drug Deliv Rev 174:53–86
Babensee JE, Anderson JM, McIntire LV, Mikos AG (1998) Host response to tissue engineered devices. Adv Drug Deliv Rev 33:111–139
Batrakova EV, Gendelman HE, Kabanov AV (2011) Cell-mediated drug delivery. Expert Opin Drug Deliv 8:415–433
Billiet T, Vandenhaute M, Schelfhout J et al (2012) A review of trends and limitations in hydrogel-rapid prototyping for tissue engineering. Biomaterials 33:6020–6041
Cao H, Chen M-M, Liu Y et al (2015) Fish collagen-based scaffold containing PLGA microspheres for controlled growth factor delivery in skin tissue engineering. Colloids Surf B: Biointerfaces 136:1098–1106
Chen G, Ushida T, Tateishi T (2002) Scaffold design for tissue engineering. Macromol Biosci 2:67–77
Chen HI, Song H, Ming G (2019a) Applications of human brain organoids to clinical problems. Dev Dyn 248:53–64
Chen RR, Mooney DJ (2003) Polymeric growth factor delivery strategies for tissue engineering. Pharm Res 20:1103–1112
Chen Y, Wu T, Huang S et al (2019b) Sustained release SDF-1α/TGF-β1-loaded silk fibroin-porous gelatin scaffold promotes cartilage repair. ACS Appl Mater Interfaces 11:14608–14618
Chen F-M, Zhang M, Z.-F. (2010) Wu, Toward delivery of multiple growth factors in tissue engineering. Biomaterials 31(24):6279–6308. https://www.sciencedirect.com/science/article/abs/pii/S0142961210005697?via%3Dihub
Christy PN, Basha SK, Kumari VS et al (2020) Biopolymeric nanocomposite scaffolds for bone tissue engineering applications–a review. J Drug Deliv Sci Technol 55:101452
Chutipakdeevong J, Ruktanonchai U, Supaphol P (2015) Hybrid biomimetic electrospun fibrous mats derived from poly (ε-caprolactone) and silk fibroin protein for wound dressing application. J Appl Polym Sci 132(132):41653
Cui C, Kim D-O, Pack MY et al (2020) 4D printing of self-folding and cell-encapsulating 3D microstructures as scaffolds for tissue-engineering applications. Biofabrication 12:018–045
Das D, Noh I (2018) Overviews of biomimetic medical materials. Biomimetic Medical Materials 1064:3–24
Dehghani F, Annabi N (2011) Engineering porous scaffolds using gas-based techniques. Curr Opin Biotechnol 22:661–666
Deng A, Yang Y, Du S et al (2021) Preparation of a recombinant collagen-peptide (RHC)-conjugated chitosan thermosensitive hydrogel for wound healing. Mater Sci Eng C 119:111555
Deville S, Saiz E, Tomsia AP (2006) Freeze casting of hydroxyapatite scaffolds for bone tissue engineering. Biomaterials 27:5480–5489
Dhandayuthapani B, Yoshida Y, Maekawa T (2011) Polymeric scaffolds in tissue engineering application: a review. Int J Polym Sci 2011:1–19
Doostmohammadi M, Forootanfar H, Ramakrishna S (2020) Regenerative medicine and drug delivery: Progress via electrospun biomaterials. Mater Sci Eng C 109:110521
Duffy DJ (2016) Problems, challenges and promises: perspectives on precision medicine. Brief Bioinform 17:494–504
El-FiqiA MN, Jo SB et al (2021) Nanotherapeutics for regeneration of degenerated tissue infected by bacteria through the multiple delivery of bioactive ions and growth factor with antibacterial/angiogenic and osteogenic/odontogenic capacity. Bioactive Material 6:123–136
Fiorillo L, Cervino G, Galindo-Moreno P, Herford AS, Spagnuolo G, Cicciù M (2021) Growth factors in oral tissue engineering: new perspectives and current therapeutic options. BioMed Research International. https://www.hindawi.com/journals/bmri/2021/8840598/

Green JD, Tollemar V, Dougherty M et al (2015) Multifaceted signaling regulators of chondrogenesis: implications in cartilage regeneration and tissue engineering. Genes & Diseases 2:307–327
Gu L, Shan T, Ma Y-X et al (2019) Novel biomedical applications of crosslinked collagen. Trends Biotechnol 37:464–491
Gu Q, Hao J, Lu Y et al (2015) Three-dimensional bio-printing. Sci China Life Sci 58:411–419
Guimarães CF, Gasperini L, Marques AP et al (2020) The stiffness of living tissues and its implications for tissue engineering. Nature Reviews Materials 5:351–370
Guvendiren M, Molde J, Soares RM et al (2016) Designing biomaterials for 3D printing. ACS Biomater Sci Eng 2:1679–1693
Han J, Li G, Yuan L (2019) Preparation and characterization of nanostructured hollow MgO spheres. Materials 12:1–8
Haraguchi Y, Shimizu T, Yamato M et al (2012) Concise review: cell therapy and tissue engineering for cardiovascular disease. Stem Cells Transl Med 1:136–141
He J, Chen G, Liu M et al (2020) Scaffold strategies for modulating immune microenvironment during bone regeneration. Mater Sci Eng C 108:110411
Hollister SJ (2005) Porous scaffold design for tissue engineering. Nat Mater 4:518–524
Horch RE, Boos AM, Quan Y et al (2013) Cancer research by means of tissue engineering–is there a rationale? J Cell Mol Med 17:1197–1206
Hull KL, Harvey S (2014) Growth hormone and reproduction: a review of endocrine and autocrine/paracrine interactions. International journal of endocrinology. https://www.hindawi.com/journals/ije/2014/234014/
Ivica A, Ghayor C, Zehnder M et al (2020) Pulp-derived exosomes in a fibrin-based regenerative root filling material. J Clin Med 9:1–12
Jacob J, Haponiuk JT, Thomas S et al (2018) Biopolymer based nanomaterials in drug delivery systems: a review. Materials Today Chem 9:43–55
Kamjula V, Kanneganti A, Metla R et al (2020) Decoding the vital segments in human ATP-dependent RNA helicase. Bioinformation 16(2):160
Khalil H, Jummaat F, Yahya EB et al (2020) A review on micro-to nanocellulose biopolymer scaffold forming for tissue engineering applications. Polymers 12:2043
Kjær M, Magnusson P, Krogsgaard M et al (2006) Extracellular matrix adaptation of tendon and skeletal muscle to exercise. J Anat 208:445–450
Ko HC, Milthorpe BK, McFarland CD (2007) Engineering thick tissues-the vascularisation problem. European Cells and Materials 14:1–19
Koons GL, Diba M, Mikos AG (2020) Materials design for bone-tissue engineering. Nature Reviews Materials 5:584–603
Kumari S, Singh BN, Srivastava P (2019) Effect of copper nanoparticles on physico-chemical properties of chitosan and gelatin-based scaffold developed for skin tissue engineering application. 3 Biotech 9:1–14
Kweon D-K, Song S-B, Park Y-Y (2003) Preparation of water-soluble chitosan/heparin complex and its application as wound healing accelerator. Biomaterials 24:1595–1601
Lebourg M, Serra RS, Estellés JM et al (2008) Biodegradable polycaprolactone scaffold with controlled porosity obtained by modified particle-leaching technique. J Mater Sci Mater Med 19:2047–2053
Liao CJ, Chen CF, Chen JH et al (2002) Fabrication of porous biodegradable polymer scaffolds using a solvent merging/particulate leaching method. J Biomed Mater Res 59:676–681
Ma PX, Elisseeff J (2005) Scaffolding in tissue engineering, vol 656. CRC Press, pp 1–241
Mabrouk M, Beherei HH, Das DB (2020) Recent progress in the fabrication techniques of 3D scaffolds for tissue engineering. Mater Sci Eng C 110:110716
MaHam A, Tang Z, Wu H et al (2009) Protein-based nanomedicine platforms for drug delivery. Small 5:1706–1721

Mallick S, Tripathi S, Srivastava P (2015) Advancement in Scaffolds for Bone Tissue Engineering: A Review. IOSR Journal of Pharmacy and Biological Sciences (IOSR-JPBS) 10(1):37–54. https://www.iosrjournals.org/iosr-jpbs/papers/Vol10-issue1/Version-4/H010143754.pdf

Mallick S, Beyene Z, Suman DK et al (2019) Strategies towards Orthopaedic tissue engineered graft generation: current scenario and application. Biotechnol Bioprocess Eng 24:854–869

Melchels FP, Domingos MA, Klein TJ et al (2012) Additive manufacturing of tissues and organs. Prog Polym Sci 37:1079–1104

Mercante LA, Iwaki LE, Scagion VP et al (2021) Electrochemical detection of bisphenol a by Tyrosinase immobilized on electrospun nanofibers decorated with gold nanoparticles. Electrochem 2:41–49

Mizuno K, Yamamura K, Yano K et al (2003) Effect of chitosan film containing basic fibroblast growth factor on wound healing in genetically diabetic mice. J Biomed MaterRes Part A 64: 177–181

Mizutani T, Haga H, Kato K et al (2009) Wide range scanning probe microscopy for probing mechanical effects on cellular function. Arch Histol Cytol 72:235–243

Modo M (2019) Bioscaffold-induced brain tissue regeneration. Front Neurosci 13:1156

Nam YS, Park TG (1999) Porous biodegradable polymeric scaffolds prepared by thermally induced phase separation. J Biomed Mater Res 47:8–17

Nam YS, Yoon JJ, Park TG (2000) A novel fabrication method of macroporous biodegradable polymer scaffolds using gas foaming salt as a porogen additive. J Biomed Mater Res 53:1–7

Narayanan N, Calve S (2021) Extracellular matrix at the muscle–tendon interface: functional roles, techniques to explore and implications for regenerative medicine. Connect Tissue Res 62:53–71

Nath SD, Abueva C, Kim B et al (2015) Chitosan–hyaluronic acid polyelectrolyte complex scaffold crosslinked with genipin for immobilization and controlled release of BMP-2. Carbohydr Polym 115:160–169

Niemczyk-Soczynska B, Gradys A et al (2020) Hydrophilic surface functionalization of electrospun Nanofibrous scaffolds in tissue engineering. Polymers 12:26–36

Niu X, Feng Q, Wang M et al (2009) Porous nano-HA/collagen/PLLA scaffold containing chitosan microspheres for controlled delivery of synthetic peptide derived from BMP-2. J Control Release 134:111–117

O'Brien FJ, Harley BA, Yannas IV et al (2005) The effect of pore size on cell adhesion in collagen-GAG scaffolds. Biomaterials 26:433–441

Oh SH, Kang SG, Kim ES et al (2003) Fabrication and characterization of hydrophilic poly (lactic-co-glycolic acid)/poly (vinyl alcohol) blend cell scaffolds by melt-molding particulate-leaching method. Biomaterials 24:4011–4021

Oliveira ÉR, Nie L, Podstawczyk D et al (2021) Advances in growth factor delivery for bone tissue engineering. Int J Mol Sci 22:903

Owen SC, Shoichet MS (2010) Design of three-dimensional biomimetic scaffolds. J Biomed Mater Res A 94:1321–1331

Patel JM, Saleh KS, Burdick JA et al (2019) Bioactive factors for cartilage repair and regeneration: improving delivery, retention, and activity. Actabiomaterialia 93:222–238

Phulara SC, Rajput VS, Mazumdar B et al (2020) Metabolic and enzyme engineering for the microbial production of anticancer Terpenoids. In: Masood N, Shakil S (eds) Essentials of cancer genomic, Computational approaches and precision medicine. Springer, Singapore, pp 237–259

Puppi D, Chiellini F, Piras AM et al (2010) Polymeric materials for bone and cartilage repair. Prog Polym Sci 35:403–440

Rajab TK, O'Malley TJ, Tchantchaleishvili V (2020) Decellularized scaffolds for tissue engineering: current status and future perspective. Artif Organs 44:1031–1043

Ren X, Zhao M, Lash B et al (2020) Growth factor engineering strategies for regenerative medicine applications. Front Bioeng Biotechnol 7:469

Rico-Llanos GA, Borrego-González S, Moncayo-Donoso M et al (2021) Collagen type I biomaterials as scaffolds for bone tissue engineering. Polymers 13:599

Riha GM, Lin PH, Lumsden AB, Yao Q, Chen C (2005) Application of stem cells for vascular tissue engineering. Tissue engineering 11(9-10):1535–1552. https://www.liebertpub.com/doi/10.1089/ten.2005.11.1535?url_ver=Z39.88-2003&rfr_id=ori:rid:crossref.org&rfr_dat=cr_pub%20%200pubmed

Runthala A (2021) Probabilistic divergence of a template-based modelling methodology from the ideal protocol. J Mol Model 27(2):25

Runthala A, Chowdhury S (2019) Refined template selection and combination algorithm significantly improves template-based modeling accuracy. J Bioinforma Comput Biol 17(2):1950006

Runthala A, Sai TH, Kamjula V et al (2020) Excavating the functionally crucial active-site residues of the DXS protein of Bacillus subtilis by exploring its closest homologues. J Genetic Eng Biotechnol 18(1):76

Saha S, Kundu B, Kirkham J, Wood D, Kundu SC, Yang XB (2013) Osteochondral tissue engineering in vivo: a comparative study using layered silk fibroin scaffolds from mulberry and nonmulberry silkworms. PloS one 8(11):e80004. https://journals.plos.org/plosone/article?id=10.1371/journal.pone.0080004

San Antonio JD, Jacenko O, Fertala A et al (2021) Collagen structure-function mapping informs applications for regenerative medicine. Bioengineering 8:1–23

Schneible JD, Daniele MA, Menegatti S (2021) Natural and synthetic biopolymers in drug delivery and tissue engineering. Biopolymers for Biomedical and Biotechnological Applications:265–356

Shi W, Nie D, Jin G et al (2012) BDNF blended chitosan scaffolds for human umbilical cord MSC transplants in traumatic brain injury therapy. Biomaterials 33:3119–3126

Shoichet MS (2010) Polymer scaffolds for biomaterials applications. Macromolecules 43:581–591

Shukla V, Runthala A, Rajput VS et al (2022) Computational and synthetic biology approaches for the biosynthesis of antiviral and anticancer Terpenoids from Bacillus subtilis. Bentham Med Chem 18(3):307–322

Silva DS, Wallace DB, Cooley PW et al (2007) An inkjet printing station for neuroregenerative tissue engineering. IEEE Xplore INSPEC Accession Number 9903125:71–73

Singh B, Pramanik K (2018) Fabrication and evaluation of non-mulberry silk fibroin fiber reinforced chitosan based porous composite scaffold for cartilage tissue engineering. Tissue Cell 55:83–90

Singh BN, Veeresh V, Mallick SP et al (2019) Design and evaluation of chitosan/chondroitin sulfate/nano-bioglass based composite scaffold for bone tissue engineering. Int J Biol Macromol 133:817–830

Singh BN, Veeresh V, Mallick SP et al (2020) Generation of scaffold incorporated with nanobioglass encapsulated in chitosan/chondroitin sulfate complex for bone tissue engineering. Int J Biol Macromol 153:1–16

Skop NB, Calderon F, Levison SW et al (2013) Heparin crosslinked chitosan microspheres for the delivery of neural stem cells and growth factors for central nervous system repair. Acta Biomater 9:6834–6843

Sodhi H, Panitch A (2021) Glycosaminoglycans in tissue engineering: a review. Biomol Ther 11:29

Soker S, Machado M, Atala A (2000) Systems for therapeutic angiogenesis in tissue engineering. World J Urol 18:10–18

Soto-Gutierrez A, Yagi H et al (2010) Cell delivery: from cell transplantation to organ engineering. Cell Transplant 19:655 665

Talebian S, Foroughi J, Wade SJ et al (2018) Biopolymers for antitumor implantable drug delivery systems: recent advances and future outlook. Adv Mater 30:1706665

Tan GZ, Zhou Y (2020) Electrospinning of biomimetic fibrous scaffolds for tissue engineering: a review. Int J Polym Mater Polym Biomater 69:947–960

Teo WE, Ramakrishna S (2006) A review on electrospinning design and nanofibre assemblies. Nanotechnology 17:R89

Thomson RC, Shung AK, Yaszemski MJ et al (2000) Polymer scaffold processing. Prin Tissue Engg 2:251–262

Tiğli RS, Gümüşderelioğlu M (2008) Evaluation of RGD-or EGF-immobilized chitosan scaffolds for chondrogenic activity. Int J Biol Macromol 43:121–128

Tong Z, Jin L, Oliveira JM et al (2021) Adaptable hydrogel with reversible linkages for regenerative medicine: dynamic mechanical microenvironment for cells. Bioactive Material 6:1375–1387

Tran RT, Naseri E, Kolasnikov A et al (2011) A new generation of sodium chloride porogen for tissue engineering. Biotechnol Appl Biochem 58:335–344

Vacanti JP, Langer R (1999) Tissue engineering: the design and fabrication of living replacement devices for surgical reconstruction and transplantation. Lancet 354:S32–S34

Vasile C, Pamfil D, Stoleru E et al (2020) New developments in medical applications of hybrid hydrogels containing natural polymers. Molecules 25:1539

Wang S, Chi J, Jiang Z et al (2021) A self-healing and injectable hydrogel based on water-soluble chitosan and hyaluronic acid for vitreous substitute. Carbohydr Polym 256:117519

Wishart DS (2016) Emerging applications of metabolomics in drug discovery and precision medicine. Nat Rev Drug Discov 15:473

Yu QH, Zhang CM, Jiang ZW et al (2020) Mussel-inspired adhesive Polydopamine-functionalized hyaluronic acid hydrogel with potential bacterial inhibition. Global Chall 4:1900068

Zhang L, Hu J, Athanasiou KA (2009) The role of tissue engineering in articular cartilage repair and regeneration. Crit Rev Biomed Eng 37(1–2):1–57

Zhang M, Wang G, Wang D et al (2021) Ag@ MOF-loaded chitosan nanoparticle and polyvinyl alcohol/sodium alginate/chitosan bilayer dressing for wound healing applications. Int J Biol Macromol 175:481–494

Zhao X, Kim J, Cezar CA et al (2011) Active scaffolds for on-demand drug and cell delivery. Proc Natl Acad Sci 108:67–72

Perspectives on Anti-Tuberculosis Drug Discovery

Shashikanta Sau and Nitin Pal Kalia

Abstract *Mycobacterium tuberculosis (Mtb)* is the causative agent of tuberculosis and is the second leading reason of death worldwide from infectious diseases after HIV. Available anti-TB chemotherapy requires a cocktail of several drugs for more than 6 months. Due to the rise of extensively and multidrug-resistant strains of *Mtb* along with HIV co-infection, there is an urgent need of new molecules for the TB treatment. These new drug molecules must have a novel mode of action and low toxicity profile, is active against both susceptible and drug-resistant strains of *Mtb*, should be well tolerated in combination with antiretroviral drugs, and more importantly shall require lesser time to clear the infection. The pathology of *Mtb* and its ability to adapt and survive in heterogeneous states inside the host during infection are the major challenges in anti-TB drug discovery. Therefore, various high-throughput screening (HTS) methods that mimic closely to the host conditions during *Mtb* infection can be used to identify novel drug molecules. This chapter summarizes these HTS methods for discovering novel anti-TB molecules.

1 Introduction

Tuberculosis (TB) is a communicable disease with the highest mortality rate from a single infectious agent (ranking above HIV/AIDS). According to the WHO report, there were an estimated 1.4 million TB deaths in 2020 (WHO 2020). With time, drug treatment for TB has also evolved. Chemotherapy for susceptible TB consists of four drugs, i.e., isoniazid, rifampicin, ethambutol, and pyrazinamide, for over 6 months (WHO 2018). Efforts to control the disease faced a big jolt with the emergence of drug-resistant strains. Incomplete antibiotic treatment and HIV co-infection are the root causes for the emergence of multidrug-resistant (MDR) and extremely drug-resistant (XDR) TB (Sotgiu et al. 2015a, 2015b; Mayer and Hamilton 2010). Every

S. Sau · N. P. Kalia (✉)
Department of Biological Sciences (Pharmacology and Toxicology), National Institute of Pharmaceutical Education and Research, Hyderabad, Telangana, India
e-mail: nitin.kalia@niperhyd.ac.in

V. S. Rajput, A. Runthala (eds.), *Drugs and a Methodological Compendium*,
https://doi.org/10.1007/978-981-19-7952-1_13

year approximately half a million of new cases of multidrug-resistant TB occur, and the treatment duration with second-line anti-TB agents is about 18–24 months. In the last few decades, only three new drugs (delamanid and pretomanid by the EMA, bedaquiline by the FDA) have been approved for use against drug-resistant TB (Ryan and Lo 2014; FDA 2012). This is not because strong efforts are not being made, but anti-TB drug discovery itself is very challenging due to the number of technical issues such as the requirement of a high level of containment facilities (BSL-3) to work, slow growth of the microorganism, and cell wall composition (lipid-rich and less permeable to hydrophilic molecules). Inside the host, *Mtb* confronts varied environments such as hypoxia and limited nutrients and hence exists as a heterogeneous population. In addition, there are several other factors as well that affect the development of new anti-tubercular agents such as new drug molecules that ideally need to be cost-effective, compatible with other companion drugs, and effective in shortening the treatment time along with a low mutation rate (TB Alliance 2019; Parish 2020). Therefore, to counter the resistant strains and effectively control the disease, there is a pressing need for drugs with a novel mode of action that is effective at killing *Mtb* prevalent in all the physiological states under various conditions and which may shorten the therapy. Strong and sincere efforts have been put into the high-throughput screening (HTS) of chemical libraries to identify novel molecules for anti-TB drug discovery. As a component of HTS drive, a major focus of researchers is on increasing the number and refining the in vitro assay systems that are needed to identify the efficacy of the compounds (Sotgiu et al. 2015a, 2015b; Abrahams and Besra 2020). A significant progress has been made in evolving robust and reproducible in vitro assays for the complete evaluation of the efficacy of the drug molecules. With time, improvements have been made in developing in vitro assays, and those assays, which are close to mimicking conditions similar to host microenvironments available to bacteria during infection, are preferred for a complete evaluation of the efficacy of drug candidates to establish structure-activity relationships (Zuniga et al. 2015). This chapter focuses on the most relevant, robust, and reproducible assays for high-throughput screening of anti-TB molecules.

2 Pathogenesis of Tuberculosis

Mtb enters the human lungs through the aerosol route where they prefer the alveolar macrophage for their existence (Fig. 1). The bacillus survives, gets away from killing, and continues to multiply by avoiding the phagosome-lysosome fusion (Sturgill-Koszycki et al. 1994; Queval et al. 2017; Russell et al. 2010). Briefly, after infection, there is a macrophage-induced inflammatory response and a localized chemokine gradient, which leads to the gathering of various immune cells (mononuclear cells and other immune cells) at the site of infection. The whole process is the building block for granuloma formation.

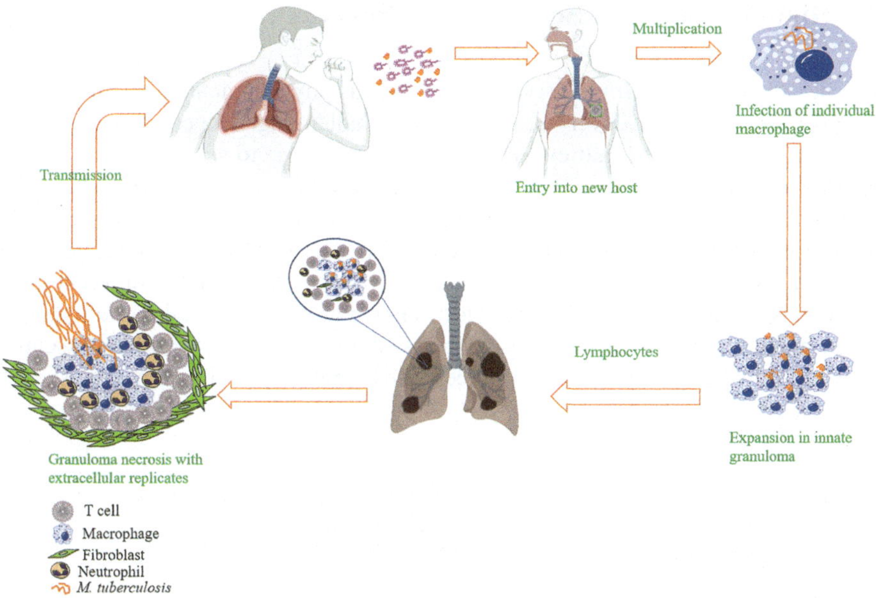

Fig. 1 Different phases of life cycle of *Mycobacterium tuberculosis*: Transmission of bacilli from infected to healthy humans. Bacilli enter the lungs of the healthy humans after transmission and then invade the macrophages. Immune cells surround the infected macrophages and then formation of granuloma (a hallmark for tuberculosis infection)

The granuloma formation restricts the bacilli replication and prevents the spread of bacteria, but later as the disease progresses, the breakdown of granuloma cells starts which leads to caseous necrosis, which may result in cavitation. The caseous cavities of granuloma play a crucial role in the containment of bacilli. The bacillus inside these lesions remains under control in persons with a healthy immune system, but those with a compromised immune system are prone to get active TB. Furthermore, caseous necrosis leads to the elimination of the adjoining host tissues, and if this eradication continues, the granuloma collapses into the lung resulting in the release of *Mtb* bacilli into the airway.

The adaptable nature of the bacilli helps it in surviving inside the host under different conditions ranging from simple vascular cellular aggregations to necrotic lesions where it faces situations similar to nutrient deprivation and hypoxia (Barry et al. 2009; Dartois 2014). These variable host conditions and the ability of *Mtb* to adapt to these numerous assorted and dynamic conditions through infection are responsible for heterogeneous bacterial populations, which are responsible for long-term chemotherapy for TB. Therefore, drug regimens targeting the bacilli residing inside the host under all these niches are likely to offer the greatest chance to lessen the treatment period and no chance of relapse.

3 High-Throughput Screening Methods

The flexible nature of *Mtb* to survive under various physiological conditions within the host makes drug discovery programs quite challenging. The conditions such as low oxygen and limited nutrient supply force the bacteria to slow down its metabolism and acquire ATP homeostasis, which is responsible for enhanced bacterial doubling time and generation of non-replicating but viable bacilli. Therefore, those high-throughput screening methods need to be adopted that can identify novel chemical entities that are effective against *Mtb* surviving under different host conditions. There are several approaches ranging from whole cell-based assay to target-based biochemical assays based on essentiality of the gene product (essential enzyme) which are being used to screen active molecules effective against *Mtb* (Fig. 2).

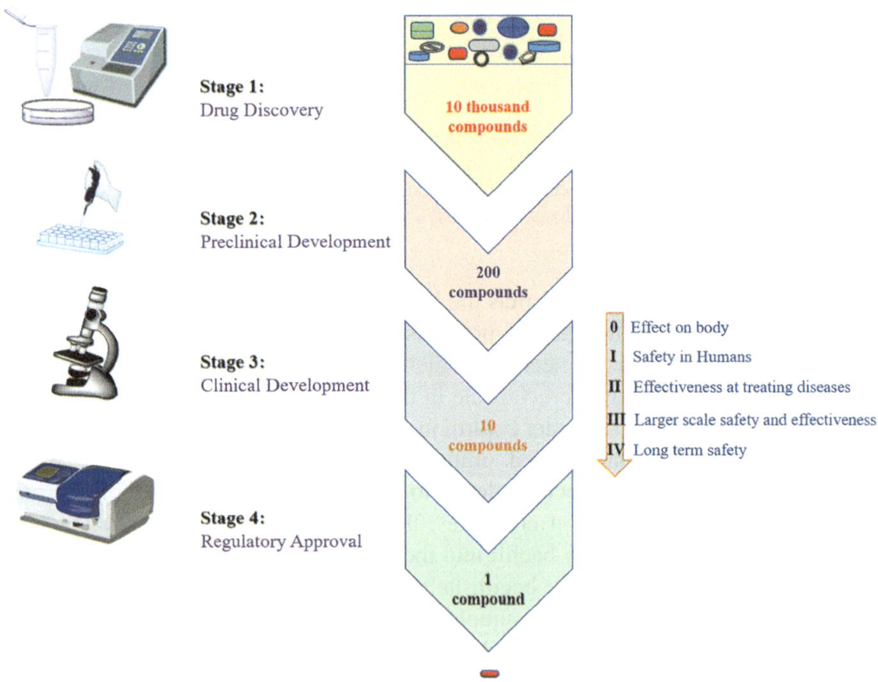

Fig. 2 High-throughput screening of chemical libraries against *Mycobacterium tuberculosis*: Screening of chemical libraries under various conditions mimicking the host environment using different in vitro assays. Success rate is very low due to the flexible nature of the bacteria and its ability to survive under various host conditions such as low oxygen and restricted or limited nutrient supply

3.1 *Whole Cell-Based Assays: Mimicking Host Conditions for Bacterial Replication (Phenotypic Screening)*

Actively Replicating Bacteria for Whole-Cell Screening

Mtb coexist in both replicating and non-replicating states during infection within the host. The primary high-throughput screens developed in the beginning focused on actively replicating bacteria growing under standard aerobic culture conditions. Under normal standard aerobic conditions, *Mtb* growth is maintained in 7H9 liquid media complemented with albumin, catalase, and dextrose along with Tween 80 (0.05%). The liquid broth assay in a 96-well plate format is the most common assay used to access the efficacy of the compound in the form of inhibition of bacterial growth. This primary assay is in common use to address the replicating bacteria where results in terms of minimum inhibition concentration (MIC) depict the efficacy of the compounds tested (TB Alliance 2019).

Fatty Acids as a Carbon Source

At the time of infection, *Mtb* translocate itself between macrophages, granulomas, mucosal epithelia, and necrotic cells (Flynn et al. 2011). The elasticity of metabolism helps it adapt to various stress conditions. The principal carbon source for *Mtb* is pyruvate or lactate as compared to fatty acids and glucose (Serafini et al. 2019). During bacterial growth, it utilizes carbon sources for growth and metabolism. When the main carbon source is exhausted, it can metabolize other carbon sources at a time. So, it shifts its metabolism to use fatty acids and cholesterol which are derived from host lipids for survival. That's why, a lot of in vitro studies have been performed utilizing fatty acids such as cholesterol, butyrate, palmitate, and acetate as a carbon source for the screening assays of the compounds.

Early et al. used butyrate media to lead a high-throughput screening for the recognition of *Mtb* inhibitors (Early et al. 2016). They verified the growth proportion in the medium using different length-chain fatty acids as a prime carbon source, for example, isovaleric acid, acetic acid, palmitic acid, and butyric acid. Among the used fatty acids, butyrate provided a vigorous growth. They screened around 87,000 compounds in butyrate media, and the initial compounds were counter-screened in glucose media for the identification of the compounds that showed activity in butyrate media. Hits obtained were classified, and six compounds of oxadiazole series with specific anti-TB activity with no toxicity to mammalian cells were identified.

It has been reported that *Mtb* catabolizes other carbon sources in vitro and macrophages (Zimmermann et al. 2017). Bloch and Segal investigated that *Mtb* prefers fatty acids for metabolism in a lung-infected mouse (Segal and Bloch 1956). Rodriguez et al. developed a model, in which *Mtb* was supplemented with 11-length long-chain (Rodríguez et al. 2014) fatty acids as the only carbon source and found the sluggish growth linked to dormancy (Grant et al. 2013). Different assays were performed to screen the activity of compounds with the use of non-sugar carbon sources which have also been established.

Nutrient-Starved Model

Mtb exists in a non-replicating state and shows tolerance to antibiotics upon exposure to nutrient starvation conditions. In this in vitro model, *Mtb* bacilli were resuspended in phosphate-buffered saline (PBS) medium or in 7H9 medium without any carbon source. But, *Mtb* forms clusters in PBS which may limit its use in the high-throughput screening methods. It has also been reported that minimum salt media added with 0.05% Tyloxapol is appropriate for high-throughput screening of *Mtb* in the medium. Around 300,000 compounds were screened using the carbon starvation model, out of which 116 compounds were found to be active against *Mtb* (Chang and Guan 2021).

Mtb adapts to the nutrient-starved conditions within 48–96 hours, with a broad change in the genetic expression, and remains viable for months to a year (Chang and Guan 2021). The new chemical entities can be identified using this simple model by measuring the viability of non-replicating *Mtb* under nutrient-starved conditions. The bacilli maintained in liquid media without any carbon source for 96 hours were used as an inoculum and incubated with test compounds in 384-well plates. The viability of the cells was measured using Alamar Blue-based fluorescence assay, and the whole process takes 7 days of incubation at 37°C. Another model was also developed with the help of luminescent strain using luciferase in a 96-well plate, but the drawback of this model is that it is less reproducible (Grant et al. 2013).

pH Homeostasis

Mtb adapts to different adverse conditions. Under the acidic condition, it keeps the internal pH neutral via redox homeostasis with the participation of the electron transport chain (Zhao et al. 2015). Under low pH conditions, non-replicating bacteria are produced. *Mtb* is highly sensitive to low pH and unable to multiply under pH 5.4, but it can continue to survive in this condition (Early et al. 2019; Darby et al. 2013). This feature helped in designing the protocols so that the viability and maintenance of internal pH can be measured. Both of these principles are utilizing the drug discovery process. "Lopicidal" assay is used to test the bactericidal effect of new molecules and measured with the help of luminescent strain of *Mtb* which is incubated at pH 4.5 in a 96-well plate. Darby et al. established a model to screen the compounds that can alter the intracellular homeostasis of *Mtb*. This helps in the identification of the pathways followed by *Mtb* to maintain the homeostasis.

Various important measures are taken to ensure that the compounds bind to the target only. Phosphate citrate buffer (pH 4.5) in addition to Tyloxapol was used as a medium just to avoid toxic fatty acid generation from the hydrolysis of a tween, which was present in 7H11 conventional medium. Another assay was also used to measure the pH homeostasis of *Mtb*. In this assay, the samples were incubated with the compounds at pH 4.5 for 2 days in which a reporter strain having a pHLUOR (radiometric GFP responsive to pH) was applied for the measurement of intracellular pH. This method can be performed on a 96- and 384-well plate (Early et al. 2018).

The *phoPR* regulon system is very much important in detecting the external pH and helps *Mtb* adapt acidic conditions within macrophages (Abramovitch et al. 2011). This regulon system is also helpful in maintaining the replication of *Mtb* in animal models as well but with no significant role in multiplication of bacilli under normal replicating conditions (Johnson et al. 2015).

The expression of a well-known locus, i.e., an acid- and phagosome-regulated (*aprABC*) locus, is linked to *phoPR* and is induced during low pH conditions. The promoter regulating the expression of this locus, i.e., *aprA*, was used to control the expression of green fluorescent protein (GFP) to generate a pH-induced fluorescent strain of *Mtb*. Further, Johnson et al. used this strain for high-throughput screening of around 22,000 molecules to recognize the hits with efficacy to inhibit the pH-inducible fluorescence. This assay was performed at pH 5.7 using 7H9 media.

Bacterial Biofilm

During persistence conditions, *Mtb* generates biofilm that comprise of drug-tolerant cells. Therefore, targeting *Mtb* biofilm may bring a dramatic change in drug discovery to eliminate persistence. Wang et al. established an in vitro model for the high-throughput screening of new molecules that are active against drug-resistant and non-replicating *Mtb*. *Mycobacterium smegmatis* was used as a model microorganism for *Mtb* biofilm development. The test compounds were incubated with *M. smegmatis* at 37 °C for 3 days; absorbance was measured to determine the cell vitality, and minimum inhibitory concentration (MIC_{50}) was determined based on absorbance. The active compounds were further verified in a secondary screen to check their capability to inhibit the biofilm formation against *M. smegmatis*. It has been reported that a small molecule TCA1 with high bactericidal activity against susceptible and multidrug-resistant strains of *Mtb* inhibits the non-replicating persisters during acute and chronic TB infection. The molecule was further evaluated in biofilm assay and showed significant inhibition in the biofilm formation (Wang et al. 2013). DprE1, an important enzyme of *Mtb* with role in cell wall synthesis, is the target for TCA1 where it binds and inhibits the cell wall synthesis in *Mtb* under nutrient starvation conditions. Apart from DprE1, TCA1 also binds to MoeW enzyme involved in molybdenum cofactor biosynthesis. The genetic and affinity-based assays confirmed the binding of TCA1 to DprE1 and MoeW enzymes which resulted inhibition of cell wall synthesis.

Hypoxia Model

Wayne and Hayes first established a hypoxic model of *Mtb*. This model is based on the principle of low dissolved oxygen saturation (0.1–0.06%) which brings about a hypoxic change in the granuloma of persistent *Mtb*. The low-oxygen culture was obtained by stirring the *Mtb* culture in Dubos media supplemented with tween and albumin incubated at 37 °C in a tightly sealed screw-capped conical flask. These cultures in sealed flasks and with continuous agitation utilize the dissolved oxygen due to efficient equilibration between gas and liquid phases and enter growth termination (non-replicating phase) after 95 to 140 hours of incubation when more 80% O_2 has been depleted.

But this technique is not appropriated for the high-throughput screening as long as duration is needed to achieve hypoxic condition and colony-forming units for the determination of MICs (Wayne 1977).

To measure the hypoxic conditions, various models were established. Out of this low-oxygen recovery assay (LORA) is the most preferable method for the drug discovery process (Sohaskey and Voskuil 2015). We can also use the Wayne model to screen the activity of few molecules. The Wayne model of hypoxia is based on the replication process of bacteria to reduce the oxygen from the tubes with a head-space ratio of 0.5. The steady decrease in the O_2 level permits the bacteria to adjust to a less oxygen condition in which they can survive but are not able to replicate for a long period. The bacteria pass through two detectable states named non-replicating persistent stage 1 (NRP1) and non-replicating persistent stage 2 (NRP2), in which the gene expression alters. In NRP1, *Mtb* exists in an immobile state with thickening of the cell wall. In the NRP2 condition, *Mtb* enters into a complete dormant condition. This model can be preferred for the screening of the compounds. LORA depends on producing bacteria in NRP while cultured under less oxygen condition, and then bacteria are exposed to new compounds for at least 10 days under hypoxic conditions in a 96-well plate. The luciferase reporter strain is used to quantify bacterial viability. The drawback of this model is the requirement of a fermenter, but this can be replaced by using the glass tubes of the Wayne model. However, an extension period is needed to produce an adequate signal and can confound explanation if the exposed molecules are active in this phase. A positive result can be established by measuring the colony-forming units. Another method of hypoxic assay has been followed in which the cultures are added to a 96-well plate and should be covered with paraffin. The compounds are added after 15–20 days followed by incubation, and the survival rate is measured using an RFP reporter 9 (Yeware and Sarkar 2018).

Further, Grant et al. investigated that only a 20% decrease in the dissolved oxygen level may result in the formation of persister *Mtb* residents (Grant et al. 2012). The efficacy of the antibiotics is reduced against this type of population. They also reported that when they are exposed to bactericidal antibiotics, the dissolved oxygen affects the existence of the persister subpopulations via the production of hydroxyl radicals. This reflection provides a new method to inhibit the *Mtb* hydroxyl formation system for the sterilization of persisters.

Macrophage-Based Intracellular Model

The granuloma has a closer resemblance with the animal model infection with human-like lesions and is a better predictor for the determination of the efficacy of drugs in vivo. So, this granuloma model is used for the screening of the compounds as it bridges between the in vivo models and in vitro extracellular assays (Sarathy et al. 2016).

In *Mtb* infection, granuloma is an inflammatory mononuclear cell infiltrate which apart from limiting the growth provides a survival niche to *Mtb*. This complex structure brings an additional obstacle in the drug discovery as the activity of test compounds depends on their ability to cross into the granuloma. Thus, simple granuloma models which mimic the pathological conditions of the disease were established to screen the compounds (Elkington et al. 2019). To evaluate the molecules for anti-TB activity, two granuloma-based models were :established one was in vitro model of granuloma like structure, and other one is in vivo model where granuloma formation was analyzed in infected animals. Both of these approaches are applied to test the new molecules.

It has been reported that Silva-Miranda et al. established a model, in which they used peripheral blood mononuclear cells (PBMCs) infected with a GFP reporter strain of *Mtb* run over a 384-well plate for 2 days after granulomas are formed, which is suitable for measuring the MIC_{50} of the compounds. This model validates the struggle of attaining lesion sterilization since no tested drugs are capable of inhibiting 100% bacterial growth (Silva-Miranda et al. 2015).

Huang et al. explained the deconstructive method, where isolated granulomas from the lungs of mice infected with the fluorescent reporter strain of *Mtb* cells were segregated into different cells and then test compounds were incubated with isolated infected cells in a 384-well microplate, by which IC_{50} was measured (Huang et al. 2018). They used a fluorescent strain for the determination of MIC_{50} of the test compounds which further can be used to distinguish whether the bacteria are susceptible or resistant to the ex vivo granuloma environment (Cronan et al. 2018; Braian et al. 2015). Further, more complex organ systems like lung tissue models are underway, but they are not fully established yet.

Zebrafish Model for Granuloma

To well understand the pathogenicity and complications of infection, the *Mycobacterium marinum*-infected zebrafish model is a suitable choice. This model closely mimics the pathologic conditions of *Mtb*-infected humans with macrophage aggregation and granuloma formation. Another advantage of the zebrafish model is the close resemblance of the histology of mature granuloma with TB granuloma in humans (Myllymäki et al. 2016). So, this model can be used to understand the pathways involved during the infection.

Two characteristics such as host survival and ease of measurement of bacterial burden are mainly considered during high-throughput screening suitability. A lot of multicellular models are established for the screening of the new chemical compounds, but they don't have the whole-body physiology, complexity of the system, and interaction between the host immunity and pathogen. However, different compounds show different activities after metabolism, and therefore rodent models are better suited as compared to vertebrate models which are not the best choice for the high-throughput screening of the compounds. Nowadays zebrafish models are used to reduce the gap. The larvae of zebrafish were infected with fluorescent strain of *M. marinum* to monitor the effect of test molecules on bacteria and its survival within the host. The efficacy of molecules can be established over a span of 10 days. Robotic insertion of larvae surges the throughput for producing more infested embryos. In this model, it is very easy to deliver the compounds, since it is added to the water; the molecules with low bioavailability can be identified, and the quantity of dose required is highly reduced.

Additionally, zebrafish have a complex immune system which mimics the humans (Yuan and Sampson 2018). Thus, zebrafish are a better model for the determination of host-derived pathways that are very much essential for infection in humans (Takaki et al. 2012). Due to the easy farming of the zebrafish larvae, and an effortless model for the in vivo screening of the compounds, the model is a preferable alternative. This method also enables the assortment of host-targeting compounds that facilitate the interaction between pathogen and host during the disease progression.

This model can be used as a proof of concept for efficacy before testing against the rodent models on a long-term basis. The actual protocol for *M. marinum* zebrafish model needs vein injection. Dalton et al. established a model to infect larvae by suspending them in an *M. marinum* solution, which is highly expressed with luciferase, followed by identification of larvae by fluorescence (Dalton et al. 2016). However, this model did not achieve good results, so it is not commonly used.

3.2 *Target-Based Assays: Based on Essentiality of Enzyme*

Target-based anti-TB drug discovery plays an important role in the identification of new chemical entities. It starts with identifying and studying enzymes or proteins which are essential for survival of the pathogen. Identification of new drug targets is important for the advancement in anti-TB drug discovery programs, especially due to a worldwide increase in the cases of resistance to available anti-TB regimen. This procedure is helpful for researchers in establishing the mechanism of action and avoid following the "false" clues about the identified drug candidates. This approach uses various biochemical assays and innovative virtual screening methods to identify anti-TB molecules. Till date a large number of targets have been identified with established assay systems, but this method has very high attrition rate and little success in the discovery of hit molecules when compared to whole cell-based screening methods for a different reason such as due to lack of permeability, efflux, and low target susceptibility (Grzelak et al. 2019).

A whole-cell target-based screen can use the particular activity of the target protein for the development of a cellular assay. Moreira et al. established a target-based whole screening method to recognize the caseinolytic protease (CLpP1P2) inhibitors present inside the cell. CLpP1P2 is a serine-based protease present inside the bacteria, and its main function is to degrade the misfolded and partially synthesized proteins which also includes transcription factor WhiB3. A designed *M. smegmatis* strain producing a caseinolytic protease-specific (SsrA) signal peptide attached to GFP was created. The SsrA labeled was fused with GFP gene which was under the regulation of tetracycline-induced promoter. Under normal conditions, ClpP1P2 degraded the SsrA-GFP with a decrease in fluorescence. When the ClpP1P2 is inhibited, the accumulation of SsrA-GFP occurs resulting in the production of more fluorescence. Therefore, the catalytic activity of ClpP1P2 is stopped and can be measured with the help of fluorescence in the high-throughput screening method (Moreira et al. 2016).

Using a whole cell-based screening, Rybniker et al. targeted the regulatory and structural EXS1 machineries by taking whole mycobacterial cell along with a mammalian cell. EXS1 is a virulent protein of type II secretion system which is composed of structural and regulatory proteins and ATPases which help in the cellular invasion and replication process. A phenotypic screen based on ESX-1 secretion-dependent toxicity was used to screen >10,000 small molecules, followed by another selectivity assay using *Mtb*-infected MRC-5 lung fibroblasts to quantify *Mtb*-induced cell death. They reported that more than 90% of the molecules are protective against lung fibrosis but are not effective against the *Mtb* growth. This assay has one more advantage of directly identifying the target that expresses the cellular toxicity and can be easily screened (Rybniker et al. 2014).

Interestingly, DprE1, decaprenylphosphoryl-β-D-ribose 2′-epimerase, brings the attention of the researcher to target this enzyme which plays a vital role in cell wall synthesis in *Mtb*. Previously, several compounds were screened against DprE1 and displayed inhibition. These compounds inhibit this enzyme by both covalent and non-covalent competitive mechanisms. Since it is found in the periplasmic space of the bacteria, it is easy to target this enzyme. The explanations for the merging of the new targets are very unclear. This target may be important in the survival of the *Mtb* and prone to inhibition by the small molecules. Otherwise, the occurrence of its selection may arise from resemblances of the compounds that are already screened. The screened compounds have been repurposed against other targets for pharmaceutical purposes with priority to the human disease targets. Further, Makarov et al. evaluated the activity of around 60,000 compounds against *Mtb*. They have reported that out of these screened compounds only the dinitrobenzamide derivative shows potency against *Mtb* by inhibiting DprE1 enzyme and establishes it as a promising target against drug-resistant TB treatment (Makarov et al. 2009).

Shikimate kinase enzyme belongs to nucleoside monophosphate class, which plays a vital role in the conversion of shikimate to shikimate-3-phosphate with the use of ATP to yield aromatic amino acids required for the survival of the mycobacterium. Interestingly the target is equally expressed in susceptible as well as resistant strains of *Mtb* and is absent in humans which makes it an attractive target anti-TB drug discovery.

Simithy et al. (2014) screened around 404 compounds having antimycobacterial activity against H37Rv, and these molecules were further analyzed for mass spectrometry-based functional assay to confirm the enzymatic target. The study revealed that these screened compounds closely bind to the binding domain of the shikimic synthase and can be used in development of rational drug combination for the TB (Simithy et al. 2014).

Further Mehra et al. screened around 20,000 compounds against the shikimic kinase activity using in silico approaches. Shortlisted molecules were evaluated using in vitro enzymatic assay developed for shikimate kinase activity followed by whole cell-based assays in *Mtb*. The lead molecules inhibiting shikimate kinase showed a synergistic effect with first-line anti-TB drugs which further proved it as a potent target for anti-TB drug discovery with the potential of targeting both susceptible and resistant strains of *Mtb* (Mehra et al. 2016; Rajput et al. 2016).

Limitations

The limitation of the target-based screening is that it utilizes low expression strains in which the primary hits are not efficacious against the wild *Mtb*. The advanced expression of the target genes in the wild type can revoke the activity. So, utilizing over- and under-expressed genes may not give the correct result. The main worry is that the target susceptibility is not evaluated before the selection of the targets. Hence, bactericidal efficacy and target susceptibility are to be checked before the whole-cell screening. In most cases, the molecule may bind to the target intracellularly, but this may not happen in vitro. Selection of hits without structure-based designing is much more difficult. Though the phenotype of the compounds depends on their mechanism of action, target-based screening does not provide any idea about the specificity of the compound (Palmer et al. 2018).

4 Animal Models for Evaluation of Hits

4.1 Mouse Model for Acute Tuberculosis Infection

The mouse has been considered as the most preferred animal model for TB infection due to the complete availability of information of their genetic make-up, ease to handle, low cost, and less space requirement. The acute model for TB infection is in common use because it takes less time for the evaluation of drug molecules compared to other models of infection. In the acute model of TB, animals are infected using the intranasal or intratracheal route with 10^5 bacilli (high inoculum), and treatment starts immediately after 48 hours of infection. After 28 days of dosing, the bacterial load can be assessed by sacrificing the animals and homogenizing the lungs. It can be used to evaluate chemotherapeutic agents, but chronic models or latent TB infection models are preferred for immune agents that target the host as acute model lacks in mimicking the complete disease conditions such as lesion formation within the lungs (Zhan et al. 2017).

4.2 *Mouse Model for Chronic Tuberculosis Infection*

The chronic mouse model is the best considered and more widely used model to understand the critical pathogenesis of *Mtb*. Contrasting from the human latent TB infection, murine models for TB infection are categorized by a heavy bacillary load, with advanced lung physiology and primary cause for death of animals (Adams et al. 1999). In general, the B6 mouse infected with *Mtb* via aerosol route is the most preferred model to evaluate a drug candidate for its anti-TB efficacy in a macrophage-dominated lesion in the lung within an immunocompetent host (Cooper 2015). Nalbandian et al. infected the mice with 5×10^4 CFU of H37Rv intravenously. The bacterial load was calculated by using 7H10 agar plates supplemented with 10% OADC enrichment for 21 to 28 days incubated at 37 °C. After 12 weeks of infection, C57BL/6J and B6.C3H-*sst1* mice were treated with isoniazid (10 mg per 100 ml) for 3 months. In the chronic infection stage, *Mtb* showed antibiotic tolerance to the bactericidal activity of isoniazid when the drug is administered after 42 days of infection, whereas it is effective when drug administration was started in the acute phase of infection (Nalbandian et al. 2009).

Further, Plumlee et al. reported that ultra-low-dose (ULD) administration of bacilli more closely resembles the human TB infection. For that, C57BL/6J and C3HeB/FeJ mice were infected with 1–3 CFUs of H37Rv having OD = 1 cultured in a Middlebrook 7H9 medium supplemented with OADC. The lungs of mice infected at ULD of bacilli exhibited heterogeneous bacterial burden and well-established granulomas with similar features as of human granulomas (Plumlee et al. 2021).

Cornell's Model for Non-Replicating Tuberculosis

At Cornell University, McCUNE and McDermott established a model of paucibacillary infection and named it Cornell's model of TB (Dutta and Karakousis 2014; McCune et al. 1966). This model is a widely accepted experimental model for latent TB. Briefly, mice were infected with 5×10^3 CFU of *M. tuberculosis* H37Rv and then treated with isoniazid and pyrazinamide resulting in no detectable bacilli. Immunosuppressive agents were administered to reactivate the latent infection. This model of latency is commonly used to study the complicated immunological pathways in latent TB, and this model is most correctly considered as a model of persistence TB (Scanga et al. 1999).

4.3 *Guinea Pig Model*

As compared to mice, guinea pig's granuloma is closer and similar to the granuloma structure in humans with respect to cellular composition and caseous necrosis. Though TB granuloma in typical mouse models is not hypoxic, the tissue hypoxic condition is well observed in guinea pigs (Ly et al. 2009). Additionally, this model differentiates between the primary granuloma and secondary lesions, which seem to be the consequences of hematogenous dissemination (Via et al. 2008; Lin et al. 2012). It has been observed that TB-infected guinea pigs generate a high burden of disease and proceed to infection. Despite the current developments, the mechanism of TB-infected guinea pig model has been difficult due to a shortage of proper reagents (Scanga and Flynn 2014).

4.4 Marmoset Model

Currently, a new non-human primate (NHP), the marmoset model, was used for the anti-*Mtb* drug screening. Marmosets are smaller monkeys (300–500 g) as compared to macaques (old primates). Via et al. confirmed the susceptibility of marmosets upon exposure to different virulent strains of mycobacteria like CDC1551, K04, Erdman strain, etc., using a low dose of aerogenic infection with the bacterial strain resulting in fulminant TB that proceeded at a rate compared to the virulence of the strain used. Infection of these NHPs with Erdman strain through intratracheal route resulted in extensive disease. Thus, marmosets showed a prominent susceptibility to the bacilli and produced clinical manifestations similar to humans. Apart from their small size, the marmoset model has another advantage that they have a high frequency of dizygotic twinning, and hence less number of animals are required (Via et al. 2013). This small NHP model of TB is good for immunologic studies as well.

4.5 Non-Human Primates

These are the advanced model used in the testing of the compounds against *Mtb* as they imitate the human characteristics in both histological and pathological conditions (Lin et al. 2009). The necrotic granuloma of this model is the best for hypoxia study as compared to mouse and guinea pig granuloma. Cynomolgus macaques were administered with a high dose of Mtb, i.e., 200–400 CFU, and successful infection leads to acute progressive TB disease, whereas no signs of disease were observed at 6 months of postinfection when these animals were latently infected by administering a low dose of Mtb, i.e., 25 CFU. Infected animals were said to have active disease only when they showed clinical, microbiologic, and immunologic signs, and approx. 45% of those animals developed symptomatic disease, while the remaining were clinically asymptomatic (Lin et al. 2014). Apart from the symptomatic and asymptomatic form of disease, the infected animals were further categorized as "percolators," a third category with no pathological evidence, but bacilli can be cultured occasionally from bronchoalveolar or gastric lavage aspirates. Though as compared to mice, it is easy to study the immunological responses in non-human primates, due to its intensive nature, the use of this model has been restricted.

5 Conclusion

Drug discovery against *Mtb* always remained the most challenging task to scientific community due to the pathogenic nature of the bacteria and requirement of a high level of containment facilities to handle the bacteria. Even after these difficulties, researchers around the globe managed to develop a wide range of assay systems including both in vitro and in vivo screening methods to test the efficacy of the compounds for their anti-TB properties. The screening capabilities have made a tremendous progress when compared to initial days of anti-TB drug discovery which was limited to phenotypic screening only. Now the screening of molecules can be done in a 384-well plate format under aerobic conditions, and several well-defined assays have been established for evaluation of molecules for their anti-TB properties under non-replicating, intracellular, and extracellular conditions. Multicellular models with high complexity which mimic the host conditions and features close to granuloma lesions have been developed to establish the in vitro assays for anti-TB screening. Conditions such as low oxygen concentration, limited nutrient supply, and fatty acids as a carbon source are the major factors which need to be considered while establishing an assay for anti-TB drug discovery as this may help in shortening of treatment and reducing the chances of failure of molecules during clinical trials. It displays a great deal of utility with these recent developments in the number and quality of available in vitro assays, but still there is no consent on as to which assays are the most extrapolative in in vivo models.

In 1998, the complete genome sequence of *Mtb* was made available to the public domain and since then drug discovery process has focused on targeting the vital enzymes of *Mtb* that are selected based upon their gene essentiality. The protein sequences were analyzed, and the proper binding sites were identified. Based on their structure and binding sites, the inhibitors are being designed by the researchers to block the activity of the essential enzymes. This strategy of inhibiting enzyme function did not show promising results and hence is not a preferred approach in drug discovery for high-throughput screening programs. This is due to the reason that the molecules inhibiting the function of the target enzyme in enzymatic assay may fail to penetrate the bacteria as these enzyme assays don't focus on the target vulnerability, permeability, and stability. All these issues bring more difficulties in the designing of the compound while keeping the inhibitory efficacy. Therefore, many researchers shift their attention from the enzyme-targeted assay to whole-cell screening for drug discovery. Furthermore, various in vitro assays with *Mtb* or its other non-pathogenic strains as substitutes are being used to identify anti-TB hits, but the chances of failure of such molecules under in vivo conditions are high which may be due to reasons such as target availability during in vivo conditions. Therefore, phenotypic screening assays mimicking the host environments are the most preferred models for the screening of anti-TB molecules. Here we summarize that the major drug discovery against TB infection was done between 2007 and 2017, based on the whole-cell screening. Thus, the screening of the compounds shifted from target-based to whole-cell screening as they provide a better environment with respect to different pathological conditions.

References

Abrahams KA, Besra GS (2020) Mycobacterial drug discovery. RSC Med Chem 11(12): 1354–1365

Abramovitch RB, Rohde KH, Hsu FF, Russell DG (2011) aprABC: a Mycobacterium tuberculosis complex-specific locus that modulates pH-driven adaptation to the macrophage phagosome. Mol Microbiol 80(3):678–694

Adams LB, Sinha I, Franzblau SG, Krahenbuhl JL, Mehta RT (1999) Effective treatment of acute and chronic murine tuberculosis with liposome-encapsulated clofazimine. Antimicrob Agents Chemother 43(7):1638–1643

Barry CE, Boshoff HI, Dartois V, Dick T, Ehrt S, Flynn J, Schnappinger D, Wilkinson RJ, Young D (2009) The spectrum of latent tuberculosis: rethinking the biology and intervention strategies. Nat Rev Microbiol 7(12):845–855

Braian C, Svensson M, Brighenti S, Lerm M, Parasa VR (2015) A 3D human lung tissue model for functional studies on Mycobacterium tuberculosis infection. J Vis Exp 104:e53084

Chang DPS, Guan XL (2021) Metabolic versatility of Mycobacterium tuberculosis during infection and dormancy. Metabolites 11(2):88

Cooper AM (2015) Mouse model of tuberculosis. Cold Spring Harb Perspect Med 5(2):a018556

Cronan MR, Matty MA, Rosenberg AF, Blanc L, Pyle CJ, Espenschied ST, Rawls JF, Dartois V, Tobin DM (2018) An explant technique for high-resolution imaging and manipulation of mycobacterial granulomas. Nat Methods 15(12):1098–1107

Dalton JP, Uy B, Okuda KS, Hall CJ, Denny WA, Crosier PS, Swift S, Wiles S (2016) Screening of anti-mycobacterial compounds in a naturally infected zebrafish larvae model. J Antimicrob Chemother. https://doi.org/10.1093/jac/dkw421

Darby CM, Ingólfsson HI, Jiang X, Shen C, Sun M, Zhao N, Burns K, Liu G, Ehrt S, Warren JD, Anderson OS (2013) Whole cell screen for inhibitors of pH homeostasis in Mycobacterium tuberculosis. PLoS One 8(7):e68942

Dartois V (2014) The path of anti-tuberculosis drugs: from blood to lesions to mycobacterial cells. Nat Rev Microbiol 12(3):159–167

Dutta NK, Karakousis PC (2014) Latent tuberculosis infection: myths, models, and molecular mechanisms. Microbiol Mol Biol Rev 78(3):343–371

Early J, Ollinger J, Darby C, Alling T, Mullen S, Casey A, Gold B, Ochoada J, Wiernicki T, Masquelin T, Nathan C (2018) Identification of compounds with pH-dependent bactericidal activity against Mycobacterium tuberculosis. ACS Infect Dis 5(2):272–280

Early JV, Casey A, Martinez-Grau MA, Gonzalez Valcarcel IC, Vieth M, Ollinger J, Bailey MA, Alling T, Files M, Ovechkina Y, Parish T (2016) Oxadiazoles have butyrate-specific conditional activity against Mycobacterium tuberculosis. Antimicrob Agents Chemother 60(6):3608–3616

Early JV, Mullen S, Parish T (2019) A rapid method to determine the bactericidal activity of compounds against non-replicating Mycobacterium tuberculosis at low pH. bioRxiv, p 578195

Elkington P, Lerm M, Kapoor N, Mahon R, Pienaar E, Huh D, Kaushal D, Schlesinger LS (2019) In vitro granuloma models of tuberculosis: potential and challenges. J Infect Dis 219(12): 1858–1866

FDA (2012). https://www.fda.gov/news-events/press-announcements/fda-approves-new-drugtreatment-resistant-forms-tuberculosis-affects-lungs. cited 2019 Aug 15

Flynn JL, Chan J, Lin PL (2011) Macrophages and control of granulomatous inflammation in tuberculosis. Mucosal Immunol 4(3):271–278

Grant SS, Kaufmann BB, Chand NS, Haseley N, Hung DT (2012) Eradication of bacterial persisters with antibiotic-generated hydroxyl radicals. Proc Natl Acad Sci 109(30):12147–12152

Grant SS, Kawate T, Nag PP, Silvis MR, Gordon K, Stanley SA, Kazyanskaya E, Nietupski R, Golas A, Fitzgerald M, Cho S (2013) Identification of novel inhibitors of nonreplicating Mycobacterium tuberculosis using a carbon starvation model. ACS Chem Biol 8(10): 2224–2234

Grzelak EM, Choules MP, Gao W, Cai G, Wan B, Wang Y, McAlpine JB, Cheng J, Jin Y, Lee H, Suh JW (2019) Strategies in anti-Mycobacterium tuberculosis drug discovery based on phenotypic screening. J Antibiot 72(10):719–728

Huang L, Kushner NL, Theriault ME, Pisu D, Tan S, McNamara CW, Petrassi HM, Russell DG, Brown AC (2018) The deconstructed granuloma: a complex high-throughput drug screening platform for the discovery of host-directed therapeutics against tuberculosis. Front Cell Infect Microbiol 8:275

Johnson BK, Colvin CJ, Needle DB, MbaMedie F, Champion PAD, Abramovitch RB (2015) The carbonic anhydrase inhibitor ethoxzolamide inhibits the Mycobacterium tuberculosis PhoPR regulon and Esx-1 secretion and attenuates virulence. Antimicrob Agents Chemother 59(8): 4436–4445

Lin PL, Dartois V, Johnston PJ, Janssen C, Via L, Goodwin MB, Klein E, Barry CE, Flynn JL (2012) Metronidazole prevents reactivation of latent Mycobacterium tuberculosis infection in macaques. Proc Natl Acad Sci 109(35):14188–14193

Lin PL, Ford CB, Coleman MT, Myers AJ, Gawande R, Ioerger T, Sacchettini J, Fortune SM, Flynn JL (2014) Sterilization of granulomas is common in active and latent tuberculosis despite within-host variability in bacterial killing. Nat Med 20(1):75–79

Lin PL, Rodgers M, Smith LK, Bigbee M, Myers A, Bigbee C, Chiosea I, Capuano SV, Fuhrman C, Klein E, Flynn JL (2009) Quantitative comparison of active and latent tuberculosis in the cynomolgus macaque model. Infect Immun 77(10):4631–4642

Ly LH, Jeevan A, McMurray DN (2009) Neutralization of TNFα alters inflammation in guinea pig tuberculous pleuritis. Microbes Infect 11(6-7):680–688

Makarov V, Manina G, Mikusova K, Möllmann U, Ryabova O, Saint-Joanis B, Dhar N, Pasca MR, Buroni S, Lucarelli AP, Milano A (2009) Benzothiazinones kill Mycobacterium tuberculosis by blocking arabinan synthesis. Science 324(5928):801–804

Mayer KH, Hamilton CD (2010) Synergistic pandemics: confronting the global HIV and tuberculosis epidemics. Clin Infect Dis 50(Supplement_3):S67–S70

McCune RM, Feldmann FM, Lambert HP, McDermott W (1966) Microbial persistence: I. The capacity of tubercle bacilli to survive sterilization in mouse tissues. J Exp Med 123(3):445–468

Mehra R, Rajput VS, Gupta M, Chib R, Kumar A, Wazir P, Khan IA, Nargotra A (2016) Benzothiazole derivative as a novel Mycobacterium tuberculosis shikimate kinase inhibitor: identification and elucidation of its allosteric mode of inhibition. J Chem Inf Model 56(5): 930–940

Moreira W, Lim JJ, Yeo SY, Ramanujulu PM, Dymock BW, Dick T (2016) Fragment-based whole cell screen delivers hits against M. tuberculosis and non-tuberculous mycobacteria. Front Microbiol 7:1392

Myllymäki H, Bäuerlein CA, Rämet M (2016) The zebrafish breathes new life into the study of tuberculosis. Front Immunol 7:196

Nalbandian A, Yan BS, Pichugin A, Bronson RT, Kramnik I (2009) Lung carcinogenesis induced by chronic tuberculosis infection: the experimental model and genetic control. Oncogene 28(17):1928–1938

Palmer AC, Chait R, Kishony R (2018) Nonoptimal gene expression creates latent potential for antibiotic resistance. Mol Biol Evol 35(11):2669–2684

Parish T (2020) In vitro drug discovery models for Mycobacterium tuberculosis relevant for host infection. Expert Opin Drug Discovery 15(3):349–358

Plumlee CR, Duffy FJ, Gern BH, Delahaye JL, Cohen SB, Stoltzfus CR, Rustad TR, Hansen SG, Axthelm MK, Picker LJ, Aitchison JD (2021) Ultra-low dose aerosol infection of mice with Mycobacterium tuberculosis more closely models human tuberculosis. Cell Host Microbe 29(1):68–82

Queval CJ, Brosch R, Simeone R (2017) The macrophage: a disputed fortress in the battle against Mycobacterium tuberculosis. Front Microbiol 8:2284

Rajput VS, Mehra R, Kumar S, Nargotra A, Singh PP, Khan IA (2016) Screening of antitubercular compound library identifies novel shikimate kinase inhibitors of Mycobacterium tuberculosis. Appl Microbiol Biotechnol 100(12):5415–5426

Rodríguez JG, Hernández AC, Helguera-Repetto C, Aguilar Ayala D, Guadarrama-Medina R, Anzóla JM, Bustos JR, Zambrano MM, González-y-Merchand J, García MJ, Del Portillo P (2014) Global adaptation to a lipid environment triggers the dormancy-related phenotype of Mycobacterium tuberculosis. MBio 5(3):e01125–e01114

Russell DG, Barry CE 3rd, Flynn JL (2010) Tuberculosis: what we don't know can, and does, hurt us. Science 328(5980):852–856

Ryan NJ, Lo JH (2014) Delamanid: first global approval. Drugs 74(9):1041–1045

Rybniker J, Chen JM, Sala C, Hartkoorn RC, Vocat A, Benjak A, Boy-Röttger S, Zhang M, Székely R, Greff Z, Őrfi L (2014) Anticytolytic screen identifies inhibitors of mycobacterial virulence protein secretion. Cell Host Microbe 16(4):538–548

Sarathy JP, Zuccotto F, Hsinpin H, Sandberg L, Via LE, Marriner GA, Masquelin T, Wyatt P, Ray P, Dartois V (2016) Prediction of drug penetration in tuberculosis lesions. ACS Infect Dis 2(8):552–563

Scanga CA, Flynn JL (2014) Modeling tuberculosis in nonhuman primates. Cold Spring Harb Perspect Med 4(12):a018564

Scanga CA, Mohan VP, Joseph H, Yu K, Chan J, Flynn JL (1999) Reactivation of latent tuberculosis: variations on the Cornell murine model. Infect Immun 67(9):4531–4538

Segal W, Bloch H (1956) Biochemical differentiation of Mycobacterium tuberculosis grown in vivo and in vitro. J Bacteriol 72(2):132–141

Serafini A, Tan L, Horswell S, Howell S, Greenwood DJ, Hunt DM, Phan MD, Schembri M, Monteleone M, Montague CR, Britton W (2019) Mycobacterium tuberculosis requires glyoxylate shunt and reverse methylcitrate cycle for lactate and pyruvate metabolism. Mol Microbiol 112(4):1284–1307

Silva-Miranda M, Ekaza E, Breiman A, Asehnoune K, Barros-Aguirre D, Pethe K, Ewann F, Brodin P, Ballell-Pages L, Altare F (2015) High-content screening technology combined with a human granuloma model as a new approach to evaluate the activities of drugs against Mycobacterium tuberculosis. Antimicrob Agents Chemother 59(1):693–697

Simithy J, Reeve N, Hobrath JV, Reynolds RC, Calderón AI (2014) Identification of shikimate kinase inhibitors among anti-Mycobacterium tuberculosis compounds by LC-MS. Tuberculosis 94(2):152–158

Sohaskey CD, Voskuil MI (2015) In vitro models that utilize hypoxia to induce non-replicating persistence in Mycobacteria. In: Mycobacteria protocols. Humana Press, New York, NY, pp 201–213

Sotgiu G, Centis R, D'ambrosio L, Migliori GB (2015a) Tuberculosis treatment and drug regimens. Cold Spring Harb Perspect Med 5(5):a017822

Sotgiu G, D'Ambrosio L, Centis R, Mura I, Castiglia P, Spanevello A, Migliori GB (2015b) The multidrug-resistant tuberculosis threat: old problems and new solutions. J Thorac Dis 7(9):E354

Sturgill-Koszycki S, Schlesinger PH, Chakraborty P, Haddix PL, Collins HL, Fok AK, Allen RD, Gluck SL, Heuser J, Russell DG (1994) Lack of acidification in Mycobacterium phagosomes produced by exclusion of the vesicular proton-ATPase. Science 263(5147):678–681

Takaki K, Cosma CL, Troll MA, Ramakrishnan L (2012) An in vivo platform for rapid high-throughput antitubercular drug discovery. Cell Rep 2(1):175–184

TB Alliance (2019). https://www.tballiance.org/news/fda-approves-new-treatment-highly-drugresistant-forms-tuberculosis. cited 2019 Aug 15

Via LE, Lin PL, Ray SM, Carrillo J, Allen SS, Eum SY, Taylor K, Klein E, Manjunatha U, Gonzales J, Lee EG (2008) Tuberculous granulomas are hypoxic in guinea pigs, rabbits, and nonhuman primates. Infect Immun 76(6):2333–2340

Via LE, Weiner DM, Schimel D, Lin PL, Dayao E, Tankersley SL, Cai Y, Coleman MT, Tomko J, Paripati P, Orandle M (2013) Differential virulence and disease progression following

Mycobacterium tuberculosis complex infection of the common marmoset (Callithrix jacchus). Infect Immun 81(8):2909–2919

Wang F, Sambandan D, Halder R, Wang J, Batt SM, Weinrick B, Ahmad I, Yang P, Zhang Y, Kim J, Hassani M (2013) Identification of a small molecule with activity against drug-resistant and persistent tuberculosis. Proc Natl Acad Sci 110(27):E2510–E2517

Wayne LG (1977) Synchronized replication of Mycobacterium tuberculosis. Infect Immun 17(3): 528–530

WHO G (2018) Global tuberculosis report 2018. Glob. Tuberc. Rep., 2020

WHO G (2020) Global tuberculosis report 2020. Glob. Tuberc. Rep., 2020

Yeware A, Sarkar D (2018) Novel red fluorescence protein-based microplate assay for drug screening against dormant Mycobacterium tuberculosis by using paraffin. Tuberculosis 110: 15–19

Yuan T, Sampson NS (2018) Hit generation in TB drug discovery: from genome to granuloma. Chem Rev 118(4):1887–1916

Zhan L, Tang J, Sun M, Qin C (2017) Animal models for tuberculosis in translational and precision medicine. Front Microbiol 8:717

Zhao N, Darby CM, Small J, Bachovchin DA, Jiang X, Burns-Huang KE, Botella H, Ehrt S, Boger DL, Anderson ED, Cravatt BF (2015) Target-based screen against a periplasmic serine protease that regulates intrabacterial pH homeostasis in Mycobacterium tuberculosis. ACS Chem Biol 10(2):364–371

Zimmermann M, Kogadeeva M, Gengenbacher M, McEwen G, Mollenkopf HJ, Zamboni N, Kaufmann SHE, Sauer U (2017) Integration of metabolomics and transcriptomics reveals a complex diet of Mycobacterium tuberculosis during early macrophage infection. mSystems 2(4):e00057–e00017

Zuniga ES, Early J, Parish T (2015) The future for early-stage tuberculosis drug discovery. Future Microbiol 10(2):217–229

Cofactor-Receptor Interaction-Based Pharmacophore Design for Development of Novel Inhibitors: A Case Study Against Tuberculosis

V. L. S. Prasad Burra

Abstract Pharmacophore modelling has evolved over a century now to become an indispensable part of computer-aided drug design (CADD) that is quite diversified providing huge time and cost savings. The pharmacophore concept has been frequently used in the rational design of new drugs. In this study, we look at how the knowledge of cofactor-receptor interactions is implemented computationally and how it is used in the novel drug development process. Pharmacophores are bio-, stereo- and chemical representations of the environment for molecular recognition that dictates the binding of a ligand with a receptor. These insights lead to discovery, design and development of novel lead compounds. The pharmacophore-based virtual screening is performed as part of the protocol with many knowledge-based variations. The pharmacophore identification is not only helpful in drug discovery but also helps in ADME-Tox modelling, understanding side effects, off-target predictions and also target identification. To improve virtual screening, pharmacophores are frequently paired with molecular docking and MD simulations. This chapter offers case studies for developing new tuberculosis (TB) inhibitors based on cofactor-receptor interaction-based pharmacophore identification and analysis. TB, which is caused by *Mycobacterium tuberculosis* (*M.tb*), has a 70,000-year natural history. The TB epidemic's reservoir is latent *M.tb* infection. Among the first-line antibiotics, *M.tb* has the most multidrug resistance. Methylation, one of the bacteria's key processes, is one of the common mechanisms to enable resistance. Two simultaneous goals can be achieved by targeting the methyltransferase: (a) eliminate methylation and (b) impede translation. Yet, no drugs or lead compounds have been found to inhibit methyltransferase activity, specifically the rRNA small subunit methyltransferase D (RsmD) family from *M.tb*. S-Adenosyl-L-methionine (SAM) is a universal methyl donor (cofactor) that aids in the transfer of a methyl group to a variety of substrates such as nucleotides, amino acids and other chemical moieties. RsmD is a SAM-dependent methyltransferase. Novel inhibitors were derived from (a) in silico high-throughput screening (HTS) of compounds from

V. L. S. P. Burra (✉)
Centre for Advanced Research and Innovation in Structural Biology of Diseases, K L E F University, Vaddeswaram, Andhra Pradesh, India

V. S. Rajput, A. Runthala (eds.), *Drugs and a Methodological Compendium*,
https://doi.org/10.1007/978-981-19-7952-1_14

a variety of small molecule databases, (b) cofactor-receptor-based pharmacophore identification and (c) modifying the filtered compounds with relevant functional groups and study their interactions based on the pharmacophore derived from RsmD. The successful attempts resulting in computationally validated novel RsmD inhibitors are presented in this chapter. The top three inhibitor IDs (AS_118_122, AS_118_129 and AS_118_76) are potential RsmD inhibitors that need further validation through in vitro and in vivo studies.

1 Introduction

Computer-aided drug design (CADD) has emerged into an indispensable field for structural biologists and/or medicinal chemists providing a wide range of applications that span the drug design and development process for addressing various diseases. CADD positively and significantly influenced the economics of drug development life cycle. The major goal of CADD has been to streamline and rationalize the drug design process while lowering the time spent and expenses incurred (Taft et al. 2008). The goal during early stages of drug development was to find the first hit of compounds accomplished using high-throughput screening (HTS), which involves evaluating a huge library of compounds with relevant activity assays involving a high cost to success ratio (Bajorath 2002). The efficacy of the lead compounds and their derivatives were increased by trial-and-error method focusing on individual compounds via another round of activity assays incurring continued expenditure (Brown n.d.; Sun et al. 2012). The vision is to build compounds that are highly effective and specific while holding unique intellectual property (IP) (Boyd 1996; Bajorath 2002). This was accomplished via traditional medicinal chemistry methods, in which the design is based on the analysis of the observable structure-activity relationships (SAR) or structural data that involves huge costs (Thiel 2004).

However, virtual screening is the in silico equivalent of in vitro HTS, and it tries to filter large libraries containing millions of compounds and/or fragments using computational approaches that enable screening/filtering and prioritize (rank) the most likely active compounds/fragments for a given target using various scoring algorithms (Bajorath 2002). Computational approaches can be used to develop a variety of derivatives based on various scaffolds and then score them for efficacy. In a significant short time, the most promising derivatives can be obtained from a large chemical universe (small molecule databases and enumerations) (Brown n.d.; Sun et al. 2012). The effectiveness and potency of the compounds are not a linear or a one-parameter problem. It is a multi-parameter problem. If a chemical is to be clinically beneficial, it must have good pharmacokinetic features (absorption, distribution, metabolism and excretion (ADME)) as well as toxicity (ADME-Tox). Virtual approaches are available for profiling the ADME-Tox properties of drug-like molecules (Li, 2001, Ekins et al., 2002, Yu and Adedoyin, 2003, Thiel, 2004).

CADD technologies are developed with primary dependence on chemoinformatics which is the data and knowledge storehouse of chemical structures, respective properties and their biological activity information. Chemo-informatics include the standard molecular descriptors, the algorithms, which enable filtering compounds and characterize the molecules based on various parameters which include the chemical and physical structure of the molecules (Agrafiotis et al. 2007; Valerio and Choudhuri 2012). Molecular fingerprints are frequently used to compare and measure (dis)similarity between molecules (Hong et al. 2008; Bel'skih et al. 2015).

Paul Ehrlich observed that specific chemical groups within a molecule significantly influence biological activity and thus proposed the notion of pharmacophore in the nineteenth century (Ehrlich 1909). Pharmacophore is an abstract feature compilation of various physicochemical properties of the binding regions of the receptor and properties of common ligands and common interactions in 3D space of bioactive compounds with their targets. The interactions include hydrogen bonds, charged and metal interactions and hydrophobic, aromatic and pi-pi contacts (Guner 2002). The above feature compilation, which now has the comprehensive knowledge about the receptor, ligand and the active site, serves as the rule book for discovery and design of more potent and reliable drug/lead molecules. After a century of research, pharmacophore methods have emerged as one of the most important techniques in drug discovery. Pharmacophore modelling has been successfully used in virtual screening, de novo design and lead optimization (Jones, Willett and Glen, 1995, Goto, Kataoka and Hirayama, 2004, Wolber et al., 2008). Various ligand-based and structure-based methodologies are developed to enhance pharmacophore modelling accuracy. Despite these achievements, pharmacophore techniques have not yet reached their full potential, particularly in terms of decreasing the current high total cost of drug discovery and development (Guner 2005).

2 Caste Study

TB is an airborne infectious illness caused by *M.tb*, a bacterium with a 70,000-year evolutionary history. TB is still a global health issue with a huge disease burden Churchyard et al. 2017. *M.tb* is the world's second deadliest infectious pathogen, after HIV/AIDS, according to the WHO. *M.tb* infects around a third of the world's population, and it is quickly becoming the biggest cause of death. The TB epidemic's reservoir is latent *M.tb* infection. *M.tb* is spreading and developing multidrug resistance to first-line antibiotics (such as isoniazid, rifampicin and fluoroquinolones) as well as injectable second-line medicines (like amikacin, kanamycin, capreomycin) Witek et al. 2017. India and other South Asian countries bear the brunt of the resistant TB burden (*Pipeline | Working Group for New TB Drugs* n. d.-a). Drug resistance to antibiotics is mostly caused by highly conserved post-transcriptional alteration processes of rRNAs, such as methylation of nucleosides. Many of the known ribosomal alterations are still unknown in terms of their function.

Due to the rise of drug-resistant strains, finding new and effective medications to treat TB has been a persistent struggle. WHO's End TB Strategy, which declares that TB prevalence will be controlled and eliminated by 2035, aims to end the global TB epidemic ('WHO | The End TB Strategy' 2018). Furthermore, none of the therapeutic candidates in the current TB drug research pipeline target *M.tb* methyltransferases, from drug discovery through market approval.

Methyltransferases (EC 2.1.1.-) are a vast, diversified and biologically significant group of enzymes that methylate substrates by transferring one carbon group (–CH3) (*Pipeline | Working Group for New TB Drugs* n.d.-b). Based on their catalytic activity, they are divided into many subclasses. The most frequent methyltransferases are class I, which utilize the cofactor SAM (also known as AdoMet) as the methyl group donor (Cantoni 1953; Waddell et al. 2000). For binding SAM, most class I methyltransferases have a Rossmann fold (Wuosmaa and Hager 1990; Thomas et al. 2004). The RsmD homologs methylate N2 atom of guanine nucleotide at position 966 of the 16S rRNA of the 30S small subunit of bacterial ribosomes Lesnyak et al. 2007.

In this chapter, we present in silico designed novel inhibitors from the Asinex building blocks library of small molecules which have shown better binding affinity and stability, indicating that they are potential lead compounds that need experimental functional validation and verification, providing a novel therapeutic mode against tuberculosis.

3 Bioinformatics Tools and Available Methods for Pharmacophore-Based Drug Design

Protein and ligand preparation, structure-based high-throughput virtual screening, MM/GBSA, attrition in drug development (ADMET) and molecular dynamics simulation (MDS henceforth) studies are among the standard computational investigations. For ligand preparation and compound generation, Schrödinger's LigPrep module (Schrödinger molecular modelling suite 2020-2) was utilized. The *GLIDE* module included in the Schrödinger suite was used to identify conformational poses in the protein's active site and generate the precision scores (SP and XP). The ADME properties for the novel compounds were determined using the QikProp module. MD simulation studies were performed using Desmond molecular dynamics systems (Desmond henceforth).

3.1 Protein Preparation

The receptor for preparation, docking, MD simulation and subsequent analysis was the highest-resolution crystal structure of RsmD solved and deposited by the author's lab at 1.74 [PDB:6AIE] Venkataraman et al. 2018. Protein preparation wizard was used to prepare the methyltransferase (RsmD) Sastry et al. 2013. Adding hydrogens and ionization metals and repairing missing residues and atoms are the tasks under the protein preparation process.

3.2 Ligand Preparation

Ligand preparation involves generating 3D structures from various 1D or 2D structure formats such as SMILES, SDF and others, adding hydrogens, generating suitable ionization states and tautomers, checking the stereochemistry and the chiralities and evaluating and modifying (if necessary) the chemical structure which then is ready for downstream processing such as docking. LigPrep module is used to perform the above tasks for a large number of drug-like compounds starting with 2D or 3D structures supporting various standard chemical structure formats. For each correctly processed input structure, LigPrep provides a single, low-energy 3D structure with correct chiralities. LigPrep also generates a variety of conformers from each input structure.

The small molecule compound library (Asinex-BB-22525) from Asinex (www.asinex.com) in SDF format was downloaded and prepared using *LigPrep* module of Schrödinger Shelley et al. 2007, which removes unwanted structures, neutralizes charged groups, generates ionization states and tautomers and optimizes 3D structures. The prepared ligands were used for virtual screening and analogue development.

3.3 Pharmacophore of 16s rRNA Methyltransferase D

After examining the stereochemistry environment and common residue-level interactions between SAM and other methyltransferases, the RsmD pharmacophore was investigated (DMT, DNA methyltransferase; RMT, RNA methyltransferase; and PMT, protein methyltransferase). DMT (PDBs: 6K0X, 5HFJ, 3AV6, 2ADM), RMT (PDBs: 5IL1, 1EJ0, 1SQF) and PMT (PDBs: 3QOW, 6KMS, 5EML) were examined for interaction using a minimum of three PDB structures. The library of small molecules was chemically modified using functional groups to develop novel ligands based on the knowledge gathered from the pharmacophore of RsmD.

3.4 Binding Site Prediction

With the use of Schrödinger's SiteMap and receptor grid generation modules (Halgren 2007, 2009), the binding sites were predicted, and the grid was constructed.

For molecular docking investigations, the active sites or binding sites were predicted using the SiteMap module. The SiteMap score was used to confirm the site's proclivity for inhibitor or ligand binding. For docking and additional analysis, the top-ranked binding sites were used. Later, the prepared target protein was used for receptor grid generation using the receptor grid generation module.

3.5 Molecular Docking

Docking investigations are divided into four stages: site-point search, diameter test, subset test and scoring. The initial stage involves determining the core conformation of the ligand and possible locations of binding around the active site and identifying the favourable poses/orientations. This particular search begins by selecting a *site-point* on an evenly spaced 2Å lattice that covers the active site region. The second stage is to look at atomic positioning. If there are significant steric clashes with the protein in any orientation, the orientation will be skipped. The atomic positions which are favourable without steric clashes are used for ChemScore computations. This stage is also known as the *greedy score* method, which favours hydrophobic areas and hydrogen bonding while penalizing steric conflicts and metal-ligation interactions. Stages 3 and 4 are energy minimization and GScore calculations, respectively, where

$$\text{GScore} = 0.05^{*}\text{vdW} + 0.15^{*}\text{Coul} + \text{Lipo} + \text{Hbond} + \text{Metal} + \text{Rewards} + \text{RotB} + \text{Site}$$

and *vdW* represents van der Waals energy, similarly, *Coul*, Coulomb energy; *Lipo*, lipophilic interaction energy; *HBond*, hydrogen-bonding energy; *Metal*, metal-binding energy; *Rewards*, rewards and penalties for buried polar groups, hydrophobic enclosures, correlated hydrogen bonds and amide twists; *RotB*, penalty for freezing rotatable bonds; and *Site*, polar interactions in the active site. Energy minimization is conducted with the help of OPLS van der Waals and electrostatic grids for the protein or receptor. The virtual screening investigations were implemented using the high-throughput virtual screening (HTVS) module, which is a fast screening method for a large number of compounds Halgren et al. 2004.

The ligand was docked into the active site of the target protein (RsmD) (PDB:6AIE). For molecular docking and ranking the best ligand interactions with the receptor, the *GLIDE* module was employed. The ligands were shortlisted and ranked using the standard precision (SP) docking scores. Extra precision (XP) docking studies were also conducted on the shortlisted ligands filtered based on the SP docking scores and further analysis. The top-ranking ligand complexes were then chosen for ADME-Tox prediction, MM/GBSA calculation and MDS studies.

3.6 ADME-Tox Prediction

ADMET is a prediction tool for pharmaceutically relative properties and significant descriptors, and *PKCSM* (Pires et al. 2015) was used to analyse the ADMET predictions. Shortlisted ligands from ADMET properties were selected for MDS studies and MM/GBSA calculation.

3.7 Binding Affinity Calculations (MM/GBSA)

The binding free energies of RsmD-ligand complexes were determined using the MM/GBSA (Molecular Mechanics/Generalized Born Model and Solvent Accessibility) module. This binding energy was calculated with CHARMM force field using Discovery Studio Visualizer (Biovia 2021). The results obtained were in Kcal/mol, with a lower negative value indicating strong binding.

3.8 Molecular Dynamics Simulation (MDS) Studies

The top 3 RsmD-ligand complexes were selected for MD simulation studies using Desmond Bowers et al. 2006. The RsmD-ligand complexes were placed in an orthorhombic box with a size of 10x10x10Å, and solvation was defined explicitly with the TIP4P water model. The system was neutralized with Na+ and Cl- ions at a final concentration of 0.15M. OPLS force field was used for simulation in this system. It was equilibrated and minimized based on the equilibration protocol developed in Desmond. It starts with Brownian dynamics NVT simulation for 100 ps at 10K temperature, then it will continue with 12ps normal pressure and temperature (NPT) simulation at 10K temperature and finally it will start from 24ps NPT (normal pressure and temperature) simulation without restraints. Now the final production was conducted for 40ns using NPT ensemble at 300K, and the pressure bar is 1.01325; we have not used any relaxation model for the system. For maintaining constant temperature and pressure, Nose-Hoover chain thermostat and Martyna-Tobias-Klein barostat were used in the system. Every 10ps interval the trajectory was saved. We have applied 4000 frames in this simulation, and trajectories and simulation interaction diagram analysis were available in maestro. The RMSD and RMSF trajectories of protein and ligand were obtained from the below formula:

$$\mathrm{RMSD}_X = \sqrt{\frac{1}{N}\sum_{i=1}^{N}(r_i'(t_x)) - r_i(t_{\mathrm{ref}}))^2} \quad \& \quad \mathrm{RMSF}_i = \sqrt{\frac{1}{T}\sum_{t=1}^{T} < (r_i'(t)) - r_i(t_{\mathrm{ref}}))^2 >}$$

N = number of atoms, r'= position of the selected in atoms in the frame, T= trajectory time, r= position of the atom

The top-ranked complexes were selected for MD simulation studies using Desmond Bowers et al. 2006. The simulation was performed for 100ns, and each 10ps frame trajectory was saved. The simulation interaction diagram was used to analyse the RMSD, RMSF and other data analyses. The flowchart is shown in Fig. 1.

4 Case Studies

4.1 *Studying the Pharmacophore of RsmD*

The active site of RsmD has been discovered near the C-terminal end of the S4 and S5 strands, which are known to be part of the SAM binding site (bottom). The interaction of SAM with several methyltransferases (DMTs, RMTs, PMTs) was examined at the residue level (Schluckebier et al. 1997; Bügl et al. 2000; Foster et al. 2003; Richon et al. 2011; Takeshita et al. 2011; Duncan et al. 2016; Ma et al. 2016; Wang et al. 2016; Guo et al. 2019; Li et al. 2019). The SAM's methionine group is situated in a deep cleft of the RsmD, surrounded by roughly 25 amino acids, 12 of which are non-polar, 6 of which are polar, 3 of which are positively charged and 4 of which are negatively charged (Table 1) (Fig. 2). The Asinex compounds (labelled as AS_NNN_NNN and referred to AS henceforth) were modified with a series of functional groups such as amide, amine, acetamide, formamide, hydroxyl, iodine, hydroxyl, fluorine, iodine and bromine at appropriate places since the chemical environment around the RsmD active site is particularly hydrophobic and ideal for SAM binding.

4.2 *Molecular Docking*

The docking was performed using the *GLIDE* module with the SP docking scores of native SAM and AS compounds ranging from −6.5 to −9.2 (Table 2) with native SAM docking score being the least at −6.54. SAM is bound in a close cavity formed by the following boundary elements: –S1, S0 (SH1 sheet); H1 helix; S5, S4 and S1 strands; and the loop between S1 and H2 helix. The methionine moiety of SAM having the –CH3 (methyl) group faces the hydrophobic interior of the RsmD, whereas the adenosine part of the SAM is facing the surface of the protein and

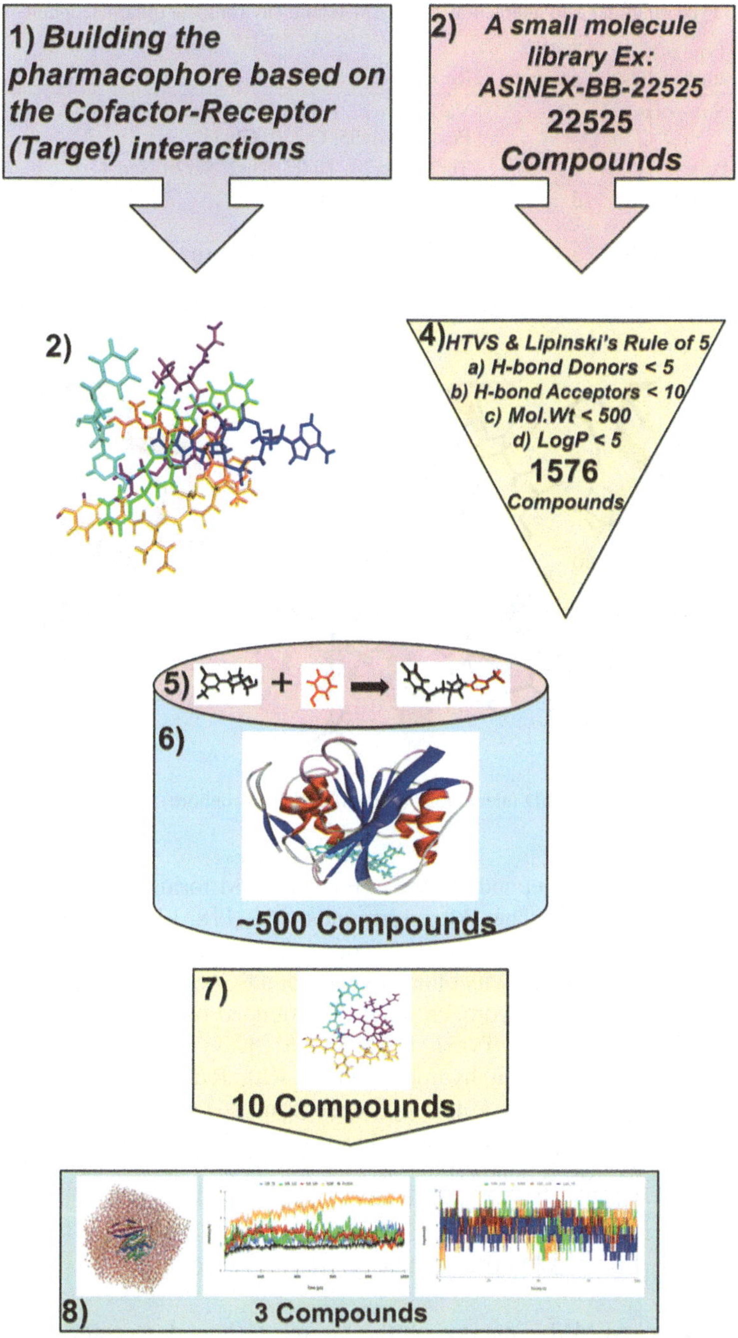

Fig. 1 Cofactor-receptor interactions based drug development flowchart: 1. Pharmacophore feature identification 2. Cofactor-receptor complex 3. Small molecule library 4. Lipinski rule based filtering 5. Pharmacophore based virtual screening 6. Molecular docking 7. ADME-Tox profiling 8. MD simulation studies

Table 1 Type of amino acid interaction with SAM-bound environment (pharmacophore of RsmD)

S. no	Type of amino acids	Total numbers	Residues
1	Non-polar	12	Val16, Pro17, Pro18, Pro23, Leu57, Ala56, Ala52, Tyr51, Pro121, Gly55, Gly20, Gly53
2	Polar	6	Thr25, Thr24, Thr21, Thr2, Ser78, Ser54
3	Positive-charged	3	Arg22, Arg19, Arg29
4	Negative-charged	4	Asp26, Asp75, Glu73, Asp119
Total residues			25

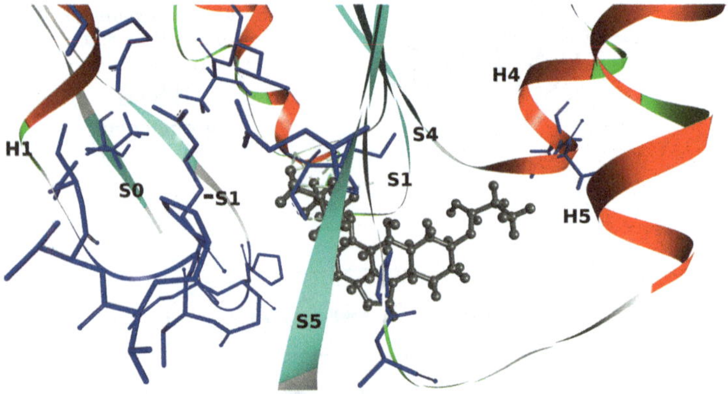

Fig. 2 Cofactor (SAM)-RsmD interactions to identify pharmacophore

interacting with SH1 sheet and H1 helix (Fig. 2). SAM forms five hydrogen bonds with RsmD (Fig. 3a). The AS_compounds, AS_118_122, AS_118_129 and AS_118_76, show favourable docking scores, minimization energies and binding free energy when compared with other AS compounds. AS_118_129 has the highest docking score with seven hydrogen bonds, surrounded by eight hydrophobic residues (Pro17, Pro18, Pro23, Tyr51, Pro120, Pro121, Tyr122, Ala 153) (Fig. 3c). AS_118_122 also has seven hydrogen bonds with RsmD surrounded by eight hydrophobic residues where a new hydrogen bond is formed with Ser78 and an existing Asp175 hydrogen bond is lost (Fig. 3d). AS_118_76 has the least docking score with five hydrogen bonds (Fig. 3b).

The binding free energies were calculated using

$$\Delta \text{Gbind} = \Delta \text{EMM} + \Delta \text{Gsolv} + \Delta \text{Gsa}$$

formula where ΔEMM= energy of the complex (bound receptor and ligand), ΔGsolv= difference GBSA solvation energy of the complex and ΔGsa= difference in surface area energy of the complex. The binding free energies varied from −34.37

Table 2 Docking scores and binding free energies of AS inhibitors

Title	Docking score	dGBind
SAM	−6.54	−35.55
AS_118_129	−9.23	−76.59
AS_118_76	−9.23	−71.64
AS_118_122	−9.18	−70.86
AS_118_78	−9.26	−70.79
AS_118_105	−9.22	−70.18
AS_118_79	−9.08	−70.10
AS_118_96	−9.22	−69.68
AS_118_75	−9.13	−68.91
AS_118_55	−9.35	−68.11
AS_118_54	−9.33	−67.85
AS_118_3	−9.24	−67.28
AS_118_9	−9.33	−62.27
AS_118_40	−9.27	−59.72
AS_118_39	−9.31	−59.44
AS_118_23	−9.19	−56.73

to −76.59 kcal/mol. The highest binding free energy was observed in AS_118_129, AS_118_122 and AS_118_76. It is interesting to note that when compared to SAM, AS_118_129, AS_118_122 and AS_118_76 have better binding free energies with significant contributions from favourable electrostatic energy component (ΔGCoul) and van der Waals energy (ΔGVdw) than the hydrogen bond energy component (Table 3). Therefore, docking and binding free energies reveal that AS_118_129, AS_118_122 and AS_118_76 inhibitors are potential lead compounds against RsmD.

4.3 ADMET Properties

ADMET predictions indicate the risk of failure of drug candidates in clinical trials, the top 14 to 16 hits were selected for ADMET properties. The predicted values are depicted in Table 4. ADMET properties reveal that AS compounds have maintained similar properties, within the admissible ranges. The top 3 ligands were selected for MD simulation based on the initial binding free energy, docking score and ADMET properties.

4.4 MDS Studies

The SAM trajectories of MDS studies are shown in Fig. 4. The average protein RMSD (black colour) is ranging from ~1.4Å to 2.2Å, indicating the protein being

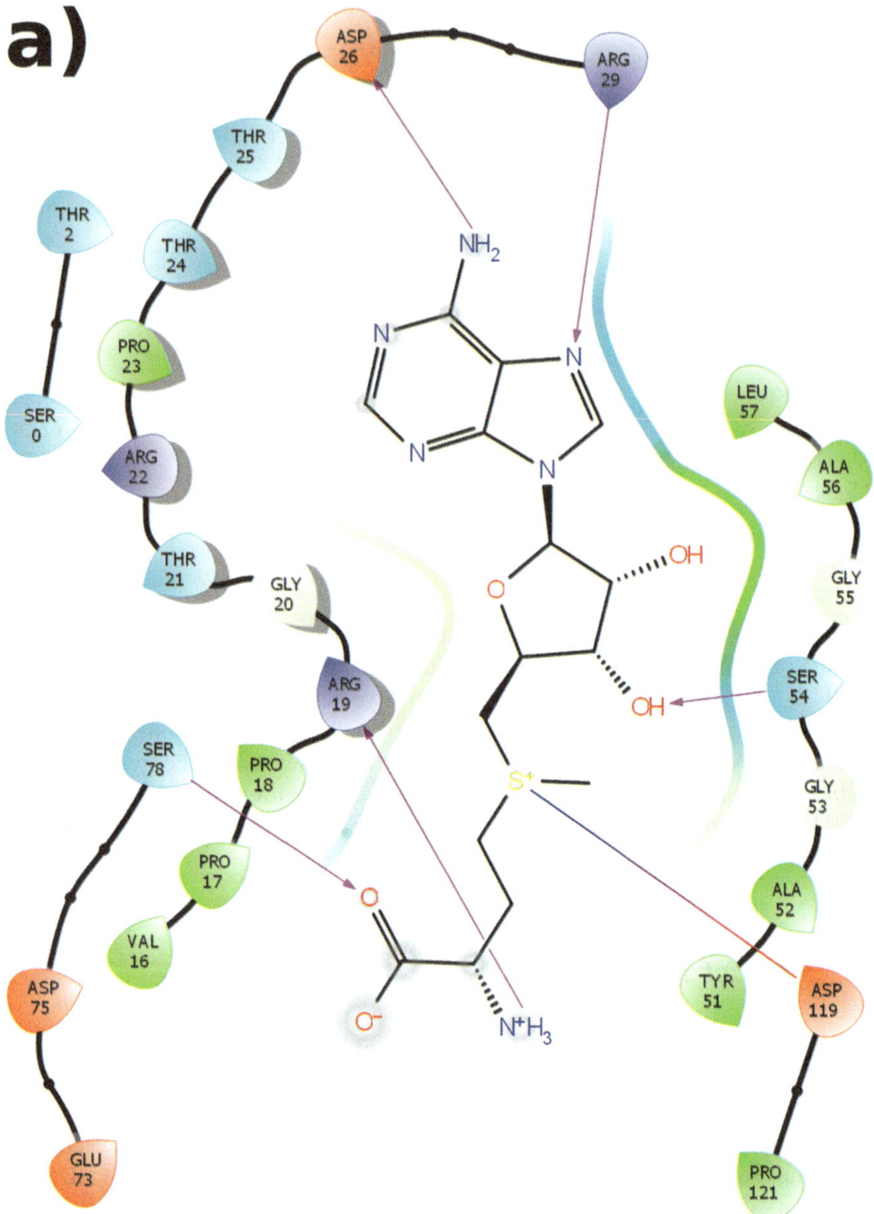

Fig. 3 (**a**) Two-dimensional (2D) interaction diagram of ligand SAM. (**b**)2D interaction diagram of ligand AS_118_76. (**c**) 2D interaction diagram of ligand AS_118_122. (**d**) 2D interaction diagram of ligand AS_118_129

b)

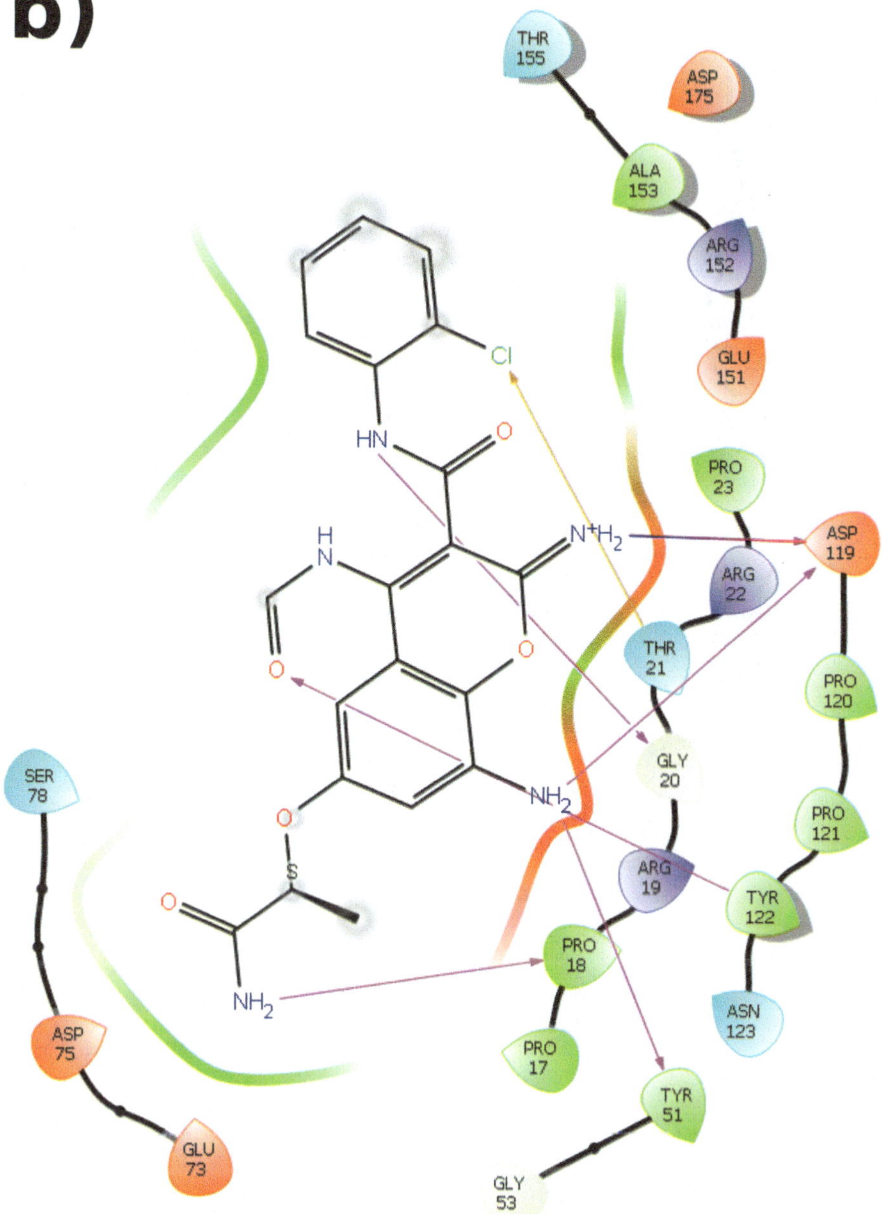

Fig. 3 (continued)

c)

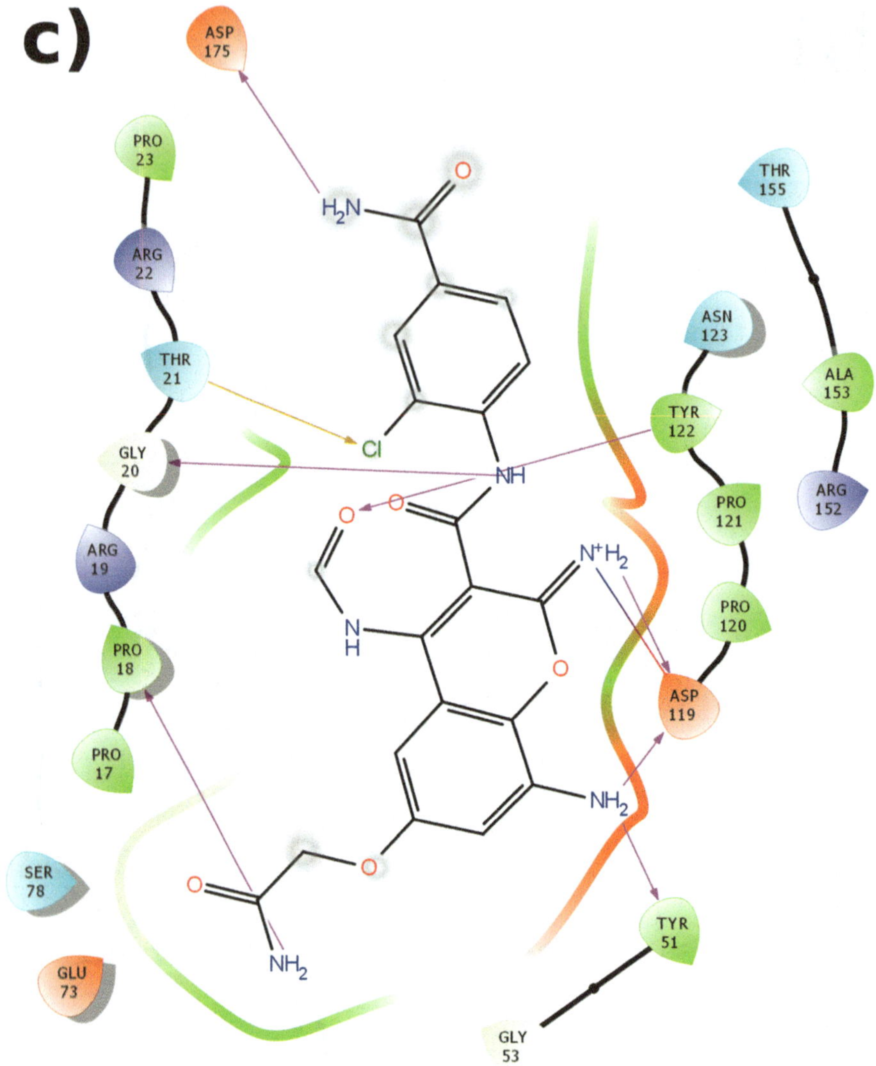

Fig. 3 (continued)

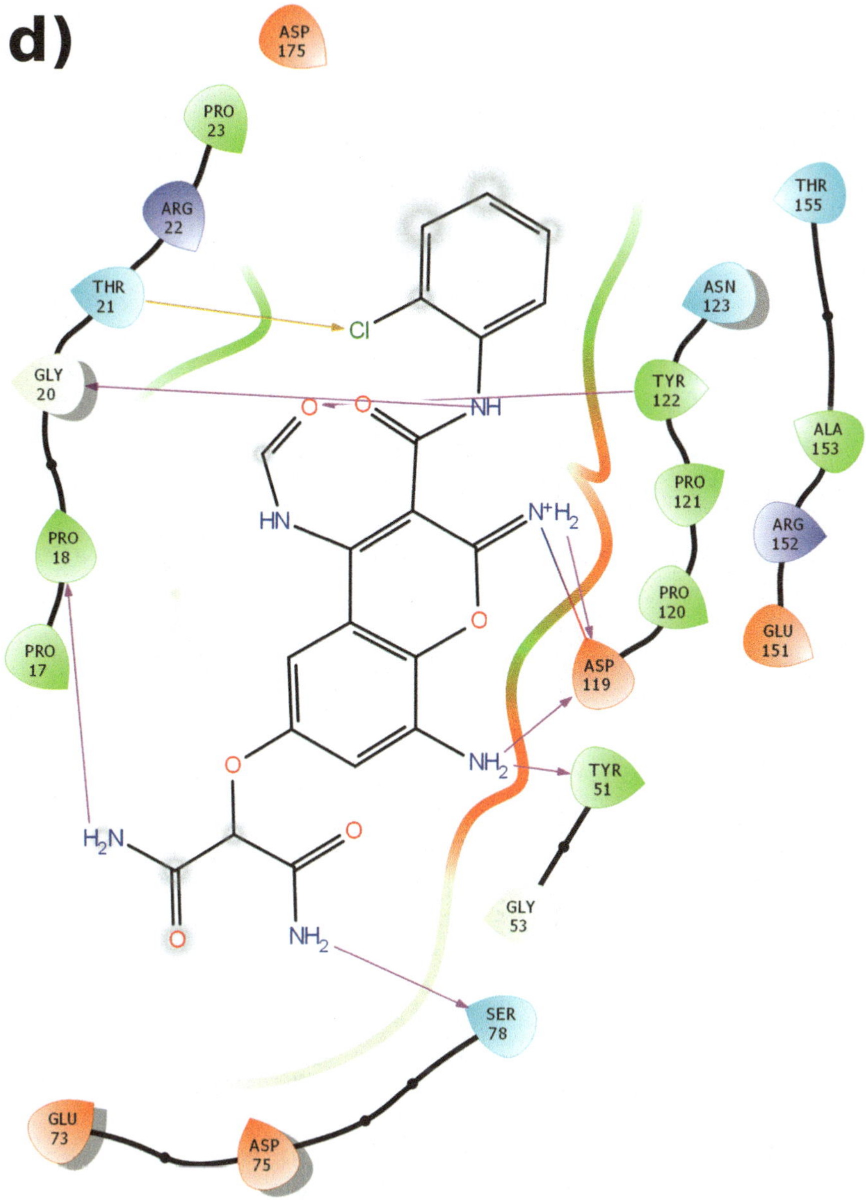

Fig. 3 (continued)

Table 3 Components of binding free energies: ΔEMM = difference in the molecular mechanical energy of the complex and the independent ligand and apo-protein structure

			ΔEMM						ΔGsolv	ΔGsa
S. no.	Title	ΔGbind (kcal/mol)	ΔGCoul	ΔGCovalent	ΔGH-bond	ΔGLipo	ΔGBind packing	ΔGVdw	Solv GB	SA GB
1	SAM	−35.55	26.60	9.10	−4.29	−10.03	−0.34	−48.99	−7.60	–
2	AS_118_129	−76.59	38.74	4.60	−4.85	−14.85	−0.57	−46.06	−53.57	–
3	AS_118_76	−71.64	43.06	5.76	−4.26	−15.58	−0.60	−44.92	−55.08	–
4	AS_118_122	−70.86	39.33	4.86	−4.83	−14.84	−0.72	−45.62	−49.02	–

Similarly, ΔGsolv = difference in the GBSA solvation energy and ΔGsa = difference in surface area energy

Table 4 List of ADMET and physicochemical properties of SAM and SAM analogues

Ligands	MW	LOGP	nROT	nON	nOHNH	TPSA	Water solubility	(LD50)	Hepatotoxicity
SAM	399.45	0.4014	7	6	5	157.26	−2.89	2.291	Yes
AS_118_129	473.853	−1.1332	8	7	6	189.667	−2.891	2.236	Yes
AS_118_76	444.855	0.4014	7	6	5	180.166	−2.946	2.167	Yes
AS_118_122	473.853	−0.8882	8	7	6	189.667	−2.911	2.17	Yes
AS_118_78	446.827	−0.2815	7	7	6	178.595	−2.912	2.24	Yes
AS_118_105	445.843	−0.4049	7	7	6	179.141	−2.88	2.16	Yes
AS_118_79	446.827	−0.2815	7	7	6	178.595	−2.913	2.273	Yes
AS_118_96	445.843	−0.0648	6	6	6	179.141	−2.894	2.221	Yes
AS_118_75	444.855	0.32132	7	6	5	180.166	−2.951	2.245	Yes
AS_118_55	556.724	0.9938	7	6	5	193.629	−2.969	2.245	Yes
AS_118_54	556.724	0.9938	7	6	5	193.629	−2.969	2.228	Yes
AS_118_3	448.818	0.2189	7	6	5	178.112	−2.952	2.238	Yes
AS_118_9	448.818	0.5283	7	6	5	178.112	−2.956	2.241	Yes
AS_118_40	465.273	0.7976	7	6	5	184.426	−2.963	2.241	Yes
AS_118_39	465.273	0.7976	7	6	5	184.426	−2.963	2.243	Yes
AS_118_23	509.724	0.9537	7	6	5	188.089	−2.967	2.291	Yes

MW molecular weight, *nROT* number of rotatable bonds, *nON* number of hydrogen bonds, *LD50* oral rat acute toxicity, *TPSA* total polar surface area. *LogP* predicted octanol/water partition coefficient

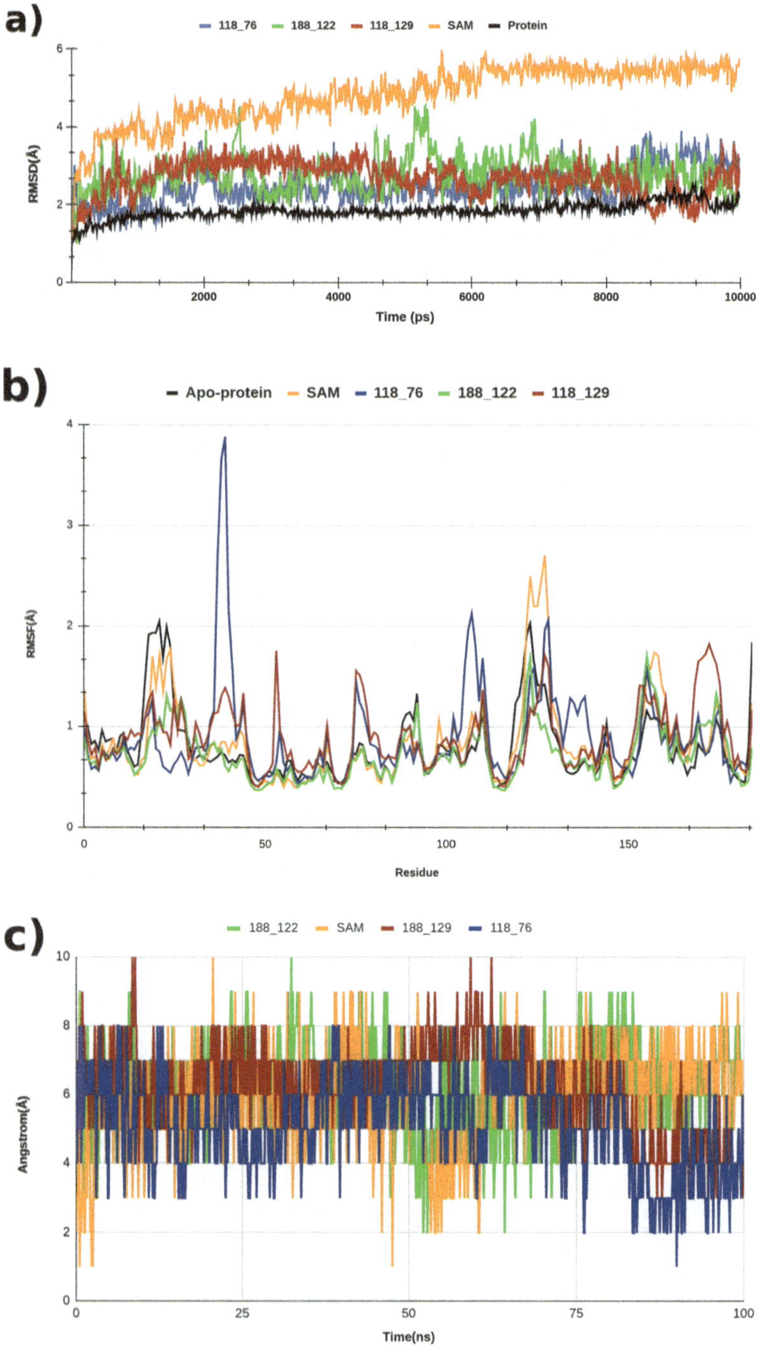

Fig. 4 MDS trajectory analysis of SAM, AS_118_122, AS_118_129 and AS_118_76: (**a**) RMSD of apo enzyme (black), SAM (blue), SAM_153 (green) and SAM_172 (pink); (**b**) RMSF of SAM (blue), SAM_153 (green) and SAM_172 (pink); and (**c**) hydrogen bonds of SAM (blue), SAM_153 (green) and SAM_172 (pink)

stable throughout the 100ns simulation (Fig. 4a). As is commonly observed, the RMSD of protein-ligand complexes ranged from 1.00Å to 5.6Å indicating certain transition states where high RMSD was observed before achieving stability. The native SAM-RsmD complex achieved stability after ~65ns (orange). The brown, blue and green colours represent AS_118_129, AS_118_76 and 118_122, respectively. From the RMSD graphs, it is clearly observed that AS compounds have reached stabilized interactions throughout the simulation period, converging faster than native SAM. AS_118-76, however, showed better and more favourable and stronger interactions as indicated by the RMSD of ~3Å when compared to the RMSD of other AS compounds. The RMSF plot shown in Fig. 4b provides the mean displacement of individual amino acids at respective positions over the simulation time span of 100ns which is interpreted for the stability of the protein and complex. In general, the higher fluctuations were observed only in the SAM binding site, 19–29 residues and 119–126 residues. Except for the active site or SAM binding environment, all other residues have less fluctuations throughout the simulation period. It is interesting to observe that the AS_118_122-RsmD complex showed the least fluctuations in the active site region with only ~1.5Å average fluctuations, whereas the SAM-RsmD and AS-RsmD complexes showed fluctuations of ~3.5Å and ~2.8Å indicating AS_118_122 analogue interactions to be stronger and stable.

The total hydrogen bonds between the RsmD and the ligands over the 100ns simulation period are depicted in Fig. 4c. It is observed that the native SAM maintained approximately five to ten hydrogen bonds (orange), with AS_118_122, AS_118_129 and AS_118_76 maintaining four to nine hydrogen bonds (green colour), four to ten hydrogen bonds (brown colour) and three to eight hydrogen bonds (blue), respectively, throughout the simulation period. Compared to native SAM, AS_188_122 shows less hydrogen bonds during the simulation period. However, AS_188_122 has maintained better binding affinity and stability during simulation. Better binding free energies of the analogues are attributed to the electrostatic and van der Waals energies. AS_188_122 has maintained continued favourable interactions at the SAM binding site throughout the 100ns simulation time.

5 Conclusions

It is a century ago that the idea of pharmacophore was first proposed as a means to understand drug/ligand-receptor interactions and develop computational mode of discovery and design of novel drugs. Pharmacophore modelling has evolved into a standard and is now a well-established CADD protocol with a variety of uses in drug discovery and design. Not all drug targets have cofactors such as RsmD and other methyltransferases. Cofactors are natural ligands with optimized interactions with the drug targets. This knowledge not only reduces the drug designing life cycle time but also adds value to the rationale of the design manifold and increases the reliability to the design. Hence, cofactor-receptor interaction-based pharmacophore design may not be possible for all varieties of drug targets; however, it is shown in this chapter that, in the case of methyltransferases (RsmD) with SAM as its cofactor (a naturally optimized ligand with the target), it is possible to design novel inhibitors against TB. The same could be extended to other drug targets with known cofactors. Cofactor receptor interaction-based pharmacophore identification can be used for (a) deriving novel leads and analogues, (b) comparative study and develop better scaffolds that could lead to new leads for similar targets, (c) faster pharmacophore-based virtual screening, (d) profiling of compounds especially repurposed drugs with known ADME-Tox properties and (e) it will be possible to understand and investigate possible off-targets as well which is a very important concern in drug designing. Given its wide applicability, cofactor-receptor-based pharmacophore modelling in particular and pharmacophore modelling in general are projected to continue to play a significant role in CADD in the foreseeable future. The advantages and prospects of pharmacophore modelling make it an indispensable computational technique in CADD that a structural biologist and/or a medicinal chemist should be aware of. The potential lead compounds which are computationally validated should be validated in vitro and in vivo in order to evaluate and validate the CADD prospects and efforts towards supporting the experimental research.

References

'WHO | The End TB Strategy' (2018). http://www.who.int/tb/strategy/en/. Accessed 24 Aug 2019

Agrafiotis DK et al (2007) Recent advances in chemoinformatics. J Chem Inf Model. https://doi.org/10.1002/chin.200743263

Bajorath J (2002) Integration of virtual and high-throughput screening. Nat Rev Drug Discov 1(11): 882–894

Bel'skih AN et al (2015) Cell engineering in nephrology: the current state and prospectives from the point of view of military medicine. Voen Med Zh 336(9):55–60

Biovia DS (2021) BIOVIA discovery studio academic research suite. Dassault Systèmes, San Diego. [Preprint].

Bowers KJ et al (2006) Molecular dynamics—Scalable algorithms for molecular dynamics simulations on commodity clusters. Proceedings of the 2006 ACM/IEEE conference on supercomputing - SC '06 [Preprint]. https://doi.org/10.1145/1188455.1188544

Boyd MR (1996) The position of intellectual property rights in drug discovery and development from natural products. J Ethnopharmacol 51(1-3):17–25; discussion 25–7.

Brown N (n.d.) In silico medicinal chemistry: computational methods to support drug design. Royal Society of Chemistry, Cambridge

Bügl H et al (2000) RNA methylation under heat shock control. Mol Cell 6(2):349–360

Cantoni GL (1953) S-Adenosylmethionine; a new intermediate formed enzymatically from L-methionine and adenosinetriphosphate. J Biol Chem 204(1):403–416

Churchyard G et al (2017) What we know about tuberculosis transmission: an overview. J Infect Dis 216(suppl_6):S629–S635

Duncan KW et al (2016) Structure and property guided design in the identification of PRMT5 tool compound EPZ015666. ACS Med Chem Lett 7(2):162–166

Ehrlich P (1909) 'Über den jetzigen Stand der Chemotherapie', Berichte der deutschen chemischen Gesellschaft, pp 17–47. https://doi.org/10.1002/cber.19090420105

Ekins S et al (2002) Towards a new age of virtual ADME/TOX and multidimensional drug discovery. Mol Divers 5(4):255–275

Foster PG et al (2003) The first structure of an RNA m5C methyltransferase, Fmu, provides insight into catalytic mechanism and specific binding of RNA substrate. Structure 11(12):1609–1620

Goto J, Kataoka R, Hirayama N (2004) Ph4Dock: pharmacophore-based protein-ligand docking. J Med Chem 47(27):6804–6811

Guner O (2002) History and evolution of the pharmacophore concept in computer-aided drug design. Curr Top Med Chem 2:1321–1332. https://doi.org/10.2174/1568026023392940

Guner OF (2005) The impact of pharmacophore modeling in drug design. IDrugs 8(7):567–572

Guo Q et al (2019) Structure of N6AMT1-TRMT112 complex with SAM. Worldwide Protein Data Bank. https://doi.org/10.2210/pdb6k0x/pdb

Halgren T (2007) New method for fast and accurate binding-site identification and analysis. Chem Biol Drug Des 69:146–148. https://doi.org/10.1111/j.1747-0285.2007.00483.x

Halgren TA (2009) Identifying and characterizing binding sites and assessing druggability. J Chem Inf Model 49:377–389. https://doi.org/10.1021/ci800324m

Halgren TA et al (2004) Glide: a new approach for rapid, accurate docking and scoring. 2. Enrichment factors in database screening. J Med Chem 47(7):1750–1759

Hong H et al (2008) Mold(2), molecular descriptors from 2D structures for chemoinformatics and toxicoinformatics. J Chem Inf Model 48(7):1337–1344

Jones G, Willett P, Glen RC (1995) A genetic algorithm for flexible molecular overlay and pharmacophore elucidation. J Comput Aided Mol Des 9(6):532–549

Lesnyak DV et al (2007) Methyltransferase that modifies guanine 966 of the 16 S rRNA. J Biol Chem 282:5880–5887. https://doi.org/10.1074/jbc.m608214200

Li AP (2001) Screening for human ADME/Tox drug properties in drug discovery. Drug Discov Today 6(7):357–366

Li W et al (2019) Structural insight into human N6amt1–Trm112 complex functioning as a protein methyltransferase. Cell Discov 5(1):1–13

Ma B et al (2016) Biochemical and structural characterization of a DNA N6-adenine methyltransferase from Helicobacter pylori. Oncotarget 7(27):40965–40977

Pipeline | Working Group for New TB Drugs (n.d.-a). https://www.newtbdrugs.org/pipeline/clinical. Accessed 17 Nov 2019

Pipeline | Working Group for New TB Drugs (n.d.-b). https://www.newtbdrugs.org/pipeline/clinical. Accessed 17 Nov 2019

Pires DEV, Blundell TL, Ascher DB (2015) pkCSM: predicting small-molecule pharmacokinetic and toxicity properties using graph-based signatures. J Med Chem 58(9):4066–4072

Richon VM et al (2011) Chemogenetic analysis of human protein methyltransferases. Chem Biol Drug Des 78(2):199–210

Sastry GM et al (2013) Protein and ligand preparation: parameters, protocols, and influence on virtual screening enrichments. J Comput Aided Mol Des 27:221–234. https://doi.org/10.1007/s10822-013-9644-8

Schluckebier G et al (1997) Differential binding of S-adenosylmethionine S-adenosylhomocysteine and Sinefungin to the adenine-specific DNA methyltransferase M.TaqI. J Mol Biol 265(1): 56–67

Shelley JC et al (2007) Epik: a software program for pK a prediction and protonation state generation for drug-like molecules. J Comput Aided Mol Des:681–691. https://doi.org/10.1007/s10822-007-9133-z

Sun H, Tawa G, Wallqvist A (2012) Classification of scaffold-hopping approaches. Drug Discov Today 17:310–324. https://doi.org/10.1016/j.drudis.2011.10.024

Taft CA, Da Silva VB, Da Silva CHTDP (2008) Current topics in computer-aided drug design. J Pharm Sci 97(3):1089–1098

Takeshita K et al (2011) Structural insight into maintenance methylation by mouse DNA methyltransferase 1 (Dnmt1). Proc Natl Acad Sci U S A 108(22):9055–9059

Thiel KA (2004) Structure-aided drug design's next generation. Nat Biotechnol 22:513–519. https://doi.org/10.1038/nbt0504-513

Thomas DJ, Waters SB, Styblo M (2004) Elucidating the pathway for arsenic methylation. Toxicol Appl Pharmacol 198(3):319–326

Valerio LG Jr, Choudhuri S (2012) Chemoinformatics and chemical genomics: potential utility of in silico methods. J Appl Toxicol 32(11):880–889

Venkataraman S et al (2018) Crystal structure of a new form of RsmD-like RNA methyl transferase from Mycobacterium tuberculosis determined at 1.74 A resolution. https://doi.org/10.2210/pdb6aie/pdb

Waddell TG et al (2000) Prebiotic methylation and the evolution of methyl transfer reactions in living cells'. Orig Life Evol Biosph 30(6):539–548

Wang X et al (2016) Structural basis of N6-adenosine methylation by the METTL3–METTL14 complex. Nature 534:575–578. https://doi.org/10.1038/nature18298

Witek MA et al (2017) A Novel Motif forS-Adenosyl-l-methionine Binding by the Ribosomal RNA Methyltransferase TlyA fromMycobacterium tuberculosis. J Biol Chem 292:1977–1987. https://doi.org/10.1074/jbc.m116.752659

Wolber G et al (2008) Molecule-pharmacophore superpositioning and pattern matching in computational drug design. Drug Discov Today 13(1-2):23–29

Wuosmaa AM, Hager LP (1990) Methyl chloride transferase: a carbocation route for biosynthesis of halometabolites. Science 249(4965):160–162

Yu H, Adedoyin A (2003) ADME-Tox in drug discovery: integration of experimental and computational technologies. Drug Discov Today 8(18):852–861

In Silico Identification of Potential Antivirals Against SARS-CoV-2 Main Protease and RBD of Spike Protein: A Drug Repurposing Approach

Vijayakumar Rajendran, Saravanan Kandasamy, Ankita Gupta, Killivalavan Asaithambi, Ashish Runthala, Jagannathan Selvaraj, and Shivanandappa Kukkaler Channappa

Abstract SARS-CoV-2 is the etiological agent liable for the viral pneumonia outbreak since later 2019 that originated in Wuhan. It has been declared as a pandemic by the World Health Organization. The virus is mainly spread from person to person mostly through airborne, fomite, contact, and droplet from the infected persons. Although there are latest vaccines available against the virus, it challenges the community with various mutants such as alpha variant (UK), beta variant (South Africa), gamma variant (Brazil), delta variant (India), eta variant (UK and Nigeria), iota variant (USA), kappa variant (India), and lambda variant (Peru). The WHO declares these as variants of concern, and only a few vaccines were reported to be effective against them. The SARS-CoV-2 enters the host cell by binding the viral surface spike glycoprotein (S-protein) to angiotensin-converting enzyme 2 (ACE2), and M^{pro} is a key enzyme that plays a crucial role in facilitating viral replication and transcription. Considering the significance of these two proteins in the viral infection, these were chosen as the potential drug target against SARS-CoV-2 infection. This chapter emphasizes the need for an additional antiviral drug, and here, through the PubChem database, we have analyzed 15 potential antiviral drugs along with

Vijayakumar Rajendran, Saravanan Kandasamy and Ankita Gupta contributed equally with all other contributors.

V. Rajendran (✉) · S. Kandasamy (✉) · K. Asaithambi
Molecular Biophysics Unit, Indian Institute of Science, Bangalore, India

A. Gupta
Department of Zoology, Gauhati University, Gauhati, India

A. Runthala
Department of Biotechnology, Koneru Lakshmaiah Education Foundation, Vijayawada, India

J. Selvaraj
Tissue culture anti-rabies vaccine section, Pasteur Institute of India, Coonoor, India

S. K. Channappa
Pertussis Vaccine Section, Pasteur Institute of India, Coonoor, India

V. S. Rajput, A. Runthala (eds.), *Drugs and a Methodological Compendium*,
https://doi.org/10.1007/978-981-19-7952-1_15

natural antiviral agent quercetin. The best hit is screened through the induced-fit docking against these two key proteins, and its credibility is further tested through molecular dynamics simulations. Our results suggest that the antiviral drugs indinavir and famciclovir could be the potential drugs against COVID-19 infection.

Keywords COVID-19 · Virtual screening · Induced-fit docking · Molecular dynamics · Quercetin · Indinavir · Famciclovir

1 Introduction

The severe acute respiratory syndrome coronavirus 2 (SARS-CoV-2) is a novel virus strain that causes a significant outbreak of coronavirus disease (COVID-19), a respiratory illness (World Health Organization 2020). SARS-CoV-2 is reported first in Wuhan, China, in late December 2019, and it then spread from country to country through a new travel route, later getting declared a global pandemic by the World Health Organization (World Health Organization 2020; Yang et al. 2020). WHO has recommended frequent hand washing and maintenance of social distance to control COVID-19 disease. However, these measures are not able to prevent the infection, spread by an infected person. India first reported the COVID-19 case in Kerala on 30 January 2020, and currently 14,043,176 patients are infected globally, including 1,077,618 Indians (on 20 July 2020) (Lathika et al. 2020).

Coronaviruses are positive-stranded RNA viruses that cause respiratory illness in vertebrates including humans. The International Committee on Taxonomy of Viruses (ICTV) is responsible for the classification and naming of the virus (Cui et al. 2019), and SARS-CoV-2 is categorized under the *Coronaviridae* family, genus *Betacoronavirus*. These are susceptible to mutation and recombination which makes the virus more vulnerable. Currently, the Global Initiative on Sharing All Influenza Data (GISAID), a database of all influenza viruses, has 7,288,649 genome sequences submitted in SARS-Cov-2 alone. The SARS-CoV-2 virus is a human-infecting *Betacoronavirus*, divergent from SARS-CoV that caused epidemic in 2003 (Gorbalenya et al. 2020).

Based on structural and biochemical studies, SARS-CoV-2 binds to human receptor angiotensin-converting enzyme 2 (ACE2), through densely glycosylated spike protein (S) as an initial step to enter the human cells. The binding of the spike protein of SARS-CoV-2 with the human receptor ACE2, through the receptor-binding domain (RBD) present in spike protein, is the most variable part of the coronavirus genome and form an RBD-ACE2 complex (Andersen et al. 2020). The amino acid sequence of RBD of SARS-CoV-2 is highly divergent from SARS-CoV-2, and the variations are shown to stabilize its atomic interactions with the human ACE2 protein (Chen and Hotez 2020). Recently, a 2.45 Å resolution crystal structure of SARS-CoV-2 S-protein has been published (PDB ID: 6M0J); through its solvent-accessible surface area analysis, several virtual screening and drug discovery reports have been published (Lan et al. 2020; Jakhmola et al. 2020). However, the highly variant RBD structure is not at all considered in several such studies, although its experimental structure is available in the Protein Data Bank.

Protease enzyme (M^{pro}, also called $3CL^{pro}$) is one of the best enzymes for drug targeting among coronavirus, due to its essential role in polyprotein processing, which is translated from viral RNA (Khan et al. 2020; Chowdhury 2020). Recently, the X-ray crystal structure of SARS-CoV-2 M^{pro} has been solved, and it provides an excellent platform for structure-based drug discovery (Jin et al. 2020; Douangamath et al. 2020). A considerable effort has been given to studying this protein for finding a therapeutic drug against SARS-CoV-2. The purpose of this study is to use the X-ray structure of protease enzyme M^{pro} and RBD of S-protein for screening the 16 selected antiviral drugs to identify the most promising therapeutic compound.

2 Methods

2.1 Selection of Antiviral Drugs

There are various databases available for drug discovery such as PubChem, ZINC, DrugBank, ChEMBL, etc. We used the PubChem database which is the repository of over 100 million drugs accessed through https://pubchem.ncbi.nlm.nih.gov/, and 15 antiviral drugs were selected along with natural compound quercetin. The 3D structure of the selected drugs was downloaded, and the drugs were optimized using the LigPrep tool using Maestro suite with default settings.

2.2 Molecular Docking Analysis

To understand the binding affinity and the intermolecular interaction of selected drug molecules with the active site amino acids of the main protease enzyme M^{pro} and RBD of S-protein, molecular docking has been performed. Before docking, selected drug molecules were prepared by the LigPrep module of Schrodinger software. Further, these molecules were docked with the main protease enzyme M^{pro}(PDB: 5RFW) and RBD of S-protein (PDB: 6M0J) by extra-precision (XP) mode incorporated in the induced-fit docking (IFD) method (Mirza et al. 2016). The IFD resulted in different conformations for each ligand-main protease enzyme M^{pro} and RBD of S-protein complexes. Based on the top score with the lowest energy conformation and intermolecular interactions, the ligand-protein complexes were selected for each case and used for further study. PyMOL software was used to analyze the intermolecular interactions between ligand and protein for the individual complex.

2.3 Molecular Dynamics Simulation

To understand the stability, intermolecular interactions, and the binding energy of the selected two complexes of main protease enzyme M^{pro} and RBD of S-protein, the molecular dynamics (MD) simulation has been performed using the OPLS3e force field implemented in the Desmond v5 package. Further, the system was built with the pre-defined TIP4P water model and orthorhombic periodic boundary conditions at the distances 10 Å. Then, the counter-ions were used to neutralize the charges of these complexes with the balancing Na+/Cl − ions. Further, the constructed system for each ligand with the main protease enzyme M^{pro} and RBD of S-protein complexes was energy minimized by heating and equilibrium processes before the MD simulation. In the minimization step, the minimization and heating protocols were fixed based on the steepest descent method, annealing temperature at 0–300 K and 2000 steps with the time steps of 0.001 ps. Further, the system was normalized in an equilibrium state at 1000 steps with a time step of 0.001 ps. Finally, the production step of the systems was continued up to 100 ns with the time steps of 0.001 ps, 300 K, 1 atm pressure, and applied using the Nose-Hoover method with NPT ensemble (Prachanronarong et al. 2016; Rajendran et al. 2018; Chandra et al. 2020). Intermolecular interactions and conformation of each ligand-protein complex were analyzed from the final results of MD simulation. Among the 1000 fractions, 10, 20, 30, 40, 50, 60, 70, 80, 90, and 100 ns fractions were used to determine the binding free energy (MMGBSA) of the protein-ligand complexes by Prime application available in the Schrodinger software package.

3 Results and Discussion

3.1 Binding Mode and Intermolecular Interactions of M^{pro} Protein with Indinavir

Several drugs have been predicted for the treatment of COVID-19 recently including antivirals such as lopinavir, ritonavir, ribavirin, telbivudine, remidesvir, etc. (Cao et al. 2020a, b; Liu et al. 2020; Dong et al. 2020; Muralidharan et al. 2020; Loustaud et al. 2016; Kandeel and Al-Nazawi 2020; Wang et al. 2020). The available PDB structures of M^{pro} show the hydrogen bond interaction between the ligand T8M 405 and amino acid residues C145, G143, and H41 of the protein. H41 also showed pi-cation interaction with the benzyl ring of the ligand (Jin et al. 2020; Douangamath et al. 2020). Ribavirin formed two hydrogen bonds with Thr25 and Gln 189, whereas the residues Gln189 and Ser46 interact with the antiviral drug telbivudine. Based on these data, the amino acid residues Cys145, Gly143, His41, Gln189, Ser46, and Thr25 are predicted as the key residues for the binding of antiviral compounds (Kandeel and Al-Nazawi 2020) (Fig. 1).

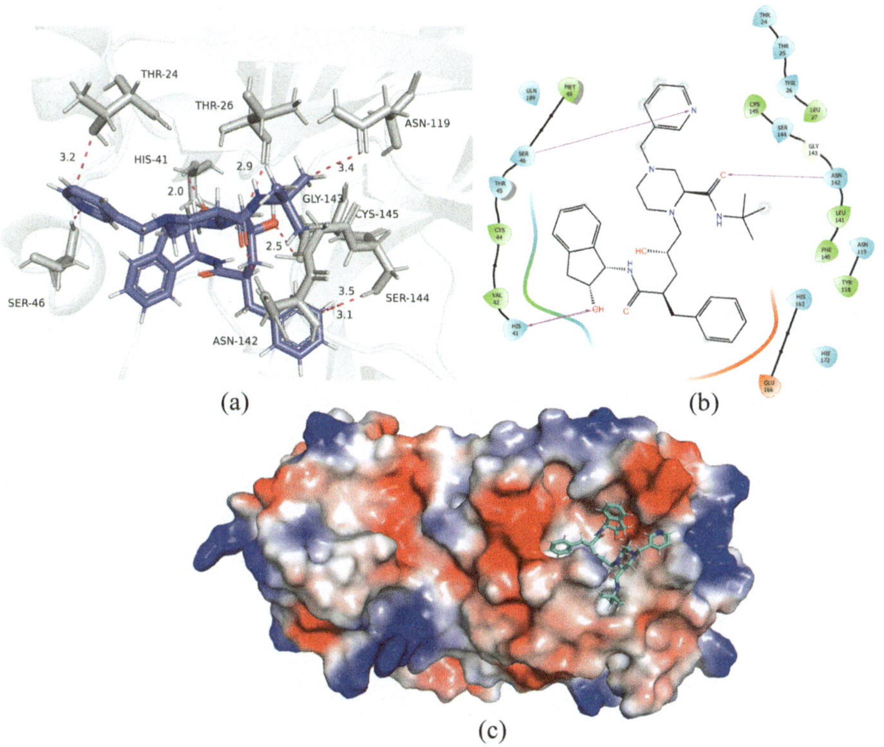

Fig. 1 (**a**) Intermolecular interactiosns between indinavir and active site residues of M^{pro}, (**b**) 2D planer view of M^{pro}-indinavir complex, and (**c**) electrostatic potential map of M^{pro} with indinavir molecule present in the active site cavity

Fifteen antiviral compounds and the natural compound quercetin were docked using the induced-fit docking method in the Schrodinger maestro suite. The drug indinavir showed a high binding affinity of −683.7 forming three hydrogen bonds between the ligand and His41, Ser46, and Asn142 residues of the M^{pro} protein. Similarly, oseltamivir formed hydrogen bonds with Gln189, His 164, Asn142, and Gly143. Famciclovir formed six hydrogen bonds with Asn142, Gly143, His41, Thr25, and Glu166. The IFD score of oseltamivir and famciclovir were −681.909 and −683.411, respectively. The natural phytochemical quercetin formed five hydrogen bonds. Glu166 formed with the hydroxyl group of the benzyl ring, Gly143 interacts with the keto and hydroxyl group of the aromatic ring, Cys145 with the keto group, and Thr26 with the hydroxyl group. Also, His41 forms pi-cation interaction. Similarly, all the compounds with the hydrogen bond-forming residues were mentioned in Table 1. The antiviral drug tromantadine didn't show any significant binding with the M^{pro} enzyme. Figure 2 shows the superimposed form of all the drug molecules in the active site of the main protease which reveals the orientation of the molecules and the differences (Fig. 3).

Table 1 Different score values obtained from molecular docking of selected viral compounds with the main protease of SARS-CoV-2

	Docking score	Glide energy	Prime energy	IFD score
Indinavir	−9.04	−68.275	−13493.2	−683.7
Oseltamivir	−8.641	−56.189	−13495.4	−683.411
Famciclovir	−6.895	−49.458	−13522.9	−683.038
Penciclovir	−9.86	−45.828	−13460.1	−682.865
Nelfinavir	−8.198	−61.213	−13488.5	−682.625
Cidofovir	−7.698	−47.802	−13494.6	−682.426
Quercetin	−8.856	−42.788	−13412.5	−679.483
Nevirapine	−5.512	−38.471	−13469.2	−678.971
Idoxuridine	−6.93	−42.994	−13427.6	−678.308
Zidovudine	−5.29	−43.774	−13458.4	−678.211
Prochloraz	−6.725	−44.421	−13426.8	−678.065
Edoxudine	−6.836	−31.541	−13,422	−677.935
Vidarabine	−8.19	−39.261	−13,393	−677.84
Ribavirin	−7.102	−44.935	−13406.1	−677.406
Rimantadine	−4.551	−21.166	−13384.9	−673.797

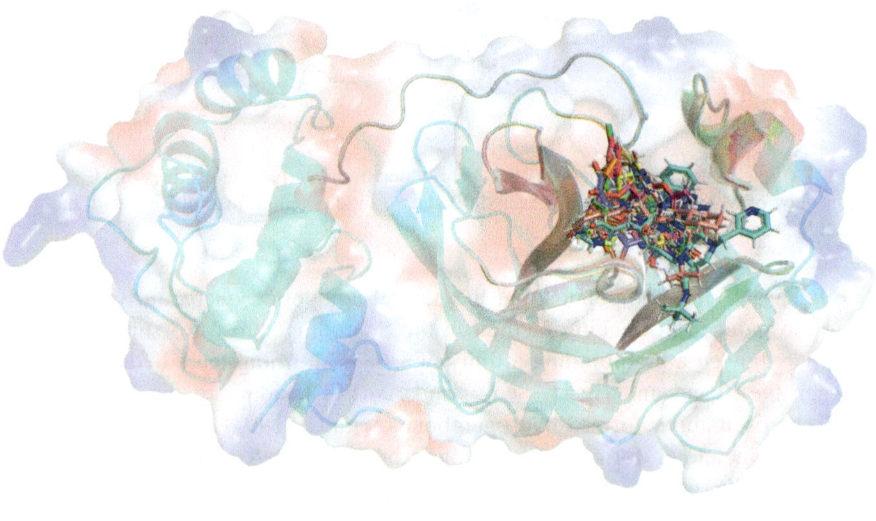

Fig. 2 The superimposed form of all the selected drug molecules in the active site of main protease which reveals the orientation of the molecules and the differences

3.2 *Binding Mode and Intermolecular Interactions of RBD of S-Protein with Famciclovir*

The S-protein of SARS-CoV-2 is most responsible for the virus entry through the binding of the receptor-binding domain of the S-protein with the ACE2 enzyme

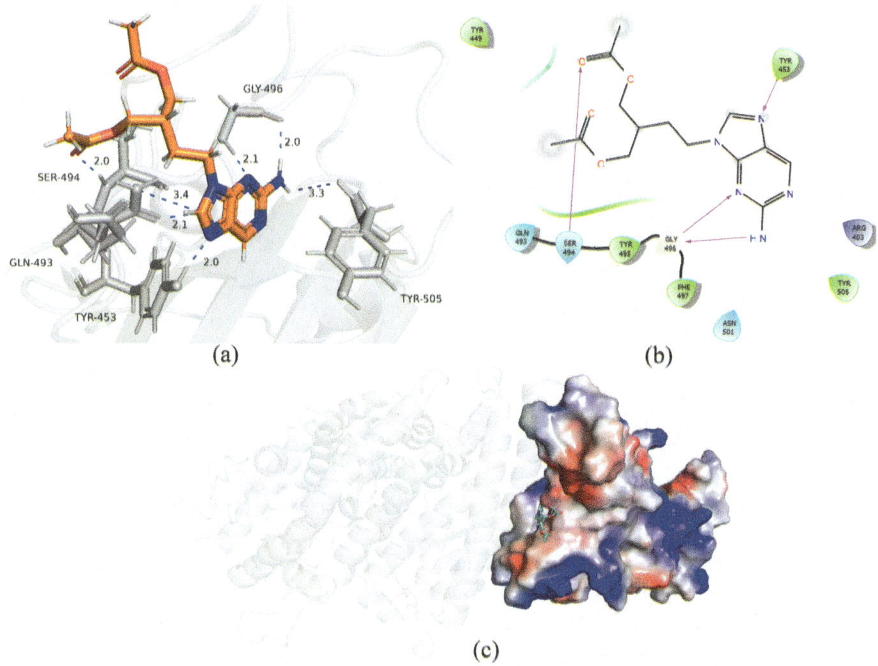

Fig. 3 (**a**) Intermolecular interactions between famciclovir and interface residues of RBD of spike protein, (**b**) 2D planer view of RBD-famciclovir complex, and (**c**) electrostatic potential map of RBD-famciclovir with ACE2 (gray) protein

(Yan et al. 2020). Reportedly, by inhibiting the RBD of S-protein, the SARS-CoV-2 fusion can be stopped. In RBD of S-protein, the protein intermediate residues are playing a major role in viral entry. Notably, the surface residues like Gln493, Gly496, Gln498, and Asn502 are forming strong interactions with the ACE2 enzyme. Hence, to inhibit the function of this enzyme, the focused inhibitors should be interacting with these key residues (Hoffmann et al. 2020). The molecular docking of selected antiviral drugs with RBD of S-protein has been performed using the induced-fit docking (IFD) method. From the docking results, the docking score and IFD score values of famciclovir show a high binding affinity toward RBD of S-protein, and the values are −7.79 kcal/mol and −430.20 kcal/mol, respectively. The difference of docking scores and glide energy confirms the nature of their intermolecular interactions with the neighboring amino acids present in the active site of RBD of S-protein. All the selected antiviral drugs are forming strong interactions with intermediate surface residues; particularly, the Gly496 is forming strong hydrogen-bonding interaction with all compounds except indinavir and nevirapine which confirms all the selected compounds are binding with the target region. Moreover, the residues Tyr453, Gln498, and Asn501 are also shown hydrogen bonding with selected compounds. In the famciclovir-RBD of S-protein complex, Tyr453, Gln493, Ser494, Gly496, and Tyr505 are showing strong interactions,

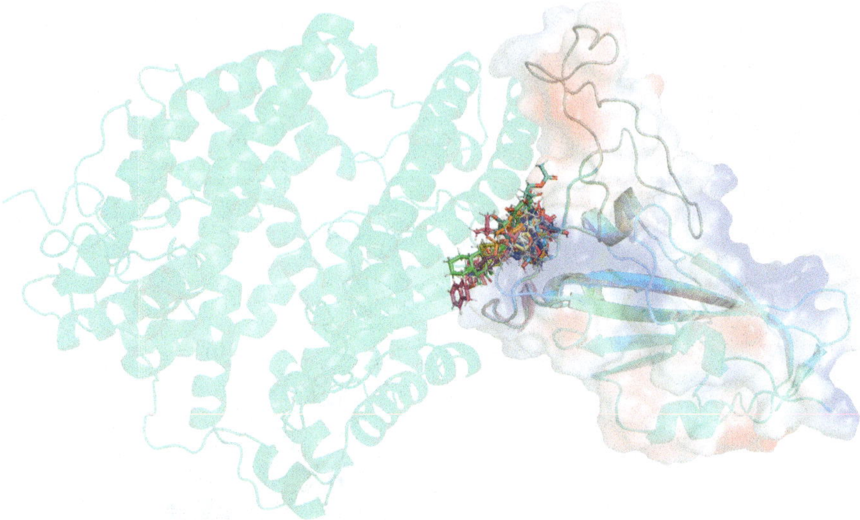

Fig. 4 The superimposed form of all the drug molecules in the active site of RBD of S-protein which reveals the orientation of the molecules and the differences

and the distances are 2.0, 2.1, 2.0, 2.1, and 3.3 Å, respectively. Based on the docking score, IFD score, and intermolecular interactions, the famciclovir-RBD of the S-protein complex has been selected for further study. Figure 4 shows the superimposed form of all the drug molecules in the active site of RBD of S-protein which reveals the orientation of the molecules and the differences. Further, we have probed the existence of these interactions and binding stability of the famciclovir drug molecules with the RBD of S-protein during the MD simulations.

3.3 RMSD, RMSF, and Intermolecular Interactions from MD Simulation

Molecular dynamics help understand the biological function of proteins and protein-ligand complexes (Rajendran et al. 2018). Here we have performed 100 ns MD simulations for indinavir-M^{pro} enzyme and famciclovir-RBD of S-protein complexes which is also helping to explore the stability and the binding affinity of the drug molecules in the corresponding protein environments. Figure 5 displays the statistical parameters (RMSD and RMSF) of the two complexes during the MD simulations. The RMSD values of the indinavir-M^{pro} complex remains within 2.5 Å (Fig. 5a), whereas the RMSD of famciclovir-RBD of S-protein complex shows 2.25 Å. Notably, the complex famciclovir-RBD of S-protein was stabilized after 30 ns of MD simulation (Fig. 5c). Figure 5b and d shows the variations of RMSF of amino acids of two proteins; among these, high fluctuations are found in the loop

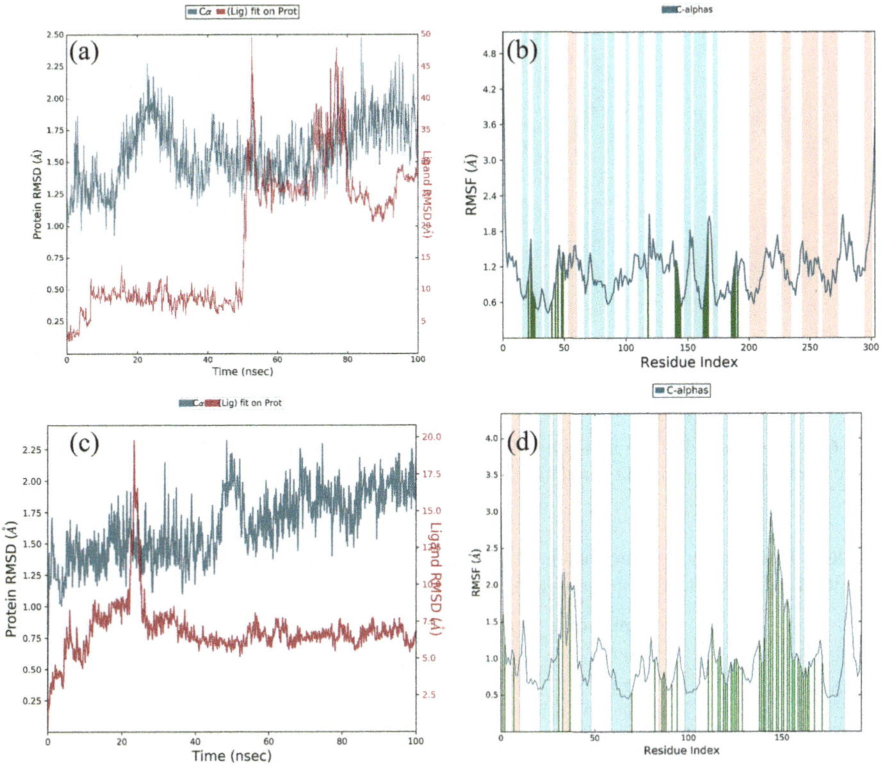

Fig. 5 RMSD and RMSF of **(a, b)** indinavir-M^{pro} complex and **(c, d)** famciclovir-RBD of S-protein complex, respectively. The cyan and orange colors in RMSF denote beta strands and alpha helix, respectively, whereas green color denotes the residues interacting with the ligand

region than the α-helix and β-sheet. The superimposed form of the three ligand-protein complexes (Fig. 6) obtained from the docking and MD simulations is showing the conformational difference of both ligands and proteins in the respective complexes; specifically, this allows us to visualize how the conformation and the orientation of the ligands and proteins are altered during the MD simulation. Overall, the fluctuation of the indinavir-M^{pro} complex is found to be very less, and it is normal.

Intermolecular interactions were analyzed after 100 ns of MD simulation, in which intermolecular interactions such as hydrogen bonds and hydrophobic and water bridge interactions exist in the indinavir-M^{pro} complex after 100 ns of MD simulations. Notably, the residues Thr24 and Thr26 are forming strong hydrogen-bonding interactions with indinavir molecule during the simulation which are highly stable (92 and 86%) throughout the simulation. Moreover, the amino acid Asn119 is forming a water bridge interaction with an indinavir molecule, whereas in famciclovir-RBD of S-protein complex, the residue Tyr453 has maintained their interaction with famciclovir molecule up to 100 ns of MD simulation. Furthermore,

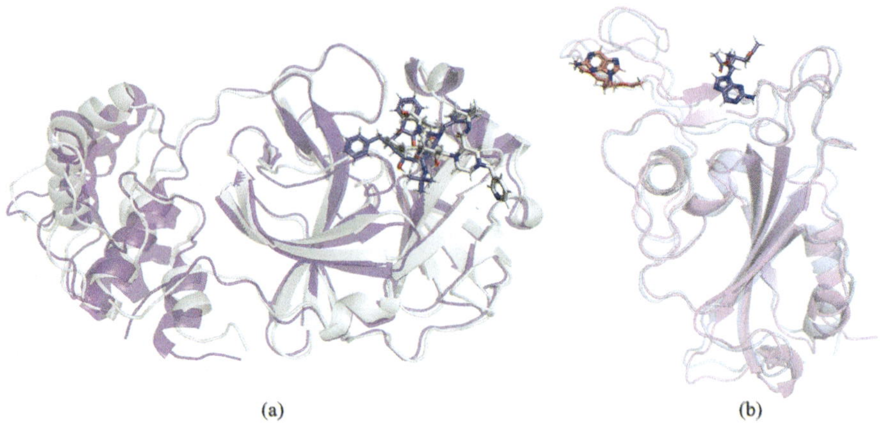

Fig. 6 Superimposed of (a) indinavir-M^{pro} complex and (b) famciclovir-RBD of S-protein complexes from docking and MD simulation

Table 2 Different score values obtained from molecular docking of selected viral compounds with receptor binding domain of spike protein

	Docking score	Glide energy	Prime energy	IFD score
Famciclovir	−7.793	−40.518	−8448.2	−430.203
Cidofovir	−7.943	−35.938	−8412.8	−428.586
Indinavir	−7.14	−56.33	−8383.3	−426.306
Edoxudine	−7.983	−35.759	−8363.8	−426.175
Penciclovir	−6.602	−40.879	−8362	−424.703
Oseltamivir	−4.155	−32.188	−8398.4	−424.076
Idoxuridine	−6.122	−32.999	−8354.7	−423.857
Nevirapine	−4.531	−27.816	−8385.5	−423.806
Nelfinavir	−4.943	−42.28	−8367.5	−423.317
Quercetin	−6.533	−32.259	−8332.4	−423.155
Zidovudine	−4.493	−32.356	−8370.2	−423.004
Ribavirin	−7.578	−36.097	−8303.3	−422.743
Prochloraz	−4.793	−32.146	−8346.7	−422.129
Vidarabine	−5.606	−32.579	−8314	−421.305
Rimantadine	−3.39	−18.111	−8323.6	−419.57
Tromantadine	−3.34	−27.093	−8312.2	−418.951

MMGBSA was calculated for these two complexes that are −74.949 kcal/mol and −31.776 kcal/mol, respectively. Therefore, from these results, we concluded that these two molecules with M^{pro} and RBD of S-protein were highly stable during the simulation and they could be potential inhibitors, and we can also design new compounds from these known drugs (Table 2).

4 Conclusion

The major source of disaster in the twenty-first century is the 2019 novel coronavirus disease (COVID-19). In the past few months, the world is eagerly waiting for a solution and put a strong effort to find drug compounds against different targets responsible for COVID-19. Since recent vaccine developments are mostly used for pre-exposure, drugs like remidesvir and 2-DG also have shown some positive indication after the infection. At this current point in time, the need for specific drugs to either treat or prevent this disease is essential since the vaccines are less effective against a variant of concern. Therefore, we have selected 16 antiviral drugs and performed molecular docking and MD simulation to put a finger on against M^{pro} and RBD of the spike protein. From our results, the drug indinavir showed high binding affinity and strong hydrogen bonding with M^{pro} protein, whereas in the RBD of S-protein, the famciclovir shows high binding affinity and strong interactions. On comparing these results, we propose these two drugs (indinavir and famciclovir) could be helpful against COVID-19.

Acknowledgments We acknowledge Mr. Vinod Devaraji from Schrodinger, India, for providing the Desmond software.

References

Andersen KG, Rambaut A, Lipkin WI et al (2020) The proximal origin of SARS-CoV-2. Nat Med 26(4):450–452
Cao B, Wang Y, Wen D et al (2020a) A Trial of Lopinavir–Ritonavir in Adults Hospitalized with Severe Covid-19. N Engl J Med 382(19):1787–1799
Cao Y-c, Deng Q-x, Dai S-x (2020b) Remdesivir for severe acute respiratory syndrome coronavirus 2 causing COVID-19: an evaluation of the evidence. Travel Med Infect Dis 35:101647
Chandra A, Goyal N, Qamar I et al (2020) Identification of hot spot residues on serine-arginine protein kinase-1 by molecular dynamics simulation studies. J Biomol Struct Dyn:1–9
Chen W-H, Hotez PJ (2020) Potential for developing a SARS-CoV receptor-binding domain (RBD) recombinant protein as a heterologous human vaccine against coronavirus infectious disease (COVID)-19. Hum Vaccin Immunother 16(6):1239–1242
Chowdhury P (2020) In silico investigation of phytoconstituents from Indian medicinal herb 'Tinospora cordifolia (giloy)' against SARS-CoV-2 (COVID-19) by molecular dynamics approach. J Biomol Struct Dyn:1–18
Cui J, Li F, Shi Z-L (2019) Origin and evolution of pathogenic coronaviruses. Nat Rev Microbiol 17(3):181–192
Dong L, Hu S, Gao J (2020) Discovering drugs to treat coronavirus disease 2019 (COVID-19). Drug Discoveries & Therapeutics 14(1):58–60
Douangamath A, Fearon D, Gehrtz P et al (2020) Crystallographic and electrophilic fragment screening of the SARS-CoV-2 main protease. Nat Commun 11(1):5047
Gorbalenya AE, Baker SC, Baric RS et al (2020) The species severe acute respiratory syndrome-related coronavirus: classifying 2019-nCoV and naming it SARS-CoV-2. Nat Microbiol 5(4):536–544

Hoffmann M, Kleine-Weber H, Schroeder S et al (2020) SARS-CoV-2 cell entry depends on ACE2 and TMPRSS2 and is blocked by a clinically proven protease inhibitor. Cell 181(2):271–280.e278

Jakhmola MR, Sehgal N, Dogra N et al (2020) Deciphering underlying mechanism of Sars-CoV-2 infection in humans and revealing the therapeutic potential of bioactive constituents from Nigella sativa to combat COVID19: in-silico study. J Biomol Struct Dyn:1–13

Jin Z, Du X, Xu Y et al (2020) Structure of Mpro from SARS-CoV-2 and discovery of its inhibitors. Nature 582(7811):289–293

Kandeel M, Al-Nazawi M (2020) Virtual screening and repurposing of FDA approved drugs against COVID-19 main protease. Life Sci 251:117627

Khan S, Fakhar Z, Hussain A et al (2020) Structure-based identification of potential SARS-CoV-2 main protease inhibitors. J Biomol Struct Dyn:1–14

Lan J, Ge J, Yu J et al (2020) Structure of the SARS-CoV-2 spike receptor-binding domain bound to the ACE2 receptor. Nature 581(7807):215–220

Lathika RA, Thakarakkattil NNA, Nangia C et al (2020) Epidemic landscape and forecasting of SARS-CoV-2 in India. J Epidemiol Glob Health 11(1):55–59

Liu F, Xu A, Zhang Y et al (2020) Patients of COVID-19 may benefit from sustained Lopinavir-combined regimen and the increase of eosinophil may predict the outcome of COVID-19 progression. Int J Infect Dis 95:183–191

Loustaud RV, Debette GM, Jacques J et al (2016) Ribavirin: past, present and future. World J Hepatol 8(2):123–130

Mirza SB, Salmas RE, Fatmi MQ et al (2016) Virtual screening of eighteen million compounds against dengue virus: combined molecular docking and molecular dynamics simulations study. J Mol Graph Model 66:99–107

Muralidharan N, Sakthivel R, Velmurugan D et al (2020) Computational studies of drug repurposing and synergism of lopinavir, oseltamivir and ritonavir binding with SARS-CoV-2 protease against COVID-19. J Biomol Struct Dyn:1–6

Prachanronarong KL, Özen A, Thayer KM et al (2016) Molecular basis for differential patterns of drug resistance in influenza N1 and N2 neuraminidase. J Chem Theory Comput 12(12):6098–6108

Rajendran V, Shukla R, Shukla H et al (2018) Structure-function studies of the asparaginyl-tRNA synthetase from Fasciola gigantica: understanding the role of catalytic and non-catalytic domains. Biochem J 475(21):3377–3391

Wang Y, Zhang D, Du G et al (2020) Remdesivir in adults with severe COVID-19: a randomised, double-blind, placebo-controlled, multicentre trial. Lancet 395(10236):1569–1578

World Health Organization (2020) Coronavirus disease 2019 (COVID-19): situation report, 72. World Health Organisation

Yan R, Zhang Y, Li Y et al (2020) Structural basis for the recognition of SARS-CoV-2 by full-length human ACE2. Science 367(6485):1444–1448

Yang X, Yu Y, Xu J et al (2020) Clinical course and outcomes of critically ill patients with SARS-CoV-2 pneumonia in Wuhan, China: a single-centered, retrospective, observational study. Lancet Respir Med 8(5):475–481

Antimalarial Drug Discovery and Development: From Bench to Bedside

Harvinder Kour Khera, Amit Kumar Srivastava, and Subhash Singh

Abstract Malaria is a mosquito-borne disease that wreaks devastation all over the world. *Plasmodium*, the disease's causative agent, is spread via female *Anopheles* mosquito bites. Despite the fact that malaria is preventable and treatable, the crisis of antimalarial drug resistance is growing, necessitating active research for new and affordable medications that target the parasite's specific pathways. Several new therapeutic targets have been discovered since the genome sequencing was completed in 2004. Through computational research of *Plasmodium falciparum* metabolism, several prospective drug development starting points have also been identified. As a result, drug development has shifted from serendipity/whole-cell screening of vast libraries to a new era of target-focused investigations, in which the method is more systematic and based on parasite genome knowledge. In this chapter, we discuss the current methodologies in antimalarial drug discovery.

1 Introduction

Malaria is a devastating mosquito-borne disease that kills and harms millions of people every year all over the world (WHO, World Malaria Report 2020). Despite the fact that malaria has been eradicated from 100 nations in the last century (Newby et al. 2016), the illness remains a serious health issue throughout the world's tropical regions, which are home to 3.2 billion people (WHO World Malaria Report 2020). Pregnant women and children are more vulnerable to malaria in endemic areas,

H. K. Khera (✉)
Tata Institute for Genetics and Society - Centre at inStem, Bengaluru, Karnataka, India
e-mail: harvinder.khera@tigs.res.in

A. K. Srivastava
Indian Institute for Technology, Roorkee, Uttarakhand, India

S. Singh
ICMR-RMRC, Bhubaneswar, Odisha, India

V. S. Rajput, A. Runthala (eds.), *Drugs and a Methodological Compendium*,
https://doi.org/10.1007/978-981-19-7952-1_16

whereas outside of endemic areas, due to the absence of sufficient immunity to the disease, all age groups are equally vulnerable.

P. falciparum (the deadliest of all species), *P. vivax*, *P. ovale*, *P. malariae* and *P. knowlesi* are the five *Plasmodium* species that cause human malaria. *P. falciparum* and *P. vivax*, two of these species, are the most dangerous. In Africa, *P. falciparum* is more prevalent and causes the most fatalities, whereas *P. vivax* is more frequent and dominant outside of sub-Saharan Africa (Lucchi et al. 2013). *P. vivax*, which was previously thought to be a benign species, has recently been discovered to produce severe clinical illness (Bartoloni and Zammarchi 2012). In humans, *P. knowlesi* produces zoonotic infection, which can be severe or fatal (Mita and Tanabe 2012).

The parasite is transmitted to humans during blood feeding by female *Anopheles* mosquitoes. Out of 400 species of *Anopheles* that are known to exist, 30 species can act as a vector for *Plasmodium* (White et al. 2014). In India, nine species of *Anopheles*, namely, *A. fluviatilis*, *A. culicifacies*, *A. minimus*, *A. dirus*, *A. philippinensis*, *A. stephensi*, *A. annularis*, *A. sundaicus*, and *A. varuna*, are known to act as vectors for the malaria parasite.

2 Life Cycle of *Plasmodium*

With more than 5000 genes, *Plasmodium* has got a sophisticated life cycle that takes place in two hosts: mosquito (definitive host) and the human (intermediate host). Sporozoites are injected into the bloodstream of humans during the blood feeding of *Anopheles*. Within an hour, the sporozoites reach hepatocytes after passing through Kupffer cells. The parasites undergo exo-erythrocytic schizogony inside the liver cells (Fig. 1).

A single sporozoite can produce up to 40,000 daughter cells (called merozoites) in 2 weeks, which are discharged into the bloodstream when the hepatocyte ruptures. The hepatocytes which are infected with schizont discharge merosomes (vesicles containing merozoites) into the bloodstream. These merosomes burst to release merozoites which later infect erythrocytes. The parasite remains inside the erythrocyte in a parasitophorous vesicle and begins the erythrocytic cycle. Within this vacuole, parasites mature from ring-form trophozoites to schizonts over the next 48 h. Asexual multiplication causes the red blood cell (RBC) to rupture, releasing roughly 20 daughter merozoites. Rather than penetrating red blood cells, some trophozoites remain in the bloodstream and grow into sexual forms known as gametocytes. In most *Plasmodium* species, gametocytes emerge after the onset of symptoms, with the exception of *P. vivax* (Baker 2010). The female *Anopheles* mosquito ingests the gametocytes, which allows the parasite to reproduce sexually. A single gametocyte can produce up to eight male microgametes or one female macrogamete. The gametes combine to produce a zygote, which further transforms into an ookinete competent of piercing the midgut wall and generating oocysts on the midgut epithelium's exterior surface. In a naturally infected mosquito, there are

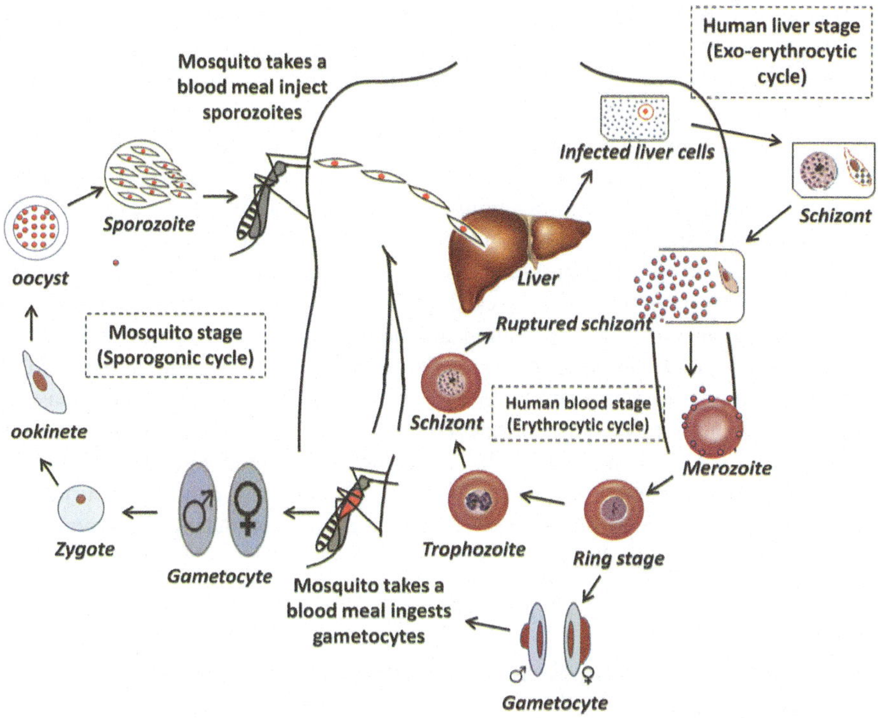

Fig. 1 Life cycle of human malaria parasite (*Plasmodium falciparum)*

approximately one to ten oocysts. The oocyst is a fast-growing structure that produces lots of sporozoites which are discharged into the mosquito's hemolymph. These sporozoites subsequently navigate their way to the mosquito's salivary glands, where they would infect them (Stone et al. 2013). Except for the zygote, all stages of parasites are haploid; meiosis occurs in the developing ookinete shortly after fertilisation. The parasite matures to the sporozoite stage and reaches the mosquito's salivary glands in 10–18 days (White et al. 2014). A percentage of the hepatocellular parasites in *P. vivax* and *P. ovale* remains inactive as hypnozoites. These hypnozoites can be latent for a few weeks or several years, and they are capable of delayed development and relapse onset. The hypnozoite stage is poorly understood, which makes malaria control difficult in *P. vivax*- and *P. ovale*-endemic areas. The only drug that is now effective against these latent stages is primaquine.

3 Current Treatment for Malaria

Malaria can be treated with a variety of medications. The World Health Organization (WHO) has been issuing recommendations for the treatment of malaria on a regular basis. Artemisinin-based combination therapies (ACTs) are prescribed for the treatment of uncomplicated *P. falciparum* infections (WHO Guidelines for Malaria Treatment 2015). ACT is a fixed-dose combination of a 4-aminoquinoline or amino alcohol plus an artemisinin derivative (artemether, dihydroartemisinin or artesunate). If chloroquine-resistant *P. vivax* has been detected in the area, chloroquine is suggested for *P. vivax* infections, while ACT is recommended for *P. vivax* cases. Primaquine should be given to the medication to prevent relapse. Injectable artesunate (intramuscular or intravenous) should be given for at least 24 h in severe malaria patients, accompanied by a full course of an ACT for 3 days.

National malaria control programmes should check the efficacy of antimalarial medications on a regular basis, according to the WHO. In India, patients with *P. vivax* are given 3 days of chloroquine and then 14 days of primaquine (primaquine prevents relapse). On the second day of *falciparum* malaria treatment, the ACT is given in combination with a single dose of primaquine. Except for the Northeastern States, where artemether-lumefantrine (ACT-AL) is to be used due to ACT-SP resistance, the ACT combination of artesunate with sulfadoxine-pyrimethamine (ACT-SP) is administered in all parts of India. Quinine salt is advised during the first trimester of pregnancy, whereas ACT is administered during the second and third trimesters (Dhillon 2008).

4 Antimalarial Drug Resistance and the Need for New Antimalarials

Resistance to many antimalarials has spread in many regions endemic to malaria in recent decades. In 1957, chloroquine-resistant strain of *P. falciparum* was found in Thailand, and it soon spread throughout Southeast Asia, sub-Saharan Africa, and South America (Packard 2014). Chloroquine-resistant strain of *P. falciparum* was found in India in 1973 in Diphu, Karbi-Anglong district, Assam state (Operational Manual for Malaria Elimination in India 2016). Alternative synthetic antimalarial medicines, such as sulfadoxine-pyrimethamine, mefloquine and artemisinin-based combination therapy, were used to combat drug resistance in the therapeutics and prevention of this disease. Antimalarial drug resistance has already been observed for practically all antimalarials, and it has become a major challenge in the global malaria eradication effort. Resistance to known therapies is confirmed in five nations of the Greater Mekong Subregion (GMS): the Lao People's Democratic Republic, Cambodia, Myanmar, Vietnam, and Thailand for the most potent antimalarial artemisinin (Hassett and Roepe 2019).

Antimalarial drug resistance is an unavoidable result of broad drug use and the manner in which medications are utilised (White 2004). It also depends on the biology of the vector and parasites, as well as pharmacokinetics and economics. Resistance develops when parasites develop spontaneous mutations that diminish their susceptibility to a particular treatment or family of drugs. Quinine resistance is linked to SNPs in transporters such as the multidrug resistance-1 gene, the chloroquine resistance transporter gene or gene amplification in Pfmdr1 gene (Cui et al. 2015). Mutations in the folate biosynthesis pathway genes induce sulfadoxine and pyrimethamine resistance (Cowman et al. 1988). *P. falciparum* artemisinin-resistant strains have mutations in the Kelch (K13) propeller domain. SNPs in the cytochrome b gene induce atovaquone resistance (Korsinczky et al. 2000).

Thus, innovative, inexpensive and safe medications (having unique modes of action and belonging to an unique physical and chemical series) for the treatment of malaria by interrupting the *Plasmodium* life cycle, inhibiting transmission to mosquitoes and preventing disease by targeting *Plasmodium* species are urgently needed. Despite rising evidence of antimalarial drug resistance, particularly to ACT (the WHO's first-line treatment), in many malaria-endemic areas, little progress has been made in the arena of antimalarial medication development. Only four drugs for malaria treatment were approved in the last 25 years, out of 1400 medicines worldwide (Trouiller et al. 2002). With public-private partnerships (e.g., MMV) assisting in the development of antimalarial medications, additional research in this area is being facilitated.

5 Antimalarial Drug Discovery

Antimalarial drug discovery and development programme typically include the in vitro screening of compounds against *P. falciparum*, along with an in vivo screening against rodent malaria or primate models. Biochemical screens can be done if purified, active target enzyme is available. The phases in the exploration and design of targeted antimalarial therapies are outlined in Fig. 2.

5.1 In Vitro Antimalarial Drug Discovery

The ability to culture *P. falciparum* in a lab is crucial for in vitro antimalarial drug development. The growth inhibitory impact of a substance on the growth and multiplication of malarial parasites is measured in a conventional in vitro antimalarial drug sensitivity test. In general, the parasites are given the medicine, and their growth is compared to that of a drug-free control group. Cut-off values for in vitro and in vivo screenings will vary based on the compound family and might be in the range of 1–5 micromolar for in vitro effective compounds and 5–25 mg/kg for in vivo effective compounds. The most promising scaffolds from the screening trials are chosen based on criteria such as efficacy, toxicity, pricing, synthesis simplicity and innovation. The derivatives of the original scaffold are then created and assessed to produce structure-activity relationships (SAR), which gives an estimate for the impact of chemical changes on the molecule's properties. To find the lead compound, several rounds of iterative SAR analysis are performed. Compounds with an

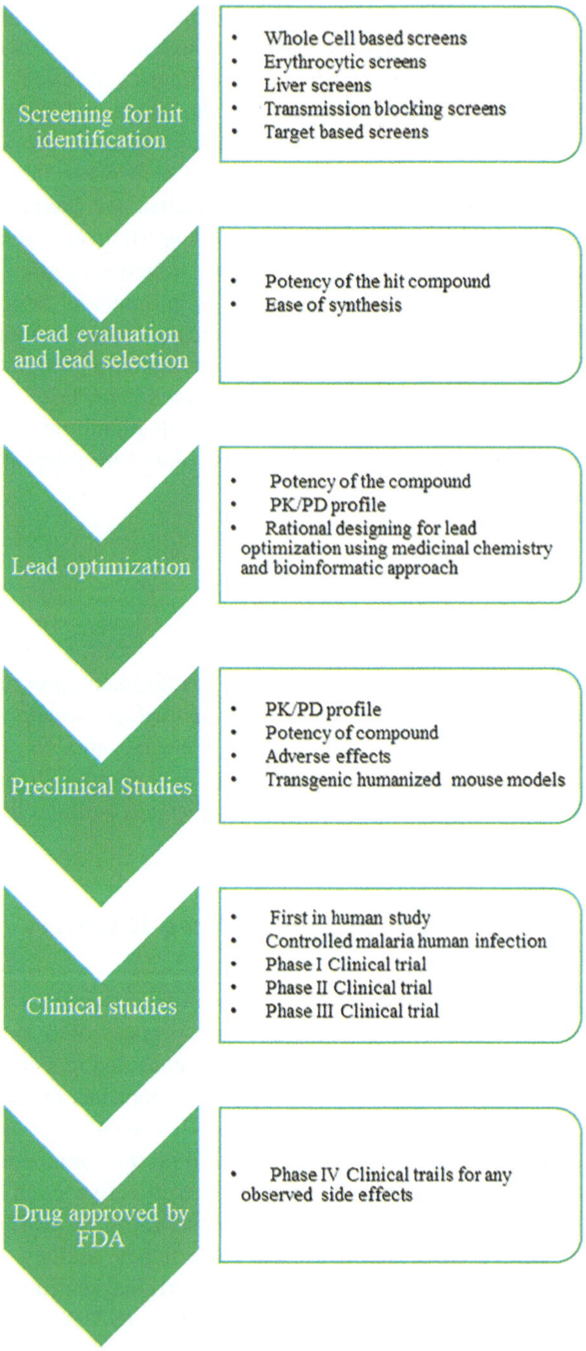

Fig. 2 The phases in the exploration and design of targeted antimalarial therapies

inhibitory concentration (IC50) of less than 1 micromolar that are successful in in vitro screening assays are chosen for in vivo testing (Flannery et al. 2013).

Because of the complex life cycle of the *Plasmodium*, the whole cell-based assays can be divided based on which stage we are targeting. Various methods have been employed to check the anti-*Plasmodium* activity of compounds or natural extracts, and some of them are given below (Sinha et al. 2017).

Microscopy Based

It is a cheap but labour-intensive antimalarial drug screening technique. The parasites are seeded at 1–2% parasitaemia and treated with the test compounds in a 96-well plate format. After 72 h of incubation, a smear is made for each well and stained with DNA binding dye like Giemsa, and an increase in parasitaemia is assessed by light microscopy (Nguyen-Dinh and Payne 1980).

Isotopic Methods

Hypoxanthine uptake: The parasite utilises radiolabelled hypoxanthine present in the media for DNA synthesis. The wells are seeded with *Plasmodium* culture at 0.25–0.5% parasitaemia. Hypoxanthine is added after the test substances have been fed to the parasites for 24 h. The incorporation of hypoxanthine is tested using a liquid scintillation counter after an additional 18-h incubation period.

Fluorescent DNA Dye Intercalation Assays

SYBR Green I, DAPI (4, 6-diamidino-2-phenylindole), YOYO-1 and other DNA dye-based dyes are used to measure parasitaemia. Fluorescence microplate readers or fluorescence-activated cell sorters are used to quantify dye fluorescence in these tests (FACS).

Enzyme Based

This is done mostly by looking at parasite lactate dehydrogenase (pLDH) and histidine-rich protein 2 inhibition (HRP2). The inhibition patterns are assessed by leveraging upon the enzymatic activity of pLDH. pLDH activity can be differentiated from host LDH activity using the 3-acetyl pyridine adenine dinucleotide (APAD)-analog of NAD. L-Lactase is transformed to pyruvate by the availability of LDH and the APAD coenzyme. Reduced APAD is produced as a result of this reaction, which further reduces blue tetrazolium to make a blue formazan compound which can be verified by spectrophotometry.

HRP2 production is measured using a simple, readily available double-site sandwich ELISA test kit to detect parasite proliferation and development. The HRP2 test requires a longer culture time (72 h instead of 48 h) allowing it to evaluate slow-acting medicines.

In vitro screens for malaria drug discovery have primarily focused on finding compounds active against asexual *P. falciparum*, but cellular screens that identify inhibitors active against other parasite stages have also been developed (Peatey et al. 2011; Dembele et al. 2011; Buchholz et al. 2011; Tanaka and Williamson 2011). Antimalarial drugs' efficacy and stage specificity can be determined using a variety of tests. Transgenic *P. falciparum* lines overexpressing the specific antigens or its human orthologue have also been produced to test drug specificity against the targeted plasmodial target enzyme. Using imaging for *P. falciparum*, researchers were able to set high-throughput screening of drugs active against liver stages (Flannery et al. 2013).

The membrane-feeding assay is still the Holy Grail for evaluating antimalarial medicines that inhibit transmission. The drug is mixed with infected blood meal and fed to mosquitos after 10 days, and the drug's effectiveness is verified by the number of oocysts in the mosquito's midgut (Wadi et al. 2019).

5.2 Target-Based Drug Discovery

When using a target-based drug discovery approach, the most important stage is target selection. Only if there are structural distinctions between the parasite and the human host can targets be exploited. Pyrimethamine and proguanil (dihydrofolate reductase inhibitors) are two examples of plasmodial enzyme inhibitors with relative selectivity. If the host target is a therapeutic target for another disease, this is advantageous. For example, knowledge gained from drug discovery projects targeting human cysteine protease cathepsin K inhibitors (target for osteoporosis therapeutics) and human farnesyl transferases (target for cancer therapeutics) can be applied to antimalarial drug development projects targeting parasite cysteine proteases and farnesyl transferases (Rosenthal et al. 2002; Gelb and Hol 2002).

Inhibitor studies for the enzymes of type II fatty acid biosynthesis pathway, mevalonate-independent isoprenoid synthesis pathway, prokaryotic protein synthesis inhibitors (tetracyclines and clindamycin) and other targets that are missing in the human host but present in the parasite can be beneficial for antimalarial drug discovery projects (Huthmacher et al. 2010).

Targets can also be identified utilising the 'reverse' drug discovery technique, which entails elucidating previously unknown targets of existing antimalarial medications as a foundation for antimalarial drug discovery. Chloroquine, for example, works by interfering with hemozoin formation, making hemozoin inhibition a promising target for antimalarial drug development (Biagini et al. 2003).

The parasite's erythrocytic stages and a few numbers of recognised targets (notably the folate and haemoglobin degradation pathways) remain the focus of

Table 1 Antimalarial drug targets for therapeutic intervention

Target	Target name	Key investigative molecules	Clinically validated?
*Pf*ATP4	Na+-ATPase 4	KAE609, SJ557733, 21A092	Yes
*Pf*PI4K	Phosphatidylinositol-4 kinase	MMV390048	Yes
*Pf*DHFR	Dihydrofolate reductase	P218	Yes
*Pf*DXR	DXP reductoisomerase	Fosmidomycin	Yes
*Pf*DHODH	Dihydroorotate dehydrogenase	DSM265	Yes
*Pf*CYT*bc*1	Cytochrome *bc*1	Decoquinate, GSK932121, ELQ300	Yes
*Pf*FP2-3	Falcipain cysteine proteases 2–3	Falcitidin	No
*Pf*HDAC1	Histone deacetylase	SB939, trichostatin A	No
*Pf*CHT1,2,4	Aspartic protease plasmepsins I, II, IV	CWHM-117, TCMDC-134674	No
*Pf*CARL	Cyclic amine resistance locus	KAF156	Yes

antimalarial drug development. With the entire genome sequencing of *P. falciparum* and *P. vivax* and subsequent bioinformatics discoveries, essential parasite enzymes have been proposed as possible starting sites for antimalarial drug development (Huthmacher et al. 2010). In recent years, several new targets have been reported which includes protein kinases (Canduri et al. 2007), plasmodial proteases, the malaria parasite mitochondrial system (Jana and Paliwal 2007), *Plasmodium* protein farnesyltransferase (Nallan et al. 2005) and *P. falciparum* enoyl reductase (Oliveira et al. 2007), as well as targets in the host nucleoside transport system (Baldwin et al. 2007) and *P. falciparum* chorismate synthase/shikimate pathway (Khera et al. 2016; Khera et al. 2017; Khera et al. 2019). As a result, there are a wealth of untapped targets in the parasite's numerous growth phases that might be investigated. For the drug discovery programme, proper validation of these drug targets is necessary. Some of the recently validated antimalarial targets with key investigative molecules are given in Table 1.

5.3 *Role of Bioinformatics in Antimalarial Drug Discovery*

Bioinformatics has become indispensable for all drug discovery projects. Right from shortlisting of the lead candidates from the genomics and metabolic-network data of target-based drug discovery projects, it includes several studies, viz. docking studies for the estimation of drug-target antigen affinity, ADME scoring analysis even before the synthesis of drugs and identification of chemical structures critical for activity using quantitative structure-activity relationship models. In vitro studies reveal more about the parasite's metabolic parameters, transporter function and potency during various stages of its life cycle. The data is then related to in vivo endpoints using in vitro and in vivo extrapolation methods. Computational approaches like mechanistic pharmacokinetic modelling and simulation (PBPK) models can be used with a PD model to predict clinical outcomes. These methods are frequently used in drug development to predict medication-drug interactions and to help with trial design (Vieira et al. 2014; Wagner et al. 2015; Wagner et al. 2016).

5.4 Role of Medicinal Chemistry in Antimalarial Drug Discovery

Medicinal chemistry plays a significant role in lead optimisation. Once a hit has been found through the preliminary high-throughput screening of the chemical libraries, the drugs are changed to improve their therapeutic characteristics through SAR and whole-cell or biochemical tests (e.g. resistance and bioavailability). Antimalarial drugs with known chemical structures that have been successful in the clinic can also be used as a basis for screening novel compounds using directed, chemistry-based approaches. Such a strategy has worked in the past, for example, synthetic ozonides are based on artemisinin (Flannery et al. 2013).

5.5 In Vivo Testing of Antimalarial Compounds

The first in vivo test is usually a four-day suppressive test. It is done using rodent malaria parasites: *P. berghei* or *P. chabaudi*. The parasite is inoculated into the mice and separated into control and test groups. After 3 h, the compound of interest is given, and the treatment is given for 4 days. The effectiveness of the chemical is determined by comparing blood parasitaemia on day 4 and the time a mouse lives in treated vs. untreated mice. Secondary testing is then carried out on the active substances. Oral bioavailability and effective dose (ED) values such as ED50 and ED90 are determined after compounds are evaluated at various dosages via subcutaneous and/or oral routes. Because rodent plasmodial species varies greatly in their degree of infection, synchrony, lethality and susceptibility to certain drugs, the choice of rodent malaria species and mouse strains is critical when utilising rodent models. *P. chabaudi* and *P. vinckei* cause synchronous infections with high parasitaemia, allowing researchers to study the parasite's stage specificity. Iron chelators and lipid biosynthesis inhibitors are more toxic to *P. chabaudi* than they are to *P. berghei* (Wengelnik et al. 2002; Peters and Robinson 1999). The use of mouse models in optimisation and development of the lead compounds for therapy can be questioned because the drug sensitivity of specified rodent plasmodial species is not equivalent to *P. falciparum*. The infection of *P. falciparum* in monkeys (Aotus and Saimiri) has been widely studied (Gysin 1998), and primate models have thus played an essential role in the preclinical development of antimalarials. Primate models predict efficacy and pharmacokinetics more accurately than mouse models, adding rationality to clinical trials of active drugs (Wengelnik et al. 2002).

6 Preclinical Studies of Antimalarials

During the initial phases of antimalarial drug discovery, pharmacological action against parasite species was often inferred from therapeutic activity against *P. berghei*, and results could often be extrapolated poorly. Later, immunodeficient nude and SCID mice models were developed that could keep *P. falciparum*- or *P. vivax*-infected human RBCs alive in the blood. The malaria SCID huMouse (humanised mouse) model has become popular for preclinical research. The model has been used to evaluate the efficacy of new therapeutics with a variety of modes of action in vivo. Survival, parasitaemia reduction, medication effect on erythrocytic stages, parasite clearance time, relapse and recrudescence are all common aims of preclinical research in the SCID huMouse model. Earlier to this, the main objective of preclinical trials was to ensure early parasite clearance, with treatments designated as either fast-acting or long-acting, instead of a numerical threshold of cured animals. There is a weak correlation between in vivo parasite reduction ratios observed in mice models and those recorded in infected individuals (McCarthy et al. 2016). Calculating the IC50 using the SCID huMouse model, on the other hand, has demonstrated promise as a human IC50 surrogate (McCarthy et al. 2016). Setting up a real quantitative link that enables preclinical data to determine parasite count in patients throughout time could aid in the prediction of clinical outcomes. SCID huMouse models will be at the forefront of malaria treatment research if this goal is met, demonstrating their potential to provide early information on new medications and assist in drug development.

7 Clinical Trials for Antimalarial Drugs

If a therapeutic candidate passes preclinical testing, it will be evaluated in clinical trials, such as first-in-human (FiH), controlled human malaria infection (CHMI), Phase 2 dose response, and Phase 3 confirmatory investigations. These studies aim to find out about the drug's safety and efficacy, as well as the most effective treatment schedule. After the drug has been approved, Phase 4 continues market surveillance. Each of these trial data sets can be used for PK/PD modelling and simulation analysis, which can help with the design of future trials to establish the compound's therapeutic potential (Andrews et al. 2018).

FiH research is conducted in healthy volunteers, both fed and fasted, using individual and periodic dose administrations, along with different quantities and compositions. This type of research provides critical PK data, such as drug exposure and half-life, including the identification and characterisation of metabolites (Andrews et al. 2018).

CHMI studies are becoming more popular as a way to investigate the efficacy of medications against *Plasmodium* parasites in the human host at an early stage. The two types of CHMI investigations are sporozoite-induced malaria (SIM) and induced blood-stage malaria (IBSM). Before receiving a single dose of the drug under research, healthy subjects are injected with sporozoites (SIM) or ring-stage parasites (IBSM). After that, a parasite count in the blood is done. Then, using qPCR, the drug concentrations and parasite numbers are tracked over time. The main purpose of the CHMI study is to assess PK/PD parameters in healthy volunteers who have been infected with the *Plasmodium* parasite. Despite the fact that the FDA does not require CHMI data for drug approval, the importance of such research is becoming more generally recognised (Burrows et al. 2017).

Parallel to other pharmaceutical research programmes, later-stage development comprises Phase 2 daily dosage trials and confirmatory Phase 3 trials in clinically diagnosed malaria patients. Phase 2a studies concentrate on monotherapy, whereas Phase 2b studies concentrate on combination therapy. In Phase 2 trials, 300–450 patients are recruited from specific geographical locations where malaria is endemic; in Phase 3 trials, the patient sample is usually expanded to encompass thousands of patients recruited from various geographical areas. These studies are essential because they allow researchers to investigate the effects of a variety of parameters on efficacy, including body size, sickness state, age, pregnancy, immunity, ethnicity, infection with several *Plasmodium* strains and species, comorbidities and natural immunity. This study gives important PK/PD data that aids in determining the efficacy and dose of a drug (Andrews et al. 2018).

For young infants and pregnant women, the determination of safe and efficacious dose is difficult. It took almost 20 years for artemether-lumefantrine to get approval from the WHO for malaria in the first trimester of pregnancy (McCarthy et al. 2016). PK/PD modelling for pregnancy and paediatrics is becoming an essential tool for efficient drug development for these vulnerable groups (Andrews et al. 2018).

8 Concluding Remarks

Malaria leads to significant death and morbidity; hence new treatments are clearly needed. Due to advancements in technology and methodologies that have been improved over several years, it is an exciting period in the arena of malaria drug discovery. This is the start of the artificial intelligence and machine learning age, which will surely help with medication development. Antibody-based treatments, in addition to small molecule medicines, are showing promise in clinical trials. This is the ideal time to take advantage of these opportunities and identify medications that will help eradicate malaria on a global scale.

Acknowledgements HKK thank Tata Institute for Genetics and Society-Centre at inStem, Bengaluru for its continued support and providing a wonderful research environment.

Conflict of Interest The authors declare no conflict of interest.

Authors Contribution HKK conceived and wrote the manuscript. AKS contributed to figures. SS provided valuable inputs for the chapter.

References

Andrews KA, Wesche D, McCarthy J et al (2018) Model-informed drug development for malaria therapeutics. Annu Rev Pharmacol Toxicol 58:567–582
Baker DA (2010) Malaria gametocytogenesis. Mol Biochem Parasitol 172(2):57–65
Baldwin SA, McConkey GA, Cass CE et al (2007) Nucleoside transport as a potential target for chemotherapy in malaria. Curr Pharm Des 13(6):569–580
Bartoloni A, Zammarchi L (2012) Clinical aspects of uncomplicated and severe malaria. Mediterr J Hematol Infect Dis 4(1):e2012026
Biagini GA, ONeill PM, Nzila A et al (2003) Antimalarial chemotherapy: young guns or back to the future? Trends Parasitol 19(11):479–487
Buchholz K, Burke TA, Williamson KC et al (2011) A high-throughput screen targeting malaria transmission stages opens new avenues for drug development. J Infect Dis 203(10):1445–1453
Burrows JN, Duparc S, Gutteridge WE et al (2017) New developments in anti-malarial target candidate and product profiles. Malar J 16:26
Canduri F, Perez PC et al (2007) Protein kinases as targets for antiparasitic chemotherapy drugs. Curr Drug Targets 8(3):389–398
Cowman AF, Morry MJ, Biggs BA et al (1988) Amino acid changes linked to pyrimethamine resistance in the dihydrofolate reductase-thymidylate synthase gene of *Plasmodium falciparum*. Proc Natl Acad Sci U S A 85(23):9109–9113
Cui L, Mharakurwa S et al (2015) Antimalarial drug resistance: literature review and activities and findings of the ICEMR Network. Am J Trop Med Hyg 93(3 Suppl):57–68
Dembele L, Gego A, Zeeman AM et al (2011) Towards an in vitro model of *Plasmodium* hypnozoites suitable for drug discovery. PLoS One 3:e18162
Dhillon GP (2008) National Vector Borne Disease Control Programme—a glimpse. Directorate of National Vector Borne Disease Control Programme Directorate General of Health Services Ministry of Health & Family Welfare Government of India
Flannery EL, Chatterjee AK, Winzeler EA (2013) Antimalarial drug discovery - approaches and progress towards new medicines. Nat Rev Microbiol 11(12):849–862
Gelb MH, Hol WG (2002) Parasitology. Drugs to combat tropical protozoan parasites. Science 297(5580):343–344
Gysin J (1998) In: Sherman I (ed) Malaria: parasite biology, pathogenesis and protection, vol 419. ASM, Washington DC, p 441
Hassett MR, Roepe PD (2019) Origin and spread of evolving artemisinin-resistant *Plasmodium* falciparum malarial parasites in Southeast Asia. Am J Trop Med 101(6):1204
Huthmacher C, Hoppe A et al (2010) Antimalarial drug targets in *Plasmodium* falciparum predicted by stage-specific metabolic network analysis. BMC Syst Biol 4:120
Jana S, Paliwal J (2007) Novel molecular targets for antimalarial chemotherapy. Int J Antimicrob Agents 30(1):4–10
Khera HK, Singh SK, Singh S (2019) Chorismate synthase from malaria parasites is bifunctional enzyme. Mol Biochem Parasitol 233:111202

Khera HK, Singh SK et al (2016) Conserved cysteine residues in malaria Chorismate synthase indicate their important role in protein structure and function. Indian J Biochem Biophys 53: 161–168

Khera HK, Singh SK et al (2017) A HRMS-based method for determination of chorismate synthase activity. Protein Pept Lett 23(3):229–234

Korsinczky M, Chen N, Kotecka B et al (2000) Mutations in *Plasmodium* falciparum cytochrome b that are associated with atovaquone resistance are located at a putative drug-binding site. Antimicrob Agents Chemother 44(8):2100–2108

Lucchi NW, Oberstaller J, Kissinger JC et al (2013) Malaria diagnostics and surveillance in the post-genomic era. Public Health Genomics 16(1-2):37–43

McCarthy JS, Marquart L, Sekuloski S et al (2016) Linking murine and human *Plasmodium* falciparum challenge models in a translational path for antimalarial drug development. Antimicrob Agents Chemother 60:3669–3675

Mita T, Tanabe K (2012) Evolution of *Plasmodium* falciparum drug resistance: implications for the development and containment of artemisinin resistance. Jpn J Infect Dis 65(6):465–475

Nallan L, Bauer KD, Bendale P et al (2005) Protein farnesyltransferase inhibitors exhibit potent antimalarial activity. J Med Chem 48(11):3704–3713

Newby G, Bennett A, Larson E et al (2016) The path to eradication: a progress report on the malaria-eliminating countries. Lancet 387(10029):1775–1784

Nguyen-Dinh P, Payne D (1980) Pyrimethamine sensitivity in *Plasmodium* falciparum: determination in vitro by a modified 48-hour test. Bull World Health Organ 58(6):909–912

Oliveira JS, Vasconcelos IB, Moreira IS et al (2007) Enoyl reductases as targets for the development of anti-tubercular and anti-malarial agents. Curr Drug Targets 8(3):399–411

Operational Manual for Malaria Elimination in India (2016) (Version 1) Directorate of National Vector Borne Disease Control Programme Directorate General of Health Services Ministry of Health & Family Welfare Government of India

Packard RM (2014) The origins of antimalarial-drug resistance. N Engl J Med 371:397–399

Peatey CL, Spicer TP, Hodder PS et al (2011) A high-throughput assay for the identification of drugs against late-stage *Plasmodium* falciparum gametocytes. Mol Biochem Parasitol 180(2): 127–131

Peters W, Robinson BL (1999) In: Zak O, Sande M (eds) Handbook of animal models of infection. Academic Press, London, pp 757–773

Rosenthal PJ, Sijwali PS, Singh A et al (2002) Cysteine proteases of malaria parasites: targets for chemotherapy. Curr Pharm Des 8(18):1659–1672

Sinha S, Sarma P, Sehgal R et al (2017) Development in assay methods for in vitro antimalarial drug efficacy testing: a systematic review. Front Pharmacol 23(8):754

Stone WJ, Eldering M, van Gemert GJ et al (2013) The relevance and applicability of oocyst prevalence as a read-out for mosquito feeding assays. Sci Rep 3:3418

Tanaka TQ, Williamson KC (2011) A malaria gametocytocidal assay using oxidoreduction indicator, alamarBlue. Mol Biochem Parasitol 177(2):160–163

Trouiller P, Olliaro P, Torreele E (2002) Drug development for neglected diseases: a deficient market and a public-health policy failure. Lancet 359(9324):2188–2194

Vieira MD, Kim MJ, Apparaju S et al (2014) PBPK model describes the effects of comedication and genetic polymorphism on systemic exposure of drugs that undergo multiple clearance pathways. Clin Pharmacol Ther 95:550–557

Wadi I, Nath M, Anvikar AR et al (2019) Recent advances in transmission-blocking drugs for malaria elimination. Future Med Chem 11(23):3047–3089

Wagner C, Pan Y, Hsu V et al (2015) Predicting the effect of cytochrome P450 inhibitors on substrate drugs: analysis of physiologically based pharmacokinetic modeling submissions to the US Food and Drug Administration. Clin Pharmacokinet 54:117–127

Wagner C, Pan Y, Hsu V et al (2016) Predicting the effect of CYP3A inducers on the pharmacokinetics of substrate drugs using physiologically based pharmacokinetic (PBPK) modeling: an analysis of PBPK submissions to the US FDA. Clin Pharmacokinet 55:475–483

Wengelnik K, Vidal V, Ancelin ML et al (2002) A class of potent antimalarials and their specific accumulation in infected erythrocytes. Science 295(5558):1311–1314
White NJ (2004) Antimalarial drug resistance. J Clin Invest 113(8):1084–1092
White NJ, Pukrittayakamee S, Hien TT et al (2014) Malaria. Lancet 383(9918):723–735
World Health Organization (2015) Guidelines for the treatment of malaria
World Health Organization (2020) World malaria report

Retraction Note to: Computational and Informatics Methodologies in Drug Discovery, with Focus on Natural Products

Anchala Kumari and Vikrant Singh Rajput

Retraction Note to:
Chapter 1 in: V. S. Rajput, A. Runthala (eds.), *Drugs and a Methodological Compendium*, https://doi.org/10.1007/978-981-19-7952-1_1

The authors have retracted this chapter because the content overlaps substantially with a previously published article (Romano and Tatonetti, 2019). The authors agree with this retraction.

Romano JD and Tatonetti NP. Informatics and Computational Methods in Natural Product Drug Discovery: A Review and Perspectives. Front. Genet. 2019, 10, https://doi.org/10.3389/fgene.2019.00368

The retracted version of this chapter can be found at
https://doi.org/10.1007/978-981-19-7952-1_1

V. S. Rajput, A. Runthala (eds.), *Drugs and a Methodological Compendium*,
https://doi.org/10.1007/978-981-19-7952-1_17

The manufacturer's authorised representative in the EU is Springer Nature Customer Service Centre GmbH, Europaplatz 3, 69115 Heidelberg, Germany. If you have any concerns regarding our products, please contact ProductSafety@springernature.com

Printed and bound by CPI Group (UK) Ltd, Croydon, CR0 4YY
30/09/2026
02225319-0017